Integrable Systems

To my Professor Pierre van Moerbeke

Integrable Systems

Ahmed Lesfari

WILEY

First published 2022 in Great Britain and the United States by ISTE Ltd and John Wiley & Sons, Inc.

Apart from any fair dealing for the purposes of research or private study, or criticism or review, as permitted under the Copyright, Designs and Patents Act 1988, this publication may only be reproduced, stored or transmitted, in any form or by any means, with the prior permission in writing of the publishers, or in the case of reprographic reproduction in accordance with the terms and licenses issued by the CLA. Enquiries concerning reproduction outside these terms should be sent to the publishers at the undermentioned address:

ISTE Ltd
27-37 St George's Road
London SW19 4EU
UK

www.iste.co.uk

John Wiley & Sons, Inc.
111 River Street
Hoboken, NJ 07030
USA

www.wiley.com

© ISTE Ltd 2022

The rights of Ahmed Lesfari to be identified as the author of this work have been asserted by him in accordance with the Copyright, Designs and Patents Act 1988.

Any opinions, findings, and conclusions or recommendations expressed in this material are those of the author(s), contributor(s) or editor(s) and do not necessarily reflect the views of ISTE Group.

Library of Congress Control Number: 2022932445

British Library Cataloguing-in-Publication Data
A CIP record for this book is available from the British Library
ISBN 978-1-78630-827-6

Contents

Preface ix

Chapter 1. Symplectic Manifolds 1

    1.1. Introduction 1
    1.2. Symplectic vector spaces 2
    1.3. Symplectic manifolds 3
    1.4. Vectors fields and flows 7
    1.5. The Darboux theorem 22
    1.6. Poisson brackets and Hamiltonian systems 25
    1.7. Examples 31
    1.8. Coadjoint orbits and their symplectic structures 35
    1.9. Application to the group $SO(n)$ 37
        1.9.1. Application to the group $SO(3)$ 39
        1.9.2. Application to the group $SO(4)$ 41
    1.10. Exercises 47

Chapter 2. Hamilton–Jacobi Theory 49

    2.1. Euler–Lagrange equation 49
    2.2. Legendre transformation 52
    2.3. Hamilton's canonical equations 53
    2.4. Canonical transformations 55
    2.5. Hamilton–Jacobi equation 57
    2.6. Applications 60
        2.6.1. Harmonic oscillator 60
        2.6.2. The Kepler problem 61
        2.6.3. Simple pendulum 62
    2.7. Exercises 64

Chapter 3. Integrable Systems 67

3.1. Hamiltonian systems and Arnold–Liouville theorem 67
3.2. Rotation of a rigid body about a fixed point 75
   3.2.1. The Euler problem of a rigid body 78
   3.2.2. The Lagrange top 82
   3.2.3. The Kowalewski spinning top 83
   3.2.4. Special cases 90
3.3. Motion of a solid through ideal fluid 91
   3.3.1. Clebsch's case 91
   3.3.2. Lyapunov–Steklov's case 93
3.4. Yang–Mills field with gauge group $SU(2)$ 93
3.5. Appendix (geodesic flow and Euler–Arnold equations) 95
3.6. Exercises 100

Chapter 4. Spectral Methods for Solving Integrable Systems 103

4.1. Lax equations and spectral curves 103
4.2. Integrable systems and Kac–Moody Lie algebras 104
4.3. Geodesic flow on $SO(n)$ 107
4.4. The Euler problem of a rigid body 108
4.5. The Manakov geodesic flow on the group $SO(4)$ 109
4.6. Jacobi geodesic flow on an ellipsoid and Neumann problem 114
4.7. The Lagrange top 115
4.8. Quartic potential, Garnier system 115
4.9. The coupled nonlinear Schrödinger equations 118
4.10. The Yang–Mills equations 119
4.11. The Kowalewski top 119
4.12. The Goryachev–Chaplygin top 121
4.13. Periodic infinite band matrix 122
4.14. Exercises 122

Chapter 5. The Spectrum of Jacobi Matrices and Algebraic Curves 129

5.1. Jacobi matrices and algebraic curves 129
5.2. Difference operators 135
5.3. Continued fraction, orthogonal polynomials and Abelian integrals . . . 137
5.4. Exercises 140

Chapter 6. Griffiths Linearization Flows on Jacobians 143

6.1. Spectral curves 143
6.2. Cohomological deformation theory 144
6.3. Mittag–Leffler problem 148
6.4. Linearizing flows 149

6.5. The Toda lattice 150
6.6. The Lagrange top 153
6.7. Nahm's equations 154
6.8. The n-dimensional rigid body 155
6.9. Exercises 156

Chapter 7. Algebraically Integrable Systems 159

7.1. Meromorphic solutions 159
7.2. Algebraic complete integrability 164
7.3. The Liouville–Arnold–Adler–van Moerbeke theorem 171
7.4. The Euler problem of a rigid body 173
7.5. The Kowalewski top 175
7.6. The Hénon–Heiles system 191
7.7. The Manakov geodesic flow on the group $SO(4)$ 200
7.8. Geodesic flow on $SO(4)$ with a quartic invariant 206
7.9. The geodesic flow on $SO(n)$ for a left invariant metric 210
7.10. The periodic five-particle Kac–van Moerbeke lattice 212
7.11. Generalized periodic Toda systems 213
7.12. The Gross–Neveu system 214
7.13. The Kolossof potential 215
7.14. Exercises 215

Chapter 8. Generalized Algebraic Completely Integrable Systems 221

8.1. Generalities 221
8.2. The RDG potential and a five-dimensional system 225
8.3. The Hénon–Heiles problem and a five-dimensional system 229
8.4. The Goryachev–Chaplygin top and a seven-dimensional system 231
8.5. The Lagrange top 236
8.6. Exercises 237

Chapter 9. The Korteweg–de Vries Equation 241

9.1. Historical aspects and introduction 241
9.2. Stationary Schrödinger and integral Gelfand–Levitan equations 243
9.3. The inverse scattering method 255
9.4. Exercises 269

Chapter 10. KP–KdV Hierarchy and Pseudo-differential Operators 275

10.1. Pseudo-differential operators and symplectic structures 275
10.2. KdV equation, Heisenberg and Virasoro algebras 279

10.3. KP hierarchy and vertex operators 281
10.4. Exercises 290

References 293

Index 305

Preface

This book is intended for a wide readership of mathematicians and physicists: students pursuing graduate, masters and higher degrees in mathematics and mathematical physics. It is devoted to some geometric and topological aspects of the theory of integrable systems and the presentation is clear and well-organized, with many examples and problems provided throughout the text. Integrable Hamiltonian systems are nonlinear ordinary differential equations that are described by a Hamiltonian function and possess sufficiently many independent constants of motion in involution. The problem of finding and integrating Hamiltonian systems has attracted a considerable amount of attention in recent decades. Besides the fact that many integrable systems have been the subject of powerful and beautiful theories of mathematics, another motivation for their study is the concepts of integrability that are applied to an increasing number of physical systems, biological phenomena, population dynamics and chemical rate equations, to mention but a few applications. However, it still seems hopeless to describe, or even to recognize with any facility, the Hamiltonian systems which are integrable, even though they are exceptional.

Chapter 1 is devoted to the study of symplectic manifolds and their connection with Hamiltonian dynamical systems. We review some interesting properties of one-parameter groups of diffeomorphisms or of flow, Lie derivative, interior product or Cartan's formula, as well as the study of a central theorem of symplectic geometry, namely, Darboux's theorem. We also show how to determine explicitly symplectic structures on adjoint and coadjoint orbits of a Lie group, with particular attention given to the group $SO(n)$.

Chapter 2 deals with the study of some notions concerning the Hamilton–Jacobi theory in the calculus of variations. We will establish the Euler–Lagrange differential equations, Hamilton's canonical equations and the Hamilton–Jacobi partial differential equation and explain how it is widely used in practice to solve some problems. As an application, we will study the geodesics, the harmonic oscillator, the Kepler problem and the simple pendulum.

In Chapter 3, we study the Arnold–Liouville theorem: the regular compact level manifolds defined by the intersection of the constants of motion are diffeomorphic to a real torus on which the motion is quasi-periodic as a consequence of the following differential geometric fact; a compact and connected n-dimensional manifold on which there exist n vector fields that commute and are independent at every point is diffeomorphic to an n-dimensional real torus, and there is a transformation to so-called action-angle variables, mapping the flow into a straight line motion on that torus. We give a proof as direct as possible of the Arnold–Liouville theorem and we make a careful study of its connection with the concept of completely integrable systems. Many problems are studied in detail: the rotation of a rigid body about a fixed point, the motion of a solid in an ideal fluid and the Yang–Mills field with gauge group $SU(2)$.

In Chapter 4, we give a detailed study of the integrable systems that can be written as Lax equations with a spectral parameter. Such equations have no *a priori* Hamiltonian content. However, through the Adler–Kostant–Symes (AKS) construction, we can produce Hamiltonian systems on coadjoint orbits in the dual space to a Lie algebra whose equations of motion take the Lax form. We outline an algebraic-geometric interpretation of the flows of these systems, which are shown to describe linear motion on a complex torus. The relationship between spectral theory and completely integrable systems is a fundamental aspect of the modern theory of integrable systems. This chapter surveys a number of classical and recent results and our purpose here is to sketch a motivated overview of this interesting subject. We present a Lie algebra theoretical schema leading to integrable systems based on the Kostant–Kirillov coadjoint action. Many problems on Kostant–Kirillov coadjoint orbits in subalgebras of infinite dimensional Lie algebras (Kac–Moody Lie algebras) yield large classes of extended Lax pairs. A general statement leading to such situations is given by the AKS theorem, and the van Moerbeke–Mumford linearization method provides an algebraic map from the complex invariant manifolds of these systems to the Jacobi variety (or some subabelian variety of it) of the spectral curve. The complex flows generated by the constants of the motion are straight line motions on these varieties. This chapter describes a version of the general scheme, and shows in detail how several important classes of examples fit into the general framework. Several examples of integrable systems of relevance in mathematical physics are carefully discussed: geodesic flow on $SO(n)$, the Euler problem of a rigid body, Manakov geodesic flow on the group $SO(4)$, Jacobi geodesic flow on an ellipsoid, the Neumann problem, the Lagrange top, a quartic potential or Garnier system, coupled nonlinear Schrödinger equations, Yang–Mills equations, the Kowalewski spinning top, the Goryachev–Chaplygin top and the periodic infinite band matrix.

The aim of Chapter 5 is to describe some connections between spectral theory in infinite dimensional Lie algebras, deformation theory and algebraic curves. We study infinite continued fractions, isospectral deformation of periodic Jacobi matrices, general difference operators, Cauchy–Stieltjes transforms and Abelian integrals from an algebraic geometrical point of view. These results can be used to obtain insight into integrable systems.

In Chapter 6, we present in detail the Griffiths' approach and his cohomological interpretation of the linearization test for solving integrable systems without reference to Kac–Moody algebras. His method is based on the observation that the tangent space to any deformation lies in a suitable cohomology group and on algebraic curves, higher cohomology can always be eliminated using duality theory. We explain how results from deformation theory and algebraic geometry can be used to obtain insight into the dynamics of integrable systems. These conditions are cohomological and the Lax equations turn out to have a natural cohomological interpretation. Several nonlinear problems in mathematical physics illustrate these results: the Toda lattice, Nahm's equations and the n-dimensional rigid body.

In Chapter 7, the notion of algebraically completely integrable Hamiltonian systems in the Adler–van Moerbeke sense is explained, and techniques to find and solve such systems are presented. These are integrable systems whose trajectories are straight line motions on Abelian varieties (complex algebraic tori). We make, via the Kowalewski–Painlevé analysis, a study of the level manifolds of the systems, which are described explicitly as being affine part of Abelian varieties and the flow can be solved by quadrature, that is to say their solutions can be expressed in terms of Abelian integrals. We describe an explicit embedding of these Abelian varieties that complete the generic invariant surfaces into projective spaces. Many problems are studied in detail: the Euler problem of a rigid body, the Kowalewski top, the Hénon–Heiles system, Manakov geodesic flow on the group $SO(4)$, geodesic flow on $SO(4)$ with a quartic invariant, geodesic flow on $SO(n)$ for a left invariant metric, the periodic five-particle Kac–van Moerbeke lattice, generalized periodic Toda systems, the Gross–Neveu system and the Kolossof potential.

In Chapter 8, we discuss the study of generalized algebraic completely integrable systems. There are many examples of differential equations that have the weak Painlevé property that all movable singularities of the general solution have only a finite number of branches, and some interesting integrable systems appear as coverings of algebraic completely integrable systems. The invariant varieties are coverings of Abelian varieties and these systems are called algebraic completely integrable in the generalized sense. These systems are Liouville integrable and by the Arnold–Liouville theorem, the compact connected manifolds invariant by the real flows are tori, the real parts of complex affine coverings of Abelian varieties. Most of these systems of differential equations possess solutions that are Laurent series of $t^{1/n}$ (t being complex time) and whose coefficients depend rationally on certain algebraic parameters. We discuss some interesting examples: Ramani–Dorizzi–Grammaticos (RDG) potential, the Hénon–Heiles system, the Goryachev–Chaplygin top, a seven-dimensional system and the Lagrange top.

Chapter 9 covers the stationary Schrödinger equation, the integral Gelfand–Levitan equation and the inverse scattering method used to solve exactly the Korteweg–de Vries (KdV) equation. The latter is a universal mathematical model for the description of weakly nonlinear long wave propagation in dispersive media. The study of this equation is the archetype of an integrable system and is one of the most fundamental equations of soliton phenomena.

In Chapter 10, we study some generalities on the algebra of infinite order differential operators. The algebras of Virasoro, Heisenberg and nonlinear evolution equations such as the KdV, Boussinesq and Kadomtsev–Petviashvili (KP) equations play a crucial role in this study. We make a careful study of some connection between pseudo-differential operators, symplectic structures, KP hierarchy and tau functions based on the Sato–Date–Jimbo–Miwa–Kashiwara theory. A few other connections and ideas concerning the KdV and Boussinesq equations and the Gelfand–Dickey flows, the Heisenberg and Virasoro algebras are given. The study of the KP and KdV hierarchies, the use of tau functions related to infinite dimensional Grassmannians, Fay identities, vertex operators and the Hirota's bilinear formalism led to obtaining remarkable properties concerning these algebras such as, for example, the existence of an infinite family of first integrals functionally independent and in involution.

It is well known that when studying integrable systems, elliptic functions and integrals, compact Riemann surfaces or algebraic curves, Abelian surfaces (as well as the basic techniques to study two-dimensional algebraic completely integrable systems) play a crucial role. These facts, which may be well known to the algebraic reader, can be found, for example, in Adler and van Moerbeke (2004); Fay (1973); Griffiths and Harris (1978); Lesfari (2015b) and Vanhaecke (2001).

I would like to thank and am grateful to P. van Moerbeke and L. Haine, from whom I learned much of this subject through conversations and remarks. I would also like to thank the editors for their interest, seriousness and professionalism. Finally my thanks go to my wife and our children for much encouragement and undeniable support, who helped bring this book into being.

Ahmed LESFARI
September 2021

1

Symplectic Manifolds

1.1. Introduction

This chapter is devoted to the study of symplectic manifolds and their connection with Hamiltonian systems. It is well known that symplectic manifolds play a crucial role in classical mechanics, geometrical optics and thermodynamics, and currently have conquered a rich territory, asserting themselves as a central branch of differential geometry and topology. In addition to their activity as an independent subject, symplectic manifolds are strongly stimulated by important interactions with many mathematical and physical specialties, among others. The aim of this chapter is to study some properties of symplectic manifolds and Hamiltonian dynamical systems, and to review some operations on these manifolds.

This chapter is organized as follows. In the second section, we begin by briefly recalling some notions about symplectic vector spaces. The third section defines and develops explicit calculation of symplectic structures on a differentiable manifold and studies some important properties. The forth section is devoted to the study of some properties of one-parameter groups of diffeomorphisms or flow, Lie derivative, interior product and Cartan's formula. We review some interesting properties and operations on differential forms. The fifth section deals with the study of a central theorem of symplectic geometry, namely Darboux's theorem: the symplectic manifolds (M, ω) of dimension $2m$ are locally isomorphic to $(\mathbb{R}^{2m}, \omega)$. The sixth section contains some technical statements concerning Hamiltonian vector fields. The latter form a Lie subalgebra of the space vector field and we show that the matrix associated with a Hamiltonian system forms a symplectic structure. Several properties concerning Hamiltonian vector fields, their connection with symplectic manifolds, Poisson manifolds or Hamiltonian manifolds as well as some interesting examples are studied in the seventh section. We will see in the eight section how to determine a symplectic structure on the orbit of the coadjoint representation of a Lie group. Section nine is dedicated to the explicit determination of symplectic structures on adjoint and coadjoint orbits of a Lie group $SO(n)$. Some exercises are proposed in the last section.

1.2. Symplectic vector spaces

DEFINITION 1.1.– *A symplectic space (E, ω) is a finite dimensional real vector space E with a bilinear form $\omega : E \times E \longrightarrow \mathbb{R}$, which is alternating (or antisymmetric), that is, $\omega(x, y) = -\omega(y, x)$, $\forall x, y \in E$, and non-degenerate, that is, $\omega(x, y) = 0$, $\forall y \in E \implies x = 0$. The form ω is referred to as symplectic form (or symplectic structure).*

The dimension of a symplectic vector space is always even. We show (using a reasoning similar to the Gram–Schmidt orthogonalization process) that any symplectic vector space (E, ω) has a base $(e_1, ..., e_{2m})$ called symplectic basis (or canonical basis), satisfying the following relations: $\omega(e_{m+i}, e_j) = \delta_{ij}$ and $\omega(e_i, e_j) = \omega(e_{m+i}, e_{m+j}) = 0$. Note that each e_{m+i} is orthogonal to all base vectors except e_i. In terms of symplectic basic vectors $(e_1, ..., e_{2m})$, the matrix (ω_{ij}) where $\omega_{ij} \equiv \omega(e_i, e_j)$ has the form

$$\begin{pmatrix} \omega_{11} & \cdots & \omega_{1\,2m} \\ \vdots & \ddots & \vdots \\ \omega_{2m\,1} & \cdots & \omega_{2m\,2m} \end{pmatrix} = \begin{pmatrix} 0 & -I_m \\ I_m & 0 \end{pmatrix},$$

where I_m denotes the $m \times m$ unit matrix.

EXAMPLE 1.1.– $\mathbb{R}^{2m}$ with the form $\omega(x, y) = \sum_{k=1}^{m}(x_{m+k} y_k - x_k y_{m+k})$, $x \in \mathbb{R}^{2m}$, $y \in \mathbb{R}^{2m}$, is a symplectic vector space. Let $(e_1, ..., e_m)$ be an orthonormal basis of $\mathbb{R}^m$. Then, $((e_1, 0), ..., (e_m, 0), (0, e_1), ..., (0, e_m))$ is a symplectic basis of $\mathbb{R}^{2m}$.

Let (E, ω) be a symplectic vector space and F a vector subspace of E. Let $F^\perp = \{x \in E : \forall y \in F, \omega(x, y) = 0\}$ be the orthogonal (symplectic) of F.

DEFINITION 1.2.– *The subspace F is isotropic if $F \subset F^\perp$, coisotropic if $F^\perp \subset F$, Lagrangian if $F = F^\perp$ and symplectic if $F \cap F^\perp = \{0\}$.*

If F, F_1 and F_2 are subspaces of a symplectic space (E, ω), then

$$\dim F + \dim F^\perp = \dim E, \qquad (F^\perp)^\perp = F,$$

$$F_1 \subset F_2 \implies F_2^\perp \subset F_1^\perp, \quad (F_1 \cap F_2)^\perp = F_1^\perp + F_2^\perp, \quad F_1^\perp \cap F_2^\perp = (F_1 + F_2)^\perp,$$

F is coisotropic if and only if $F^\perp$ is isotropic and F is Lagrangian if and only if F is isotropic and coisotropic.

1.3. Symplectic manifolds

DEFINITION 1.3.– *Let M be an even-dimensional differentiable manifold. A symplectic structure (or symplectic form) on M is a closed non-degenerate differential 2-form ω on M. The non-degeneracy condition means that: $\forall x \in M$, $\forall \xi \neq 0$, $\exists \eta$: $\omega(\xi, \eta) \neq 0$, $(\xi, \eta \in T_x M)$. The pair (M, ω) (or simply M) is called a symplectic manifold.*

At a point $p \in M$, we have a non-degenerate antisymmetric bilinear form on the tangent space $T_p M$, which explains why the dimension of the manifold M is even.

EXAMPLE 1.2.– $\mathbb{R}^{2m}$ with the 2-form $\omega = \sum_{k=1}^{m} dx_k \wedge dy_k$ is a symplectic manifold. The vectors $\left(\frac{\partial}{\partial x_1}\right)_p, ..., \left(\frac{\partial}{\partial x_m}\right)_p, \left(\frac{\partial}{\partial y_1}\right)_p, ..., \left(\frac{\partial}{\partial y_m}\right)_p$, $p \in \mathbb{R}^{2m}$, constitute a symplectic basis of the tangent space $T\mathbb{R}^{2m} = \mathbb{R}^{2m}$. Similarly, $\mathbb{C}^m$ with the form $\omega = \frac{i}{2} \sum_{k=1}^{m} dz_k \wedge d\bar{z}_k$ is a symplectic manifold. This form coincides with the previous form by means of the identification $\mathbb{C}^m \simeq \mathbb{R}^{2m}$, $z_k = x_k + iy_k$. Riemann surfaces, Kählerian manifolds and complex projective manifolds are symplectic manifolds. Another class of symplectic manifolds consists of the coadjoint orbits (see section 1.8).

We will see that the cotangent bundle T^*M (i.e. the union of all cotangent spaces of M) admits a natural symplectic structure. The phase spaces of the Hamiltonian systems studied below are symplectic manifolds and often they are cotangent bundles equipped with the canonical structure.

THEOREM 1.1.– *Let M be a differentiable manifold of dimension m and T^*M its cotangent bundle. Then T^*M possesses a symplectic structure and in a local coordinate $(x_1, ..., x_m, y_1, ..., y_m)$, the form ω is given by $\omega = \sum_{k=1}^{m} dx_k \wedge dy_k$.*

PROOF.– Let (U, φ) be a local chart in the neighborhood of $p \in M$, and consider the application $\varphi : U \subset M \longrightarrow \mathbb{R}^m$, $p \longmapsto \varphi(p) = \sum_{k=1}^{m} x_k e_k$, where e_k are the vectors basis of $\mathbb{R}^m$. Consider the canonical projections $TM \longrightarrow M$, and $T(T^*M) \longrightarrow T^*M$, of tangent bundles, respectively, to M and T^*M on their bases. We note $\pi^* : T^*M \longrightarrow M$, the canonical projection and $d\pi^* : T(T^*M) \longrightarrow TM$, its linear tangent application. We have $\varphi^* : T^*M \longrightarrow \mathbb{R}^{2m}$, $\alpha \longmapsto \varphi^*(\alpha) = \sum_{k=1}^{m}(x_k e_k + y_k \varepsilon_k)$, where ε_k are the basic forms of $T^*\mathbb{R}^m$ and α denotes $\alpha_p \in T^*M$. So, if α is a 1-form on M and ξ_α is a vector tangent to T^*M, then $d\varphi^* : T(T^*M) \longrightarrow T\mathbb{R}^{2m} = \mathbb{R}^{2m}$, $\xi_\alpha \longmapsto d\varphi^*(\xi_\alpha) = \sum_{k=1}^{m}(\beta_k e_k + \gamma_k \varepsilon_k)$, where β_k, γ_k are the components of ξ_α in the local chart of $\mathbb{R}^{2m}$. Consider $\lambda_\alpha(\xi_\alpha) = \alpha(d\pi^* \xi_\alpha) = \alpha(\xi)$, where ξ is a tangent vector to M. Let $(x_1, ..., x_m, y_1, ..., y_m)$ be a system of local coordinates compatible with a local trivialization of the tangent bundle T^*M. Let us show that:

$$\lambda_\alpha(\xi_\alpha) = \alpha\left(\sum_{k=1}^m \beta_k e_k\right) = \sum_{k=1}^m (x_k e_k + y_k \varepsilon_k)\left(\sum_{j=1}^m \beta_j e_j\right) = \sum_{k=1}^m \beta_k y_k.$$

Indeed, let $(x_1, ..., x_m)$ be a system of local coordinates around $p \in M$. Since $\forall \alpha \in T^*M$, $\alpha = \sum_{k=1}^m \alpha_k dx_k$, then by defining local coordinates $y_1, ..., y_m$ by $y_k(\alpha) = y_k$, $k = 1, ..., m$, the 1-form λ is written as $\lambda = \sum_{k=1}^m y_k dx_k$. The form λ on the cotangent bundle T^*M (doing correspondence λ_α to α) is called Liouville form. We have $\lambda(\alpha) = \sum_{k=1}^m y_k(\alpha) dx_k(\alpha)$, $\lambda(\alpha)(\xi_\alpha) = \sum_{k=1}^m y_k(\alpha) dx_k(\alpha)$ $\left(\sum_{j=1}^m \beta_j e_j + \gamma_j \varepsilon_j\right) = \sum_{k=1}^m y_k \beta_k = \lambda_\alpha(\xi_\alpha)$, $\lambda = \sum_{k=1}^m y_k dx_k$. The symplectic structure of T^*M is given by the exterior derivative of λ, that is, the 2-form $\omega = -d\lambda$. The forms λ and ω are called canonical forms on T^*M. We can visualize all this with the help of the following diagram:

$$\begin{array}{ccccccc}
 & & & T^*(T^*M) & & & \\
 & & & \uparrow \lambda & & & \\
\mathbb{R} & \stackrel{\lambda_\alpha(\xi)}{\longleftarrow} & T(T^*M) & \longrightarrow & T^*M & \stackrel{\varphi^*}{\longrightarrow} & \mathbb{R}^{2m} \\
 & & \downarrow d\pi^* & & \downarrow \pi^* & & \\
\mathbb{R} & \stackrel{\alpha(\xi)}{\longleftarrow} & TM & \longrightarrow & M & \stackrel{\varphi}{\longrightarrow} & \mathbb{R}^m
\end{array}$$

The form ω is closed: $d\omega = 0$ since $d \circ d = 0$ and it is non-degenerate. To show this last property, just note that the form is well defined independently of the chosen coordinates but we can also show it using a direct calculation. Indeed, let $\xi = (\xi_1, ..., \xi_{2m}) \in T_p M$ and $\eta = (\eta_1, ..., \eta_{2m}) \in T_p M$. We have

$$\omega(\xi, \eta) = \sum_{k=1}^m dx_k \wedge dy_k(\xi, \eta) = \sum_{k=1}^m (dx_k(\xi) dy_k(\eta) - dx_k(\eta) dy_k(\xi)).$$

Since $dx_k(\xi) = \xi_{m+k}$ is the $(m+k)$th component of ξ and $dy_k(\xi) = \xi_k$ is the kth component of ξ, then

$$\omega(\xi, \eta) = \sum_{k=1}^m (\xi_{m+k} \eta_k - \eta_{m+k} \xi_k) = (\xi_1 ... \xi_{2m}) \begin{pmatrix} O & -I \\ I & O \end{pmatrix} \begin{pmatrix} \eta_1 \\ \vdots \\ \eta_{2m} \end{pmatrix},$$

with O the null matrix and I the unit matrix of order m. Then, for all $x \in M$ and for all $\xi = (\xi_1, ..., \xi_{2m}) \neq 0$, it exists $\eta = (\xi_{m+1}, ..., \xi_{2m}, -\xi_1, ..., -\xi_m)$ such that: $\omega(\xi, \eta) = \sum_{k=1}^m (\xi_{m+k}^2 - \xi_k^2) \neq 0$, because $\xi_k \neq 0$, $\forall k = 1, ..., 2m$. In the local coordinate system $(x_1, ..., x_m, y_1, ..., y_m)$, this symplectic form is written as $\omega = \sum_{k=1}^{n} dx_k \wedge dy_k$, which completes the proof. $\square$

A manifold M is said to be orientable if there exists on M an atlas such that the Jacobian of any change of chart is strictly positive or if M has a volume form (i.e. a differential form that does not vanish anywhere). For example, $\mathbb{R}^n$ is oriented by the volume form $dx_1 \wedge ... \wedge dx_n$. The circle S^1 is oriented by $d\theta$. The torus $T^2 = S^1 \times S^1$ is oriented by the volume form $d\theta \wedge d\varphi$. All holomorphic manifolds are orientable.

THEOREM 1.2.– (a) A closed differential 2-form ω on a differentiable manifold M of dimension $2m$ is symplectic, if and only if, ω^m is a volume form. (b) Any symplectic manifold is orientable. (c) Any orientable manifold of dimension two is symplectic. However, in even dimensions larger than 2, this is no longer true.

PROOF.– (a) This is due to the fact that the non-degeneracy of ω is equivalent to the fact that ω^m is never zero. (b) We have $\omega = dx_1 \wedge dx_{m+1} + \cdots + dx_m \wedge dx_{2m}$ in a system of symplectic charts $(x_1, ..., x_{2m})$. Therefore, $\omega^m = dx_1 \wedge dx_{m+1} \wedge ... \wedge dx_m \wedge dx_{2m} = (-1)^{\frac{m(m-1)}{2}} dx_1 \wedge dx_2 \wedge ... \wedge dx_{2m}$, which means that the $2m$-form ω^m is a volume form on the manifold M and therefore this one is orientable. The orientation associated with the differential form ω is the canonical orientation of $\mathbb{R}^{2m}$. (c) This results from the fact that any differential 2-form on a 2-manifold is always closed. $\square$

THEOREM 1.3.– Let α be a differential 1-form on the manifold M and $\alpha^*\lambda$ the reciprocal image of the Liouville form λ on the cotangent bundle T^*M. Then, $\alpha^*\lambda = \alpha$.

PROOF.– Since $\alpha : M \longrightarrow T^*M$, we can consider the reciprocal image that we note $\alpha^* : T^*T^*M \longrightarrow T^*M$, of $\lambda : T^*M \longrightarrow T^*T^*M$ (Liouville form), such that, for any vector ξ tangent to M, we have the relation $\alpha^*\lambda(\xi) = \lambda(\alpha)(d\alpha\xi)$. Since $d\alpha$ is an application $TM \longrightarrow TT^*M$, then $\alpha^*\lambda(\xi) = \lambda(\alpha)(d\alpha\xi) = \lambda_\alpha(d\alpha\xi) = \alpha d\pi^* d\alpha(\xi) = \alpha d(\pi^*\alpha)(\xi) = \alpha(\xi)$, because $\pi^*\alpha(p) = p$ where $p \in M$ and the result follows. $\square$

A submanifold $\mathcal{N}$ of a symplectic manifold M is called Lagrangian if for all $p \in \mathcal{N}$, the tangent space $T_p\mathcal{N}$ coincides with the following configuration space $\{\eta \in T_pM : \omega_p(\xi, \eta) = 0, \forall \xi \in T_p\mathcal{N}\}$. On this space, the 2-form $\sum dx_k \wedge dy_k$ that defines the symplectic structure is identically zero. Lagrangian submanifolds are considered among the most important submanifolds of symplectic manifolds. Note that $\dim \mathcal{N} = \frac{1}{2} \dim M$ and that for all vector fields X, Y on $\mathcal{N}$, we have $\omega(X, Y) = 0$.

EXAMPLE 1.3.– If $(x_1, ..., x_m, y_1, ..., y_m)$ is a local coordinate system on an open $U \subset M$, then the subset of U defined by $y_1 = \cdots = y_m = 0$ is a Lagrangian submanifold of M. The submanifold $\alpha(M)$ is Lagrangian in T^*M if and only if the form α is closed because $0 = \alpha^*\omega = \alpha^*(-d\lambda) = -d(\alpha^*\lambda) = -d\alpha$.

Let M be a differentiable manifold, T^*M its cotangent bundle with the symplectic form ω, $s_\alpha : U \longrightarrow T^*M$, $p \longmapsto \alpha(p)$, a section on an open $U \subset M$. From the local expression of ω (theorem 1.6), we deduce that the null section of the bundle T^*M is a Lagrangian submanifold of T^*M. If $s_\alpha(U)$ is a Lagrangian submanifold of T^*M, then s_α is called the Lagrangian section. We have (theorem 1.8), $s_\alpha^*\lambda = \alpha$, and according to the previous example, $s_\alpha(U)$ is a Lagrangian submanifold of T^*M if and only if the form α is closed. Let (M,ω), (N,η) be two symplectic manifolds of the same dimension and $f : M \longrightarrow N$, a differentiable application. We say that f is a symplectic morphism if it preserves the symplectic forms, that is, if f satisfies $f^*\eta = \omega$. When f is a diffeomorphism, we say that f is a symplectic diffeomorphism or f is a symplectomorphism.

THEOREM 1.4.– (a) A symplectic morphism is a local diffeomorphism. (b) A symplectomorphism preserves the orientation.

PROOF.– (a) Indeed, since the 2-form ω is non-degenerate, then the differential $df(p) : T_pM \longrightarrow T_pN$, $p \in M$, is a linear isomorphism and according to the local inversion theorem, f is a local diffeomorphism. Another proof is to note that $f^*\eta^m = (f^*\eta)^m = \omega^m$. The map f has constant rank $2m$ because ω^m and η^m are volume forms on M and N, respectively. And the result follows. (b) It is deduced from (a) that the symplectic diffeomorphisms or symplectomorphisms preserve the volume form and therefore the orientation. The Jacobian determinant of the transformation is $+1$. $\square$

REMARK 1.1.– Note that the inverse $f^{-1} : N \longrightarrow M$ of a symplectomorphism $f : M \longrightarrow N$ is also a symplectomorphism.

Let (M,ω), (N,η) be two symplectic manifolds, $pr_1 : M \times N \longrightarrow M$, $pr_2 : M \times N \longrightarrow N$, the projections of $M \times N$ on its two factors. The forms $pr_1^*\omega + pr_2^*\eta$ and $pr_1^*\omega - pr_2^*\eta$ on the product $M \times N$ are symplectic forms. Take the case where $\dim M = \dim N = 2m$ and consider a differentiable map $f : M \longrightarrow N$, as well as its graph defined by the set $A = \{(x,y) \in M \times N : y = f(x)\}$. The application g defined by $g : M \longrightarrow A$, $x \longmapsto (x, f(x))$ is a diffeomorphism. The set A is a $2m$-dimensional Lagrangian submanifold of $(M \times N, pr_1^*\omega - pr_2^*\eta)$ if and only if the reciprocal image of $pr_1^*\omega - pr_2^*\eta$ by application g is the identically zero form on M. For the differentiable map f to be a symplectic morphism, it is necessary and sufficient that the graph of f is a Lagrangian submanifold of the product manifold $(M \times N, pr_1^*\omega - pr_2^*\eta)$.

THEOREM 1.5.– (a) Let $f : M \longrightarrow M$ be a diffeomorphism. Then, the application $f^* : T^*M \longrightarrow T^*M$ is a symplectomorphism. (b) Let $g : T^*M \longrightarrow T^*M$ be a diffeomorphism such that: $g^*\lambda = \lambda$. Then, there is a diffeomorphism $f : M \longrightarrow M$ such that: $g = f^*$.

PROOF.– (a) Let us show that $f^{**}\omega = \omega$. We have

$$f^{**}\lambda(\alpha)(\xi_\alpha) = \lambda(f^*(\alpha))(df^*\xi_\alpha) = f^*(\alpha)d\pi^*df^*(\xi_\alpha) = \alpha(df d\pi^* df^*(\xi_\alpha)),$$

and therefore, $f^{**}\lambda(\alpha)(\xi_\alpha) = \alpha(d(f \circ \pi^* \circ f^*)(\xi_\alpha))$. Since $f^*\alpha = \alpha_{f^{-1}(p)}$ and $\pi^* f^* \alpha = f^{-1}(p)$, then $f \circ \pi^* \circ f^*(\alpha) = p = \pi^*\alpha$, that is,

$$f \circ \pi^* \circ f^* = \pi^* \qquad [1.1]$$

and $f^{**}\lambda(\alpha)(\xi_\alpha) = \alpha(d\pi^*(\xi_\alpha)) = \lambda_\alpha(\xi_\alpha) = \lambda(\alpha)(\xi_\alpha)$. Consequently, $f^{**}\lambda = \lambda$ and $f^{**}\omega = \omega$. (b) Since $g^*\lambda = \lambda$, then $g^*\lambda(\eta) = \lambda(dg\eta) = \omega(\xi, dg\eta) = \lambda(\eta) = \omega(\xi, \eta)$. Moreover, we have $g^*\omega = \omega$, hence $\omega(dg\xi, dg\eta) = \omega(\xi, \eta) = \omega(\xi, dg\eta)$ and $\omega(dg\xi - \xi, dg\eta) = 0, \forall \eta$. Since the form ω is non-degenerate, we deduce that $dg\xi = \xi$ and that g preserves the integral curves of ξ. On the null section of the tangent bundle (i.e. on the manifold), we have $\xi = 0$ and then $g|_M$ is an application $f : M \longrightarrow M$. Let us show that: $f \circ \pi^* \circ g = \pi^* = f \circ \pi^* \circ f^*$. Indeed, taking the differential, we get $df \circ d\pi^* \circ dg(\xi) = df \circ d\pi^*(\xi) = df(\xi_p)$, because $dg(\xi) = \xi$ and $\xi_p \equiv d\pi^*(\xi)$), hence $df \circ d\pi^* \circ dg(\xi) = \xi_p = d\pi^*(\xi)$. Therefore, $df \circ d\pi^* \circ dg = d\pi^*$, $f \circ \pi^* \circ g = \pi^*$. Since $f \circ \pi^* \circ f^* = \pi^*$ (according to [1.1]), so $g = f^*$. □

THEOREM 1.6.– Let $I : T^*_x M \longrightarrow T_x M$, $\omega^1_\xi \longmapsto \xi$, where $\omega^1_\xi(\eta) = \omega(\eta, \xi)$, $\forall \eta \in T_x M$. Then I is an isomorphism generated by the symplectic form ω.

PROOF.– Denote by I^{-1} the map $I^{-1} : T_x M \longrightarrow T^*_x M$, $\xi \longmapsto I^{-1}(\xi) \equiv \omega^1_\xi$, with $I^{-1}(\xi)(\eta) = \omega^1_\xi(\eta) = \omega(\eta, \xi), \forall \eta \in T_x M$. The form ω being bilinear, then we have $I^{-1}(\xi_1 + \xi_2)(\eta) = I^{-1}(\xi_1)(\eta) + I^{-1}(\xi_2)(\eta), \forall \eta \in T_x M$. To show that I^{-1} is bijective, it suffices to show that it is injective (because $\dim T_x M = \dim T^*_x M$). The form ω is non-degenerate, and it follows that $Ker I^{-1} = \{0\}$. Hence, I^{-1} is an isomorphism and consequently I is also an isomorphism. □

1.4. Vectors fields and flows

Let M be a differentiable manifold of dimension m. Let $TM = \bigcup_{x \in M} T_x M$, be the bundle tangent to M (union of spaces tangent to M at all its points x). This bundle has a structure of a differentiable manifold of dimension $2m$ and allows us to immediately transfer to manifolds the theory of ordinary differential equations.

DEFINITION 1.4.– *A vector field (also called a tangent bundle section) on M is an application, noted by X, which at any point $x \in M$ associates a tangent vector $X_x \in T_x M$. In other words, it is an application: $X : M \longrightarrow TM$, such that if $\pi : TM \longrightarrow M$, is the natural projection, then we have $\pi \circ X = id_M$.*

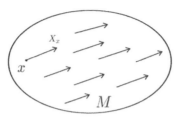

Figure 1.1. *Vector field*

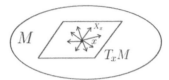

Figure 1.2. *Tangent space*

Note that the diagram

$$\begin{array}{ccc} M & \xrightarrow{X} & TM \\ & \searrow{id_M} & \downarrow{\pi} \\ & & M \end{array}$$

is commutative. In a local coordinate system $(x_1, ..., x_m)$ in a neighborhood $U \subset M$, the vector field X is written in the form $X = \sum_{k=1}^{m} f_k(x) \frac{\partial}{\partial x_k}$, $x \in U$, where the functions $f_1, ..., f_m : U \longrightarrow \mathbb{R}$, are the components of X with respect to $(x_1, ..., x_m)$. A vector field X is differentiable if its components $f_k(x)$ are differentiable functions. This definition of differentiability does not obviously depend on the choice of the local coordinate system. Indeed, if $(y_1, ..., y_m)$ is another local coordinate system in U, then $X = \sum_{k=1}^{m} h_k(x) \frac{\partial}{\partial y_k}$, $x \in U$, where $h_1, ..., h_m : U \longrightarrow \mathbb{R}$, are the components of X in relation to $(y_1, ..., y_m)$ and the result follows from the fact that $h_k(x) = \sum_{l=1}^{m} \frac{\partial y_k}{\partial x_l f_l(x)}$, $x \in U$. To the vector field X corresponds to a system of differential equations

$$\frac{dx_1}{dt} = f_1(x_1, ..., x_m), ..., \frac{dx_m}{dt} = f_m(x_1, ..., x_m). \qquad [1.2]$$

DEFINITION 1.5.– *A differentiable vector field X over M is called a dynamical system.*

A vector field is written locally in the form [1.2].

Symplectic Manifolds 9

DEFINITION 1.6.– *An integral curve (or trajectory) of a vector field X is a differentiable curve $\gamma : I \longrightarrow M$, $t \longmapsto \gamma(t)$, such that $\forall t \in I$, $\frac{d\gamma(t)}{dt} = X(\gamma(t))$, where I is an interval of $\mathbb{R}$.*

If $\sum_{k=1}^{m} f_k(x) \frac{\partial}{\partial x_k}$ is the local expression of X, then the integral curves (or trajectories) of X are the solutions $\gamma(t) = \{x_k(t)\}$ of [1.2]. We assume in the following that the vector field X is differentiable (of class $\mathcal{C}^\infty$) and with compact support (i.e. X is zero outside of a compact of M). This will especially be the case if the manifold M is compact. Given a point $x \in M$, we denote by $g_t^X(x)$ (or quite simply $g_t(x)$) the position of x after a displacement of a duration $t \in \mathbb{R}$.

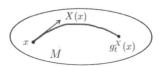

Figure 1.3. *Flow*

We therefore have an application $g_t^X : M \longrightarrow M$, $t \in \mathbb{R}$, which is a diffeomorphism (a one-to-one differentiable mapping with a differentiable inverse), by virtue of the theory of differential equations. More precisely, to the vector field X we associate a one-parameter group of diffeomorphisms g_t^X on M, that is, a differentiable application (of class $\mathcal{C}^\infty$): $M \times \mathbb{R} \longrightarrow M$, verifying a group law: (i) $\forall t \in \mathbb{R}$, $g_t^X : M \longrightarrow M$ is a diffeomorphism of M on M. (ii) $\forall t, s \in \mathbb{R}$, $g_{t+s}^X = g_t^X \circ g_s^X$. Condition (ii) means that the correspondence $t \longmapsto g_t^X$ is a homomorphism of the additive group $\mathbb{R}$ in the group of diffeomorphisms from M to M. It implies that $g_{-t}^X = \left(g_t^X\right)^{-1}$, because $g_0^X = id_M$ is the identical transformation that leaves each point invariant.

DEFINITION 1.7.– *The one-parameter group of diffeomorphism g_t^X on M is called flow. It admits the vector field X for velocity field $\frac{d}{dt} g_t^X(x) = X\left(g_t^X(x)\right)$, with the initial condition: $g_0^X(x) = x$.*

Obviously, $\left.\frac{d}{dt} g_t^X(x)\right|_{t=0} = X(x)$. So through these formulas $g_t^X(x)$ is the curve on the manifold, which passes through x such that the tangent at each point is the vector $X\left(g_t^X(x)\right)$. We will now see how to construct the flow g_t^X over the whole variety M.

THEOREM 1.7.– *The vector field X generates a unique one-parameter group of diffeomorphism of M.*

PROOF.– a) Construction of g_t^X for small t. For x fixed, the differential equation $\frac{d}{dt} g_t^X(x) = X\left(g_t^X(x)\right)$, function of t with the initial condition: $g_0^X(x) = x$, admits a unique solution g_t^X defined in the neighborhood of the point x_0 and depending on the initial condition $\mathcal{C}^\infty$. So g_t^X is locally a diffeomorphism. Therefore, for each point $x_0 \in M$, we can find a neighborhood $U(x_0) \subset M$, a positive real number $\varepsilon \equiv \varepsilon(x_0)$ such that for all $t \in \,]-\varepsilon,\varepsilon[$, the differential equation in question with its initial condition admits a unique solution differentiable $g_t^X(x)$ defined in $U(x_0)$ and verifying the group relation $g_{t+s}^X(x) = g_t^X \circ g_s^X(x)$, with $t, s, t+s \in \,]-\varepsilon,\varepsilon[$. Indeed, let us pose $x_1 = g_t^X(x)$, t fixed and consider the solution of the differential equation satisfying in the neighborhood of the point x_0 to the initial condition $g_{s=0}^X = x_1$. This solution satisfies the same differential equation and coincides at a point $g_t^X(x) = x_1$, with the function g_{t+s}^X. Therefore, by uniqueness of the solution of the differential equation, the two functions are locally equal. Therefore, the application g_t^X is locally a diffeomorphism. We recall that the vector field X is supposed to be differentiable (of class $\mathcal{C}^\infty$) and with compact support K. From the open cover of K formed by $U(x)$, we can extract a finite subcover (U_i), since K is compact. Let us denote ε_i by the numbers ε corresponding to U_i and put $\varepsilon_0 = \inf(\varepsilon_i)$, $g_t^X(x) = x$, $x \notin K$. Therefore, the differential equation in question admits a unique solution g_t^X on $M \times \,]-\varepsilon_0, \varepsilon_0[$ verifying the group relation: $g_{t+s}^X = g_t^X \circ g_s^X$, the inverse of g_t^X being g_{-t}^X and so g_t^X is a diffeomorphism for t small enough.

b) Construction of g_t^X for all $t \in \mathbb{R}$. According to (a), it suffices to construct g_t^X for $t \in \,]-\infty, -\varepsilon_0[\cup]\varepsilon_0, \infty[$. We will see that the applications g_t^X are defined according to the multiplication law of the group. Note that t can be written in the form $t = k\frac{\varepsilon_0}{2} + r$, with $k \in \mathbb{Z}$ and $r \in \left[0, \frac{\varepsilon_0}{2}\right[$. Let us consider, for $t \in \mathbb{R}_+^*$ and for $t \in \mathbb{R}_-^*$,

$$g_t^X = \underbrace{g_{\frac{\varepsilon_0}{2}}^X \circ \cdots \circ g_{\frac{\varepsilon_0}{2}}^X}_{k-\text{times}} \circ g_r^X, \qquad g_t^X = \underbrace{g_{-\frac{\varepsilon_0}{2}}^X \circ \cdots \circ g_{-\frac{\varepsilon_0}{2}}^X}_{k-\text{times}} \circ g_r^X,$$

respectively. The diffeomorphisms $g_{\pm\frac{\varepsilon_0}{2}}^X$ and g_r^X were defined in (a), and we deduce that for any real t, g_t^X is a diffeomorphism defined globally on the manifold M. $\square$

COROLLARY 1.1.– Every solution of the differential equation $\frac{dx(t)}{dt} = X(x(t))$, $x \in M$, with the initial condition x (for $t = 0$), can be extended indefinitely. The value of the solution $g_t^X(x)$ at the instant t is differentiable with respect to t and x.

With a slight abuse of notation, we can write the preceding differential equation in the form of the system of differential equations [1.2] with the initial conditions $x_1, ..., x_m$ for $t = 0$. With the vector field X, we associate the first-order differential operator L_X. We refer here to the differentiation of functions in the direction of the field X. We have $L_X : \mathcal{C}^\infty(M) \longrightarrow \mathcal{C}^\infty(M)$, $F \longmapsto L_X F(x) = \frac{d}{dt} F\left(g_t^X(x)\right)\big|_{t=0}$, $x \in M$. Here, $\mathcal{C}^\infty(M)$ designates the set of functions $F : M \longrightarrow \mathbb{R}$ of class $\mathcal{C}^\infty$. The operator L_X is linear: $L_X(\alpha_1 F_1 + \alpha_2 F_2) = \alpha_1 L_X F_1 + \alpha_2 L_X F_2$, $(\alpha_1, \alpha_2 \in \mathbb{R})$, and satisfies Leibniz's formula: $L_X(F_1 F_2) = F_1 L_X F_2 + F_2 L_X F_1$. Since $L_X F(x)$ only depends on the values of F in the neighborhood of x, we can therefore apply the operator L_X without the need to extend them to the whole manifold M. Let $(x_1, ..., x_m)$ be local coordinates on M. In this coordinate system, the vector X is given by its components $f_1, ..., f_m$ and the flow g_t^X is given by the system of differential equations [1.2]. So the derivative of the function $F = F(x_1, ..., x_m)$ in the direction X is $L_X F = f_1 \frac{\partial F}{\partial x_1} + \cdots + f_m \frac{\partial F}{\partial x_m}$. In other words, in the coordinates $(x_1, ..., x_m)$ the operator L_X has the form $L_X = f_1 \frac{\partial}{\partial x_1} + \cdots + f_m \frac{\partial}{\partial x_m}$.

DEFINITION 1.8.– *We say that two vector fields X_1 and X_2 on a manifold M commute (or are commutative) if and only if the corresponding flows commute,*

$$g_{t_1}^{X_1} \circ g_{t_2}^{X_2}(x) = g_{t_2}^{X_2} \circ g_{t_1}^{X_1}(x), \quad \forall x \in M.$$

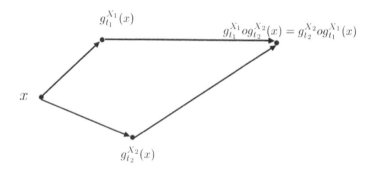

Figure 1.4. *Commutative flows*

THEOREM 1.8.– Two vector fields X_1 and X_2 on a manifold M commute if and only if, $[L_{X_1}, L_{X_2}] \equiv L_{X_1} L_{X_2} - L_{X_2} L_{X_1} = 0$.

PROOF.– a) Let us first show that the condition is necessary. Note that,

$$\frac{\partial^2}{\partial t_1 \partial t_2} \left(F\left(g_{t_2}^{X_2} \circ g_{t_1}^{X_1}(x) \right) - F\left(g_{t_1}^{X_1} \circ g_{t_2}^{X_2}(x) \right) \right) \bigg|_{t_2 = t_1 = 0}$$

$$= (L_{X_1} L_{X_2} - L_{X_2} L_{X_1}) F(x), \quad \forall F \in \mathcal{C}^\infty(M), \quad \forall x \in M.$$

Indeed, according to the definition of L_{X_2}, we find,

$$\frac{\partial}{\partial t_2} F\left(g_{t_2}^{X_2} \circ g_{t_1}^{X_1}(x) \right) \bigg|_{t_2 = 0} = L_{X_2} F\left(g_{t_1}^{X_1}(x) \right).$$

Thus,

$$\frac{\partial^2}{\partial t_1 \partial t_2} F\left(g_{t_2}^{X_2} \circ g_{t_1}^{X_1}(x)\right)\bigg|_{t_2=t_1=0} = \frac{\partial}{\partial t_1} L_{X_2} F\left(g_{t_1}^{X_1}(x)\right)\bigg|_{t_1=0},$$

$$= \frac{\partial}{\partial t_1} G\left(g_{t_1}^{X_1}(x)\right)\bigg|_{t_1=0} \quad \text{where } G \equiv L_{X_2} F,$$

$$= L_{X_1} G(x) \text{ by definition of } L_{X_1},$$

$$= L_{X_1} L_{X_2} F(x).$$

Likewise, we have

$$\frac{\partial^2}{\partial t_2 \partial t_1} F\left(g_{t_1}^{X_1} \circ g_{t_2}^{X_2}(x)\right)\bigg|_{t_1=0} = L_{X_1} F\left(g_{t_2}^{X_2}(x)\right),$$

and

$$\frac{\partial^2}{\partial t_2 \partial t_1} F\left(g_{t_1}^{X_1} \circ g_{t_2}^{X_2}(x)\right)\bigg|_{t_2=t_1=0} = L_{X_2} L_{X_1} F(x).$$

Therefore,

$$\frac{\partial^2}{\partial t_2 \partial t_1} F\left(g_{t_1}^{X_1} \circ g_{t_2}^{X_2}(x)\right)\bigg|_{t_1=0} - \frac{\partial^2}{\partial t_2 \partial t_1} F\left(g_{t_1}^{X_1} \circ g_{t_2}^{X_2}(x)\right)\bigg|_{t_2=t_1=0}$$

$$= \frac{\partial^2}{\partial t_1 \partial t_2}\left(F\left(g_{t_2}^{X_2} \circ g_{t_1}^{X_1}(x)\right) - F\left(g_{t_1}^{X_1} \circ g_{t_2}^{X_2}(x)\right)\right)\bigg|_{t_2=t_1=0},$$

$$= L_{X_1} L_{X_2} F(x) - L_{X_2} L_{X_1} F(x).$$

So if X_1 and X_2 commute on M, that is, $g_{X_1}^{t_1} \circ g_{t_2}^{X_2}(x) = g_{t_2}^{X_2} \circ g_{t_1}^{X_1}(x)$, $\forall x \in M$, then according to the above formula, $(L_{X_1} L_{X_2} - L_{X_2} L_{X_1}) F(x) = 0$, $\forall F \in C^\infty(M)$, $\forall x \in M$. Consequently, $L_{X_1} L_{X_2} = L_{X_2} L_{X_1}$.

b) Let us show that $g_{t_1}^{X_1} \circ g_{t_2}^{X_2}(x) = g_{t_2}^{X_2} \circ g_{t_1}^{X_1}(x)$, $\forall x \in M$, that is, that the condition is sufficient, or that: $F\left(g_{t_1}^{X_1} \circ g_{t_2}^{X_2}(x)\right) = F\left(g_{t_2}^{X_2} \circ g_{t_1}^{X_1}(x)\right)$, $\forall F \in C^\infty(M)$, $\forall x \in M$. Let us pose $\xi = g_{t_1}^{X_1} \circ g_{t_2}^{X_2}(x)$, $\zeta = g_{t_2}^{X_2} \circ g_{t_1}^{X_1}(x)$, and develop in Taylor series the function $F(\xi) - F(\zeta)$ at the neighborhood of $t_1 = t_2 = 0$. We have,

$$F(\xi) - F(\zeta) = F(x) - F(x) + t_1 \left(\frac{\partial}{\partial t_1}(F(\xi) - F(\zeta))\right)\bigg|_{t_1=t_2=0}$$

$$+ t_2 \left(\frac{\partial}{\partial t_2}(F(\xi) - F(\zeta))\right)\bigg|_{t_1=t_2=0} + \frac{t_1^2}{2}\left(\frac{\partial^2}{\partial t_1^2}(F(\xi) - F(\zeta))\right)\bigg|_{t_1=t_2=0}$$

$$+ \frac{t_2^2}{2}\left(\frac{\partial^2}{\partial t_2^2}(F(\xi) - F(\zeta))\right)\bigg|_{t_1=t_2=0} + t_1 t_2 \left(\frac{\partial^2}{\partial t_1 \partial t_2}(F(\xi) - F(\zeta))\right)\bigg|_{t_1=t_2=0}$$

$$+ o\left(t_1^3, t_2^3, t_1^2 t_2, t_1 t_2^2\right).$$

Let us calculate the different terms. We have

$$\frac{\partial}{\partial t_1} F(\xi)\bigg|_{t_1=t_2=0} = \frac{\partial}{\partial t_1} F\left(g_{t_1}^{X_1} \circ g_{t_2}^{X_2}(x)\right)\bigg|_{t_1 t_2=0}$$

$$= = L_{x_1} F\left(g_{t_2}^{X_2}(x)\right)\bigg|_{t_2=0} = L_{x_1} F(x),$$

and

$$\frac{\partial}{\partial t_1} F(\zeta)\bigg|_{t_1=t_2=0} = \frac{\partial}{\partial t_1} F\left(g_{t_2}^{X_2} \circ g_{t_1}^{X_1}(x)\right)\bigg|_{t_1=t_2=0},$$

$$= \frac{\partial}{\partial t_1} G\left(g_{t_1}^{X_1}(x)\right)\bigg|_{t_1=0} \text{ where } G = F g_{t_2}^{X_2}\bigg|_{t_2=0},$$

$$= L_{x_1} G(x) = L_{x_1} F\left(g_{t_2}^{X_2}\right)\bigg|_{t_2=0} = L_{x_1} F(x).$$

Therefore, we have $\frac{\partial}{\partial t_1}(F(\xi) - F(\zeta))\bigg|_{t_1=t_2=0} = 0$. By symmetry, we also have $\frac{\partial}{\partial t_2}(F(\xi) - F(\zeta))\bigg|_{t_1=t_2=0} = 0$. Likewise, we have

$$\frac{\partial^2}{\partial t_1^2}(F(\xi) - F(\zeta))\bigg|_{t_1=t_2=0} = \frac{\partial^2}{\partial t_1^2}\left(F\left(g_{t_1}^{X_1} \circ g_{t_2}^{X_2}(x)\right)\right.$$

$$\left. - F\left(g_{t_2}^{X_2} \circ g_{t_1}^{X_1}(x)\right)\right)\bigg|_{t_1=t_2=0}.$$

Now $\frac{\partial}{\partial t_1} F\left(g_{t_1}^{X_1} \circ g_{t_2}^{X_2}(x)\right) = \frac{\partial}{\partial t_1} F\left(g_{t_1}^{X_1}(y)\right) = L_{X_1} F\left(g_{t_1}^{X_1}(y)\right)$, where $y = g_{t_2}^{X_2}(x)$, so

$$\frac{\partial^2}{\partial t_1^2} F\left(g_{t_1}^{X_1} \circ g_{t_2}^{X_2}(x)\right) = \frac{\partial}{\partial t_1} L_{X_1} F\left(g_{t_1}^{X_1}(y)\right) = L_{X_1} L_{X_1} F\left(g_{t_1}^{X_1}(y)\right),$$

$$= L_{X_1} L_{X_1} F\left(g_{t_1}^{X_1} \circ g_{t_2}^{X_2}(x)\right) \underset{t_1 = t_2 = 0}{\longrightarrow} L_{X_1} L_{X_1} F(x).$$

Likewise, we have $\frac{\partial}{\partial t_1} F\left(g_{t_2}^{X_2} \circ g_{t_1}^{X_1}(x)\right) = \frac{\partial}{\partial t_1} G\left(g_{t_1}^{X_1}(x)\right) = L_{X_1} G\left(g_{t_1}^{X_1}(x)\right)$, where $G = F g_{t_2}^{X_2}$, hence

$$\frac{\partial^2}{\partial t_1^2} F\left(g_{t_2}^{X_2} \circ g_{t_1}^{X_1}(x)\right) = \frac{\partial}{\partial t_1} L_{X_1} G\left(g_{t_1}^{X_1}(x)\right) = L_{X_1} L_{X_1} G\left(g_{t_1}^{X_1}(x)\right),$$

$$= L_{X_1} L_{X_1} F\left(g_{t_2}^{X_2} \circ g_{t_1}^{X_1}(x)\right) \underset{t_1 = t_2 = 0}{\longrightarrow} L_{X_1} L_{X_1} F(x).$$

Thus, $\frac{\partial^2}{\partial t_1^2}(F(\xi) - F(\zeta))\Big|_{t_1 = t_2 = 0} = 0$. It follows, by symmetry, that $\frac{\partial^2}{\partial t_2^2}(F(\xi) - F(\zeta))\Big|_{t_1 = t_2 = 0} = 0$. Moreover, we deduce from the necessary condition and from the fact that the vector fields X_1 and X_2 commute the following relation:

$$\frac{\partial^2}{\partial t_1 \partial t_2}(F(\xi) - F(\zeta))|_{t_1 = t_2 = 0}$$

$$= \frac{\partial^2}{\partial t_1 \partial t_2}\left(F\left(g_{t_1}^{X_1} \circ g_{t_2}^{X_2}(x)\right) - F\left(g_{t_2}^{X_2} \circ g_{t_1}^{X_1}(x)\right)\right)\Big|_{t_1 = t_2 = 0},$$

$$= \frac{\partial^2}{\partial t_1 \partial t_2}\left(L_{X_2} L_{X_1} - L_{X_1} L_{X_2}\right) F(x) = 0.$$

Therefore, $F\left(g_{t_1}^{X_1} \circ g_{t_2}^{X_2}(x)\right) - F\left(g_{t_2}^{X_2} \circ g_{t_1}^{X_1}(x)\right) = o\left(t_1^3, t_2^3, t_1^2 t_2, t_1 t_2^2\right)$. Consider the times t_1 and t_2 of the order ε. We find a difference between the two new points of the manifold, depending on whether we apply the field X_1 before the field X_2 or the inverse, of the order of ε^3. $F\left(g_{t_1}^{X_1} \circ g_{t_2}^{X_2}(x)\right) - F\left(g_{t_2}^{X_2} \circ g_{t_1}^{X_1}(x)\right) = o\left(\varepsilon^3\right)$. Now, if t_1 and t_2 are arbitrary fixed times, let us square the space between the two paths with squares of sides ε. Each square represents the small space traveled during a small time ε, either according to the field X_1 or according to the field X_2. We have found that when the space between two paths differs from that of a square, we get a difference ε^3. By modifying the path traveled by a square in successive stages, we obtain

$$F\left(g_{t_1}^{X_1} \circ g_{t_2}^{X_2}(x)\right) - F\left(g_{t_2}^{X_2} \circ g_{t_1}^{X_1}(x)\right) \leq \frac{t_1 t_2}{\varepsilon^2} o\left(\varepsilon^3\right),$$

by the fact that we have $\frac{t_1}{\varepsilon} \times \frac{t_2}{\varepsilon}$ steps intermediaries. This is valid for all ε; just take ε small enough, tending to zero, so that $\frac{t_1 t_2}{\varepsilon^2} o\left(\varepsilon^3\right) = t_1 t_2 o\left(\varepsilon\right) \xrightarrow[\varepsilon \to 0]{} 0.$ □

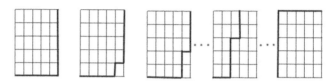

Figure 1.5. *Proof of the commutativity of flows*

Since every first-order linear differential operator is given by a vector field, $L_{X_2} L_{X_1} - L_{X_1} L_{X_2}$ being a first-order linear differential operator, the latter also corresponds to some vector field that we denote by X_3.

DEFINITION 1.9.– *The Poisson bracket or commutator of two vector fields X_1 and X_2 on the manifold M, denoted by $X_3 = \{X_1, X_2\}$ or $X_3 = [X_1, X_2]$, is the vector field X_3 for which $L_{X_3} = L_{X_2} L_{X_1} - L_{X_1} L_{X_2}$.*

EXAMPLE 1.4.– The Poisson bracket transforms the vector space of vector fields over a manifold into Lie algebra.

Let X be a vector field on a differentiable manifold M. We have shown (theorem 1.7) that X generates a unique one-parameter group of diffeomorphism g_t^X (which we also denote by g_t) on M, solution to the differential equation: $\frac{d}{dt} g_t^X(p) = X(g_t^X(p))$, $p \in M$, with the initial condition $g_0^X(p) = p$. Let ω be a k-form differential on M.

DEFINITION 1.10.– *The Lie derivative of ω with respect to X is the k-form differential defined by $L_X \omega = \frac{d}{dt} g_t^* \omega \big|_{t=0} = \lim_{t \to 0} \frac{g_t^*(\omega(g_t(p))) - \omega(p)}{t}$.*

In general, for $t \neq 0$, we have

$$\frac{d}{dt} g_t^* \omega = \frac{d}{ds} g_{t+s}^* \omega \bigg|_{s=0} = g_t^* \frac{d}{ds} g_s^* \omega \bigg|_{s=0} = g_t^*(L_X \omega). \qquad [1.3]$$

For all $t \in \mathbb{R}$, the application $g_t : \mathbb{R} \longrightarrow \mathbb{R}$ being a diffeomorphism then dg_t and dg_{-t} are the applications, $dg_t : T_p M \longrightarrow T_{g_t(p)} M$, $dg_{-t} : T_{g_t(p)} M \longrightarrow T_p M$.

DEFINITION 1.11.– *The Lie derivative of a vector field Y in the direction X is defined by $L_X Y = \frac{d}{dt} g_{-t} Y \big|_{t=0} = \lim_{t \to 0} \frac{g_{-t}(Y(g_t(p))) - Y(p)}{t}$.*

In general, for $t \neq 0$, we have

$$\frac{d}{dt} g_{-t} Y = \frac{d}{ds} g_{-t-s} Y \bigg|_{s=0} = g_{-t} \frac{d}{ds} g_{-s} Y \bigg|_{s=0} = g_{-t}(L_Y).$$

DEFINITION 1.12.– *The interior product of a k-form differential ω by a vector field X on the differentiable manifold M is a $(k-1)$-form differential, $i_X \omega$, defined by $(i_X \omega)(X_1, ..., X_{k-1}) = \omega(X, X_1, ..., X_{k-1})$, where $X_1, ..., X_{k-1}$ are vector fields.*

It is easy to show that if ω is a k-differential form, λ a differential form of any degree, X and Y two vector fields, f a linear map and a a constant, then

$$i_{X+Y} \omega = i_X \omega + i_Y \omega, \qquad i_{aX} \omega = a i_X \omega, \qquad i_X i_Y \omega = -i_Y i_X \omega,$$

$$i_X i_X \omega = 0, \qquad i_X(f\omega) = f(i_X \omega),$$

$$i_X(\omega \wedge \lambda) = (i_X \omega) \wedge \lambda + (-1)^k \omega \wedge (i_X \lambda), \qquad i_X f^* \omega = f^*(i_{fX} \omega),$$

where $f^* \omega$ denote the pull-back by f.

EXAMPLE 1.5.– Let us calculate the expression of the interior product in local coordinates. If $X = \sum_{j=1}^{m} X_j(x) \frac{\partial}{\partial x_j}$ is the local expression of the vector field on the manifold M of dimension m and $\omega = \sum_{i_1 < i_2 < ... < i_k} f_{i_1...i_k}(x) dx_{i_1} \wedge ... \wedge dx_{i_k}$ is a k-differential form, then

$$i_X \omega = \omega(X, \cdot) = \sum_{i_2 < i_3 < ... < i_k} \sum_{j=1}^{m} f_{j i_2 ... i_k} X_j dx_{i_2} \wedge ... \wedge dx_{i_k}$$

$$- \sum_{i_1 < i_3 < ... < i_k} \sum_{j=1}^{m} f_{i_1 j ... i_k} X_j dx_{i_1} \wedge dx_{i_3} \wedge ... \wedge dx_{i_k} + \cdots$$

$$+ (-1)^{k-1} \sum_{i_1 < i_2 < ... < i_{k-1}} \sum_{j=1}^{m} f_{i_1 i_2 ... j} X_j dx_{i_1} \wedge dx_{i_2} \wedge ... \wedge dx_{i_{k-1}},$$

$$= k \sum_{i_2 < i_3 < ... < i_k} \sum_{j=1}^{m} f_{j i_2 ... i_k} X_j dx_{i_2} \wedge ... \wedge dx_{i_k}.$$

Hence, $i_{\frac{\partial}{\partial x_j}} \omega = \frac{\partial}{\partial (dx_j)} \omega$, where we put dx_j in the first position in ω.

The following properties are often involved in solving practical problems using Lie derivatives.

PROPOSITION 1.1.– a) If $f : M \longrightarrow \mathbb{R}$ is a differentiable function, then the Lie derivative of f is the image of X by the differential of f, $L_X f = df(X) = X.f$.
b) L_X and d commute, $L_X \circ d = d \circ L_X$. c) Let $X, X_1, ..., X_k$ be vector fields on M and ω a k-form differential. So

$$(L_X \omega)(X_1, ..., X_k) = L_X(\omega(X_1, ..., X_k)) - \sum_{j=1}^{k} \omega(X_1, ..., L_X X_j, ..., X_k).$$

d) For all differential forms ω and λ, $L_X(\omega \wedge \lambda) = L_X \omega \wedge \lambda + \omega \wedge L_X \lambda$.

PROOF.– a) Indeed, we have

$$L_X f = \frac{d}{dt} g_t^* f \bigg|_{t=0} = \frac{d}{dt} f \circ g_t \bigg|_{t=0} = df \left(\frac{dg_t}{dt} \right) \bigg|_{t=0} = df(X),$$

and (see theorem 1.9), $L_X f = i_X df = X.f$, hence the result.

b) Indeed, as the differential and the inverse image commute, then

$$d \circ L_X \omega = d \circ \frac{d}{dt} g_t^* \omega \bigg|_{t=0} = \frac{d}{dt} g_t^* \circ d\omega \bigg|_{t=0} = L_X \circ d\omega.$$

c) We have

$$(L_X \omega)(X_1, ..., X_k) = \frac{d}{dt} g_t^* \omega(X_1, ..., X_k) \bigg|_{t=0},$$

$$= \frac{d}{dt} \omega(g_t)(dg_t X_1, ..., dg_t X_k) \bigg|_{t=0},$$

$$= L_X \omega(g_t)(dg_t X_1, ..., dg_t X_k)|_{t=0}$$

$$+ \sum_{j=1}^{k} \omega(g_t) \left(dg_t X_1, ..., \frac{d}{dt} dg_t X_j, ..., dg_t X_k \right) \bigg|_{t=0},$$

and the result is deduced from the fact that

$$\frac{d}{dt} dg_t X_j \bigg|_{t=0} = -\frac{d}{dt} dg_{-t} X_j \bigg|_{t=0} = -L_X X_j.$$

d) Just consider $\omega = f dx_{i_1} \wedge ... \wedge dx_{i_k}$ and $\lambda = g dx_{j_1} \wedge ... \wedge dx_{j_l}$. We have $\omega \wedge \lambda = fg dx_{i_1} \wedge ... \wedge dx_{i_k} \wedge dx_{j_1} \wedge ... \wedge dx_{j_l}$, and

$$L_X(\omega \wedge \lambda)(X_1, ..., X_k, X_{k+1}, ..., X_{k+l})$$
$$= (L_X f).g dx_{i_1} \wedge ... \wedge dx_{i_k} \wedge dx_{j_1} \wedge ... \wedge dx_{j_l}(X_1, ..., X_k, X_{k+1}, ..., X_{k+l})$$
$$+ f(L_X g) dx_{i_1} \wedge ... \wedge dx_{i_k} \wedge dx_{j_1} \wedge ... \wedge dx_{j_l}(X_1, ..., X_k, X_{k+1}, ..., X_{k+l}),$$
$$= ((L_X \omega) \wedge \lambda + \omega \wedge (L_X \lambda))(X_1, ..., X_k, X_{k+1}, ..., X_{k+l}),$$

and the result follows. $\square$

PROPOSITION 1.2.– Let X and Y be two vector fields on M. Then, the Lie derivative of $L_X Y$ is the Lie bracket $[X, Y]$.

PROOF.– We have $L_X Y(f) = \lim_{t \to 0} \frac{dg_{-t} Y - Y}{t}(f) = \lim_{t \to 0} dg_{-t} \frac{Y - dg_t Y}{t}(f)$, hence, $L_X Y(f) = \lim_{t \to 0} \frac{Y(f) - dg_t Y(f)}{t} = \lim_{t \to 0} \frac{Y(f) - Y(f \circ g_t) \circ g_t^{-1}}{t}$. Put $g_t(x) \equiv g(t, x)$, and apply to $g(t, x)$ the Taylor formula with integral remainder. So there is $h(t, x)$ such that: $f(g(t, x)) = f(x) + th(t, x)$, with $h(0, x) = \frac{\partial}{\partial t} f(g(t, x))(0, x)$. According to the definition of the tangent vector, we have $X(f) = \frac{\partial}{\partial t} f \circ g_t(x)(0, x)$, hence $h(0, x) = X(f)(x)$. Therefore,

$$L_X Y(f) = \lim_{t \to 0} \left(\frac{Y(f) - Y(f) \circ g_t^{-1}}{t} - Y(h(t, x)) \circ g_t^{-1} \right),$$
$$= \lim_{t \to 0} \left(\frac{(Y(f) \circ g_t - Y(f)) \circ g_t^{-1}}{t} - Y(h(t, x)) \circ g_t^{-1} \right).$$

Since $\lim_{t \to 0} g_t^{-1}(x) = g_0^{-1}(x) = id.$, we deduce that:

$$L_X Y(f) = \lim_{t \to 0} \left(\frac{Y(f) \circ g_t - Y(f)}{t} - Y(h(0, x)) \right),$$
$$= \frac{\partial}{\partial t} Y(f) \circ g_t(x) - Y(X(f)),$$
$$= X(Y(f)) - Y(X(f)) = [X, Y],$$

which completes the proof. $\square$

We will now establish a fundamental formula for the Lie derivative, which can be used as a definition.

THEOREM 1.9.– Let X be a vector field on M and ω a differential k-form. Then $L_X \omega = d(i_X \omega) + i_X(d\omega)$. We have the Cartan homotopy formula $L_X = d \circ i_X + i_X \circ d$.

PROOF.– We will reason by induction on the degree k of the differential form ω. Let $D_X \equiv d \circ i_X + i_X \circ d$. For a differential 0-form, that is, a function f, we have $D_X f = d(i_X f) + i_X(df)$, or $i_X f = 0$, hence $d(i_X f) = 0$, $i_X df = df(X)$, and so $D_X f = df(X)$. We know (proposition 1.1, (a)) that $L_X f = df(X) = X.f$, so $D_X f = L_X f$. Assume that the formula in question is true for a differential $(k-1)$-form and is proved to be true for a differential k-form. Let λ be a differential $(k-1)$-form and let $\omega = df \wedge \lambda$, where f is a function. We have

$$L_X \omega = L_X(df \wedge \lambda) = L_X df \wedge \lambda + df \wedge L_X \lambda, \quad \text{(proposition 1.1, (d))},$$
$$= dL_X f \wedge \lambda + df \wedge L_X \lambda, \quad \text{(because } L_X df = dL_X f, \text{proposition 1.1, (b))},$$
$$= d(df(X)) \wedge \lambda + df \wedge L_X \lambda, \quad \text{(because } L_X f = df(X), \text{proposition 1.1, (a))}.$$

Hypothetically, we have $L_X \lambda = d(i_X \lambda) + i_X(d\lambda)$. Since $i_X df = df(X)$, then

$$L_X \omega = d(i_X df) \wedge \lambda + df \wedge d(i_X \lambda) + df \wedge i_X(d\lambda). \qquad [1.4]$$

Moreover, we have

$$i_X d\omega = i_X d(df \wedge \lambda) = -i_X(df \wedge d\lambda) = -(i_X df) d \wedge d\lambda + df \wedge i_X(d\lambda),$$

and

$$d(i_X \omega) = d i_X(df \wedge \lambda) = d\left((i_X df) \wedge \lambda - df \wedge (i_X \lambda)\right),$$
$$= d(i_X df) \wedge \lambda + (i_X df) \wedge d\lambda + df \wedge d(i_X \lambda),$$

because $d(df) = 0$. Hence,

$$d i_X(df \wedge \lambda) + i_X d(df \wedge \lambda) = d(i_X df) \wedge \lambda + df \wedge d(i_X \lambda) + df \wedge i_X(d\lambda).$$

Comparing this expression with that obtained in [1.4] above, we finally obtain $L_X \omega = d(i_X \omega) + i_X(d\omega)$, and the theorem is proved. □

PROPOSITION 1.3.– For a differential form ω, we have $i_X L_X \omega = L_X i_X \omega$.

PROOF.– Taking into account the previous theorem and the fact that $i_X i_X = 0$, we obtain

$$i_X L_X \omega = i_X(d i_X \omega) + i_X(i_X d\omega) = i_X(d i_X \omega),$$
$$L_X i_X \omega = (d \circ i_X + i_X \circ d) i_X \omega = i_X(d i_X \omega),$$

hence $i_X L_X \omega - L_X i_X \omega = 0$ and the proof ends. □

PROPOSITION 1.4.– Let X and Y be two vector fields on M and ω a differential form. Then, $L_{X+Y}\omega = L_X\omega + L_Y\omega$, $L_{fX}\omega = fL_X\omega + df \wedge i_X\omega$, where $f : M \longrightarrow \mathbb{R}$ is a differentiable function.

PROOF.– Indeed, just use theorem 1.9,

$$L_{X+Y}\omega = d(i_{X+Y}\omega) + i_{X+Y}(d\omega) = d(i_X\omega + i_Y\omega) + i_X(d\omega) + i_Y(d\omega),$$
$$= d(i_X\omega) + i_X(d\omega) + d(i_Y\omega) + i_Y(d\omega) = L_X\omega + L_Y\omega.$$

Similarly, we have

$$L_{fX}\omega = d(i_{fX}\omega) + i_{fX}(d\omega) = d(fi_X\omega) + fi_X(d\omega),$$
$$= df \wedge i_X\omega + fd(i_X\omega) + fi_X(d\omega) = df \wedge i_X\omega + fL_X\omega,$$

which completes the proof. □

EXAMPLE 1.6.– The expression of the Lie derivative of the differential form $\omega = \sum_{i_1<...<i_k} f_{i_1...i_k} dx_{i_1} \wedge ... \wedge dx_{i_k}$ in local coordinates is given by

$$L_X\omega = \sum_{i_1<...<i_k} \sum_{j=1}^{m} \left(\frac{\partial f_{i_1...i_k}}{\partial x_j} X_j + k f_{ji_2...i_k} \frac{\partial X_j}{\partial x_{i_1}} \right) dx_{i_1} \wedge ... \wedge dx_{i_k}.$$

Indeed, if $X = \sum_{j=1}^{m} X_j(x) \frac{\partial}{\partial x_j}$ is the local expression of the vector field on the m-dimensional manifold M, then

$$L_X\omega = \sum_{j=1}^{m} L_{X_j \frac{\partial}{\partial x_j}} \omega = \sum_{j=1}^{m} \left(dX_j \wedge i_{\frac{\partial}{\partial x_j}} \omega + X_j L_{\frac{\partial}{\partial x_j}} \omega \right).$$

According to example 1.5, we know that

$$i_{\frac{\partial}{\partial x_j}}\omega = \frac{\partial}{\partial(dx_j)}\omega = k \sum_{i_2<i_3<...<i_k} f_{ji_2...i_k} dx_{i_2} \wedge dx_{i_3} \wedge ... \wedge dx_{i_k},$$

hence, $dX_j \wedge i_{\frac{\partial}{\partial x_j}}\omega = k \sum_{i_1<i_2<...<i_k} f_{ji_2...i_k} \frac{\partial X_j}{\partial x_{i_1}} dx_{i_1} \wedge ... \wedge dx_{i_k}$. Similarly, using proposition 1.1, (c), we obtain $L_{\frac{\partial}{\partial x_j}}\omega = \sum_{i_1<...<i_k} \frac{\partial f_{i_1...i_k}}{\partial x_j} dx_{i_1} \wedge ... \wedge dx_{i_k}$. Since $\left[\frac{\partial}{\partial x_j}, \frac{\partial}{\partial x_l}\right] = 0$, we finally get the result.

PROPOSITION 1.5.– We have $[L_X, i_Y] = i_{[X,Y]}$, $[L_X, L_Y] = L_{[X,Y]}$, where X and Y are two vector fields on M.

PROOF.– The proof is to show that for a differential k-form ω, we have $[L_X, i_Y]\omega = i_{[X,Y]}\omega$ and $[L_X, L_Y]\omega = L_{[X,Y]}\omega$. We reason by induction assuming first that $k = 1$, that is, $\omega = df$. We have $[L_X, i_Y]df = L_X i_Y df - i_Y L_X df = L_X(Y.f) - i_Y dL_X f$, because $L_X \circ d = d \circ L_X$). Since $L_X f = X.f$, we have $[L_X, i_Y]df = X.(Y.f) - i_Y d(X.f) = X.(Y.f) - Y.(X.f), = [X, Y].f = i_{[X,Y]}df$. Suppose the formula in question is true for a form ω of degree less than or equal to $k - 1$. Let λ and θ be two forms of degree less than or equal to $k - 1$, so that $\omega = \lambda \wedge \theta$ is a form of degree k. We have

$$\begin{aligned}
[L_X, i_Y]\omega &= L_X i_Y \omega - i_Y L_X \omega = L_X i_Y (\lambda \wedge \theta) - i_Y L_X (\lambda \wedge \theta), \\
&= L_X(i_Y \lambda \wedge \theta + (-1)^{deg\lambda} \lambda \wedge i_Y \theta) - i_Y(L_X \lambda \wedge \theta + \lambda \wedge L_X \theta), \\
&= L_X i_Y \lambda \wedge \theta + i_Y \lambda \wedge L_X \theta + (-1)^{deg\lambda} L_X \lambda \wedge i_Y \theta \\
&\quad + (-1)^{deg\lambda} \lambda \wedge L_X i_Y \theta - i_Y L_X \lambda \wedge \theta - (-1)^{deg\lambda} L_X \lambda \wedge i_Y \theta \\
&\quad - i_Y \lambda \wedge L_X \theta - (-1)^{deg\lambda} \lambda \wedge i_Y L_X \theta, \\
&= (L_X i_Y \lambda - i_Y L_X \lambda) \wedge \theta + (-1)^{deg\lambda} \lambda \wedge (L_X i_Y \theta - i_Y L_X \theta), \\
&= i_{[X,Y]} \lambda \wedge \theta + (-1)^{deg\lambda} \lambda \wedge i_{[X,Y]} \theta, \\
&= i_{[X,Y]} (\lambda \wedge \theta) = i_{[X,Y]} \omega.
\end{aligned}$$

Similarly, we have

$$\begin{aligned}
[L_X, L_Y]\omega &= L_X L_Y \omega - L_Y L_X \omega = L_X d i_Y \omega + L_X i_Y d\omega - d i_Y L_X \omega - i_Y dL_X \omega, \\
&= dL_X i_Y \omega + L_X i_Y d\omega - d i_Y L_X \omega - i_Y L_X d\omega, (L_X d\omega = dL_X \omega) \\
&= d i_{[X,Y]} \omega + i_{[X,Y]} d\omega, \text{ (according to } [L_X, i_Y] = i_{[X,Y]}) = L_{[X,Y]}\omega,
\end{aligned}$$

and the proof ends. $\square$

REMARK 1.2.– Using the results above, we give a quick proof of the Poincaré lemma: in the neighborhood of a point of a manifold, any closed differential form is exact. Consider the differential equation in $\mathbb{R}^n$, $\dot{x} = X_t(x) = \frac{x}{t}$, whose solution $g_t(x_0) = x_0 t$ is a one-parameter group of diffeomorphisms. That proof is based on the properties of this differential equation, whose solution is a family of maps $g_t(x) = x_0 t, t \in]0, 1]$. It can also be regarded as a time dependent family of vector fields $X(t)$. Note that they are only defined for $t > 0$. For $t = 0$, the right part of that equation is undefined. Nevertheless, that family $g_t(x) = x_0 t$ can be defined for all $t \in [0, 1]$, while for $t = 0$, the map $g_0(x) = 0$ is a constant map. We have $g_0(x_0) = 0$, $g_1(x_0) = x_0$, $g_0^* \omega = 0$, $g_1^* \omega = \omega$, and $\omega = g_1^* \omega - g_0^* \omega = \int_0^1 \frac{d}{dt} g_t^* \omega \, dt = \int_0^1 g_t^* (L_X \omega) dt$, according to [1.3]. By theorem 1.9 and the fact that $d\omega = 0$, we have $\omega = \int_0^1 g_t^* (di_X \omega) dt = \int_0^1 dg_t^* i_X \omega \, dt$, because $df^* \omega = f^* d\omega$). We can, therefore, find a differential form λ such that: $\omega = d\lambda$, where $\lambda = \int_0^1 g_t^* i_X \omega \, dt$, which completes the proof.

1.5. The Darboux theorem

We will study a central theorem of symplectic geometry, namely Darboux's theorem, which states that every point in a symplectic manifold has a neighborhood with the so-called Darboux coordinates. The Darboux theorem plays a central role in symplectic geometry; the symplectic manifolds (M, ω) of dimension $2m$ are locally isomorphic to $(\mathbb{R}^{2m}, \omega)$. More precisely, if (M, ω) is a symplectic manifold of dimension $2m$, then in the neighborhood of each point of M, there exist local coordinates $(x_1, ..., x_{2m})$ such that: $\omega = \sum_{k=1}^{m} dx_k \wedge dx_{m+k}$. We call such coordinates Darboux coordinates. In particular, there is no local invariant in symplectic geometry, analogous to the curvature in Riemannian geometry. The classical proof given by Darboux is by induction on the dimension of the manifold and is accessible in different versions in the basic literature (see, for example, Arnold 1989; Sternberg 1983). We will give an overview in remark 1.3. The proof presented in this section, however, is due to J. Moser (consult Guillemin and Sternberg 1984) and is based on standard differential calculus. The idea of proof, known as the Moser's trick, works in many situations. The Moser's trick involves constructing an appropriate isotopy g_t generated by a time-dependent vector field X_t on M such that $g_t^* \omega_t = \omega_0$. Let us also point out the information that Givental (see Arnold and Givental 1990) proved, that a germ of a submanifold in a symplectic manifold is defined, up to a symplectomorphism, by the restriction of the symplectic structure to the tangent bundle of the submanifold. This theorem constitutes a generalization of Darboux's theorem. The latter corresponds to the particular case when the submanifold is a point. Givental's theorem is closely related to the following generalization of Darboux's theorem obtained by Weinstein (1971): a submanifold of a symplectic space is defined, up to a symplectomorphism of a neighborhood of it, by the restriction of the symplectic form to the bundle of tangent vectors of the ambient space at the point of the submanifold. Weinstein's theorem is global in the sense that it gives a conclusion on the whole submanifold. Givental's theorem is local; however, it requires only interior information, whereas in order to be able to apply Weinstein's theorem we must know some exterior information (namely, the values of the symplectic form on non-tangential elements). We will first give a proof of Moser (1965) and then prove Darboux's theorem based on this lemma.

LEMMA 1.1.– Let $\{\omega_t\}$, $0 \leq t \leq 1$, be a family of symplectic forms, differentiable in t. Then, for all $p \in M$, there exists a neighborhood $\mathcal{U}$ of p and a function $g_t : \mathcal{U} \longrightarrow \mathcal{U}$, such that: $g_0^* = $ identity and $g_t^* \omega_t = \omega_0$.

PROOF.– Let us first remember that we have shown in theorem 1.7 that the vector field generates a unique one-parameter group of diffeomorphism of M. Looking for a family of vector fields X_t on $\mathcal{U}$ such that these fields generate locally a one-parameter group of diffeomorphisms g_t with $\frac{d}{dt} g_t(p) = X_t(g_t(p))$, $g_0(p) = p$. First note that the form ω_t is closed (i.e. $d\omega_t = 0$) as the form $\frac{d}{dt}\omega_t$ (since $d\frac{d}{dt}\omega_t = \frac{d}{dt}d\omega_t = 0$). Therefore, by deriving the relationship $g_t^* \omega_t = \omega_0$ and using the Cartan homotopy formula $L_{X_t} = i_{X_t}d + di_{X_t}$, taking into account that ω_t depends on time, we obtain the expression $\frac{d}{dt}g_t^*\omega_t = g_t^*\left(\frac{d}{dt}\omega_t + L_{X_t}\omega_t\right) = g_t^*\left(\frac{d}{dt}\omega_t + di_{X_t}\omega_t\right)$. By Poincaré's lemma (in the neighborhood of a point, any closed differential form is exact), the form $\frac{\partial}{\partial t}\omega_t$ is exact in the neighborhood of p. In other words, we can find a form λ_t such that: $\frac{d}{dt}\omega_t = d\lambda_t$. Hence,

$$\frac{d}{dt}g_t^*\omega_t = g_t^*d(\lambda_t + i_{X_t}\omega_t). \qquad [1.5]$$

We want to show that for all $p \in M$, there exists a neighborhood $\mathcal{U}$ of p and a function $g_t : \mathcal{U} \longrightarrow \mathcal{U}$, such that: $g_0^* = $ identity and $g_t^*\omega_t = \omega_0$, therefore, $\frac{d}{dt}g_t^*\omega_t = 0$. By [1.5], the problem amounts to finding X_t such that: $\lambda_t + i_{X_t}\omega_t = 0$. Since the form ω_t is non-degenerate, the above equation is solvable with respect to the vector field X_t and (according to theorem 1.7) defines the family $\{g_t\}$ for $0 \leq t \leq 1$. In local coordinates (x_k) of the $2m$-dimensional manifold M, with $\left(\frac{\partial}{\partial x_k}\right)$ a basis of TM and (dx_k) the dual basis of $\left(\frac{\partial}{\partial x_k}\right)$, $k = 1,...,2m$, we have $\lambda_t = \sum_{k=1}^{2m}\lambda_k(t,x)dx_k$, $X_t = \sum_{k=1}^{2m}X_k(t,x)\frac{\partial}{\partial x_k}$, $\omega_t = \sum_{\substack{k,l=1\\k<l}}^{2m}\omega_{k,l}(t,x)dx_k \wedge dx_l$, $i_{X_t}\omega_t = 2\sum_{l=1}^{2m}\left(\sum_{k=1}^{2m}\omega_{k,l}X_k\right)dx_l$. We, therefore, solve the system of equations in $x_k(t,x)$ according to: $\lambda_l(t,x) + 2\sum_{k=1}^{2m}\omega_{k,l}(t,x)X_k(t,x) = 0$. The form ω_t is non-degenerate and the matrix $(\omega_k(t,x))$ is non-singular. Then the above system has a unique solution. This determines the vector field X_t and thus functions g_t^* such that: $g_t^*\omega_t = \omega_0$, which completes the proof. $\square$

Using the above lemma, we give a proof (we will proceed with "Moser's trick") of the Darboux theorem (sometimes also referred to as the Darboux–Weinstein theorem).

THEOREM 1.10.– Any symplectic form on a manifold M of dimension $2m$ is locally diffeomorphic to the standard form on $\mathbb{R}^{2m}$. In other words, if (M,ω) is a symplectic manifold of dimension $2m$, then in the neighborhood of each point p of M, there exist local coordinates $(x_1,...,x_{2m})$ such that: $\omega = \sum_{k=1}^{m}dx_k \wedge dx_{m+k}$.

PROOF.– In theorem 1.7, we have seen that the vector field X generates a unique one-parameter group of diffeomorphism g_t^X of M with $\frac{d}{dt}g_t(p) = X_t(g_t(p))$, $g_0(p) = p$. We have given a detailed proof on the construction of the flow g_t^X for small t and for all $t \in \mathbb{R}$. In the proof of Darboux's theorem, one can be satisfied with the symplectic forms defined on an open neighborhood of $0 \in \mathbb{R}^{2m}$. Indeed, since the statement is local, without loss of generality, we may assume that $M = \mathbb{R}^{2m}$ and $p = 0$. If we have a symplectic manifold (M,ω) and a point p, we can take a smooth coordinate chart about p and then use the coordinate function to push ω forward to a symplectic form on a neighborhood of 0 in R^{2m}. If the result holds on $\mathbb{R}^{2m}$, we can compose the coordinate chart with the resulting symplectomorphism to get the theorem in general. Let $\{\omega_t\}$, $0 \leq t \leq 1$, be a family of 2-differential forms that depends differentiably on t and let $\omega_t = \omega_0 + t(\omega - \omega_0)$, $\omega_0 = \sum_{k=1}^{m} dx_k \wedge d_{m+k}$, where $(x_1, ..., x_{2m})$ are local coordinates on $\mathbb{R}^{2m}$. These ω_t's are volume forms. Note that these 2-forms are closed. We have $\omega_0(0) = \omega(0)$, so $\omega_t(0) = \omega_0(0)$. Since the function $\det : [0,1] \times \mathbb{R}^{2m} \longrightarrow \mathbb{R}$, $(t, p) \longmapsto \det \omega_t(p)$ is continuous and as ω_0 is non-degenerate, we can find a small neighborhood of $[0, 1] \times \{0\}$ in $\mathbb{R} \times \mathbb{R}^{2m}$ where the form $\omega_t(0)$ is non-degenerate for all $t \in [0, 1]$ (in short, ω_t is non-degenerate in a neighborhood of 0 and independent of t at 0). Since the interval $[0, 1]$ is compact and $\mathbb{R}^{2m}$ is locally compact, we can find a neighborhood of 0 on which all the ω_t are symplectic forms. In other words, ω_t are symplectic forms and by lemma 1.1, for all $0 \in \mathbb{R}^{2m}$, there exists a neighborhood $\mathcal{U}$ of 0 and a function $g_t : \mathcal{U} \longrightarrow \mathcal{U}$ such that: $g_t^* = $ identity and $g_t^* \omega_t = \omega_0$. Differentiating this relation with respect to t, we obtain (as in the proof of lemma 1.1) $\frac{d}{dt} g_t^* \omega_t = 0$, $g_t^* \left(\frac{d}{dt} \omega_t + L_{X_t} \omega_t \right) = 0$, $g_t^* \left(\frac{d}{dt} \omega_t + di_{X_t} \omega_t \right) = 0$. Therefore, $di_{X_t} \omega_t = -\frac{d}{dt} \omega_t$ and since the form $\frac{d}{dt} \omega_t$ is exact in the neighborhood of 0 (indeed, the forms ω_0 and ω are closed, so it is the same for $\omega_0 - \omega$ and we can therefore, according to Poincaré's lemma, find in the neighborhood of 0 a 1-form θ_t such that: $\frac{d}{dt} \omega_t = \omega_0 - \omega = d\theta_t$ on $\mathcal{U}$, and by adding an exact 1-form if necessary, we can achieve that $\theta_t(0) = 0$). Since $\theta_t(0) = 0$, we see that $X_t(0) = 0$, that is 0 is a fixed point of g_t. Then $di_{X_t} \omega_t = d\theta_t$, where θ_t is a 1-differential form. In addition, with ω_t being non-degenerate, the equation $i_{X_t} \omega_t = \theta_t$ is solvable and determines solely the vector field X_t depending on t. Note that for $t = 1$, $\omega_1 = \omega$ and for $t = 0$, $\omega_0 = \omega_0$, and also we can find g_1^* such that: $g_1^* \omega = \omega_0$. According to theorem 1.7, vector fields X_t generate one-parameter families of diffeomorphisms $\{g_t\}$, $0 \leq t \leq 1$. In other words, you can make a change of coordinates such as: $\omega = \sum_{k=1}^{m} dx_k \wedge d_{m+k}$, and the proof is completed. $\square$

REMARK 1.3.– As we pointed out at the beginning of this section, we will give an overview on the classical proof given by Darboux of his theorem. We proceed by induction on m. Suppose the result is true for $m - 1 \geq 0$ and show that it is also for m. Fix x and let x_{m+1} be a differentiable function on M whose differential dx_m is a non-zero point x. Let X be the unique differentiable vector field satisfying the relation $i_X \omega = dx_{m+1}$. As this vector field does not vanish at x, we can find a function x_1 in a neighborhood $\mathcal{U}$ of x such that $X(x_1) = 1$. Consider a vector field Y on $\mathcal{U}$ satisfying the relation $i_Y \omega = -dx_1$. Since $d\omega = 0$, then $L_X \omega = L_Y \omega = 0$, according to the Cartan homotopy formula (theorem 1.9). Therefore, $i_{[X,Y]} \omega = L_X i_Y \omega = L_X(i_Y \omega) - i_Y(L_X \omega) = L_X(-dx_1) = -d(X(x_1)) = 0$, from which we have $[X, Y] = 0$, since any point in the form ω is of rank equal to $2m$. By the Recovery theorem[1], it follows that there exist local coordinates $x_1, x_{m+1}, z_1, z_2, \ldots, z_{2m-2}$ on a neighborhood $\mathcal{U}_1 \subset \mathcal{U}$ of x such that: $X = \frac{\partial}{\partial x_1}$, $Y = \frac{\partial}{\partial x_{m+1}}$. Consider the differential form $\lambda = \omega - dx_1 \wedge dx_{m+1}$. We have $d\lambda = 0$ and $i_X \lambda = L_X \lambda = i_Y \lambda = L_Y \lambda = 0$. So λ is expressed as a 2-differential form based only on variables $z_1, z_2, \ldots, z_{2m-2}$. In particular, we have $\lambda^{m+1} = 0$. Furthermore, we have $0 \neq \omega^m = m dx_1 \wedge dx_{m+1} \wedge \lambda^{m-1}$. The 2-form λ is closed and of maximal rank (rank half) $m - 1$ on an open set of $\mathbb{R}^{2m-2}$. It is, therefore, sufficient to apply the induction hypothesis to λ.

1.6. Poisson brackets and Hamiltonian systems

As a result of what was previously mentioned, the symplectic form ω induces a Hamiltonian vector field $IdH : M \longrightarrow T_x M$, $x \longmapsto IdH(x)$, where $H : M \longrightarrow \mathbb{R}$ is a differentiable function (Hamiltonian). In others words, the differential system defined by $\dot{x}(t) = X_H(x(t)) = IdH(x)$ is a Hamiltonian vector field associated with the function H. The Hamiltonian vector fields form a Lie subalgebra of the vector field space. The flow g_X^t leaves invariant the symplectic form ω.

DEFINITION 1.13.– *A Hamiltonian system is a triple (M, ω, H) where (M, ω) is a symplectic manifold and $H \in C^\infty(M)$ a smooth function (Hamiltonian).*

THEOREM 1.11.– The matrix that is associated with a Hamiltonian system determines a symplectic structure.

PROOF.– Let $(x_1, \ldots, x_m)$ be a local coordinate system on M, $(m = \dim M)$. We have

$$\dot{x}(t) = \sum_{k=1}^{m} \frac{\partial H}{\partial x_k} I(dx_k) = \sum_{k=1}^{m} \frac{\partial H}{\partial x_k} \xi^k, \qquad [1.6]$$

where $I(dx_k) = \xi^k \in T_x M$ is defined such that: $\forall \eta \in T_x M$, $\eta_k = dx_k(\eta) = \omega(\eta, \xi^k)$ (kth-component of η). Define $(\eta_1, \ldots, \eta_m)$ and $(\xi_1^k, \ldots, \xi_m^k)$ to be, respectively, the components of η and ξ^k, then

$$\eta_k = \sum_{i=1}^{m} \eta_i \left(\frac{\partial}{\partial x_i}, \frac{\partial}{\partial x_j} \right) \xi_j^k = (\eta_1, \ldots, \eta_m) J^{-1} \begin{pmatrix} \xi_1^k \\ \vdots \\ \xi_m^k \end{pmatrix},$$

---

[1] Let $X_1, \ldots, X_r$ be differentiable vector fields on a manifold M and $x \in M$. Assume that for all $k, l = 1, \ldots, r$, $[X_k, X_l] = 0$ and $X_1(x), \ldots, X_r(x)$ are linearly independent. We show that there is an open $\mathcal{U}$ of M containing x and a local coordinate system on $\mathcal{U}$ such that: $X_1|_\mathcal{U} = \frac{\partial}{\partial x_1}, \ldots, X_r|_\mathcal{U} = \frac{\partial}{\partial x_r}$.

where J^{-1} is the matrix defined by $J^{-1} \equiv \left(\omega\left(\frac{\partial}{\partial x_i}, \frac{\partial}{\partial x_j}\right)\right)_{1 \leq i,j \leq m}$. Note that this matrix is invertible. Indeed, it suffices to show that the matrix J^{-1} has maximal rank. Suppose this were not possible, that is, we assume that $rank(J^{-1}) \neq m$. Hence $\sum_{i=1}^{m} a_i \omega\left(\frac{\partial}{\partial x_i}, \frac{\partial}{\partial x_j}\right) = 0$, $\forall 1 \leq j \leq m$, with a_i not all null and $\omega\left(\sum_{i=1}^{m} a_i \frac{\partial}{\partial x_i}, \frac{\partial}{\partial x_j}\right) = 0$, $\forall 1 \leq j \leq m$. In fact, since ω is non-degenerate, we have $\sum_{i=1}^{m} a_i \frac{\partial}{\partial x_i} = 0$. Now $\left(\frac{\partial}{\partial x_1}, \ldots, \frac{\partial}{\partial x_m}\right)$ is a basis of $T_x M$, then $a_i = 0$, $\forall i$, contradiction. Since this matrix is invertible, we can search ξ^k such that:

$$J^{-1}\begin{pmatrix} \xi_1^k \\ \vdots \\ \xi_m^k \end{pmatrix} = \begin{pmatrix} 0 \\ \vdots \\ 0 \\ 1 \\ 0 \\ \vdots \\ 0 \end{pmatrix} \rightsquigarrow k^{th}\text{-place} \implies \begin{pmatrix} \xi_1^k \\ \vdots \\ \xi_m^k \end{pmatrix} = J\begin{pmatrix} 0 \\ \vdots \\ 0 \\ 1 \\ 0 \\ \vdots \\ 0 \end{pmatrix},$$

so $\xi^k = (k$th-column of $J)$, that is, $\xi_i^k = J_{ik}$, $1 \leq i \leq m$ and $\xi^k = \sum_{i=1}^{m} J_{ik} \frac{\partial}{\partial x_i}$. The matrix J is skew-symmetric[2]. From [1.6], we deduce that $\dot{x}(t) = \sum_{k=1}^{m} \frac{\partial H}{\partial x_k} \sum_{i=1}^{m} J_{ik} \frac{\partial}{\partial x_i} = \sum_{i=1}^{m}\left(\sum_{k=1}^{m} J_{ik} \frac{\partial H}{\partial x_k}\right)\frac{\partial}{\partial x_i}$. Writing $\dot{x}(t) = \sum_{i=1}^{m} \frac{dx_i(t)}{dt}\frac{\partial}{\partial x_i}$, it is seen that $\dot{x}_i(t) = \sum_{k=1}^{m} J_{ik} \frac{\partial H}{\partial x_k}$, where $1 \leq i \leq j \leq m$, which can be written in more compact form $\dot{x}(t) = J(x)\frac{\partial H}{\partial x}$, this is the Hamiltonian vector field associated with the function H. $\square$

Let (M, ω) be a symplectic manifold. To any pair of differentiable functions (F, G) over M, we associate the function $\{F, G\} = d_u F(X_G) = X_G F(u) = \omega(X_G, X_F)$, where X_F and X_G are the Hamiltonian vector fields associated with the functions F and G, respectively. We say that $\{F, G\}$ is a Poisson bracket (or Poisson structure) of the functions F and G. It is easily verified that the Poisson bracket on the space $\mathcal{C}^\infty(M)$, that is, the bilinear application $\{,\} : \mathcal{C}^\infty(M) \times \mathcal{C}^\infty(M) \longrightarrow \mathcal{C}^\infty(M)$, $(F, G) \longmapsto \{F, G\}$, defined above (where $\mathcal{C}^\infty(M)$ is the commutative algebra of regular functions on M) is skew-symmetric $\{F, G\} = -\{G, F\}$, obeys the Leibniz rule $\{FG, H\} = F\{G, H\} + G\{F, H\}$, and satisfies the Jacobi identity $\{\{H, F\}, G\} + \{\{F, G\}, H\} + \{\{G, H\}, F\} = 0$, for all $F, G, H \in \mathcal{C}^\infty(M)$.

---

[2] Indeed, since $\omega\left(\frac{\partial}{\partial x_i}, \frac{\partial}{\partial x_j}\right) = -\omega\left(\frac{\partial}{\partial x_j}, \frac{\partial}{\partial x_i}\right)$, that is, ω is symmetric, it follows that J^{-1} is skew-symmetric. Then, $I = J.J^{-1} = \left(J^{-1}\right)^\top.J^\top = -J^{-1}.J$ and consequently $J^\top = J$.

DEFINITION 1.14.– *A Poisson structure on a manifold M is a Lie algebra structure $\{.,.\}$ on the space $C^\infty(M)$ that satisfies the Leibniz rule. The pair $(M, \{.,.\})$ is called a Poisson manifold.*

The Leibniz formula ensures that the mapping $G \longmapsto \{G, F\}$ is a derivation. The antisymmetry and identity of Jacobi ensure that $\{,\}$ is a Lie bracket, and they provide $C^\infty(M)$ an infinite-dimensional Lie algebra structure. When this Poisson structure is non-degenerate, we obtain the symplectic structure discussed above. So any symplectic manifold is a Poisson manifold.

EXAMPLE 1.7.– Consider $M = \mathbb{R}^{2n}$ with symplectic form $\omega = \sum_{k=1}^m dx_k \wedge dy_k$ and let H be the Hamiltonian function. We have $dH = \sum_{i=1}^n \left(\frac{\partial H}{\partial x_i} dx_i + \frac{\partial H}{\partial y_i} dy_i \right)$, and it follows that in the local coordinate system $(x_1, \ldots, x_n, y_1, \ldots, y_n)$, the vector field X_H is given by $X_H = \sum_{i=1}^n \left(\frac{\partial H}{\partial x_i} \frac{\partial}{\partial y_i} - \frac{\partial H}{\partial y_i} \frac{\partial}{\partial x_i} \right)$, and $X_H F = \{H, F\}$, $\forall F \in C^\infty(M)$. In particular, the Poisson brackets of the Darboux coordinate functions are $\{x_i, x_j\} = \{y_i, y_j\} = 0$, $\{x_i, y_j\} = [\delta_{ij}$. The manifold M with the local coordinates and the canonical Poisson bracket mentioned above is a Poisson manifold. The Hamiltonian systems form a Lie algebra.

A non-constant function F is called a first integral (or constant of motion) of X_F, if $X_H F = 0$; F is constant on the trajectories of X_H. In particular, H is integral. Two functions F, G are said to be in involution or to commute if $\{F, G\} = 0$. An interesting result is given by the following Poisson theorem:

THEOREM 1.12.– *If F and G are two first integrals of a Hamiltonian system, then $\{F, G\}$ is also a first integral.*

PROOF.– Jacobi's identity is written as $\{\{H, F\}, G\} + \{\{F, G\}, H\} + \{\{G, H\}, F\} = 0$, where H is the Hamiltonian. Since $\{H, F\} = \{H, G\} = 0$, then we have $\{\{F, G\}, H\} = 0$, which shows that $\{F, G\}$ is a first integral. □

REMARK 1.4.– If we know two first integrals, we can, according to Poisson's theorem, find new integrals, but we often fall back on known first integrals or a constant.

Let M and N be two differentiable manifolds and $f \in C^\infty(M, N)$. The linear tangent map to f at the point p is the induced mapping between the tangent spaces $T_p M$ and $T_{f(p)} N$, defined by $f_* : T_p M \longrightarrow T_{f(p)} N$, $f_* v(\varphi) = v(\varphi \circ f)$, where $v \in T_p M$ and $\varphi \in C^\infty(N, \mathbb{R})$. Let $L : TM \longrightarrow \mathbb{R}$ be a differentiable function (Lagrangian) on the tangent bundle TM. We say that (M, L) is invariant under the differentiable application $g : M \longrightarrow M$ if for all $v \in TM$, we have $L(g_* v) = L(v)$. The Noether theorem below expresses the existence of a first integral associated with a symmetry of the Lagrangian. In other words, each parameter of a group of transformations corresponds to a conserved quantity. One of the consequences of the invariance of the Lagrangian with respect to a group of transformations is the conservation of the generators of the group. For example, the first integral associated with rotation invariance is the kinetic moment. Similarly, the first integral associated with the invariance with respect to the translations is the velocity. The Noether theorem applies to certain classes of theories, described either by a Lagrangian or a Hamiltonian. We will give below the theorem in its original version, which applies to the theories described by a Lagrangian. There is also a version (see Adler and van Moerbeke 2004) that applies to theories described by a Hamiltonian.

THEOREM 1.13.– If (M, L) is invariant under a parameter group of diffeomorphisms $g_s : M \longrightarrow M$, $s \in \mathbb{R}$, $g_0 = E$, then the system of Lagrange equations, $\frac{d}{dt}\frac{\partial L}{\partial \dot{q}} - \frac{\partial L}{\partial q} = 0$, corresponding to L admits a first integral $I : TM \longrightarrow \mathbb{R}$ with $I(q, \dot{q}) = \frac{\partial L}{\partial \dot{q}} \frac{dg_s(q)}{ds}\Big|_{s=0}$, the q being local coordinates on M.

PROOF.– The first integral I is independent of the choice of local coordinates q over M and so we consider the case $M = \mathbb{R}^n$. Let $f : \mathbb{R} \longrightarrow M$, $t \longmapsto q = f(t)$, be a solution of the system of Lagrange equations above. By hypothesis, g_{*s} leaves L invariant, so $g_s \circ f : \mathbb{R} \longrightarrow M$, $t \longmapsto g_s \circ f(t)$, also satisfies the system of Lagrange equations. We translate the solution $f(t)$ considering the application $F : \mathbb{R} \times \mathbb{R} \longrightarrow \mathbb{R}^n$, $(s, t) \longmapsto q = g_s(f(t))$. The fact that g_s leaves invariant L implies that:

$$0 = \frac{\partial L(F, \dot{F})}{\partial s} = \frac{\partial L}{\partial q}\frac{\partial F}{\partial s} + \frac{\partial L}{\partial \dot{q}}\frac{\partial \dot{F}}{\partial s} \Longrightarrow \frac{\partial L}{\partial q}\frac{\partial q}{\partial s} + \frac{\partial L}{\partial \dot{q}}\frac{\partial \dot{q}}{\partial s} = 0. \quad [1.7]$$

Since F is also a solution of the system of Lagrange equations, that is,

$$\frac{d}{dt}\left(\frac{\partial L}{\partial \dot{q}}\left(F(s,t), \dot{F}(s,t)\right)\right) = \frac{\partial L}{\partial q}\left(F(s,t), \dot{F}(s,t)\right),$$

so noting that: $\frac{\partial \dot{q}}{\partial s} = \frac{d}{dt}\frac{\partial q}{\partial s}$, and equation [1.7] is written in the form

$$0 = \frac{\partial q}{\partial s}\frac{d}{dt}\left(\frac{\partial L}{\partial \dot{q}}\left(F(s,t), \dot{F}(s,t)\right)\right) + \frac{\partial L}{\partial \dot{q}}\frac{d}{dt}\frac{\partial q}{\partial s}$$

$$= \frac{d}{dt}\left(\frac{\partial L}{\partial \dot{q}}\left(F(s,t), \dot{F}(s,t)\right)\frac{\partial q}{\partial s}\right),$$

which completes the proof of the theorem. □

We now give the following definition of the Poisson bracket: $\{F, G\} = \left\langle \frac{\partial F}{\partial x}, J\frac{\partial G}{\partial x} \right\rangle = \sum_{i,j} J_{ij}\frac{\partial F}{\partial x_i}\frac{\partial G}{\partial x_j}$. We will look for conditions on the matrix J for Jacobi's identity to be satisfied. This is the purpose of the following theorem.

THEOREM 1.14.– The matrix J satisfies the Jacobi identity, if

$$\sum_{k=1}^{2n}\left(J_{kj}\frac{\partial J_{li}}{\partial x_k}+J_{ki}\frac{\partial J_{jl}}{\partial x_k}+J_{kl}\frac{\partial J_{ij}}{\partial x_k}\right)=0,\ \forall 1\leq i,j,l\leq 2n.$$

PROOF.– Consider the Jacobi identity: $\{\{H,F\},G\}+\{\{F,G\},H\}+\{\{G,H\},F\}=0$. We have

$$\{\{H,F\},G\}=\left\langle\frac{\partial\{H,F\}}{\partial x},J\frac{\partial G}{\partial x}\right\rangle=\sum_{k,l}J_{kl}\frac{\partial\{H,F\}}{\partial x_k}\frac{\partial G}{\partial x_l},$$

$$=\sum_{k,l}\sum_{i,j}J_{kl}\frac{\partial J_{ij}}{\partial x_k}\frac{\partial H}{\partial x_i}\frac{\partial F}{\partial x_j}\frac{\partial G}{\partial x_l}+\sum_{k,l}\sum_{i,j}J_{kl}J_{ij}\frac{\partial^2 H}{\partial x_k\partial x_i}\frac{\partial F}{\partial x_j}\frac{\partial G}{\partial x_l}$$

$$+\sum_{k,l}\sum_{i,j}J_{kl}J_{ij}\frac{\partial H}{\partial x_i}\frac{\partial^2 F}{\partial x_k\partial x_j}\frac{\partial G}{\partial x_l}.$$

By symmetry, we have immediately $\{\{F,G\},H\}$ and $\{\{G,H\},F\}$. Then

$$\{\{H,F\},G\}+\{\{F,G\},H\}+\{\{G,H\},F\}$$

$$=\sum_{k,l}\sum_{i,j}J_{kl}\frac{\partial J_{ij}}{\partial x_k}\frac{\partial H}{\partial x_i}\frac{\partial F}{\partial x_j}\frac{\partial G}{\partial x_l}$$

$$+\sum_{k,l}\sum_{i,j}J_{kl}J_{ij}\frac{\partial^2 H}{\partial x_k\partial x_i}\frac{\partial F}{\partial x_j}\frac{\partial G}{\partial x_l} \qquad [1.8]$$

$$+\sum_{k,l}\sum_{i,j}J_{kl}J_{ij}\frac{\partial H}{\partial x_i}\frac{\partial^2 F}{\partial x_k\partial x_j}\frac{\partial G}{\partial x_l} \qquad [1.9]$$

$$+\sum_{k,l}\sum_{i,j}J_{kl}\frac{\partial J_{ij}}{\partial x_k}\frac{\partial G}{\partial x_i}\frac{\partial H}{\partial x_j}\frac{\partial F}{\partial x_l}$$

$$+\sum_{k,l}\sum_{i,j}J_{kl}J_{ij}\frac{\partial^2 G}{\partial x_k\partial x_i}\frac{\partial H}{\partial x_j}\frac{\partial F}{\partial x_l} \qquad [1.10]$$

$$+\sum_{k,l}\sum_{i,j}J_{kl}J_{ij}\frac{\partial G}{\partial x_i}\frac{\partial^2 H}{\partial x_k\partial x_j}\frac{\partial F}{\partial x_l} \qquad [1.11]$$

$$+\sum_{k,l}\sum_{i,j}J_{kl}\frac{\partial J_{ij}}{\partial x_k}\frac{\partial F}{\partial x_i}\frac{\partial G}{\partial x_j}\frac{\partial H}{\partial x_l}$$

$$+ \sum_{k,l} \sum_{i,j} J_{kl} J_{ij} \frac{\partial^2 F}{\partial x_k \partial x_i} \frac{\partial G}{\partial x_j} \frac{\partial H}{\partial x_l} \qquad [1.12]$$

$$+ \sum_{k,l} \sum_{i,j} J_{kl} J_{ij} \frac{\partial F}{\partial x_i} \frac{\partial^2 G}{\partial x_k \partial x_j} \frac{\partial H}{\partial x_l}. \qquad [1.13]$$

Note that the indices i, j, k and l play a symmetric roll. Applying in the term [1.11] the permutation $i \leftarrow l, j \leftarrow k, k \leftarrow i, l \leftarrow j$, and add the term [1.8], with the understanding that $J_{lk} = -J_{kl}$, we get $\sum_{k,l} \sum_{i,j} (J_{ij} J_{lk} + J_{kl} J_{ij}) \frac{\partial G}{\partial x_l} \frac{\partial^2 H}{\partial x_i \partial x_k} \frac{\partial F}{\partial x_j} = 0$, as a consequence of Schwarz's lemma. Again, applying in the term [1.12] the permutation $i \leftarrow k, j \leftarrow l, k \leftarrow j, l \leftarrow i$, and adding the term [1.9] yields $\sum_{k,l} \sum_{i,j} (J_{ji} J_{kl} + J_{kl} J_{ij}) \frac{\partial^2 F}{\partial x_j \partial x_k} \frac{\partial G}{\partial x_l} \frac{\partial H}{\partial x_i} = 0$. By the same argument as above, applying in the term [1.13] the permutation $i \leftarrow l, j \leftarrow k, k \leftarrow i, l \leftarrow j$, and adding the term [1.10], we obtain $\sum_{k,l} \sum_{i,j} (J_{ij} J_{lk} + J_{kl} J_{ij}) \frac{\partial F}{\partial x_l} \frac{\partial^2 G}{\partial x_i \partial x_k} \frac{\partial H}{\partial x_j} = 0$, and thus

$$\{\{H, F\}, G\} + \{\{F, G\}, H\} + \{\{G, H\}, F\}$$

$$= \sum_{k,l} \sum_{i,j} J_{kl} \frac{\partial J_{ij}}{\partial x_k} \frac{\partial H}{\partial x_i} \frac{\partial F}{\partial x_j} \frac{\partial G}{\partial x_l} \qquad [1.14]$$

$$+ \sum_{k,l} \sum_{i,j} J_{kl} \frac{\partial J_{ij}}{\partial x_k} \frac{\partial G}{\partial x_i} \frac{\partial H}{\partial x_j} \frac{\partial F}{\partial x_l} \qquad [1.15]$$

$$+ \sum_{k,l} \sum_{i,j} J_{kl} \frac{\partial J_{ij}}{\partial x_k} \frac{\partial F}{\partial x_i} \frac{\partial G}{\partial x_j} \frac{\partial H}{\partial x_l}.$$

Under permuting the indices $i \leftarrow l, j \leftarrow i, k \leftarrow k, l \leftarrow j$, for [1.14] and $i \leftarrow j, j \leftarrow l, k \leftarrow k, l \leftarrow i$, for [1.15], we obtain the following:

$$\{\{H, F\}, G\} + \{\{F, G\}, H\} + \{\{G, H\}, F\}$$

$$= \sum_{i,j,l} \left[\sum_k \left(J_{kj} \frac{\partial J_{li}}{\partial x_k} + J_{ki} \frac{\partial J_{jl}}{\partial x_k} + J_{kl} \frac{\partial J_{ij}}{\partial x_k} \right) \right] \frac{\partial H}{\partial x_l} \frac{\partial F}{\partial x_i} \frac{\partial G}{\partial x_j}.$$

Since the Jacobi identity must be identically zero, the expression to prove follows immediately, ending the proof of theorem. □

CONCLUSION.– Consequently, we have a complete characterization of the Hamiltonian vector field

$$\dot{x}(t) = X_H(x(t)) = J \frac{\partial H}{\partial x}, \quad x \in M, \qquad [1.16]$$

where $H : M \longrightarrow \mathbb{R}$ is a smooth function (Hamiltonian) and $J = J(x)$ is a skew-symmetric matrix, for which the corresponding Poisson bracket satisfies the Jacobi identity: $\{\{H,F\},G\} + \{\{F,G\},H\} + \{\{G,H\},F\} = 0$, with $\{H,F\} = \langle \frac{\partial H}{\partial x}, J\frac{\partial F}{\partial x}\rangle = \sum_{i,j} J_{ij} \frac{\partial H}{\partial x_i}\frac{\partial F}{\partial x_j}$ (Poisson bracket).

1.7. Examples

The following examples will be studied in detail in other chapters.

EXAMPLE 1.8.– An important special case is when $J = \begin{pmatrix} O & I \\ -I & O \end{pmatrix}$, where O is the $n \times n$ zero matrix and I is the $n \times n$ identity matrix. The condition on J is trivially satisfied. Indeed, here the matrix J does not depend on the variable x and we have

$$\{H,F\} = \sum_{i=1}^{2n} \frac{\partial H}{\partial x_i} \sum_{j=1}^{2n} J_{ij}\frac{\partial F}{\partial x_j} = \sum_{i=1}^{n}\left(\frac{\partial H}{\partial x_{n+i}}\frac{\partial F}{\partial x_i} - \frac{\partial H}{\partial x_i}\frac{\partial F}{\partial x_{n+i}}\right).$$

Moreover, equations [1.16] are transformed into $\dot{q}_1 = \frac{\partial H}{\partial p_1}, \ldots, \dot{q}_n = \frac{\partial H}{\partial p_n}, \dot{p}_1 = -\frac{\partial H}{\partial q_1}, \ldots, \dot{p}_n = -\frac{\partial H}{\partial q_n}$, where $q_1 = x_1, \ldots, q_n = x_n, p_1 = x_{n+1}, \ldots, p_n = x_{2n}$. These are exactly the well-known differential equations of classical mechanics in the canonical form. They show that it suffices to know the Hamiltonian function H to determine the equations of motion. They are often interpreted by considering that the variables p_k and q_k are the coordinates of a point that moves in a space with $2n$ dimensions, called phase space. The flow associated with the system above obviously leaves invariant each hypersurface of constant energy $H = c$. The Hamilton equations mentioned above can still be written in the form $\dot{q}_i = \{H, q_i\} = \frac{\partial H}{\partial p_i}, \dot{p}_i = \{H, p_i\} = -\frac{\partial H}{\partial q_i}$, where $1 \leq i \leq n$. Note that the functions $1, q_i, p_i$ ($1 \leq i \leq n$) verify the following commutation relations: $\{q_i, q_j\} = \{p_i, p_j\} = \{q_i, 1\} = \{p_i, 1\} = 0$, $\{p_i, q_j\} = \delta_{ij}, 1 \leq i, j \leq n$. These functions constitute a basis of a real Lie algebra (Heisenberg algebra) of dimension $2n + 1$.

EXAMPLE 1.9.– The Hénon–Heiles differential equations are defined by

$$\dot{x}_1 = y_1, \quad \dot{y}_1 = -Ax_1 - 2x_1 x_2,$$
$$\dot{x}_2 = y_2, \quad \dot{y}_2 = -Bx_2 - x_1^2 - \varepsilon x_2^2,$$

where x_1, x_2, y_1, y_2 are the canonical coordinates and moments and A, B, ε are free constant parameters. The above equations can be rewritten as a Hamiltonian vector field, $\dot{x} = J\frac{\partial H}{\partial x}$, $x = (x_1, x_2, y_1, y_2)^\top$, where

$$H = \frac{1}{2}(y_1^2 + y_2^2 + Ax_1^2 + Bx_2^2) + x_1^2 x_2 + \frac{\varepsilon}{3}x_2^3, \quad \text{(Hamiltonian)},$$

and $J = \begin{pmatrix} 0 & I \\ -I & 0 \end{pmatrix}$ is the matrix associated with the vector field (where O is the 2×2 zero matrix and I is the 2×2 identity matrix).

EXAMPLE 1.10.– The Euler equations (usually referred to as Euler top) of motion of a rotating rigid body around a fixed point, taken as the origin of the reference bound to the solid, when no external force is applied to the system, are written as

$$\dot{m}_1 = (\lambda_3 - \lambda_2) m_2 m_3, \quad \dot{m}_2 = (\lambda_1 - \lambda_3) m_1 m_3, \quad \dot{m}_3 = (\lambda_2 - \lambda_1) m_1 m_2,$$

where (m_1, m_2, m_3) is the angular momentum of the solid and $\lambda_i \equiv I_i^{-1}$, I_1, I_2, I_3 are moments of inertia. These equations can be written in the form of a Hamiltonian vector field: $\dot{x} = J \frac{\partial H}{\partial x}$, $x = (m_1, m_2, m_3)^\top$, with

$$H = \frac{1}{2} \left(\lambda_1 m_1^2 + \lambda_2 m_2^2 + \lambda_3 m_3^2 \right), \quad \text{(Hamiltonian)}.$$

To determine the matrix $J = (J_{ij})_{1 \leq i,j \leq 3}$, we proceed as follows: since J is antisymmetric, then obviously $J_{ii} = 0$, $J_{ij} = -J_{ji}$, $1 \leq i,j \leq 3$, hence

$$J = \begin{pmatrix} 0 & J_{12} & J_{13} \\ -J_{12} & 0 & J_{23} \\ -J_{13} & -J_{23} & 0 \end{pmatrix}.$$

Therefore,

$$\begin{pmatrix} \dot{m}_1 \\ \dot{m}_2 \\ \dot{m}_3 \end{pmatrix} = \begin{pmatrix} 0 & J_{12} & J_{13} \\ -J_{12} & 0 & J_{23} \\ -J_{13} & -J_{23} & 0 \end{pmatrix} \begin{pmatrix} \lambda_1 m_1 \\ \lambda_2 m_2 \\ \lambda_3 m_3 \end{pmatrix}, \quad [1.17]$$

$$= \begin{pmatrix} (\lambda_3 - \lambda_2) m_2 m_3 \\ (\lambda_1 - \lambda_3) m_1 m_3 \\ (\lambda_2 - \lambda_1) m_1 m_2 \end{pmatrix}. \quad [1.18]$$

Comparing expressions [1.17] and [1.18], we deduce that: $J_{12} = -m_3$, $J_{13} = m_2$ and $J_{23} = -m_1$. Finally,

$$J = \begin{pmatrix} 0 & -m_3 & m_2 \\ m_3 & 0 & -m_1 \\ -m_2 & m_1 & 0 \end{pmatrix} \in so(3),$$

is the matrix of the Hamiltonian vector field. It satisfies the Jacobi identity.

EXAMPLE 1.11.– The equations of the geodesic flow[3] on the group $SO(4)$ are given as

$$\begin{aligned}
\dot{x}_1 &= (\lambda_3 - \lambda_2)\, x_2 x_3 + (\lambda_6 - \lambda_5)\, x_5 x_6, \\
\dot{x}_2 &= (\lambda_1 - \lambda_3)\, x_1 x_3 + (\lambda_4 - \lambda_6)\, x_4 x_6, \\
\dot{x}_3 &= (\lambda_2 - \lambda_1)\, x_1 x_2 + (\lambda_5 - \lambda_4)\, x_4 x_5, \\
\dot{x}_4 &= (\lambda_3 - \lambda_5)\, x_3 x_5 + (\lambda_6 - \lambda_2)\, x_2 x_6, \\
\dot{x}_5 &= (\lambda_4 - \lambda_3)\, x_3 x_4 + (\lambda_1 - \lambda_6)\, x_1 x_6, \\
\dot{x}_6 &= (\lambda_2 - \lambda_4)\, x_2 x_4 + (\lambda_5 - \lambda_1)\, x_1 x_5,
\end{aligned} \qquad [1.19]$$

where $\lambda_1, \ldots, \lambda_6$ are constants. These equations can be written in the form of a Hamiltonian vector field. We have $\dot{x}(t) = J\frac{\partial H}{\partial x}$, $x = (x_1, x_2, x_3, x_4, x_5, x_6)^\top$, with

$$H = \frac{1}{2}\left(\lambda_1 x_1^2 + \lambda_2 x_2^2 + \cdots + \lambda_6 x_6^2\right), \quad \text{(Hamiltonian).}$$

By proceeding in a similar way to the previous example, we obtain

$$J = \begin{pmatrix}
0 & -x_3 & x_2 & 0 & -x_6 & x_5 \\
x_3 & 0 & -x_1 & x_6 & 0 & -x_4 \\
-x_2 & x_1 & 0 & -x_5 & x_4 & 0 \\
0 & -x_6 & x_5 & 0 & -x_3 & x_2 \\
x_6 & 0 & -x_4 & x_3 & 0 & -x_1 \\
-x_5 & x_4 & 0 & -x_2 & x_1 & 0
\end{pmatrix}.$$

EXAMPLE 1.12.– The movement of the Kowalewski spinning top is governed by,

$$\dot{m} = m \wedge \lambda m + \gamma \wedge l, \qquad \dot{\gamma} = \gamma \wedge \lambda m,$$

where m, γ and l denote, respectively, the angular momentum, the direction cosine of the z axis (fixed in space), the center of gravity which can be reduced to $l = (1, 0, 0)$ and $\lambda m = \left(\frac{m_1}{2}, \frac{m_2}{2}, \frac{m_3}{2}\right)$. This system is written in the form of a Hamiltonian vector field $\dot{x} = J\frac{\partial H}{\partial x}$, $x = (m_1, m_2, m_3, \gamma_1, \gamma_2, \gamma_3)^\top$, with $H = \frac{1}{2}\left(m_1^2 + m_2^2\right) + m_3^2 + 2\gamma_1$, the Hamiltonian and

$$J = \begin{pmatrix}
0 & -m_3 & m_2 & 0 & -\gamma_3 & \gamma_2 \\
m_3 & 0 & -m_1 & \gamma_3 & 0 & -\gamma_1 \\
-m_2 & m_1 & 0 & -\gamma_2 & \gamma_1 & 0 \\
0 & -\gamma_3 & \gamma_2 & 0 & 0 & 0 \\
\gamma_3 & 0 & -\gamma_1 & 0 & 0 & 0 \\
-\gamma_2 & \gamma_1 & 0 & 0 & 0 & 0
\end{pmatrix}.$$

---

[3] For some information on the geodesic flow and Euler–Arnold equations, see section 3.5.

EXAMPLE 1.13.– The motion of a solid in a perfect fluid is described using the Kirchhoff equations:

$$\dot{p} = p \wedge \frac{\partial H}{\partial l}, \qquad \dot{l} = p \wedge \frac{\partial H}{\partial p} + l \wedge \frac{\partial H}{\partial l}, \qquad [1.20]$$

where $p = (p_1, p_2, p_3) \in \mathbb{R}^3$, $l = (l_1, l_2, l_3) \in \mathbb{R}^3$ and H the Hamiltonian. The problem of this movement is a limit case of the geodesic flow on $SO(4)$. In the case of Clebsch, we have $H = \frac{1}{2} \sum_{k=1}^{3} \left(a_k p_k^2 + b_k l_k^2 \right)$, with the condition: $\frac{a_2 - a_3}{b_1} + \frac{a_3 - a_1}{b_2} + \frac{a_1 - a_2}{b_3} = 0$. The system [1.20] is written in the form of a Hamiltonian vector field: $\dot{x} = J \frac{\partial H}{\partial x}$, $x = (p_1, p_2, p_3, l_1, l_2, l_3)^{\top}$, where

$$J = \begin{pmatrix} 0 & 0 & 0 & 0 & -p_3 & p_2 \\ 0 & 0 & 0 & p_3 & 0 & -p_1 \\ 0 & 0 & 0 & -p_2 & p_1 & 0 \\ 0 & -p_3 & p_2 & 0 & -l_3 & l_2 \\ p_3 & 0 & -p_1 & l_3 & 0 & -l_1 \\ -p_2 & p_1 & 0 & -l_2 & l_1 & 0 \end{pmatrix}.$$

EXAMPLE 1.14.– Let $\frac{df}{dt} = \sum_{k=1}^{n} \left(\frac{\partial f}{\partial p_k} \dot{p}_k + \frac{\partial f}{\partial q_k} \dot{q}_k \right) + \frac{\partial f}{\partial t}$ be the total derivative of a function $f(p, q, t)$ with respect to t. We will determine a necessary and sufficient condition for f to be a first integral of a system described by a Hamiltonian H. Taking into account Hamilton's equations, we obtain the expression $\frac{df}{dt} = \{f, H\} + \frac{\partial f}{\partial t}$. We deduce that f is a first integral of a system described by a Hamiltonian $H(p, q, t)$ explicitly dependent on t if and only if

$$\{f, H\} + \frac{\partial f}{\partial t} = 0, \qquad [1.21]$$

and obviously if f does not depend explicitly on t, we have $\{f, H\} = 0$.

EXAMPLE 1.15.– Consider a Hamiltonian, $H = \frac{1}{2m}(p_1^2 + p_2^2 + p_3^2) + V(r, t)$, $r = \sqrt{q_1^2 + q_2^2 + q_3^2}$, describing the motion of a particle having a mass m and immersed into a potential $V(r, t)$. We will determine three first integrals of the system described by this Hamiltonian. The two components of angular momentum are equal to $H_1 = q_2 p_3 - q_3 p_2$ and $H_2 = q_3 p_1 - q_1 p_3$. They are obviously first integrals. According to Poisson's theorem 1.12, we have $\{H_1, H_2\} = q_1 p_2 - q_2 p_1 = H_3$, which shows that H_3 is also a first integral. Note also that: $\{H_3, H_1\} = H_2$, $\{H_2, H_3\} = H_1$. If in a system two components of angular momentum are first integrals, then the third component is also a first integral.

EXAMPLE 1.16.– We have already seen that in a conservative system, the Hamiltonian $H(p, q)$ is a first integral. We will show that if $F(p, q, t)$ denotes another first integral explicitly dependent on t, then $\frac{\partial^k F}{\partial t^k}$ is also a first integral. We will apply this result to the case of the Hamiltonian of the harmonic oscillator: $H = \frac{1}{2m}p^2 + \frac{m\omega^2}{2}q^2$. According to the Poisson theorem 1.12, $\{F, H\}$ is also a first integral. Therefore, $\frac{\partial F}{\partial t} = -\{F, H\}$, is a first integral under [1.21]. Similarly, we have $\{\frac{\partial F}{\partial t}, H\} + \frac{\partial^2 F}{\partial t^2} = 0$, which shows that $\frac{\partial^2 F}{\partial t^2} = -\{\frac{\partial F}{\partial t}, H\}$ is also a first integral. Similarly, we show that $\frac{\partial^k F}{\partial t^k}$ is a first integral. For the Hamiltonian of the harmonic oscillator, we see that $F = q\cos\omega t - \frac{1}{m\omega}p\sin\omega t$ and $\frac{\partial F}{\partial t} = -\omega q\sin\omega t - \frac{1}{m\omega}p\cos\omega t$ are first integrals of the Hamiltonian system associated with H.

1.8. Coadjoint orbits and their symplectic structures

We will first define the adjoint and coadjoint orbits of a Lie group. Let G be a Lie group and g an element of G. The Lie group G operates on itself by left translation: $L_g : G \longrightarrow G, h \longmapsto gh$, and by right translation: $R_g : G \longrightarrow G, h \longmapsto hg$. By virtue of the associative law of the group, we have $L_g L_h = L_{gh}$, $R_g L_h = R_{hg}$, $L_{g^{-1}} = L_g^{-1}$, $R_{g^{-1}} = R_g^{-1}$. In particular, the applications R_g and L_g are diffeomorphisms of G. Also, because of associativity, R_g and L_g commute. Consider $R_g^{-1} L_g : G \longrightarrow G, h \longmapsto ghg^{-1}$ the automorphism of the group G. It leaves the unit e of the group G fixed, that is, $R_g^{-1} L_g(e) = geg^{-1} = e$. We can define the adjoint representation of the group G as the derivative of $R_g^{-1} L_g$ in the unit e, that is, the induced application of tangent spaces as follows $Ad_g : \mathcal{G} \longrightarrow \mathcal{G}, \xi \longmapsto \frac{d}{dt} R_g^{-1} L_g(e^{t\xi})\big|_{t=0}$, where $\mathcal{G} = T_e G$ is the Lie algebra of the group G; it is the tangent space at G in its unit e. This definition holds meaning because $R_g^{-1} L_g(e^{t\xi})$ is a curve in G and passes through the identity in $t = 0$. Therefore, $g\xi g^{-1} \in \mathcal{G}$.

THEOREM 1.15.– For any element $\xi \in \mathcal{G}$, we have $Ad_g(\xi) = g\xi g^{-1}$, $g \in G$ and $Ad_{gh} = Ad_g.Ad_h$. The application Ad_g is an algebra homomorphism, that is, $Ad_g[\xi, \eta] = [Ad_g\xi, Ad_g\eta]$, $(\xi, \eta \in \mathcal{G})$.

PROOF.– We have

$$Ad_g(\xi) = \frac{d}{dt} R_g^{-1} L_g(e^{t\xi})\bigg|_{t=0} = \frac{d}{dt} g e^{t\xi} g^{-1}\bigg|_{t=0} = \frac{d}{dt} g \left(\sum_{n=0}^{\infty} \frac{t^n \xi^n}{n!}\right) g^{-1}\bigg|_{t=0},$$

hence,

$$Ad_g(\xi) = \frac{d}{dt} \sum_{n=0}^{\infty} \frac{t^n}{n!} g\xi^n g^{-1}\bigg|_{t=0} = \frac{d}{dt} \sum_{n=0}^{\infty} \frac{t^n}{n!} \underbrace{g\xi g^{-1}.g\xi g^{-1}...g\xi g^{-1}}_{n-times}\bigg|_t = 0,$$

and finally

$$Ad_g(\xi) = \frac{d}{dt} \sum_{n=0}^{\infty} \frac{t^n}{n!} (g\xi g^{-1})^n\bigg|_{t=0} = \frac{d}{dt} e^{t(g\xi g^{-1})}\bigg|_{t=0} = g\xi g^{-1}.$$

We easily check that: $Ad_{gh} = Ad_g.Ad_h$. Indeed, we have

$$Ad_{gh}(\xi) = gh\xi(gh)^{-1} = gh\xi h^{-1}g^{-1},$$
$$Ad_g.Ad_h(\xi) = Ad_g(h\xi h^{-1}) = gh\xi h^{-1}g^{-1},$$

and

$$Ad_g[\xi,\eta] = Ad_g(\xi\eta - \eta\xi) = g(\xi\eta - \eta\xi)g^{-1} = g\xi\eta g^{-1} - g\eta\xi g^{-1},$$
$$Ad_g[\xi,\eta] = g\xi g^{-1}g\eta g^{-1} - g\eta g^{-1}g\xi g^{-1} = [g\xi g^{-1}, g\eta g^{-1}] = [Ad_g\xi, Ad_g\eta],$$

which completes the demonstration. □

The adjoint orbit of ξ is defined by $\mathcal{O}_G(\xi) = \{Ad_g(\xi) : g \in G\} \subset \mathcal{G}$. Now consider the map $Ad : G \longrightarrow \text{End}(\mathcal{G})$, $g \longmapsto Ad(g) \equiv Ad_g$, where $\text{End}(\mathcal{G})$ is the space of the linear operators on the algebra $\mathcal{G}$. The map Ad is differentiable and its derivative Ad_{*e} in the unit of the group G is a linear map from the algebra $T_eG = \mathcal{G}$ to the vector space $T_I\text{End}(\mathcal{G}) = \text{End}(\mathcal{G})$. This map will be noted $ad \equiv Ad_{*e} : \mathcal{G} \longrightarrow \text{End}(\mathcal{G})$, $\xi \longmapsto ad_\xi = \frac{d}{dt}Ad_{g(t)}\big|_{t=0}$, where $g(t)$ is a one-parameter group with $\frac{d}{dt}g(t)\big|_{t=0} = \xi$ and $g(0) = e$.

THEOREM 1.16.– Let $\xi \in \mathcal{G}$ and $\eta \in \text{End}(\mathcal{G})$. We have $ad_\xi(\eta) = [\xi, \eta]$, where $ad_\xi \equiv Ad_{*e}(\xi)$.

PROOF.– We have $ad_\xi(\eta) = Ad_{*e}(\xi)(\eta) = \frac{d}{dt}Ad_{g(t)}(\eta)\big|_{t=0} = \frac{d}{dt}(g(t)\eta g^{-1}(t))\big|_{t=0} = \dot{g}(t)\eta g^{-1}(t)\big|_{t=0} - g(t)\eta g^{-1}(t)\dot{g}(t)g^{-1}(t)\big|_{t=0} = \dot{g}(0)\eta - \eta\dot{g}(0)$, and consequently $ad_\xi(\eta) = \xi\eta - \eta\xi = [\xi, \eta]$. □

Let T_g^*G be the cotangent space to the group G at g; it is the dual space to the tangent space T_gG. Then an element $\zeta \in T_g^*G$ is a linear form on T_gG and its value on $\eta \in T_gG$ will be denoted by $\zeta(\eta) \equiv \langle \zeta, \eta \rangle$. Let $\mathcal{G}^* = T_e^*G$ be the dual vector space to the Lie algebra $\mathcal{G}$; it is the cotangent space to the group G in its unit e. The transpose operators $Ad_g^* : \mathcal{G}^* \longrightarrow \mathcal{G}^*$, where g runs through the Lie group G are defined by $\langle Ad_g^*(\zeta), \eta \rangle = \langle \zeta, Ad_g\eta \rangle$, $\zeta \in \mathcal{G}^*$, $\eta \in \mathcal{G}$. Ad_g^* is called coadjoint representation of the Lie group G. The coadjoint orbit (also called the Kostant–Kirillov orbit) is defined at the point $x \in \mathcal{G}^*$ by $\mathcal{O}_G^*(x) = \{Ad_g^*(x) : g \in G\} \subset \mathcal{G}^*$.

THEOREM 1.17.– The transpose operators Ad_g^* form a representation of the Lie group G, that is, they satisfy the relations: $Ad_{gh}^* = Ad_h^*.Ad_g^*$.

PROOF.– Indeed, let $\zeta \in \mathcal{G}^*$, $\eta \in \mathcal{G}$. We have $\langle Ad_{gh}^*(\zeta), \eta \rangle = \langle \zeta, Ad_{gh}(\eta) \rangle = \langle \zeta, Ad_h.Ad_g(\eta) \rangle = \langle Ad_g^*(\zeta), Ad_h(\eta) \rangle = \langle Ad_h^*.Ad_g^*(\zeta), \eta \rangle$. □

Consider the map $Ad^* : G \longrightarrow \text{End}(\mathcal{G}^*)$, $g \longmapsto Ad^*(g) \equiv Ad_g^*$, and its derivative in the unity of the group $ad^* \equiv (Ad^*)_{*e} : \mathcal{G} \longrightarrow \text{End}(\mathcal{G}^*)$, $\xi \longmapsto ad_\xi^*$.

THEOREM 1.18.– By setting $\langle ad_\xi^*(\zeta), \eta \rangle = \langle \zeta, [\xi, \eta] \rangle = \langle \{\xi, \zeta\}, \eta \rangle$, where $\{,\}$: $\mathcal{G} \times \mathcal{G}^* \longrightarrow \mathcal{G}^*, (\xi, \zeta) \longmapsto \{\xi, \zeta\}, (\xi, \eta \in \mathcal{G}, \zeta \in \mathcal{G}^*)$, then $ad_\xi^*(\zeta) = \{\xi, \zeta\}$.

PROOF.– We have $\langle ad_\xi^*(\zeta), \eta \rangle = \langle (Ad^*)_{*e}(\zeta), \eta \rangle = \langle \frac{d}{dt} Ad^*_{e^{t\xi}}(\zeta)\big|_{t=0}, \eta \rangle$, with $e^{t\xi}\big|_{t=0} = e$ and $\frac{d}{dt} e^{t\xi}\big|_{t=0} = \xi$. Hence, $\langle ad_\xi^*(\zeta), \eta \rangle = \frac{d}{dt} \langle Ad^*_{e^{t\xi}}(\zeta), \eta \rangle\big|_{t=0} = \frac{d}{dt} \langle \zeta, Ad_{e^{t\xi}}(\eta) \rangle\big|_{t=0}$, and therefore,

$$\langle ad_\xi^*(\zeta), \eta \rangle = \left\langle \zeta, \frac{d}{dt} Ad_{e^{t\xi}}(\eta)\bigg|_{t=0} \right\rangle = \langle \zeta, ad_\xi \eta \rangle = \langle \zeta, [\xi, \eta] \rangle = \langle \{\xi, \zeta\}, \eta \rangle,$$

which completes the proof. □

1.9. Application to the group $SO(n)$

We will show how to find the adjoint orbit and the coadjoint orbit in the case of the group $SO(n)$. Recall that $SO(n)$ is the special orthogonal group of order n, that is, the set of matrices X of order $n \times n$ such that: $X^\top . X = I$ (or $X^{-1} = X^\top$) and $\det X = 1$. $SO(n)$ is a Lie group. The tangent space to the identity of $SO(n)$, which is denoted as $so(n)$, consists of the antisymmetric matrices of order $n \times n$, that is, matrices A such that: the commutator of two antisymmetric matrices is still an antisymmetric matrix (if $A, B \in so(n)$, then $[A, B] = AB - BA \in so(n)$). This product defines a Lie algebra structure on $so(n)$; it is the Lie algebra of the group $SO(n)$. In addition, we have $\dot{X} = AX$ with $A \in so(n)$ and therefore the tangent space to the identity of $SO(n)$ is $T_I SO(n) = so(n)$. Let $R_Y^{-1} L_Y : SO(n) \longrightarrow SO(n), X \longmapsto YXY^{-1}$, $Y \in SO(n)$, be the automorphism interior of the group $SO(n)$. When looking for the coadjoint orbit, we have to use the following lemma.

LEMMA 1.2.– The Lie algebra $so(n)$ with the commutator $[,]$ of matrix is isomorphic to the vector space $\mathbb{R}^{\frac{n(n-1)}{2}}$ with the vector product $\wedge$. The isomorphism is given by $a \wedge b \longmapsto [A, B] = AB - BA$, where $a, b \in \mathbb{R}^{\frac{n(n-1)}{2}}$ and $A, B \in so(n)$.

THEOREM 1.19.– The orbit of the adjoint representation of the group $SO(n)$ is

$$\mathcal{O}_{SO(n)}(A) = \{YAY^{-1} : Y \in SO(n)\}, \quad A \in so(n).$$

Let $A \in so(n)$. The coadjoint orbit of the group $SO(n)$ is

$$\mathcal{O}^*_{SO(n)}(A) = \{Y^{-1}AY : Y \in SO(n)\},$$
$$= \{C \in so(n) : C = Y^{-1}AY, \text{spectrum of } C = \text{spectrum of } A\}.$$

With the notation of theorem 1.18, we have $\{A, B\} = [B, A], (A, B \in so(n))$.

PROOF.– Let $Y \in SO(n)$, $A \in so(n)$. By definition, the adjoint representation of the group $SO(n)$ is $Ad_Y : so(n) \longrightarrow so(n)$, $A \longmapsto YAY^{-1}$. We have $(YAY^{-1})^\top = (Y^{-1})^\top A^\top Y^\top = -YAY^\top = -YAY^{-1}$. So $YAY^{-1} \in so(n)$. Let $Ad : SO(n) \longrightarrow End(so(n))$, $Y \longmapsto Ad_Y$, where $Ad_Y(A) = YAY^{-1}$, $A \in so(n)$, and let $ad : so(n) \longrightarrow End(so(n))$, $\dot{Y}(0) \longmapsto ad_{\dot{Y}(0)}$, with $ad_{\dot{Y}(0)} \bullet = [\dot{Y}(0), \bullet] : so(n) \longrightarrow so(n)$, $A \longmapsto [\dot{Y}(0), A]$, where $Y(t)$ is a curve in $SO(n)$ with $Y(0) = I$. Since $(R^{n \times n})^* \simeq R^{n \times n}$, then according to lemma 1.2, we also have the isomorphism $(so(n))^* \simeq so(n)$. We define Ad^* by $Ad_Y^* : so(n) \longrightarrow so(n)$, with $\langle Ad_Y^*(A), B \rangle = \langle A, Ad_Y B \rangle = \langle A, YBY^{-1} \rangle = -\frac{1}{2} tr(AYBY^{-1}) = -\frac{1}{2} tr(Y^{-1}AYB) = \langle Y^{-1}AY, B \rangle$, where $A, B \in so(n)$. Hence, $Ad_Y^*(A) = Y^{-1}AY$. We check that $Y^{-1}AY \in so(n)$. Indeed, we have $(Y^{-1}AY)^\top = Y^\top A^\top (Y^{-1})^\top = -Y^{-1}AY$, because $Y \in SO(n)$ and $A \in so(n)$. Then $\mathcal{O}_{SO(n)}^*(A) = \{Y^{-1}AY : Y \in SO(n)\}$, and we can write that in the form $\mathcal{O}_{SO(n)}^*(A) = \{C \in so(n) : \exists Y \in SO(n), C = Y^{-1}AY\}$. Note that $\det(C - \lambda I) = \det(A - \lambda I)$. Then the matrices C and A have the same characteristic polynomial, and consequently they have the same spectrum.

$$\mathcal{O}_{SO(n)}^*(A) = \{C \in so(n) : C = Y^{-1}AY, \text{ spectrum of } C = \text{spectrum of } A\}.$$

We apply theorem 1.18 to the case of $SO(n)$. Let us go back to the linear form knowing that $(so(n))^* = so(n)$, $\{,\} : so(n) \times so(n) \longrightarrow so(n)$, $(A, B) \longmapsto \{A, B\}$, as well as the applications $Ad^* : SO(n) \longrightarrow End(so(n))$, $Y \longmapsto Ad_Y^*(B) = Y^{-1}BY$, $B \in so(n)$, and $ad^* : so(n) \longrightarrow End(so(n))$, $A \longmapsto ad_A^*$, where $\langle ad_A^*(B), C \rangle = \langle B, [A, C] \rangle = \langle \{A, B\}, C \rangle$. We have

$$\langle \{A, B\}, C \rangle = \langle B, [A, C] \rangle = -\frac{1}{2} tr(B.[A, C]) = -\frac{1}{2} tr(BAC - BCA),$$

hence $\langle \{A, B\}, C \rangle = -\frac{1}{2} tr([B, A].C) = \langle [B, A], C \rangle$. Then $\{A, B\} = [B, A]$. $\square$

We will see how to explicitly determine a symplectic structure on the coadjoint orbit with an application in the case of the groups $SO(3)$ and $SO(4)$. Let $x \in \mathcal{G}^*$ and ξ be the tangent vector in x to the orbit. Since $\mathcal{G}^*$ is a vector space, then obviously $\xi \in T_x \mathcal{G}^* = \mathcal{G}^*$. Let us remember that $\mathcal{O}_G^*(x) = \{Ad_g^*(x) : g \in G\} \subset \mathcal{G}^*$. For $x \in \mathcal{O}_G^*(x)$, there exists $g \in G$ such that: $x = Ad_g^*$. Let $a \in \mathcal{G}$ and e^{ta} be a group with a parameter in G with $e^{ta}|_{t=0} = g$, $\frac{d}{dt} Ad_{e^{ta}}^*(x)|_{t=0} = \xi$. Since $\frac{d}{dt} Ad_{e^{ta}}^*(x)|_{t=0} \equiv ad_a^*(x) = \{a, x\}$, therefore the vector ξ can be represented as the velocity vector of the motion of x under the action of a group e^{ta}, $a \in \mathcal{G}$. In other words, any vector ξ tangent to the orbit $\mathcal{O}_G^*(x)$ is expressed as a function of $a \in \mathcal{G}$ by

$$\xi = \{a, x\}, \quad a \in \mathcal{G}, \quad x \in \mathcal{G}^*. \qquad [1.22]$$

Therefore, we can determine the value of a 2-form Ω on the orbit $\mathcal{O}_G^*(x)$ as follows: let ξ_1 and ξ_2 be two vectors tangent to the orbit of x. From the above, we have $\xi_1 = \{a_1, x\}$, $\xi_2 = \{a_2, x\}$, $(a_1, a_2 \in \mathcal{G})$, $x \in \mathcal{G}^*$. We can easily verify that the differential 2-form

$$\Omega(\xi_1, \xi_2)(x) = \langle x, [a_1, a_2] \rangle, \quad a_1, a_2 \in \mathcal{G}, \quad x \in \mathcal{G}^*, \qquad [1.23]$$

on $\mathcal{O}_G^*(x)$ is well defined; its value does not depend on the choice of a_1, a_2. It is antisymmetric, non-degenerate and closed.

1.9.1. *Application to the group $SO(3)$*

To determine the symplectic structure on $\mathcal{O}_{SO(3)}^*(X)$, we proceed as follows: according to [1.23], we have $\Omega(\xi_1, \xi_2)(X) = \langle X, [A, B] \rangle$, $A, B \in so(3)$, $X \in (so(3))^* = so(3)$, and according to [1.22], $\xi_1 = \{A, X\}$, $\xi_2 = \{B, X\}$ are two tangent vectors to the orbit in X or what amounts to the same according to theorem 1.19, $\xi_1 = [X, A]$, $\xi_2 = [X, B]$. Using the isomorphism (lemma 1.2) between $(so(3), [,])$ and $(\mathbb{R}^3, \wedge)$, we also have $\xi_1 = x \wedge a$, $\xi_2 = x \wedge b$, $\Omega(\xi_1, \xi_2)(x) = \langle x, a \wedge b \rangle$. According to theorem 1.19, the coadjoint orbit of $SO(3)$ is $\mathcal{O}_{SO(3)}^*(A) = \{C \in so(3) : C = Y^{-1}AY$, spectrum of C = spectrum of $A\}$, where $A \in so(3)$ and $Y \in SO(3)$. Let us determine the spectrum of the matrix

$$A = \begin{pmatrix} 0 & -a_3 & a_2 \\ a_3 & 0 & -a_1 \\ -a_2 & a_1 & 0 \end{pmatrix} \in so(3).$$

We have $\det(A - \lambda I) = -\lambda(\lambda^2 + a_1^2 + a_2^2 + a_3^2) = 0$, hence $\lambda = 0$ and $\lambda = \pm i\sqrt{a_1^2 + a_2^2 + a_3^2}$. Then $\mathcal{O}_{SO(3)}^*(A) = \{C \in so(3) : c_1^2 + c_2^2 + c_3^2 = r^2\}$, with

$$C = \begin{pmatrix} 0 & -c_3 & c_2 \\ c_3 & 0 & -c_1 \\ -c_2 & c_1 & 0 \end{pmatrix} \in so(3),$$

and $r^2 = a_1^2 + a_2^2 + a_3^2$. Since the algebra $so(3)$ is isomorphic to $\mathbb{R}^3$, we deduce that the orbit $\mathcal{O}_{SO(3)}^*(A)$ is isomorphic to a sphere S^2 of radius r. Like vectors ξ_1, ξ_2 belong to the tangent plane $T_X \mathcal{O}_{SO(3)}^*$ to X; they also belong to the tangent plane $T_x S^2$ in x. Let $S^2 = \{(y_1, y_2, y_3) \in \mathbb{R}^3 : y_1^2 + y_2^2 + y_3^2 = r^2\}$ be the sphere of radius r, then the plane tangent to this sphere in x of coordinates (x_1, x_2, x_3) is

$$\begin{aligned} T_x S^2 &= \{(y_1, y_2, y_3) \in \mathbb{R}^3 : y_1 x_1 + y_2 x_2 + y_3 x_3 = 0\}, \\ &= \left\{ \left(y_1, y_2, -\frac{y_1 x_1 + y_2 x_2}{x_3} \right) \right\}. \end{aligned}$$

Let $z = (z_1, z_2, z_3) \in T_x S^2$ and determine $a = (a_1, a_2, a_3)$ such that: $x \wedge a = z$. The latter is equivalent to the system

$$\begin{pmatrix} 0 & -a_3 & a_2 \\ a_3 & 0 & -a_1 \\ -a_2 & a_1 & 0 \end{pmatrix} \begin{pmatrix} a_1 \\ a_2 \\ a_3 \end{pmatrix} = \begin{pmatrix} z_1 \\ z_2 \\ -\frac{z_1 x_1 + z_2 x_2}{x_3} \end{pmatrix},$$

whose solution is $a = \left(\frac{x_1 a_3 + z_2}{x_3}, \frac{x_2 a_3 - z_1}{x_3}, a_3 \right)$, $a_3 \in \mathbb{R}$. The symplectic form on S^2 that one wants to determine is intrinsic, that is, it does not depend on the choice of local coordinates. We choose as local coordinates x_1, x_2 and the same reasoning will be valid for the other cases, that is, x_2, x_3 and x_3, x_1. So we will calculate a and b relative to the basis $\left(\frac{\partial}{\partial x_1}, \frac{\partial}{\partial x_2} \right)$ of $T_x S^2$ with $\frac{\partial}{\partial x_1} = \left(1, 0, -\frac{x_1}{x_3} \right)$, $\frac{\partial}{\partial x_2} = \left(0, 1, -\frac{x_2}{x_3} \right)$. We have $a = (a_1, a_2, a_3) = \left(\frac{x_1 b_3 + 1}{x_3}, \frac{x_2 b_3}{x_3}, b_3 \right)$, and

$$a \wedge b = (a_2 b_3 - a_3 b_2, a_3 b_1 - a_1 b_3, a_1 b_2 - a_2 b_1) = \left(-\frac{b_3}{x_3}, \frac{a_3}{x_3}, \frac{x_1 b_3 - x_2 a_3 + 1}{x_3^2} \right).$$

Therefore, $\Omega \left(\frac{\partial}{\partial x_1}, \frac{\partial}{\partial x_2} \right) = (x, a \wedge b) = \frac{1}{x_3}$, consequently, $\Omega = \frac{dx_1 \wedge dx_2}{x_3}$. The symplectic form being intrinsic, we will finally have

$$\Omega = \frac{dx_1 \wedge dx_2}{x_3} = \frac{dx_2 \wedge dx_3}{x_1} = \frac{dx_3 \wedge dx_1}{x_2}.$$

EXAMPLE 1.17.– The symplectic structure obtained here is equivalent to that associated with the system [1.17]. Indeed, we know that $J^{-1} = \left(\omega \left(\frac{\partial}{\partial x_i}, \frac{\partial}{\partial x_j} \right) \right)_{i,j=1,2}$, so the matrix associated with the form $\Omega = \frac{dx_1 \wedge dx_2}{x_3}$ is $\begin{pmatrix} 0 & -x_3 \\ x_3 & 0 \end{pmatrix}$. Let us show that there is equivalence between

$$\dot{x}(t) = J \frac{\partial H}{\partial x}, \quad \text{where} \quad \begin{cases} x = (m_1, m_2, m_3)^\top, \\ H = \frac{1}{2} \left(\lambda_1 m_1^2 + \lambda_2 m_2^2 + \lambda_3 m_3^2 \right), \\ J = \begin{pmatrix} 0 & -m_3 & m_2 \\ m_3 & 0 & -m_1 \\ -m_2 & m_1 & 0 \end{pmatrix}, \end{cases}$$

and

$$\dot{x}(t) = \mathbf{J} \frac{\partial \mathbf{H}}{\partial x}, \quad \text{where} \quad \begin{cases} x = (m_1, m_2, m_3)^\top, \\ \mathbf{H} = H(m_1, m_2, m_3), \\ \mathbf{J} = \begin{pmatrix} 0 & -m_3 \\ m_3 & 0 \end{pmatrix}. \end{cases}$$

Indeed, we have $\dot{m}_1 = -m_3 \frac{\partial H}{\partial m_2} = -m_3 \left(\frac{\partial H}{\partial m_2} + \frac{\partial H}{\partial m_3} \frac{\partial m_3}{\partial m_2} \right)$ and $\dot{m}_2 = m_3 \frac{\partial H}{\partial m_1} = m_3 \left(\frac{\partial H}{\partial m_1} + \frac{\partial H}{\partial m_3} \frac{\partial m_3}{\partial m_1} \right)$. According to example 1.10, we have $dm_3 = -\frac{m_1 dm_1 + m_2 dm_2}{m_3}$, hence $\frac{dm_3}{dm_2} = -\frac{m_2}{m_3}$ and $\frac{dm_3}{dm_1} = -\frac{m_1}{m_3}$. Therefore, we have $\dot{m}_1 = (\lambda_3 - \lambda_2) m_2 m_3$, $\dot{m}_2 = (\lambda_1 - \lambda_3) m_1 m_3$, and the result follows.

1.9.2. Application to the group $SO(4)$

To determine the symplectic structure on the coadjoint orbit of $SO(4)$, one can easily obtain the result by using a geometric approach by observing that $so(4)$ breaks down into two copies of $so(3)$ and that the generic orbits are a product of two spheres. More precisely, from $SO(4) = SO(3) \otimes SO(3)$, it is more interesting to consider the coordinates (x_1, x_2, x_3), (x_4, x_5, x_6) with $(x_1, x_2, x_3) \oplus (x_4, x_5, x_6) \in so(4) \simeq so(3) \oplus so(3)$. On the other hand, we can follow the same method as in the previous case but the calculation is longer. We will follow what has been done above to determine the symplectic structure on the orbit $\mathcal{O}^*_{SO(4)}(X)$. According to [1.23], $\Omega(\xi_1, \xi_2)(X) = \langle X, [A, B] \rangle$, where $A, B \in so(4)$, $X \in (so(4))^* = so(4)$ and according to [1.22], we have $\xi_1 = \{A, X\}$, $\xi_2 = \{B, X\}$, are two vectors tangent to the orbit in X. According to theorem 1.7, we have $\xi_1 = [X, A]$, $\xi_2 = [X, B]$. Since $(so(4), [,])$ and $(\mathbb{R}^6, \wedge)$ are isomorphic (lemma 1.2), $\xi_1 = x \wedge a$, $\xi_2 = x \wedge b$, with $\Omega(\xi_1, \xi_2)(x) = \langle x, a \wedge b \rangle$. According to theorem 1.19, the coadjoint orbit of the group $SO(4)$ is $\mathcal{O}^*_{SO(4)}(A) = \{C \in so(4) : C = Y^{-1} AY,$ spectrum of C = spectrum of $A\}$, where $A \in so(4)$ and $Y \in SO(4)$. Let us determine the spectrum of the matrix

$$A = \begin{pmatrix} 0 & -a_3 & a_2 & -a_4 \\ a_3 & 0 & -a_1 & -a_5 \\ -a_2 & a_1 & 0 & -a_6 \\ a_4 & a_5 & a_6 & 0 \end{pmatrix} \in so(4),$$

$\det(A - \lambda I) = \lambda^4 + \lambda^2 (a_1^2 + a_2^2 + a_3^2 + a_4^2 + a_5^2 + a_6^2) + (a_1 a_4 + a_2 a_5 + a_3 a_6)^2$. So $\mathcal{O}^*_{SO(4)}(A)$ is the set of matrices

$$C = \begin{pmatrix} 0 & -c_3 & c_2 & -c_4 \\ c_3 & 0 & -c_1 & -c_5 \\ -c_2 & c_1 & 0 & -c_6 \\ c_4 & c_5 & c_6 & 0 \end{pmatrix} \in so(4),$$

such that $c_1^2 + c_2^2 + c_3^2 + c_4^2 + c_5^2 + c_6^2 = a_1^2 + a_2^2 + a_3^2 + a_4^2 + a_5^2 + a_6^2$ and $c_1 c_4 + c_2 c_5 + c_3 c_6 = a_1 a_4 + a_2 a_5 + a_3 a_6$. Since the algebra $so(4)$ is isomorphic to $\mathbb{R}^6$, we deduce that the tangent plane $T_X \mathcal{O}^*_{SO(4)}$ in X is isomorphic to the set of $(u_1, u_2, u_3, u_4, u_5, u_6) \in \mathbb{R}^6$ such that:

$$u_1 x_1 + u_2 x_2 + u_3 x_3 + u_4 x_4 + u_5 x_5 + u_6 x_6 = 0, \quad [1.24]$$

$$u_1 x_4 + u_2 x_5 + u_3 x_6 + u_4 x_1 + u_5 x_2 + u_6 x_3 = 0. \quad [1.25]$$

Note that:

$$(x_1, x_2, x_3) \wedge (a_1, a_2, a_3) = (a_3 x_2 - a_2 x_3 \quad a_1 x_3 - a_3 x_1 \quad a_2 x_1 - a_1 x_2),$$

$$(x_4, x_5, x_6) \wedge (a_4, a_5, a_6) = (a_6 x_5 - a_5 x_6 \quad a_4 x_6 - a_6 x_4 \quad a_5 x_4 - a_4 x_5),$$

$$(x_1, x_2, x_3) \wedge (a_4, a_5, a_6) = (a_6 x_2 - a_5 x_3 \quad a_4 x_3 - a_6 x_1 \quad a_5 x_1 - a_4 x_2),$$

$$(x_4, x_5, x_6) \wedge (a_1, a_2, a_3) = (a_3 x_5 - a_2 x_6 \quad a_1 x_6 - a_3 x_4 \quad a_2 x_4 - a_1 x_5).$$

Let us consider $\xi_1, \xi_2 \in T_X \mathcal{O}^*_{SO(4)}$, $\xi_1 = (u_1, u_2, u_3, u_4, u_5, u_6)$, $\xi_2 = (v_1, v_2, v_3, v_4, v_5, v_6)$ and determine $a, b \in \mathbb{R}^6$, $a = (a_1, a_2, a_3, a_4, a_5, a_6)$, $b = (b_1, b_2, b_3, b_4, b_5, b_6)$, such that: $x \wedge a = \xi_1$, $x \wedge b = \xi_2$.

i) Let us take the case where $x \wedge a = \xi_1$. We obtain the following equations:

$$a_3 x_2 - a_2 x_3 + a_6 x_5 - a_5 x_6 = u_1,$$

$$a_1 x_3 - a_3 x_1 + a_4 x_6 - a_6 x_4 = u_2,$$

$$a_2 x_1 - a_1 x_2 + a_5 x_4 - a_4 x_5 = u_3, \quad [1.26]$$

$$a_6 x_2 - a_5 x_3 + a_3 x_5 - a_2 x_6 = u_4,$$

$$a_4 x_3 - a_6 x_1 + a_1 x_6 - a_3 x_4 = u_5,$$

$$a_5 x_1 - a_4 x_2 + a_2 x_4 - a_1 x_5 = u_6. \quad [1.27]$$

We will see that this system is compatible, and more precisely equations [1.26] and [1.27] can be expressed as linear combinations of others. Indeed, according to equation [1.24], we have

$$u_3 = -\frac{1}{x_3}(u_1 x_1 + u_2 x_2 + u_4 x_4 + u_5 x_5 + u_6 x_6), \quad x_3 \neq 0.$$

By replacing this expression in equation [1.25], we express u_6 in terms of u_1, u_2, u_4, u_5 as follows:

$$u_6 = \frac{1}{x_3^2 - x_6^2}[u_1(x_1 x_6 - x_3 x_4) + u_2(x_2 x_6 - x_3 x_5)$$

$$+ u_4(x_4 x_6 - x_1 x_3) + u_5(x_5 x_6 - x_2 x_3)], \quad x_3^2 - x_6^2 \neq 0. \quad [1.28]$$

Similarly, equation [1.25] is derived from the following expression:

$$u_6 = -\frac{1}{x_6}(u_1 x_1 + u_2 x_2 + u_3 x_3 + u_4 x_4 + u_5 x_5), \quad x_6 \neq 0$$

and after replacing it in equation [1.24], we get u_3 depending on u_1, u_2, u_4, u_5:

$$u_3 = \frac{1}{x_3^2 - x_6^2}[u_1(x_4 x_6 - x_1 x_3) + u_2(x_5 x_6 - x_2 x_3)$$
$$+ u_4(x_1 x_6 - x_3 x_4) + u_5(x_2 x_6 - x_3 x_5)], \quad x_3^2 - x_6^2 \neq 0. \quad [1.29]$$

By replacing u_3 (respectively, u_6) by its expression above in equation [1.26] (respectively, [1.27]), it is easy to show that equations [1.26] and [1.27] are linear combinations of others. Therefore, the resolution of the previous system is reduced to the following system of four equations with six unknowns:

$$a_3 x_2 - a_2 x_3 + a_6 x_5 - a_5 x_6 = u_1, \quad a_6 x_2 - a_5 x_3 + a_3 x_5 - a_2 x_6 = u_4,$$
$$a_1 x_3 - a_3 x_1 + a_4 x_6 - a_6 x_4 = u_2, \quad a_4 x_3 - a_6 x_1 + a_1 x_6 - a_3 x_4 = u_5,$$

that we can write it in the form:

$$a_2 x_3 + a_5 x_6 = -u_1 + a_3 x_2 + a_6 x_5 \equiv \alpha, \qquad [1.30]$$
$$a_1 x_3 + a_4 x_6 = u_2 + a_3 x_1 + a_6 x_4 \equiv \beta, \qquad [1.31]$$
$$a_5 x_3 + a_2 x_6 = -u_4 + a_6 x_2 + a_3 x_5 \equiv \gamma, \qquad [1.32]$$
$$a_4 x_3 + a_1 x_6 = u_5 + a_6 x_1 + a_3 x_4 \equiv \delta. \qquad [1.33]$$

We fix a_3, a_6, and we draw from equations [1.31] and [1.33] the values of a_1 and a_4: $a_1 = \frac{\beta x_3 - \delta x_6}{x_3^2 - x_6^2}$, $a_4 = \frac{\delta x_3 - \beta x_6}{x_3^2 - x_6^2}$, while the values of a_2 and a_5 are obtained from equations [1.31] and [1.32] as follows: $a_2 = \frac{\alpha x_3 - \gamma x_6}{x_3^2 - x_6^2}$, $a_5 = \frac{\gamma x_3 - \alpha x_6}{x_3^2 - x_6^2}$. We determined $x \wedge a = \xi_1$, $a = (a_1, a_2, a_3, a_4, a_5, a_6)$ where a_3, a_6 are fixed and a_1, a_2, a_4, a_5 are given explicitly by the formulas above: $x = (x_1, x_2, x_3, x_4, x_5, x_6)$, $\xi_1 = (u_1, u_2, u_3, u_4, u_5, u_6)$, where u_3 is determined by [1.29] and u_6 by [1.28].

ii) As mentioned above, we determine $x \wedge b = \xi_2$, $b = (b_1, b_2, b_3, b_4, b_5, b_6)$, where b_3, b_6 are fixed and b_1, b_2, b_4, b_5 are given by formulas similar to those obtained above (for a_1, a_2, a_4, a_5), $x = (x_1, x_2, x_3, x_4, x_5, x_6)$ and $\xi_2 = (v_1, v_2, v_3, v_4, v_5, v_6)$ where v_3, v_6 are obtained in terms of v_1, v_2, v_4, v_5 (as in the case of u_3, u_6 above depending on u_1, u_2, u_4, u_5). By replacing these expressions in $\Omega(\xi_1, \xi_2)(x) = \langle x, a \wedge b \rangle$, we obtain the following expression: $\frac{(u_1 v_2 - u_2 v_1 + u_4 v_5 - u_5 v_4)x_3 - (u_1 v_5 - u_5 v_1 + u_4 v_2 - u_2 v_4)x_6}{x_3^2 - x_6^2}$.

We will calculate a and b relative to the basis $\left(\frac{\partial}{\partial x_1}, \frac{\partial}{\partial x_2}, \frac{\partial}{\partial x_4}, \frac{\partial}{\partial x_5}\right)$ of the tangent plane with

$$\frac{\partial}{\partial x_1} = \left(1, 0, \frac{x_4 x_6 - x_1 x_3}{x_3^2 - x_6^2}, 0, 0, \frac{x_1 x_6 - x_3 x_4}{x_3^2 - x_6^2}\right),$$

$$\frac{\partial}{\partial x_2} = \left(0, 1, \frac{x_5 x_6 - x_2 x_3}{x_3^2 - x_6^2}, 0, 0, \frac{x_2 x_6 - x_3 x_5}{x_3^2 - x_6^2}\right),$$

$$\frac{\partial}{\partial x_4} = \left(0, 0, \frac{x_1 x_6 - x_3 x_4}{x_3^2 - x_6^2}, 1, 0, \frac{x_4 x_6 - x_1 x_3}{x_3^2 - x_6^2}\right),$$

$$\frac{\partial}{\partial x_5} = \left(0, 0, \frac{x_2 x_6 - x_3 x_5}{x_3^2 - x_6^2}, 0, 1, \frac{x_5 x_6 - x_2 x_3}{x_3^2 - x_6^2}\right).$$

Hence, $\Omega\left(\frac{\partial}{\partial x_1}, \frac{\partial}{\partial x_2}\right) = \frac{x_3}{x_3^2 - x_6^2}$, $\Omega\left(\frac{\partial}{\partial x_1}, \frac{\partial}{\partial x_4}\right) = 0$, $\Omega\left(\frac{\partial}{\partial x_1}, \frac{\partial}{\partial x_5}\right) = -\frac{x_6}{x_3^2 - x_6^2}$, $\Omega\left(\frac{\partial}{\partial x_2}, \frac{\partial}{\partial x_4}\right) = \frac{x_6}{x_3^2 - x_6^2}$, $\Omega\left(\frac{\partial}{\partial x_2}, \frac{\partial}{\partial x_5}\right) = 0$, $\Omega\left(\frac{\partial}{\partial x_4}, \frac{\partial}{\partial x_5}\right) = \frac{x_3}{x_3^2 - x_6^2}$. Taking into account the formula $J^{-1} = \left(\omega\left(\frac{\partial}{\partial x_i}, \frac{\partial}{\partial x_j}\right)\right)$, we have

$$J = \begin{pmatrix} 0 & -x_3 & 0 & -x_6 \\ x_3 & 0 & x_6 & 0 \\ 0 & -x_6 & 0 & -x_3 \\ x_6 & 0 & x_3 & 0 \end{pmatrix}.$$

Consequently, $\Omega = -x_3 dx_1 \wedge dx_2 - x_6 dx_1 \wedge dx_5 + x_6 dx_2 \wedge dx_4 - x_3 dx_4 \wedge dx_5$.

EXAMPLE 1.18.– The symplectic structure obtained here is equivalent to that associated with the system [1.19] (see example 1.11). Indeed, we will show that there is an equivalence between $\dot{x}(t) = J\frac{\partial H}{\partial x}$, $x = (x_1, x_2, x_3, x_4, x_5, x_6)^\top$, where

$$J = \begin{pmatrix} 0 & -x_3 & x_2 & 0 & -x_6 & x_5 \\ x_3 & 0 & -x_1 & x_6 & 0 & -x_4 \\ -x_2 & x_1 & 0 & -x_5 & x_4 & 0 \\ 0 & -x_6 & x_5 & 0 & -x_3 & x_2 \\ x_6 & 0 & -x_4 & x_3 & 0 & -x_1 \\ -x_5 & x_4 & 0 & -x_2 & x_1 & 0 \end{pmatrix}.$$

and $\dot{x}(t) = \mathbf{J}\frac{\partial \mathbf{H}}{\partial x}$, $x = (x_1, x_2, \mathbf{x}_3, x_4, x_5, \mathbf{x}_6)^\top$, where

$$\mathbf{H} = H(x_1, x_2, \mathbf{x}_3, x_4, x_5, \mathbf{x}_6), \quad \mathbf{J} = \begin{pmatrix} 0 & -x_3 & 0 & -x_6 \\ x_3 & 0 & x_6 & 0 \\ 0 & -x_6 & 0 & -x_3 \\ x_6 & 0 & x_3 & 0 \end{pmatrix}.$$

Indeed, let us put $J = (J_{ij})$ and $\mathbf{J} = (\mathbf{J}_{ij})$. When calculating $\dot{x}_1, \dot{x}_2, \dot{x}_4, \dot{x}_5$, we will need the following equations (which we obtain from expressions [1.29] and [1.28]:

$$dx_3 = \frac{1}{x_3^2 - x_6^2}[(x_4x_6 - x_1x_3)dx_1 + (x_5x_6 - x_2x_3)dx_2$$
$$+ (x_1x_6 - x_3x_4)dx_4 + (x_2x_6 - x_3x_5)dx_5], \quad x_3^2 - x_6^2 \neq 0, \qquad [1.34]$$

$$dx_6 = \frac{1}{x_3^2 - x_6^2}[(x_1x_6 - x_3x_4)dx_1 + (x_2x_6 - x_3x_5)dx_2$$
$$+ (x_4x_6 - x_1x_3)dx_4 + (x_5x_6 - x_2x_3)dx_5], \quad x_3^2 - x_6^2 \neq 0. \qquad [1.35]$$

– Let us determine $\dot{x}_1$. We have $\dot{x}_1 = \sum_{\substack{j=1 \\ j \neq 3,6}}^{6} \mathbf{J}_{ij} \frac{\partial H}{\partial x_j} = -x_3 \frac{\partial H}{\partial x_2} - x_6 \frac{\partial H}{\partial x_5}$, hence,

$$\dot{x}_1 = -x_3 \left(\frac{\partial H}{\partial x_2} + \frac{\partial H}{\partial x_3}\frac{\partial x_3}{\partial x_2} + \frac{\partial H}{\partial x_6}\frac{\partial x_6}{\partial x_2} \right) - x_6 \left(\frac{\partial H}{\partial x_5} + \frac{\partial H}{\partial x_3}\frac{\partial x_3}{\partial x_5} + \frac{\partial H}{\partial x_6}\frac{\partial x_6}{\partial x_5} \right).$$

According to equations [1.34] and [1.35], we have

$$\frac{\partial x_3}{\partial x_2} = x_5x_6 - x_2x_3, \qquad \frac{\partial x_3}{\partial x_5} = x_2x_6 - x_2x_5,$$

$$\frac{\partial x_6}{\partial x_2} = x_3x_5 - x_2x_6, \qquad \frac{\partial x_6}{\partial x_5} = x_2x_3 - x_5x_6.$$

By replacing these expressions in the equation above, we get

$$\dot{x}_1 = -x_3\frac{\partial H}{\partial x_2} + x_2\frac{\partial H}{\partial x_3} - x_6\frac{\partial H}{\partial x_5} + x_5\frac{\partial H}{\partial x_6}.$$

Consequently,

$$\dot{x}_1 = \sum_{\substack{j=1 \\ j \neq 3,6}}^{6} \mathbf{J}_{1j}\frac{\partial \mathbf{H}}{\partial x_j} \iff \dot{x}_1 = \sum_{j=1}^{6} J_{1j}\frac{\partial H}{\partial x_j}.$$

– Let us determine $\dot{x}_2$. We have $\dot{x}_2 = \sum_{\substack{j=1 \\ j \neq 3,6}}^{6} \mathbf{J}_{ij} \frac{\partial H}{\partial x_j} = x_3 \frac{\partial H}{\partial x_1} + x_6 \frac{\partial H}{\partial x_4}$, hence,

$$\dot{x}_2 = x_3 \left(\frac{\partial H}{\partial x_1} + \frac{\partial H}{\partial x_3}\frac{\partial x_3}{\partial x_1} + \frac{\partial H}{\partial x_6}\frac{\partial x_6}{\partial x_1} \right) + x_6 \left(\frac{\partial H}{\partial x_4} + \frac{\partial H}{\partial x_3}\frac{\partial x_3}{\partial x_4} + \frac{\partial H}{\partial x_6}\frac{\partial x_6}{\partial x_4} \right).$$

According to equations [1.25] and [1.26], we have

$$\frac{\partial x_3}{\partial x_1} = x_4 x_6 - x_1 x_3, \qquad \frac{\partial x_3}{\partial x_4} = x_1 x_6 - x_3 x_4,$$

$$\frac{\partial x_6}{\partial x_1} = x_3 x_4 - x_1 x_6, \qquad \frac{\partial x_6}{\partial x_4} = x_1 x_3 - x_4 x_6.$$

By replacing these expressions in the equation above, we get

$$\dot{x}_2 = x_3 \frac{\partial H}{\partial x_1} - x_1 \frac{\partial H}{\partial x_3} + x_6 \frac{\partial H}{\partial x_4} - x_4 \frac{\partial H}{\partial x_6}.$$

Consequently,

$$\dot{x}_2 = \sum_{\substack{j=1 \\ j \neq 3,6}}^{6} J_{2j} \frac{\partial \mathbf{H}}{\partial x_j} \iff \dot{x}_2 = \sum_{j=1}^{6} J_{2j} \frac{\partial H}{\partial x_j}.$$

– Let us determine $\dot{x}_4$. We have $\dot{x}_4 = \sum_{\substack{j=1 \\ j \neq 3,6}}^{6} J_{ij} \frac{\partial \mathbf{H}}{\partial x_j} = -x_6 \frac{\partial H}{\partial x_2} - x_3 \frac{\partial H}{\partial x_5}$, hence,

$$\dot{x}_4 = -x_6 \left(\frac{\partial H}{\partial x_2} + \frac{\partial H}{\partial x_3} \frac{\partial x_3}{\partial x_2} + \frac{\partial H}{\partial x_6} \frac{\partial x_6}{\partial x_2} \right) - x_3 \left(\frac{\partial H}{\partial x_5} + \frac{\partial H}{\partial x_3} \frac{\partial x_3}{\partial x_5} + \frac{\partial H}{\partial x_6} \frac{\partial x_6}{\partial x_5} \right).$$

The expressions, $\frac{\partial x_3}{\partial x_2}, \frac{\partial x_3}{\partial x_5}, \frac{\partial x_6}{\partial x_2}, \frac{\partial x_6}{\partial x_5}$, have been calculated above. By replacing these in the equation above, we obtain $\dot{x}_4 = -x_6 \frac{\partial H}{\partial x_2} + x_5 \frac{\partial H}{\partial x_3} - x_3 \frac{\partial H}{\partial x_5} + x_2 \frac{\partial H}{\partial x_6}$ and consequently,

$$\dot{x}_4 = \sum_{\substack{j=1 \\ j \neq 3,6}}^{6} J_{4j} \frac{\partial \mathbf{H}}{\partial x_j} \iff \dot{x}_4 = \sum_{j=1}^{6} J_{4j} \frac{\partial H}{\partial x_j}.$$

– Let us determine $\dot{x}_5$. We have $\dot{x}_5 = \sum_{\substack{j=1 \\ j \neq 3,6}}^{6} J_{ij} \frac{\partial \mathbf{H}}{\partial x_j} = x_6 \frac{\partial H}{\partial x_1} + x_3 \frac{\partial H}{\partial x_4}$, hence,

$$\dot{x}_5 = x_6 \left(\frac{\partial H}{\partial x_1} + \frac{\partial H}{\partial x_3} \frac{\partial x_3}{\partial x_1} + \frac{\partial H}{\partial x_6} \frac{\partial x_6}{\partial x_1} \right) + x_3 \left(\frac{\partial H}{\partial x_4} + \frac{\partial H}{\partial x_3} \frac{\partial x_3}{\partial x_4} + \frac{\partial H}{\partial x_6} \frac{\partial x_6}{\partial x_4} \right).$$

The expressions, $\frac{\partial x_3}{\partial x_1}, \frac{\partial x_3}{\partial x_4}, \frac{\partial x_6}{\partial x_1}, \frac{\partial x_6}{\partial x_4}$, have been calculated above. By replacing these in the equation above, we obtain $\dot{x}_5 = x_6 \frac{\partial H}{\partial x_1} - x_4 \frac{\partial H}{\partial x_3} + x_3 \frac{\partial H}{\partial x_4} - x_1 \frac{\partial H}{\partial x_6}$. Consequently,

$$\dot{x}_5 = \sum_{\substack{j=1 \\ j \neq 3,6}}^{6} J_{5j} \frac{\partial \mathbf{H}}{\partial x_j} \iff \dot{x}_5 = \sum_{j=1}^{6} J_{5j} \frac{\partial H}{\partial x_j}.$$

1.10. Exercises

EXERCISE 1.1.– Let (M, ω) be a $2n$-dimensional symplectic manifold. Show that M is oriented by the Liouville form: $\Omega = (-1)^{n(n-1)/2} \omega \wedge \cdots \wedge \omega$, ($n$ times), and deduce that in canonical coordinates $(x_1, ..., x_n, y_1, ..., y_n)$, the form Ω is written as $\Omega = dx_1 \wedge \cdots \wedge dx_n \wedge dy_1 \wedge \cdots \wedge dy_n$.

EXERCISE 1.2.– Consider $M = \mathbb{R}^{2n}$ with coordinate system $(x_1, ..., x_n, y_1, ..., y_n)$. Show in detail that the latter is a Darboux coordinate system if and only if $\{x_i, x_j\} = \{y_i, y_j\} = 0$, $\{x_i, y_j\} = \delta_{ij}$ (hint: see example 1.7.)

EXERCISE 1.3.– Let ω_1 and ω_2 be symplectic forms on a smooth $2n$-dimensional manifold M. Let N be a compact n-dimensional submanifold and suppose it is a Lagrangian submanifold of (M, ω_1) and (M, ω_2). Show that there are neighborhoods U_1 and U_2 of N, and a symplectomorphism $\eta : (U_1, \omega_1) \longrightarrow (U_2, \omega_2)$, such that: $i_2 = \eta \circ i_1$ where $i_1 : N \longrightarrow U_1$ and $i_2 : N \longrightarrow U_2$ are the inclusion maps (hint: see Weinstein (1971)).

EXERCISE 1.4.– Let $(M, \{.,.\})$ be a Poisson manifold. Show that, for any $H \in \mathcal{C}^\infty(M)$, the map $\{., H\} : \mathcal{C}^\infty(M) \longrightarrow \mathbb{R}$ is a derivation on $\mathcal{C}^\infty(M)$ (hint: use Leibniz's law). Deduce that the above map defines a unique vector field X_H on M so that for any F, $X_H(F) = \{F, H\}$.

EXERCISE 1.5.– Let (M, ω) be a symplectic manifold, H the Hamiltonian function and consider the Poisson bracket $\{H, F\} = \omega(X_H, X_F)$ of $H, F \in \mathcal{C}^\infty(M)$. Prove that $\{H, F\} = 0$ if and only if F is constant along the integral curve of X_H (i.e. the Hamiltonian vector field X_H is tangent to the level sets $F = c$ and, in particular, the Hamiltonian vector field X_H is always tangent to the level sets $H = c$).

EXERCISE 1.6.– Let (M, ω, H) be a Hamiltonian system, $\dim M = 2n$, and suppose that $H_1 = H, H_2, ..., H_k$ are independent first integrals of (M, ω, H).

a) Show that for any i, j, we have $\omega(X_{H_i}, X_{H_j}) = 0$.

b) Show that at a point p where $dH_1, ..., dH_k$ are linearly independent, the vector fields $X_{H_1}(p), ..., X_{H_k}(p)$ span an isotropic subspace of $T_p M$.

c) Show that if the independent first integrals $H_1, ..., H_k$ are Poisson commuting (i.e. $\{H_i, H_j\} = 0$, for all $1 \leq i, j \leq k$), then $k \leq n$.

EXERCISE 1.7.– Consider the Hamiltonian $H = p_1 q_1 - p_2 q_2 - \alpha q_1^2 + \beta q_2^2$, where α and β are constants. Show that the functions $F = \frac{1}{q_1}(p_2 - \beta q_2)$, $G = q_1 q_2$, are first integrals, and that $q_1(t) = e^{t + \log q_1(0)}$.

EXERCISE 1.8.– We use the notation from section 1.5. Prove that if the variety M is compact, connected and $\int_M \omega_t = \int_M \omega_0$, where $\{\omega_t\}$, $0 \leq t \leq 1$ is a family of volume forms, then one can find a family of diffeomorphisms $g_t : M \longrightarrow M$, such that: $g_0^* =$ identity, $g_t^* \omega_t = \omega_0$ (hint: use a reasoning similar to lemma 1.1, replace Poincaré's lemma, which is local, with De Rham's theorem, which is global. This means that a volume form ω on M is exact if and only if $\int_M \omega = 0$).

EXERCISE 1.9.– Let (M, ω) be a symplectic manifold. Suppose that there exist n smooth functions $H_1, ..., H_n$ on M which Poisson bracket commute: $\{H_i, H_j\} = 0$ for all $1 \leq i, j \leq n$. Let $c \in \mathbb{R}^n$ be a regular value of the map $\varphi : M \longrightarrow \mathbb{R}^n$, $p \longmapsto \varphi(p) = (H_1(p), ..., H_n(p))$. Recall that $c \in \mathbb{R}^n$ is a regular value of the map φ, which means that every $p \in \varphi^{-1}(c)$ is a regular point provided that $\text{rank}(d\varphi_p) = \dim \mathbb{R}^n = n$. Prove that the fiber $\varphi^{-1}(c)$ is coisotropic and deduce that in particular if $n = \frac{1}{2} \dim M$, then $\varphi^{-1}(c)$ is a Lagrangian submanifold of M.

EXERCISE 1.10.– (a) Let $x \in M$ and $X_1, ..., X_r$ be differentiable vector fields on a manifold M. Show that there exists an open $\mathcal{U}$ of M containing x and $\varepsilon > 0$ such that, for any $x \in \mathcal{U}$, the function: $(t_1, ..., t_r) \longmapsto g_{t_1}^{X_1} \circ \cdots \circ g_{t_r}^{X_r}(x)$ exists for all $t_1, ..., t_r \in]-\varepsilon, \varepsilon[$. (b) Rectification theorem: prove that if $\forall k, l = 1, ..., r$, $[X_k, X_l] = 0$ and $X_1(x), ..., X_r(x)$ are linearly independent, then there is an open $\mathcal{U} \subset M$ containing x and a local coordinate system on $\mathcal{U}$ such that: $X_1|_{\mathcal{U}} = \frac{\partial}{\partial x_1}, ..., X_r|_{\mathcal{U}} = \frac{\partial}{\partial x_r}$. (c) Frobenius theorem: we assume that $X_1(x), ..., X_r(x)$ are independent everywhere and that these vector fields form a completely integrable system (i.e. for all $x \in M$, there exists a submanifold $N \supset M$ such that: $X_k(y) \in T_y N$ for any $y \in N$ and for any $k = 1..., r$). Show that this is equivalent to the existence of differentiable functions C_{kl}^j such as: $[X_k, X_l] = \sum_{j=1}^r C_{kl}^j X_j$, $(j, k, l = 1, ..., r)$.

2

Hamilton–Jacobi Theory

In this chapter we study some notions concerning the variational principle and will establish the Euler–Lagrange differential equation. We then introduce the Legendre transformation that allows to associate a system of differential equations of the first order, Hamilton's canonical equations, with this second-order equation. In the study of a Hamiltonian system and in order to simplify the system in question, it will sometimes be necessary to perform transformations. Not all transformations transform a Hamiltonian system into a Hamiltonian system, only the canonical transformations have this effect. Such transformations will be studied. We will then introduce the Hamilton–Jacobi equation; it is a nonlinear partial differential equation, and its resolution requires the knowledge of a canonical transformation, which is generally difficult to determine. As applications, we will study the geodesics, the harmonic oscillator and the Kepler problem as well as the simple pendulum.

2.1. Euler–Lagrange equation

Let $\gamma = \{(t, q) : q = q(t), t_1 \leq t \leq t_2\}$ be a curve defined on a differentiable manifold and connecting two parameter points (t_1, q_1) and (t_2, q_2). We consider the functional

$$\Phi(\gamma) = \int_{t_1}^{t_2} L(q(t), \dot{q}(t), t) dt, \quad \dot{q}(t) = \frac{dq}{dt}, \qquad [2.1]$$

where $L : \mathbb{R}^n \times \mathbb{R}^n \times \mathbb{R} \longrightarrow \mathbb{R}$, $(q, \dot{q}, t) \longmapsto L(q, \dot{q}, t)$, is a function of class $\mathcal{C}^2$. The variational problem is to find the function $q(t)$ of class $\mathcal{C}^2$ with $q(t_1) = q_1$, $q(t_2) = q_2$, (q_1, q_2 given values), such that the integral [2.1] (called integral action or simply action) is extremal (maximum or minimum). Classical trajectories can often be saddle points of the action, i.e. neither maxima nor minima. Geometrically, we search the arc of the curve joining the point $A(t_1, q_1)$ to the point $B(t_2, q_2)$ admitting a tangent that varies continuously for which the integral [2.1] is stationary. This means that if one considers the following function $t \longmapsto q(t) + \varepsilon x(t)$, with $\varepsilon \in \mathbb{R}$ small and $x(t)$ any function of class $\mathcal{C}^1$ satisfying $x(t_1) = x(t_2) = 0$, then the function Φ is stationary for q if $\int_{t_1}^{t_2} L(q(t)+\varepsilon x(t), \dot{q}(t)+\varepsilon \dot{x}(t), t)dt - \int_{t_1}^{t_2} L(q(t), \dot{q}(t), t)dt = o(\varepsilon)$.

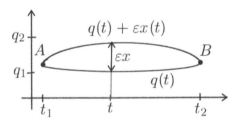

Figure 2.1. *Stationary functional*

We denote by $\delta \Phi$ the functional derivative (or variational derivative) of the functional Φ, and we symbolize the fact that the integral [2.1] is extremal by writing $\delta \Phi = 0$, during a functional variation of paths. This means that the difference between this evaluated integral along the actual trajectory and the evaluated integral along any infinitesimal neighboring virtual trajectory is an infinitesimal small of the second order.

EXAMPLE 2.1.– Let $\gamma = \{(t, q) : q = q(t), t_1 \leq t \leq t_2\}$ be a curve in the (t, q)-plane, and $\dot{q}(t) = \frac{dq}{dt}$. In the particular case where $L(a, b, c) = \sqrt{1 + b^2}$, that is, $\Phi(\gamma) = \int_{t_1}^{t_2} \sqrt{1 + \dot{q}^2} dt$, we obtain the length of the curve γ.

REMARK 2.1.– In Lagrangian and Hamiltonian mechanics (see below), L is called the Lagrangian of the variational principle and is equal to the difference between the kinetic energy and the potential energy. The Hamilton principle is $\delta \Phi = \delta \int_{t_1}^{t_2} L(q(t), \dot{q}(t), t) dt = 0$, with $\delta q(t_1) = \delta q(t_2) = 0$ for the real movement of the system to be studied.

The functional Φ is differentiable if $\Phi(\gamma + h) - \Phi(\gamma) = F + R$, where F refers to the differential of Φ and depends linearly on h (i.e. for a fixed γ, $F(\alpha h_1 + \beta h_2) = \alpha F(h_1) + \beta F(h_2)$), and $R(h, \gamma) = O(h^2)$ (in the sense that, for $|h| < \varepsilon$ and $|\frac{dh}{dt}| < \varepsilon$, we have $|R| < C\varepsilon^2$). We also say that F (respectively, h) is a variation of the functional Φ (respectively, the curve γ).

THEOREM 2.1.– The functional [2.1] is differentiable. Its differential $F(h)$ is given by $F(h) = \int_{t_1}^{t_2} \left(\frac{\partial L}{\partial q} - \frac{d}{dt} \frac{\partial L}{\partial \dot{q}} \right) h dt + \left(\frac{\partial L}{\partial \dot{q}} h \right) \Big|_{t_1}^{t_2}$.

PROOF.– We have

$$\Phi(\gamma + h) - \Phi(h) = \int_{t_1}^{t_2} (L(q + h, \dot{q} + \dot{h}, t) - L(q, \dot{q}, t)dt,$$

$$= \int_{t_1}^{t_2} \left(\frac{\partial L}{\partial q} h + \frac{\partial L}{\partial \dot{q}} \dot{h} \right) dt + o(h^2).$$

By setting $F(h) = \int_{t_1}^{t_2} \left(\frac{\partial L}{\partial q} h + \frac{\partial L}{\partial \dot{q}} \dot{h} \right) dt$, $R = o(h^2)$, we get $\Phi(\gamma + h) - \Phi(\gamma) = F + R$. By integrating by parts, we find $\int_{t_1}^{t_2} \frac{\partial L}{\partial \dot{q}} \dot{h} dt = \frac{\partial L}{\partial \dot{q}} h \Big|_{t_1}^{t_2} - \int_{t_1}^{t_2} h \frac{d}{dt} \frac{\partial L}{\partial \dot{q}} dt$. Therefore, $F(h) = \frac{\partial L}{\partial \dot{q}} h \Big|_{t_1}^{t_2} + \int_{t_1}^{t_2} h \left(\frac{\partial L}{\partial q} - \frac{d}{dt} \frac{\partial L}{\partial \dot{q}} \right) dt$, which ends the proof. □

DEFINITION 2.1.– *A curve γ is called extremal of a differentiable functional [2.1] if and only if for any h, $F(h, \gamma) = 0$.*

THEOREM 2.2.– The curve γ is the extremal of the functional [2.1] if and only if

$$\frac{d}{dt}\left(\frac{\partial L}{\partial \dot{q}}\right) - \frac{\partial L}{\partial q} = 0. \qquad [2.2]$$

PROOF.– According to the previous theorem, we have $F(h) = \int_{t_1}^{t_2} h \left(\frac{\partial L}{\partial q} - \frac{d}{dt} \left(\frac{\partial L}{\partial \dot{q}} \right) \right) dt$, because $h(t_1) = h(t_2) = 0$. The rest of the proof uses the fundamental lemma of the calculus of variations: if $f(t)$ is a continuous function on $[t_1, t_2]$ such that, for any function $g(t)$ of class $\mathcal{C}^1$ null in t_1 and t_2, we have $\int_{t_1}^{t_2} f(t)g(t)dt = 0$, then $f(t) = 0$ for $t \in [t_1, t_2]$. Therefore, according to this lemma applied to functions $f(t) = \frac{\partial L}{\partial q} - \frac{d}{dt}\left(\frac{\partial L}{\partial \dot{q}}\right)$ and $g(t) = h(t)$, we have $\frac{\partial L}{\partial q} - \frac{d}{dt}\left(\frac{\partial L}{\partial \dot{q}}\right) = 0$. The converse is obvious because if this last equation is satisfied, then we have $F(h) = 0$, which completes the proof. □

In mechanics, equation [2.2] is called the Lagrange equation, whereas in the calculus of variations it is called the Euler equation. It can be also called the Lagrange equation, Euler equation or Euler–Lagrange equation.

EXAMPLE 2.2.– Geodesics in the plane and on a surface: in a sufficiently small domain surrounding two points of a surface S, a line that realizes the shortest path between these two points is called a geodesic. Thus, in the Euclidean plane, the geodesics are the straight lines and on a sphere the geodesics are the great circles. In the plane, the length of the arc of equation $q = q(t)$ that joins the point $A(t_1, a)$ to the point $B(t_2, b)$ is $\Phi = \int_{t_1}^{t_2} \sqrt{1 + \dot{q}^2} dt$. The Euler–Lagrange equation with

$L = \sqrt{1+\dot{q}^2}$ is written as $\frac{\dot{q}}{\sqrt{1+\dot{q}^2}}$ = constant. Hence, $\dot{q}$ is constant, the extremities are straight lines and the shortest path between A and B is realized by the line segment AB. In the case of a surface S, suppose that a parametric representation is given as a function of the parameters u, v. The arc element ds is expressed as $ds^2 = Edu^2 + 2Fdudv + Gdv^2$. Therefore, we look for the shortest path on the surface S between two points $A(u = t_1, v = v_1)$ and $B(u = t_2, v = v_2)$, that is, the function $v = v(u)$ that minimizes the integral $\Phi = \int_{t_1}^{t_2} \sqrt{E + 2F\dot{v} + G\dot{v}^2} du$, and taking for t_1 the value v_1 and for t_2 the value v_2. We can deduce the differential equation of the geodesics

$$\frac{d}{du} \frac{F + G\dot{v}}{\sqrt{E + 2F\dot{v} + G\dot{v}^2}} = \frac{\frac{\partial E}{\partial v} + 2\frac{\partial F}{\partial v}\dot{v} + \frac{\partial G}{\partial v}\dot{v}^2}{2\sqrt{E + 2F\dot{v} + G\dot{v}^2}}.$$

2.2. Legendre transformation

We will see that, using a Legendre transformation, we can associate a system of $2n$ first-order differential equations called Hamilton's canonical equations to the Lagrange equations [2.2], which are second-order equations. But before that, let us first define a Legendre transformation. Let $f : \mathbb{R} \longrightarrow \mathbb{R}$ be a convex function, $f''(x) > 0$. The Legendre transformation of f is a function $G(s)$ of a new variable s defined as follows: consider in the plane of (x, y), the graph of f and the equation of a straight line $y = sx$, $s \in \mathbb{R}$. Let us determine a point $x = x(s)$ such that the vertical distance $F(s, x) \equiv sx - f(x)$ is maximal. In other words, $(x(s), f(x(s)))$ is the point farthest from the straight line with respect to the direction of the y axis. At this point (if it exists, it is unique since f is convex), we have $\frac{\partial F}{\partial x} = s - f'(x) = 0$ and the tangent $y = f'(x)$ is parallel to the straight line $y = sx$. Hence, we define the function $G(s)$ by setting $G(s) = F(s, x(s)) = \max_x (sx - f(x))$.

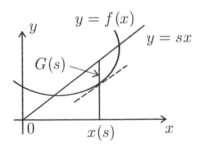

Figure 2.2. *Legendre transformation*

EXAMPLE 2.3.– Let $f(x) = x^2$. We have $F(s, x) = sx - x^2$ and $\frac{\partial F}{\partial x} = s - 2x = 0$. Then $x(s) = \frac{s}{2}$ and $G(s) = F(s, x(s)) = \frac{s^2}{4}$.

The Legendre transformation is involutive (i.e. its square is the identity). Two functions f and g are said to be dual (in the Young's sense) if they are deduced from each other by the Legendre transformation. In fact, the Legendre transformation can be considered as a process that allows us to pass from functions defined on a vector space to functions on the dual space. We now assume that $f : \mathbb{R}^n \longrightarrow \mathbb{R}$ is a convex function defined on $\mathbb{R}^n$ with $x = (x_1, ..., x_n) \in \mathbb{R}^n$; the quadratic form $\left(\frac{\partial^2 f}{\partial x^2} dx, dx\right)$ is positive definite. The Legendre transformation of f is a new function $G(s)$ of several variables $s = (s_1, ..., s_n)$ defined similarly to the previous case using formulas: $G(s) = F(s, x(s)) = \max_x F(s, x) = \max_x ((s, x) - f(x))$, $s = \frac{\partial f}{\partial x}$. Let us take the case of a function $f(x_1, ..., x_n; \lambda_1, ..., \lambda_m)$ depending on $n + m$ variables and assumed twice continuously differentiable with respect to the x_i and once with respect to the λ_i. We introduce new variables $s_1, ..., s_n$ by the transformation formulas $s_1 = \frac{\partial f}{\partial x_1}, ..., s_n = \frac{\partial f}{\partial x_n}$. It is assumed that these formulas are invertible by requiring, for example, the determinant of the Hessian matrix $\left(\frac{\partial^2 f}{\partial x_i \partial x_j}\right)_{1 \leq i,j \leq n}$ to be non-zero. The quantities $\lambda_1, ..., \lambda_m$ appear simply as parameters in the transformation. Let us now introduce the function G by $G(s_1, ..., s_n; \lambda_1, ..., \lambda_m) = \sum_{i=1}^n s_i x_i - f(x_1, ..., x_n; \lambda_1, ..., \lambda_m)$. In this equation, it is obviously assumed that the x_i are expressed in terms of the s_i. We have

$$\frac{\partial G}{\partial s_k} = \sum_{i=1}^n \frac{\partial x_i}{\partial s_k} s_i + x_k - \sum_{i=1}^n \frac{\partial f}{\partial x_i} \frac{\partial x_i}{\partial s_k} = \sum_{i=1}^n \frac{\partial x_i}{\partial s_k} \left(s_i - \frac{\partial f}{\partial x_i}\right) + x_k = x_k,$$

because $s_i = \frac{\partial f}{\partial x_i}$, $1 \leq i \leq n$. In the same way,

$$\frac{\partial G}{\partial \lambda_k} = \sum_{i=1}^n \frac{\partial x_i}{\partial \lambda_k} s_i - \sum_{i=1}^n \frac{\partial f}{\partial x_i} \frac{\partial x_i}{\partial \lambda_k} - \frac{\partial f}{\partial \lambda_k}$$

$$= \sum_{i=1}^n \frac{\partial x_i}{\partial \lambda_k} \left(s_i - \frac{\partial f}{\partial x_i}\right) - \frac{\partial f}{\partial \lambda_k} = -\frac{\partial f}{\partial \lambda_k}.$$

Consequently, the formulas of the Legendre transformation are written in the form $G(s_1, ..., s_n; \lambda_1, ..., \lambda_n) = \sum_{i=1}^n s_i x_i - f(x_1, ..., x_n; \lambda_1, ..., \lambda_n)$, $s_i = \frac{\partial f}{\partial x_i}$, $x_i = \frac{\partial G}{\partial s_i}$, and $\frac{\partial G}{\partial \lambda_k} = -\frac{\partial f}{\partial \lambda_k}$.

2.3. Hamilton's canonical equations

Let $n \in \mathbb{N}^*$, $t \in \mathbb{R}$, $q = (q_1, ..., q_n) \in \mathbb{R}^n$ and $\dot{q} = (\dot{q}_1, ..., \dot{q}_n) \in \mathbb{R}^n$. The application $L : \mathbb{R}^n \times \mathbb{R}^n \times \mathbb{R} \longrightarrow \mathbb{R}$, $(q, \dot{q}, t) \longmapsto L(q, \dot{q}, t)$, is the Lagrangian function (or Lagrangian) if L and $\frac{\partial L}{\partial q}$ are of class $\mathcal{C}^1$ and if L is a second-degree polynomial in the variables $\dot{q}_1, ..., \dot{q}_n$. We define n new variables $p = (p_1, ..., p_n)$

by setting $p = \frac{\partial L}{\partial \dot{q}}(q, \dot{q}, t)$. We consider the Hamiltonian defined by the function $H: \mathbb{R}^n \times \mathbb{R}^n \times \mathbb{R} \longrightarrow \mathbb{R}$, $(p, q, t) \longmapsto H(p, q, t)$, where

$$H(p, q, t) = \sum_{i=1}^{n} p_i \dot{q}_i - L(q_1, ..., q_n, \dot{q}_1, ..., \dot{q}_n, t),$$

with H, $\frac{\partial H}{\partial q}$ and $\frac{\partial H}{\partial p}$ of class $\mathcal{C}^1$. Let us note that this expression is nothing but a Legendre transformation applied on the Lagrangian L to the variables $\dot{q}_i$, and the variables q_i and t are passive. Here, the determinant of the Hessian matrix of L with respect to the $\dot{q}_i$ is also non-zero. By differentiating the above equation, we get

$$dH = \sum_{i=1}^{n} \left(\dot{q}_i dp_i + p_i d\dot{q}_i - \frac{\partial L}{\partial q_i} dq_i - \frac{\partial L}{\partial \dot{q}_i} d\dot{q}_i \right) - \frac{\partial L}{\partial t} dt,$$

$$= \sum_{i=1}^{n} \left(\dot{q}_i dp_i - \frac{\partial L}{\partial q_i} dq_i \right) - \frac{\partial L}{\partial t} dt,$$

because $p_i = \frac{\partial L}{\partial \dot{q}_i}$. According to the Euler–Lagrange equation, we have

$$\frac{\partial L}{\partial q_i} = \frac{dp_i}{dt} = \dot{p}_i \Longrightarrow dH = \sum_{i=1}^{n} (\dot{q}_i dp_i - \dot{p}_i dq_i) - \frac{\partial L}{\partial t} dt.$$

As the differential of the function $H = H(q_1, ..., q_n, p_1, ..., p_n, t)$ is given by

$$dH = \sum_{i=1}^{n} \frac{\partial H}{\partial q_i} dq_i + \sum_{i=1}^{n} \frac{\partial H}{\partial p_i} dp_i + \frac{\partial H}{\partial t} dt,$$

we deduce by comparison: $\dot{q}_i = \frac{\partial H}{\partial p_i}$, $\dot{p}_i = -\frac{\partial H}{\partial q_i}$, $\frac{\partial H}{\partial t} = -\frac{\partial L}{\partial t}$. Hence,

THEOREM 2.3.– The Euler–Lagrange equations [2.2] are equivalent to Hamilton's canonical equations

$$\dot{q}_i = \frac{\partial H}{\partial p_i}, \qquad \dot{p}_i = -\frac{\partial H}{\partial q_i}, \qquad i = 1, ..., n. \qquad [2.3]$$

These Hamilton equations constitute a system of $2n$ ordinary differential equations of the first order and are known when we know the (Hamiltonian) function H defined above. In Hamiltonian mechanics, H is equal to the sum of kinetic energy and potential energy. The space $\mathbb{R}^{2n}$ of the $2n$-tuples of numbers $(q_1, ..., q_n, p_1, ..., p_n)$ is called the phase space of the problem, and equations [2.3] determine a Hamiltonian vector field on $\mathbb{R}^{2n}$ (see below). Note that when the Hamiltonian does not explicitly depend on t (i.e. in the case of conservative systems), we have $\frac{dH}{dt} = \sum_{i=1}^{n} \left(\frac{\partial H}{\partial q_i} \dot{q}_i + \frac{\partial H}{\partial p_i} \dot{p}_i \right) + \frac{\partial H}{\partial t} = 0$, taking into account equations [2.3] and the fact that $\frac{\partial H}{\partial t} = 0$. Therefore, we have $H = constant$, and this equation expresses the conservation of the total energy. Geometrically, this equation represents, according to the different values attributable to the constant, a family of hypersurfaces on which the point (p, q) representative of the system remains throughout the whole movement.

NOTATION.– *We will sometimes use the following notations in other chapters*: x_i for q_i and y_i for p_i.

2.4. Canonical transformations

We have shown previously that the Euler–Lagrange equations are equivalent to Hamilton's canonical equations: $\dot{q}_i = \frac{\partial H}{\partial p_i}, \dot{p}_i = -\frac{\partial H}{\partial q_i}$, of Hamiltonian

$$H \equiv H(p, q, t) = \sum_{i=1}^{n} p_i \dot{q}_i - L, \qquad [2.4]$$

where L is the Lagrangian and $p = (p_1, ..., p_n)$, $q = (q_1, ..., q_n)$. Contrary to what one might think, the advantage of these does not lie in the form of the equations, which are, in general, as difficult to solve in one formulation as in the other. In fact, for the Euler–Lagrange equations, they can only be simplified by transformations of variables that deal only with the coordinates q_i, whereas for Hamilton's equations one can transform into both q_i and p_i. It should be noted that not all transformations transform a Hamiltonian system into a Hamiltonian system, only canonical transformations have this effect. A canonical transformation is a change of canonical coordinates $(q, p, t) \longmapsto (Q, P, t)$ that preserves the form of Hamilton's equations.

DEFINITION 2.2.– *A differentiable mapping g of $\mathbb{R}^{2n} = \{p, q\}$ to $\mathbb{R}^{2n}$ is called a canonical transformation if it preserves the 2-form*: $\omega = \sum_{i=1}^{n} dp_i \wedge dq_i$.

REMARK 2.2.– The above definition can be written in one of the following equivalent forms: the differentiable mapping g is a canonical transformation if (i) the pull-back of ω is $g^*\omega = \omega$, or (ii) for any closed contour $\mathcal{C}$, $\oint_{\mathcal{C}} p dq = \oint_{g\mathcal{C}} p dq$, or (iii) if $\int \int_S \omega = \int_{gS} \omega$, where S is any surface with boundary $\mathcal{C}$.

Let $P_i = P_i(p, q, t), Q_i = Q_i(p, q, t)$ be canonical transformations and

$$\mathcal{H} \equiv \mathcal{H}(P, Q, t) = \sum_{i=1}^{n} P_i \dot{Q}_i - \mathcal{L}, \qquad [2.5]$$

the Hamiltonian expressed in the new variables where $\mathcal{L}$ is the Lagrangian and

$$\dot{Q}_i = \frac{\partial \mathcal{H}}{\partial P_i}, \qquad \dot{P}_i = -\frac{\partial \mathcal{H}}{\partial Q_i}, \qquad [2.6]$$

with $P = (P_1, ..., P_n)$, $Q = (Q_1, ..., Q_n)$.

THEOREM 2.4.– A necessary and sufficient condition for the transformation $p, q \longmapsto P(p,q,t), Q(p,q,t)$, to be canonical is that the following differential form, $\omega = \sum_{i=1}^n (p_i dq_i - P_i dQ_i) - (H - \mathcal{H}) dt$, is exact.

PROOF.–

– *Necessary condition*. According to the principle of least action, we have $\delta \int_{t_1}^{t_2} L dt = 0$, $\delta \int_{t_1}^{t_2} \mathcal{L} dt = 0$, hence $\delta \int_{t_1}^{t_2} (L - \mathcal{L}) dt = 0$. This condition is satisfied in general by

$$L - \mathcal{L} = \frac{dS}{dt}, \qquad [2.7]$$

(S is a function called the generating function). Indeed, $\delta \int_{t_1}^{t_2} dS = \delta(S(t_2) - S(t_1)) = 0$, because the canonical variables are identical at the instants t_1 and t_2. Taking into account equations [2.4] and [2.5], equation [2.7] is written as $\left(\sum_{i=1}^n p_i \dot{q}_i - H\right) - \left(\sum_{i=1}^n P_i \dot{Q}_i - \mathcal{H}\right) = \frac{dS}{dt}$, or

$$\left(\sum_{i=1}^n p_i dq_i - P_i dQ_i\right) - (H - \mathcal{H}) dt = dS, \qquad [2.8]$$

so the form ω is exact. The function S depends only on $2n + 1$ variables because p, q, P, Q, t are related by $2n$ equations: $P_i = P_i(p,q,t)$, $Q_i = Q_i(p,q,t)$.

– *Sufficient condition*. From the expression $\delta \int_{t_1}^{t_2} L dt = 0$, we derive the relation

$$\sum_{i=1}^n p_i dq_i - H dt = dF, \qquad [2.9]$$

where F is some function. This differential form is called the Poincaré–Cartan invariant. By subtracting [2.8] from [2.9], we get

$$\sum_{i=1}^n P_i dQ_i - \mathcal{H} dt = dF - dS \equiv d\lambda, \qquad [2.10]$$

or $d\left(\sum_{i=1}^n P_i Q_i\right) - \sum_{i=1}^n Q_i dP_i - \mathcal{H} dt = d\lambda$, or

$$\sum_{i=1}^n Q_i dP_i + \mathcal{H} dt = d\left(\sum_{i=1}^n P_i Q_i\right) - d\lambda \equiv d\mu. \qquad [2.11]$$

The forms [2.10] and [2.11] are exact if and only if $\frac{\partial P_i}{\partial t} = -\frac{\partial \mathcal{H}}{\partial Q_i}$, $\frac{\partial Q_i}{\partial t} = \frac{\partial \mathcal{H}}{\partial P_i}$, respectively. The canonical variables P_i and Q_i depend only on t, whence $\dot{P}_i = -\frac{\partial \mathcal{H}}{\partial Q_i}$, $\dot{Q}_i = \frac{\partial \mathcal{H}}{\partial P_i}$, which completes the proof. $\square$

REMARK 2.3.– Another way to characterize a canonical transformation is to use a symplectic matrix, or a Poisson bracket (see exercise 2.9 for more information).

2.5. Hamilton–Jacobi equation

The basic idea of the Hamilton–Jacobi method consists of finding a canonical transformation which reduces the Hamiltonian function to a form such that the canonical equations can be integrated, which implies the integration of the original canonical equations. Let $S = S(q, Q, t)$ be a function of $2n + 1$ variables $q = (q_1, ..., q_n)$, $Q = (Q_1, ..., Q_n)$ and t. We have $dS = \sum_{i=1}^{n} \left(\frac{\partial S}{\partial q_i} dq_i + \frac{\partial S}{\partial Q_i} dQ_i \right) + \frac{\partial S}{\partial t} dt$, and by comparing with [2.8], we get

$$p_i = \frac{\partial S}{\partial q_i}, \qquad P_i = -\frac{\partial S}{\partial Q_i}, \qquad \mathcal{H} = \frac{\partial S}{\partial t} + H. \qquad [2.12]$$

Similarly, if $S = S(q, P, t)$ we obtain $p_i = \frac{\partial S}{\partial q_i}$, $Q_i = \frac{\partial S}{\partial P_i}$, $\mathcal{H} = \frac{\partial S}{\partial t} + H$. Similar formulas are obtained by assuming that S depends on other coordinates, for example, $S(p, Q, t)$, $S(p, P, t)$. In any case, whatever the canonical transformation considered, the generating function S and the Hamiltonians H, $\mathcal{H}$ are connected by equation $\mathcal{H} = \frac{\partial S}{\partial t} + H$.

Consider a Hamiltonian system $H(p, q, t)$ and look for a canonical transformation depending on t such that the new Hamiltonian $\mathcal{H}(P, Q, t)$ is zero. Thus, we define $S = S(q, Q, t)$ depending on $2n + 1$ variables so that $\mathcal{H} = 0$. This method of making $\mathcal{H} = 0$ implies that the new variables P and Q are constants. Indeed, equations [2.12] and [2.6] reduce, respectively, to $p_i = \frac{\partial S}{\partial q_i}$, $P_i = -\frac{\partial S}{\partial Q_i}$, $\mathcal{H} = \frac{\partial S}{\partial t} + H$ and $\dot{Q}_i = \frac{\partial \mathcal{H}}{\partial P_i} = 0$, $\dot{P}_i = -\frac{\partial \mathcal{H}}{\partial Q_i} = 0$, so $Q_i = a_i$, $P_i = b_i$, where a_i and b_i are constants. We can determine p_i and q_i using the canonical transformation and thus the solution of the system. To do this, we must determine the generating function S and this is obtained by solving equation $\frac{\partial S}{\partial t} + H = 0$. Since $p_i = \frac{\partial S}{\partial q_i}$, then the Hamiltonian $H(p_i, q_i, t)$ is written as $H\left(\frac{\partial S}{\partial q_i}, q_i, t\right)$ and the equation above becomes

$$\frac{\partial S}{\partial t} + H\left(\frac{\partial S}{\partial q_i}, q_i, t\right).$$

This partial differential equation is called the Hamilton–Jacobi equation. S depends on t, on n variables $q_1, ..., q_n$ and on n constants of integration $a_1, ..., a_n$. In fact, instead of $2n+1$, this equation depends only on $n+1$ constants. Now the function S appears in the Hamilton–Jacobi equation only as partial derivatives with respect to t and q_i. If S is a solution of this equation, then $S+c$ is another solution where c is an arbitrary constant. One can thus clearly replace one of the constants of integration in the (complete) solution by an additive constant. Since only the partial derivatives of the generating function are involved in the transformation equations, we can neglect this additive constant and write the solution in the form $S = S(q_1,...,q_n, a_1,...,a_n, t)$. We have $\frac{d}{dt}\left(\frac{\partial S}{\partial a_i}\right) = \frac{\partial^2 S}{\partial t \partial a_i} + \sum_{j=1}^{n} \frac{\partial^2 S}{\partial q_j \partial a_i} \cdot \frac{dq_j}{dt}$. From the Hamilton–Jacobi equation, we deduce the expression $\frac{\partial^2 S}{\partial t \partial a_i} = \sum_{j=1}^{n} \frac{\partial H}{\partial p_j} \cdot \frac{\partial^2 S}{\partial q_j \partial a_i}$. So

$$\frac{d}{dt}\left(\frac{\partial S}{\partial a_i}\right) = -\sum_{j=1}^{n} \frac{\partial H}{\partial p_j} \cdot \frac{\partial^2 S}{\partial q_j \partial a_i} + \sum_{j=1}^{n} \frac{\partial^2 S}{\partial q_j \partial a_i} \cdot \frac{dq_j}{dt},$$

$$= \sum_{j=1}^{n} \frac{\partial H}{\partial p_j} \cdot \frac{\partial^2 S}{\partial q_j \partial a_i}\left(\frac{dq_j}{dt} - \frac{\partial H}{\partial p_j}\right) = 0,$$

because $\frac{dq_j}{dt} = \frac{\partial H}{\partial p_j}$ along an extremal. Therefore, $\frac{\partial S}{\partial a_i}(q, a_i, t) = c_i$, $1 \leq i \leq n$, where a_i are constants and these equations can be solved for q_i in terms of a_i, c_i, t,

$$q_i = q_i(a_1,...,a_n, c_1,...,c_n, t), \qquad [2.13]$$

assuming that the Hessian determinant $\det\left(\frac{\partial^2 S}{\partial q_j \partial a_i}\right) \neq 0$. Then, by inserting the solution [2.13] into the relation $p_i = \frac{\partial S}{\partial q_i}(q_1,...,q_n, a_1,...,a_n, t)$, we get the p_i as functions of a_i, c_i and t,

$$p_i = p_i(a_1,...,a_n, c_1,...,c_n, t). \qquad [2.14]$$

The expressions [2.13] and [2.14] thus constitute the complete solution of Hamilton's canonical equations. The constants $a_1,...,a_n, c_1,...,c_n$ are determined by the initial conditions. We therefore have the following result (Jacobi's theorem):

THEOREM 2.5.– Let $S(q_1,...,a_1,...,a_n, t)$ be a solution of the Hamilton–Jacobi equation, dependent on n parameters $a_1,...,a_n$ and assume: $\det\left(\frac{\partial^2 S}{\partial q_j \partial a_i}\right) \neq 0$. Then Hamilton's canonical equations $\dot{p} = -\frac{\partial H}{\partial q}$, $\dot{q} = \frac{\partial H}{\partial p}$ can be integrated by quadratures and the partial derivatives $\frac{\partial S}{\partial a_i}$, $1 \leq i \leq n$ are the first integrals of these canonical equations.

We will distinguish and analyze several cases that we often encounter in practice:

1) Suppose that the Hamiltonian H does not explicitly depend on t. Let $H = H(p, q)$. From the relation $\frac{dH}{dt} = \frac{\partial H}{\partial t} = 0$, we deduce that $H = a_1 \equiv$ constant.

As mentioned before, let us look for a generating function S so that: $\mathcal{H} = 0$. The Hamilton–Jacobi equation is written as $\frac{\partial S}{\partial t} + H\left(\frac{\partial S}{\partial q_i}\right) = 0$, hence $\frac{\partial S}{\partial t} + a_1 = 0$ and, separating variables (the separated variable being t), we get an equation in the form

$$S(q_i, a_i, t) = -a_1 t + S'(q_i, a_i), \quad i \geq 2 \qquad [2.15]$$

Since $\frac{\partial S}{\partial q_i} = \frac{\partial S'}{\partial Q_i}$, the Hamilton–Jacobi equation is written as $H\left(\frac{\partial S'}{\partial q_i}, q_i\right) = a_1$, $i \geq 2$ and the problem consists of determining the complete solution $S'(q_i, a_i)$ of this equation. The generating function is given by the relation [2.15] and the new variables are written as $P_i = -\frac{\partial S}{\partial q_i} = c_i$, $i \geq 2$, because $\dot{P}_i = -\frac{\partial \mathcal{H}}{\partial Q_i} = 0$. In particular, we have $c_1 = P_1 = -\frac{\partial S}{\partial a_1} = t - \frac{\partial S'}{\partial Q_1}$.

2) Suppose that the generating function is written in the form: $S = S(q, P, t)$. Hence, the variables p and Q are defined by $p_i = -\frac{\partial S}{\partial q_i}$, $Q_i = \frac{\partial S}{\partial P_i}$, with $\mathcal{H} = \frac{\partial S}{\partial t} + H$. The Hamilton–Jacobi equation is written as $\frac{\partial S}{\partial t} + H\left(q_i, \frac{\partial S}{\partial q_i}, t\right) = 0$, and the problem consists of determining a solution $S(q_i, a_i, t)$ of this equation with $Q_i = \frac{\partial S}{\partial a_i} = c_i$, $P_i = a_i$. The solutions of the Hamilton–Jacobi equation above are obtained by repeating the same method as that used previously for the generating function $S(q, Q, t)$.

3) We have just seen that when the Hamiltonian H does not depend explicitly on t, we can conclude that t is separable in the Hamilton–Jacobi equation. Similarly, we note that when the coordinates are cyclic (remember that a coordinate q_i is cyclic if $\frac{\partial H}{\partial q_i} = 0$), then these are always separable, the new Hamiltonian $\mathcal{H}$ depends only on new impulses and the Hamilton equations can easily be solved. Let us take the case of a system for which H is a constant of the motion, say a, and that q_1 is a cyclic coordinate as an example. Therefore $\dot{p}_1 = -\frac{\partial H}{\partial q_1} = 0$, which implies: $p_1 = b$ = constant. The Hamilton–Jacobi equation in this case is written in the form

$$H\left(\frac{\partial S}{\partial q_2}, ..., \frac{\partial S}{\partial q_n}, q_2, ..., q_n, b\right) = a. \qquad [2.16]$$

In order to use the method of separation of the variables, we consider S in the form: $S = S_1(q_1, a) + S'(q_2, ..., q_n, a)$. Therefore, equation [2.16] is written as $H\left(\frac{\partial S'}{\partial q_2}, ..., \frac{\partial S'}{\partial q_n}, q_2, ..., q_n, b\right) = a$ and since $p_1 = b = \frac{\partial S_1}{\partial q_1}$, then $S_1 = bq_1$, modulo a constant. Thus, b is the separation constant, and therefore, we get $S = bq_1 + S'(q_2, ..., q_n, a)$. This is similar to equation [2.15] obtained in the case where the Hamiltonian H does not depend explicitly on t. This is due to the fact that since $H = a$, t can be assimilated to a cyclic coordinate. We thus obtain results similar to those obtained previously.

4) Suppose now that q_1 is not cyclic but that all other coordinates $q_2, ..., q_n$ are cyclic. We have $\dot{p}_i = -\frac{\partial H}{\partial q_i}$, $i = 2, ..., n$, so $p_i = b_i$, $i = 2, ..., n$. The

Hamilton–Jacobi equation is $H\left(\frac{\partial S}{\partial q_1}, q_1, b_2, ..., b_n, b\right) = a$, and we proceed as before. We consider $S = S_1(q_1, a) + S_2(q_2, a) + \cdots + S_n(q_n, a)$. Since $p_i = b_i = \frac{\partial S_i}{\partial q_i}$, $i = 2, ..., n$, then $S_i = b_i q_i$, $i = 2, ..., n$ and therefore, $S = S_1'(q_1, a) + \sum_{i=2}^{n} b_i q_i$, with $H\left(\frac{\partial S}{\partial q_1}, q_1, b_2, ..., b_n, b\right) = a$. This last differential equation in q_1 can be solved by quadratures, and we easily get its solution S_1 and the complete solution S.

5) When a coordinate q_i and the derivative $\frac{\partial S}{\partial q_i}$ can be grouped into a combination of the form $\varphi\left(\frac{\partial S}{\partial q_i}, Q_i\right)$, then q_i is separable. A coordinate q_i is separable if it is included with the impulse p_i in the Hamiltonian as a function $\varphi(p_i, q_i)$. We look for solutions of the equation in the form $S = S_i(q_i, b) + S'(q_j, b)$, $j = 1, ..., n, j \neq i$. The Hamilton–Jacobi equation is $H\left(\frac{\partial S'}{\partial q_j}, q_j, \varphi\left(\frac{\partial S}{\partial q_i}, q_i\right)\right) = a$, and it is solved for φ, $\varphi\left(\frac{\partial S}{\partial q_i}, q_i\right) = \psi\left(\frac{\partial S'}{\partial q_j}, q_j, a\right)$. Using the separation of variables method, we put $\varphi\left(\frac{\partial S}{\partial q_i}, q_i\right) =$ constant, $\psi\left(\frac{\partial S'}{\partial q_j}, q_j, a\right) =$ constant. In the first equation, the separation of the variable q_i has been performed. If in the second equation a variable is separable, the above procedure can be repeated and so on until all the variables are separated. When this is possible (i.e. all variables are separable), the Hamilton–Jacobi equation is completely separable and the corresponding Hamiltonian system is integrated by quadratures.

2.6. Applications

2.6.1. *Harmonic oscillator*

Let us apply the preceding method to the study of the harmonic oscillator defined by the Hamiltonian $H \equiv H(p, q) = \frac{1}{2m}p^2 + \frac{1}{2}m\omega^2 q^2$. The corresponding Hamilton–Jacobi equation is written in the form $\frac{\partial S}{\partial t} + \frac{1}{2m}\left(\frac{\partial S}{\partial q}\right)^2 + \frac{1}{2}m\omega^2 q^2 = 0$. Since H does not explicitly depend on t, we can conclude that t can be separated in this equation and therefore look for a solution in the form

$$S = S_1(q) + S_2(t). \qquad [2.17]$$

The above equation reduces to $\frac{1}{2m}\left(\frac{\partial S}{\partial q}\right)^2 + \frac{1}{2}m\omega^2 q^2 = -\left(\frac{dS_2}{dt}\right)$. According to the method of separation of the variables, each of these terms must be constant. Let c be this constant. So, $\frac{1}{2m}\left(\frac{dS_1}{dq}\right)^2 + \frac{1}{2}m\omega^2 q^2 = c$, $\frac{dS_2}{dt} = -c$, hence $S_1 = \sqrt{2m}\int \sqrt{c - \frac{1}{2}m\omega^2 q^2}dq$, $S_2 = -ct$, modulo a constant. Substituting these expressions in [2.17], one obtains $S = \sqrt{2m}\int \sqrt{c - \frac{1}{2}m\omega^2 q^2}dq - ct = S(q, c, t)$.

From the above equation, we identify the constant c with the new momentum P and since Q is a constant($\equiv a$). So $a = Q = \frac{\partial S}{\partial P} = \frac{\partial S}{\partial c}$, i.e. $a = \sqrt{\frac{m}{2}} \int \frac{dq}{\sqrt{c - \frac{1}{2}m\omega^2 q^2}} - t$, from which we derive the expression of q in terms of constants of integration c and a, $q = \sqrt{\frac{2c}{m\omega^2}} \sin \omega(t + a)$. For the impulse p, we have $p = \frac{\partial S_1}{\partial q} = \sqrt{2m}\sqrt{c - \frac{1}{2}m\omega^2 q^2} = \sqrt{2mc} \cos \omega(t+a)$. The constant c identifies itself with the total energy of the system. The constants c and a can be determined from the initial conditions. By denoting by p_0 and q_0 the values of p and q in $t = 0$, one obtains $c = \frac{1}{2m}p_0^2 + \frac{m\omega^2}{2}q_0, a = \frac{1}{\omega}\operatorname{arctg} m\omega \frac{q_0}{p_0}$.

2.6.2. The Kepler problem

Consider the Kepler problem on the motion of a mass point m, mobile in a fixed plane and attracted by a force (called Newtonian) to the origin (a fixed center). This force is proportional to the inverse of the square of its distance at the origin. The material point is identified by polar coordinates (r, θ) with the origin as the pole. This material point being in a central field of potential $-\frac{k}{r}$ where k is a number strictly greater than 0, the Hamiltonian is defined by $H = \frac{1}{2m}\left(p_r^2 + \frac{p_\theta^2}{r^2}\right) - \frac{k}{r}$. We will write the canonical equations of the motion, determine the solutions of the corresponding Hamilton–Jacobi equation for H and discuss the results. The canonical equations of the motion are $\dot{p}_r = -\frac{\partial H}{\partial r} = \frac{p_\theta^2}{mr^3} - \frac{k}{r^2}$, $\dot{p}_\theta = \frac{\partial H}{\partial \theta} = 0$, which implies that p_θ is constant. Here again H does not explicitly depend on t and since $p_r = \frac{\partial S}{\partial r}$, $p_\theta = \frac{\partial S}{\partial \theta}$, the corresponding Hamilton–Jacobi equation is written as $\frac{\partial S}{\partial t} + \frac{1}{2m}\left[\left(\frac{\partial S}{\partial r}\right)^2 + \frac{1}{r^2}\left(\frac{\partial S}{\partial \theta}\right)^2\right] - \frac{k}{r} = 0$. The coordinate θ being cyclic, we look for a solution in the form

$$S = S_1(r) + S_2(\theta) + S_3(t). \qquad [2.18]$$

The equation above becomes $\frac{1}{2m}\left[\left(\frac{\partial S_1}{\partial r}\right)^2 + \frac{1}{r^2}\left(\frac{\partial S_2}{\partial \theta}\right)^2\right] - \frac{k}{r} = -\frac{dS_3}{dt}$. The separation of variables method is used, $\frac{1}{2m}\left[\left(\frac{\partial S_1}{\partial r}\right)^2 + \frac{1}{r^2}\left(\frac{\partial S_2}{\partial \theta}\right)^2\right] - \frac{k}{r} = c$, $\frac{dS_3}{dt} = -c$, where c is a constant. Hence $\left(\frac{dS_2}{d\theta}\right)^2 = -r^2\left(\frac{dS_1}{dr}\right)^2 + 2mkr + 2mcr^2$, $S_3 = -ct$, modulo a constant. In the first equation above, we note that the left term is independent of r and depends only on θ. When in the right term, it depends only on r. We therefore apply separation of variables method again: $\frac{dS_2}{d\theta} = a$, $r^2\left(\frac{dS_1}{dr}\right)^2 - 2mkr - 2mcr^2 = -a^2$, where a is a constant. Hence $S_2 = a\theta$, $S_1 = \int \sqrt{-\frac{a^2}{r^2} + 2\frac{mk}{r} + 2mc}\, dr$, by considering only the positive root. By inserting the expressions of S_1, S_2 and S_3 into [2.18], one obtains $S = \int \sqrt{-\frac{a^2}{r^2} + 2\frac{mk}{r} + 2mc}\, dr + a\theta - ct$. The sequence consists of identifying the constants a and c with the new quantities of motion P_r and P_θ, respectively. We denote the constants Q_r and Q_θ by b_1 and b_2, respectively. So $b_1 = Q_r = \frac{\partial S}{\partial P_r} = \frac{\partial S}{\partial a}$, $b_2 = Q_\theta = \frac{\partial S}{\partial P_\theta} = \frac{\partial S}{\partial c}$, that is,

$$a \int \frac{dr}{r^2 \sqrt{-\frac{a^2}{r^2} + 2\frac{mk}{r} + 2mc}} = \theta - b_1, \qquad m \int \frac{dr}{\sqrt{-\frac{a^2}{r^2} + 2\frac{mk}{r} + 2mc}} = t + b_2.$$

The first equation gives the equation of the orbit, that is, it is sufficient to carry out the change of variable $r = \frac{1}{u}$, and the second equation determines r as a function of t. The constants a and c are identical to the angular momentum and the total energy. The orbit is an ellipse if $c = 0$, a parabola if $c > 0$ and a hyperbola if $c < 0$.

2.6.3. *Simple pendulum*

A simple pendulum consists of a small mass m suspended at the end of a cord in such a manner that it can swing in an arc of a circle. We denote by l the length of the cord (i.e. the radius of the circle) and by g the acceleration of gravity. The motion of the simple pendulum is described by an angle displacement x of the pendulum from its vertical equilibrium position. The kinetic and potential energies are $T = \frac{1}{2}ml^2\dot{x}^2$ and $U = mg(1 - \cos x)$. The Lagrangian is $L(x, \dot{x}) = T - U = \frac{1}{2}ml^2\dot{x}^2 - mg(1 - \cos x)$. The generalized momentum is $p = \frac{\partial L}{\partial \dot{x}} = ml^2\dot{x}$, so that $\dot{x} = \frac{p}{ml^2}$. The Hamiltonian is $H(x, p) = T + U = \frac{1}{2}ml^2\dot{x}^2 + mg(1 - \cos x)$, and the Hamiltonian equations, $\dot{p} = -\frac{\partial H}{\partial x}$, $\dot{x} = \frac{\partial H}{\partial p}$. Using these equations, we obtain, after substituting for p, the equation of motion in terms of the angle x and its derivative:

$$\frac{d^2 x}{dt^2} + \frac{g}{l} \sin x = 0. \qquad [2.19]$$

Let $\theta = \frac{dx}{dt}$, equation [2.19] is written as $\theta d\theta + \frac{g}{l} \sin x\, dx = 0$. By integrating, we get $\frac{\theta^2}{2} = \frac{g}{l} \cos x + C$, where C is a constant. To determine the latter, note that when $t = 0$, $x = x_0$ (initial angle), then $\theta = 0$ (the speed is zero), hence $C = -\frac{g}{l} \cos x_0$. Therefore

$$\frac{l}{2g}\left(\frac{dx}{dt}\right)^2 = \frac{l}{2g}\theta^2 = \cos x - \cos x_0. \qquad [2.20]$$

We will study several cases:

a) Consider the case of an oscillatory movement, that is, the case where the mass passes from $x = x_0$ (the largest angle reached by the pendulum, it corresponds to a speed $\theta = 0$) to $x = 0$ (maximum speed). Since $\cos x = 1 - 2\sin^2 \frac{x}{2}$, then equation [2.20] becomes

$$\frac{l}{4g}\left(\frac{dx}{dt}\right)^2 = \sin^2 \frac{x_0}{2} - \sin^2 \frac{x}{2}. \qquad [2.21]$$

Let $\sin\frac{x}{2} = \sin\frac{x_0}{2}\sin\varphi$, hence $\frac{1}{2}\cos\frac{x}{2}dx = \sin\frac{x_0}{2}\cos\varphi d\varphi$,

$$\frac{1}{2}\sqrt{1-\sin^2\frac{x}{2}}dx = \sin\frac{x_0}{2}\sqrt{1-\sin^2\varphi}d\varphi,$$

$$\frac{1}{2}\sqrt{1-\sin^2\frac{x_0}{2}\sin^2\varphi}dx = \sin\frac{x_0}{2}\sqrt{1-\sin^2\varphi}d\varphi,$$

and so $dx = \frac{2\sin\frac{x_0}{2}\sqrt{1-\sin^2\varphi}}{\sqrt{1-\sin^2\frac{x_0}{2}\sin^2\varphi}}d\varphi$. By substitution in [2.21], we obtain $\left(\frac{d\varphi}{dt}\right)^2 = \frac{g}{l}\left(1 - k^2\sin^2\varphi\right)$, where $k = \sin\frac{x_0}{2}$, is the module and $\frac{x_0}{2}$ the modular angle. Note that for $x = 0$, we have $\varphi = 0$ and therefore $t = \pm\sqrt{\frac{l}{g}}\int_0^\varphi \frac{d\varphi}{\sqrt{1-k^2\sin^2\varphi}}$. We have[1] $\varphi = \pm\mathbf{am}\sqrt{\frac{g}{l}}t$, $\sin\varphi = \pm\sin\mathbf{am}\sqrt{\frac{g}{l}}t = \pm\mathbf{sn}\sqrt{\frac{g}{l}}t$, and, consequently, $\sin\frac{x}{2} = \pm\sin\frac{x_0}{2}\mathbf{sn}\sqrt{\frac{g}{l}}t$.

b) Consider the case of a circular motion. Equation [2.20] is written in the form $\frac{l}{2g}\left(\frac{dx}{dt}\right)^2 = 1 - 2\sin^2\frac{x}{2} - \cos x_0 = (1 - \cos x_0)\left(1 - k^2\sin^2\frac{x}{2}\right)$, where $k^2 = \frac{2}{1-\cos x_0}$, with k positive and $0 < k < 1$. Taking into account the initial condition $x(0) = 0$, we obtain $dt = \pm\sqrt{\frac{2l}{g(1-\cos x_0)}}\int_0^\varphi \frac{d\varphi}{\sqrt{1-k^2\sin^2\varphi}}$, $\varphi = \frac{x}{2}$. Hence, $\varphi = \pm\mathbf{am}\sqrt{\frac{g(1-\cos x_0)}{2l}}t$, $x = \pm 2\mathbf{am}\sqrt{\frac{g(1-\cos x_0)}{2l}}t$.

c) Finally, consider the case of an asymptotic movement. This is the case where $x_0 = \pm\pi$ and equation [2.20] is written as $\frac{l}{2g}\left(\frac{dx}{dt}\right)^2 = \cos x + 1 = 2\cos^2\frac{x}{2}$. Hence, $t = \pm\frac{1}{2}\sqrt{\frac{l}{g}}\int_0^x \frac{dx}{\cos\frac{x}{2}} = \pm\sqrt{\frac{l}{g}}\ln\tan\left(\frac{x}{\pi} + \frac{\pi}{4}\right)$, and $x = 4\arctan e^{\pm\sqrt{\frac{g}{l}}t} - \pi$. We check that $x \to \pm\pi$ when $t \to \infty$.

REMARK 2.4.– For small oscillations, equation [2.19] can be reduced to $\frac{d^2x}{dt^2} + \frac{g}{l}x = 0$, whose general solution is immediate: $x(t) = C_1\cos\sqrt{\frac{g}{l}}t + C_2\sqrt{\frac{l}{g}}\sin\sqrt{\frac{g}{l}}t$, $C_1 = x(0)$, $C_2 = \frac{dx}{dt}(0)$. For small oscillations, the period of the pendulum (the time needed for a complete oscillation; a round trip) is $2\pi\sqrt{\frac{l}{g}}$. On the other hand, in the case of

---

[1] The function $t = \int_0^\varphi \frac{d\varphi}{\sqrt{1-k^2\sin^2\varphi}}$ defined by this integral has an inverse, called amplitude of t and is written $\varphi = \mathbf{am}\,t = \mathbf{am}(t;k)$. Note that if $k = 0$, then $t = \arcsin s$, and $s = \sin t$. For $k \neq 0$, we note by analogy the inverse function of the integral in question by $s = \mathbf{sn}\,t = \mathbf{sn}(t;k)$, called Jacobi elliptic function. When there is no ambiguity on the module k, we simply write $\mathbf{sn}\,t$ instead of $\mathbf{sn}(t;k)$. The function $\varphi = \mathbf{am}\,t$ is a strictly increasing odd function of t and we have $\mathbf{am}(0) = 0$, $\frac{\partial\mathbf{am}}{\partial t}(0) = 1$. Since $s = \sin\varphi$, we can write $s = \mathbf{sn}\,t = \sin(\mathbf{am}\,t)$ (see, for example (Lesfari 2008b, 2015b)).

oscillations that are not necessarily small, the period corresponds to what precedes $4\sqrt{\frac{l}{g}}\int_0^{\frac{\pi}{2}}\frac{dx}{\sqrt{1-k^2\sin^2 x}}$ with $k=\sin\frac{x_0}{2}$.

2.7. Exercises

EXERCISE 2.1.– Show that the composition of canonical transformations and the inverse of a canonical transformation is a canonical transformation. Deduce that the set of canonical transformations forms a group.

EXERCISE 2.2.– Are the following transformations canonical? Justify your answer (hint: use, for example, theorem 2.4).

a) $q_1 = Q_1, q_2 = -P_2, p_1 = P_1, p_2 = Q_2$.

b) $q_1 = \sqrt{Q_1}\cos P_1 + \sqrt{Q_2}\cos P_2$, $q_2 = -\sqrt{Q_1}\cos P_1 + \sqrt{Q_2}\cos P_2$, $p_1 = \sqrt{Q_1}\sin P_1 + \sqrt{Q_2}\sin P_2$, $p_2 = -\sqrt{Q_1}\sin P_1 + \sqrt{Q_2}\sin P_2$.

EXERCISE 2.3.– Prove that canonical transformations preserve the Poisson bracket and that the converse statement holds locally, and deduce that a transformation $Q_i = Q_i(q,p,t)$ and $P_i = P_i(q,p,t)$ is canonical if and only if $\{Q_i, Q_j\} = \{P_i, P_j\} = 0$, $\{Q_i, P_j\} = \delta_{ij}$ (hint: although the result is true for time-dependent transformations and so as not to make the notations heavier, it suffices to consider the case of transformations independent of time $Q_i = Q_i(q;p)$ and $P_i = P_i(q;p)$, which does not affect the generality).

EXERCISE 2.4.– Let $g: \mathbb{R}^{2n} \longrightarrow \mathbb{R}^{2n}$ be a canonical transformation. Show that g preserves the differential forms $\omega, \omega \wedge \omega, ..., \omega \wedge \cdots \wedge \omega$, n-times (these forms are the invariants of g), and deduce that a canonical transformation preserves volume in phase space (Liouville's theorem), that is, the phase flow of a Hamiltonian system is volume preserving. In other words, the volume of gD is equal to the volume of D for any region D (hint: it suffices to show that the Jacobian of the transformation is equal to 1).

EXERCISE 2.5.– Consider the ellipsoid: $\frac{x_1^2}{\alpha_1} + \frac{x_2^2}{\alpha_2} + \frac{x_3^2}{\alpha_3} = 1$, $\alpha_1 > \alpha_2 > \alpha_3$, and let u_1, u_2, u_3 be the roots of the equation: $\frac{x_1^2}{\alpha_1 - u} + \frac{x_2^2}{\alpha_2 - u} + \frac{x_3^2}{\alpha_3 - u} = 1$, $u_3 < \alpha_3 < u_2 < \alpha_2 < u_1 < \alpha_1$. It was Jacobi (1969) who introduced elliptic coordinates u_1, u_2, u_3, to study the geodesic flow on an ellipsoid. These elliptic coordinates correspond to the confocal quadrics. The Hamiltonian associated with this problem is $H = \frac{u_1 - u_2}{8}\left(\frac{u_1}{\prod_{j=1}^3(\alpha_j - u_1)}\frac{du_1}{dt} + \frac{u_2}{\prod_{j=1}^3(\alpha_j - u_2)}\frac{du_2}{dt}\right)$. Prove that the solution of the corresponding Hamilton–Jacobi equation is given by $S = \sqrt{\frac{H}{2}}\left(\int\frac{u_1 - c}{\sqrt{P(u_1)}}du_1 + \int\frac{u_2 - c}{\sqrt{P(u_2)}}du_2\right)$, and that the linearization of the equations of motion is done using genus 2 hyperelliptic functions: $\frac{du_1}{dt} = \frac{\sqrt{P(u_1)}}{u_1 - u_2}$, $\frac{du_2}{dt} = \frac{\sqrt{P(u_2)}}{u_2 - u_1}$, where $P(u) = \frac{u-c}{u} \prod_{j=1}^{3} (u - \alpha_j)$, and c is any constant such that: $\alpha_3 < c < \alpha_1$.

EXERCISE 2.6.– We consider a particle of mass m moving in two dimensions under the influence of a potential $U(r)$, which is a function of the distance from the origin, $r = \sqrt{x^2 + y^2}$. The Lagrangian is written as $L = \frac{m}{2}\left(\dot{r}^2 + r^2\dot{\theta}^2\right) - U(r)$, where r and θ are polar coordinates (see the Kepler problem, section 2.6). Show that the Hamiltonian is $H = \frac{1}{2m}\left(p_r^2 + \frac{p_\theta^2}{r^2}\right) + U(r)$, and write the Hamilton–Jacobi equation (hint: note that H does not explicitly depend on t). Separate the variables and integrate to obtain S.

EXERCISE 2.7.– Consider the central force field described by the Lagrangian in spherical coordinates r, θ, ϕ, $L = \frac{m}{2}\left(\dot{r}^2 + r^2\dot{\theta}^2 + r^2\sin^2\theta.\dot{\phi}^2\right) - U(r)$. Show that the Hamiltonian is $H = \frac{1}{2m}\left(p_r^2 + \frac{p_\theta^2}{r^2} + \frac{p_\phi^2}{r^2 \sin^2\theta}\right) + U(r)$, and the Hamilton–Jacobi equation is given by $\frac{1}{2m}\left(\left(\frac{\partial S}{\partial r}\right)^2 + \frac{1}{r^2}\left(\frac{\partial S}{\partial \theta}\right)^2 + \frac{1}{r^2 \sin^2\theta}\left(\frac{\partial S}{\partial \phi}\right)^2\right) + U(r) = E$. Separate the variables, determine the complete integral and analyze the equations describing the dynamics.

EXERCISE 2.8.– Consider the brachistochrone problem that consists of minimizing the travel time of a mass moving under its own weight between two points $A = (a, b)$ and $B = (c, d)$. The problem is to minimize the integral $\frac{1}{\sqrt{2g}} \int_a^c \sqrt{\frac{1+y'^2}{b-y}}$, subject to the conditions: $y(a) = b, y(c) = d$.

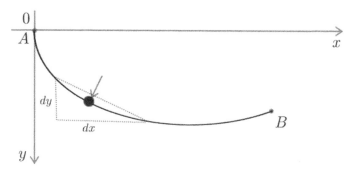

Figure 2.3. *Brachistochrone*

Show that the brachistochrone is the inverted cycloid: $x(\theta) = a + r(\theta - \sin\theta)$, $y(\theta) = b - r(1 - \cos\theta)$, where r is a parameter determined by the initial and terminal points. We put $\sqrt{2g} = 1$ which shows that the above integral leads to the Hamilton–Jacobi equation: $\left(\frac{\partial S}{\partial x}\right)^2 + \left(\frac{\partial S}{\partial y}\right)^2 = \frac{1}{y}$.

EXERCISE 2.9.– Hamilton's equations [2.3] can be written as: $\dot{z} = J\frac{\partial H}{\partial z}$, where $z = (q_1, ..., q_n, p_1, ..., p_n)^\top$ and $\frac{\partial H}{\partial z} = \left(\frac{\partial H}{\partial q_1}, ..., \frac{\partial H}{\partial q_n}, \frac{\partial H}{\partial p_1}, ..., \frac{\partial H}{\partial p_n}\right)$ are column vectors, and $J = \begin{pmatrix} O & I \\ -I & O \end{pmatrix}$, where O is the $n \times n$ zero matrix and I is the $n \times n$ identity matrix. A canonical transformation is a change of variables $z_i = z_i(z', t)$ that transforms any system of Hamiltonian equations into a system of Hamiltonian equations. It is necessary that for every function $H(z, t)$ there exists a function $K(z', t)$ such that the new equations are written as $\dot{z}' = J\frac{\partial K}{\partial z'}$. Consider the time-independent transformations: $Q_i = Q_i(q, p)$, $P_i = P_i(q, p)$, and define $z' = (Q_1, ..., Q_n, P_1, ..., P_n)^\top$, which is a function of the original coordinates, so we can write $z' = z'(z)$. This implies $K = H$. Let $\dot{z}'_i = \frac{\partial z'_i}{\partial z_j} \dot{z}_j$, the time derivative of z'.

a) Show that the transformed Hamilton's equations are written in the form: $\dot{z} = ZJZ^\top \frac{\partial H}{\partial z'}$, where Z is the Jacobian of the transformation; $Z_{ij} = \frac{\partial z'_i}{\partial z_j}$. Deduce that $z' = z'(z)$ is a canonical transformation if and only if the identity $ZJZ^\top = J$ is satisfied. (Such a matrix Z is called symplectic and for this reason the term canonical transformation is sometimes replaced by symplectic transformation).

b) Show that the results obtained above for the case of time-independent transformations remain valid for the case of time-dependent transformations and determine the corresponding formulas.

3

Integrable Systems

The aim of this chapter is to study the Arnold–Liouville theorem and its connection with completely integrable systems. Many integrable systems are studied in detail, including: the problem of the rotation of a rigid body about a fixed point – the Euler problem of a rigid body, the Lagrange top, the Kowalewski top and other special cases, such as the Hesse–Appel'rot top, Goryachev–Chaplygin top and Bobylev–Steklov top; the problem of motion of a solid in an ideal fluid – Clebsch's case and Lyapunov–Steklov's case; the Yang–Mills field with gauge group $SU(2)$. Some of these problems will be studied in detail in other chapters, using other methods.

3.1. Hamiltonian systems and Arnold–Liouville theorem

DEFINITION 3.1.– *A Hamiltonian system is a triple (M, ω, H), where (M, ω) is a $2n$-dimensional symplectic manifold (the phase space) and $H \in \mathcal{C}^\infty(M)$ is a smooth function (Hamiltonian).*

Recall section 1.5, where we showed that we have a complete characterization of Hamiltonian vector field

$$\dot{x}(t) = X_H(x(t)) = J\frac{\partial H}{\partial x}, \quad x \in M, \qquad [3.1]$$

where $H : M \longrightarrow \mathbb{R}$ is the Hamiltonian, $J = J(x)$ is a skew-symmetric matrix, possibly depending on $x \in M$, and for which the corresponding Poisson bracket $\{H, F\} = \sum_{i,j} J_{ij} \frac{\partial H}{\partial x_i} \frac{\partial F}{\partial x_j}$ satisfies the Jacobi identity: $\{\{H, F\}, G\} + \{\{F, G\}, H\} + \{\{G, H\}, F\} = 0$.

DEFINITION 3.2.– *A Hamiltonian system [3.1] is called Liouville integrable or completely integrable if it possesses n smooth functions (first integrals or constants of motion), $H_1 = H, H_2, ..., H_n$ defined on M, such that (i) $H_1, H_2, ..., H_n$ are in* *involution, that this* $\{H_i, H_j\} = 0$, $i,j = 1,...,n$, *and (ii)* $H_1, H_2, ..., H_n$ *are functionally independent. In other words, the differentials* $dH_1, ..., dH_n$ *are linearly independent on a dense open subset of* M *or* $dH_1 \wedge ... \wedge dH_n \neq 0$.

REMARK 3.1.– Note that these cannot be more than n independent first integrals in involution, otherwise the Poisson bracket would be degenerate.

We have defined a Hamiltonian system on a $2n$-dimensional symplectic manifold to be completely integrable if it has n first integrals in involution, which are functionally independent on some open and dense subset of this manifold. It is natural to extend the notion of complete integrability to systems defined on Poisson manifolds $(M, \{.,.\})$ by requiring that on each symplectic leaf, such system defines a completely integrable system in the usual sense. This generalization implies that an integrable system is associated with a maximal Abelian Poisson subalgebra of $(\mathcal{C}^\infty(M), \{.,.\})$. In the general context of Poisson manifolds, we have the following:

DEFINITION 3.3.– *Let* $(M, \{.,.\})$ *be a Poisson manifold of rank* $2r$. *A Hamiltonian system* $(M, \{.,.\}, \varphi)$ *is called completely integrable if it admits a complete set of first integrals* $H_1, ..., H_s$ *in involution (i.e.* $\{H_i, H_j\} = 0$, *for all* $i, j = 1, ..., s$) *and independent (i.e. their differentials are independent on a dense open subset of* M), *where* $r + s = \dim M$, $\varphi = (H_1, ..., H_s) : M \longrightarrow \mathbb{R}^s$ *is the so-called momentum map.*

REMARK 3.2.– The definition also makes sense when M is a complex manifold, such as a smooth affine variety, equipped with a Poisson structure on its algebra of holomorphic functions.

We will see that the overall regularity of the dynamical system in the case of complete integrability follows from the so-called Arnold–Liouville theorem (which will be discussed further). This theorem plays a crucial role in the study of the integrability of Hamiltonian systems. Before studying this theorem and its consequences, let us first show the following purely differential geometric fact:

PROPOSITION 3.1.– *A compact and connected n-dimensional manifold M, on which there exist m differential (of class $\mathcal{C}^\infty$) vector fields $X_1, ..., X_m$, which commute and are independent at every point, and diffeomorphic to an m-dimensional real torus:* $T^m = \mathbb{R}^m / lattice = \{(\varphi_1, ..., \varphi_m) \mod 2\pi\}$.

PROOF.– Indeed, let $g_{t_i}^{X_i}$ be the flow generated by X_i on M and let us define the application $g : \mathbb{R}^m \longrightarrow M$, $(t_1, ..., t_m) \longmapsto g(t_1, ..., t_m)$, where

$$g(t_1, ..., t_m) = g_{t_1}^{X_1} \circ \cdots \circ g_{t_m}^{X_m}(x) = g_{t_m}^{X_m} \circ \cdots \circ g_{t_1}^{X_1}(x), \ x \in M.$$

a) The application g is a local diffeomorphism. Indeed, let

$$g_r \equiv g\mid_U : U \longrightarrow M, \ (t_1,...,t_m) \longmapsto g_r(t_1,...,t_m) = g_{t_m}^{X_m} \circ \cdots \circ g_{t_1}^{X_1}(x),$$

be the restriction of g on a neighborhood U of $(0,...,0)$ in $\mathbb{R}^m$, with $x = g_r(0,...,0)$. Let us show that the map g_r is differentiable (of class $\mathcal{C}^\infty$). We have

$$\frac{\partial}{\partial t_1} g_{t_1}^{X_1} = X_1(x) = (\dot{x}_1, ..., \dot{x}_m) = (f_1(x_1,...,x_m), ..., f_m(x_1,...,x_m)),$$

where $f_1,...,f_m : M \longrightarrow \mathbb{R}$ are functions on M. Similarly, we have

$$\frac{\partial^2}{\partial t_1^2} g_{t_1}^{X_1} = (\ddot{x}_1, ..., \ddot{x}_m) = \left(\sum_{k=1}^m \frac{\partial f_1}{\partial x_k} \dot{x}_k, ..., \sum_{k=1}^m \frac{\partial f_m}{\partial x_k} \dot{x}_k \right),$$

$$\frac{\partial^3}{\partial t_1^3} g_{t_1}^{X_1} = (\dddot{x}_1, ..., \dddot{x}_m)$$

$$= \left(\sum_{k=1}^m \sum_{l=1}^m \frac{\partial^2 f_1}{\partial x_k \partial x_l} \dot{x}_k \dot{x}_l + \frac{\partial f_1}{\partial x_k} \ddot{x}_k, ..., \sum_{k=1}^m \sum_{l=1}^m \frac{\partial^2 f_m}{\partial x_k \partial x_l} \dot{x}_k \dot{x}_l + \frac{\partial f_m}{\partial x_k} \ddot{x}_k \right).$$

All of these expressions have a meaning because by hypothesis, all of the functions $f_1,...,f_m$ are $\mathcal{C}^\infty$. A similar reasoning shows that $g_{t_2}^{X_2}, ..., g_{t_m}^{X_m}$ are also $\mathcal{C}^\infty$. Since the composite of functions $\mathcal{C}^\infty$ is $\mathcal{C}^\infty$, we deduce that $g_r(t_1,...,t_m)$ is $\mathcal{C}^\infty$. Let us now show that the Jacobian matrix of g_r in $(0,...,0)$ is invertible. To do this, we consider $g_r(t_1,...,t_m) \equiv (G_1(t_1,...,t_m), ..., G_m(t_1,...,t_m))$. We have

$$\det \begin{pmatrix} \frac{\partial G_1}{\partial t_1} & \cdots & \frac{\partial G_m}{\partial t_1} \\ \vdots & \ddots & \vdots \\ \frac{\partial G_1}{\partial t_m} & \cdots & \frac{\partial G_m}{\partial t_m} \end{pmatrix} = \det \begin{pmatrix} \frac{\partial}{\partial t_1} g_{t_m}^{X_m} \circ \cdots \circ g_{t_1}^{X_1}(x) \\ \vdots \\ \frac{\partial}{\partial t_m} g_{t_m}^{X_m} \circ \cdots \circ g_{t_1}^{X_1}(x) \end{pmatrix} \neq 0,$$

because the vector fields $X_1,...,X_m$ are linearly independent at each point of M. According to the local inversion theorem, there exists a sufficiently small neighborhood $V \subset U$ of $(0,...,0)$ and a neighborhood W of x such that g_r induces a bijection of V on W, whose inverse $g_r^{-1} : W \to V$ is $\mathcal{C}^\infty$. In other words, g_r is a diffeomorphism of V over $g_r(V)$. This result is local because even if the above Jacobian matrix is invertible for any $(t_1,...,t_m)$, then the inverse "global" of g_r does not necessarily exist.

b) The application g is surjective. Indeed, let $(t_1,...,t_m) \in \mathbb{R}^m$ such that: $g(t_1,...,t_m) = g_{t_m}^{X_m} \circ \cdots \circ g_{t_1}^{X_1}(x) = y \in M$. In part (a) we showed that g is a local diffeomorphism. So for every point x_1 contained in a neighborhood of x, there exists $(t_1,...,t_m) \in \mathbb{R}^m$ such that: $g_{t_m}^{X_m} \circ \cdots \circ g_{t_1}^{X_1}(x) = x_1$. Since the variety M is connected, we can connect the point x to the point y by a curve $\mathcal{C}$. Let B_1 be an open ball in M containing the point x_1. This ball exists since M is compact. Let $x_2 \in \mathcal{C}$, such that x_2 be contained in the ball B_1. As before, we reason that the map g is a local diffeomorphism, then there exists $(t'_1,...,t'_m) \in \mathbb{R}^m$ such that: $\left(g_{t_m}^{X_m}\right)' \circ \cdots \circ \left(g_{t_1}^{X_1}\right)'(x_1) = x_2$. Hence, $x_2 = \left(g_{t_m}^{X_m}\right)' + t_m \circ \cdots \circ \left(g_{t_1}^{X_1}\right)' + t_1(x)$. Similarly, let B_2 be an open ball in M containing the point x_2 and let $x_3 \in \mathcal{C}$, such that x_3 be contained in the ball B_2. Since the application g is a local diffeomorphism, then there exists $(t''_1,...,t''_m) \in \mathbb{R}^m$ such that: $\left(g_{t_m}^{X_m}\right)'' \circ \cdots \circ \left(g_{t_1}^{X_1}\right)''(x_2) = x_3$. So $x_3 = \left(g_{t_m}^{X_m}\right)'' + t'_m + t_m \circ \cdots \circ \left(g_{t_1}^{X_1}\right)'' + t'_1 + t_1(x)$. Continuing this way, we show (after a finite number k of steps) the existence of a point $\left(t_1^{(k-1)},...,t_m^{(k-1)}\right) \in \mathbb{R}^m$, such that: $\left(g_{t_m}^{X_m}\right)^{(k-1)} \circ \cdots \circ \left(g_{t_1}^{X_1}\right)^{(k-1)}(x_{k-1}) = x_k$, where $x_k \in \mathcal{C}$, x_k is contained in an open ball B_{k-1} of M, with $B_{k-1} \ni x_{k-1}$. Therefore, for k finite, we have $x_k = \left(g_{t_m}^{X_m}\right)^{(k-1)} + t_m^{(k-2)} + \cdots + t'_m + t_m \circ \cdots \circ \left(g_{t_1}^{X_1}\right)^{(k-1)} + t_1^{(k-2)} + \cdots + t'_1 + t_1(x)$.

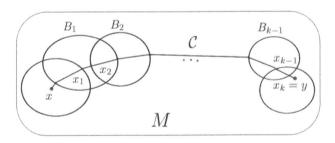

Figure 3.1. *Covering of $\mathcal{C}$ by open balls*

This construction shows that in a finite number k of steps, we can cover the curve $\mathcal{C}$ connecting the point x to the point y by neighborhoods of x; the point y plays the role of x_k. The application g cannot be injective, if it was we would have, according to part (a), a bijection between a compact M and a non-compact $\mathbb{R}^m$, which is absurd.

c) Let $\Lambda = \left\{(t_1,...,t_m) \in \mathbb{R}^m : g(t_1,...,t_m) = g_{t_m}^{X_m} \circ \cdots \circ g_{t_1}^{X_1}(x) = x\right\}$ be the stationary group. It is a discrete subgroup of $\mathbb{R}^m$ independent of point $x \in M$. Indeed, let us first note that $\Lambda \neq \emptyset$ because $(0,...,0) \in \Lambda$. Let $(t_1,...,t_m) \in \Lambda$,

$(t'_1, ..., t'_m) \in \Lambda$. We have $g(t_1, ..., t_m) = g(t'_1, ..., t'_m) = x$. As the vector fields $X_1, ..., X_m$ are commutative, then

$$g(t_1 + t'_1, ..., t_m + t'_m) = g^{X_m}_{t_m + t'_m} \circ \cdots \circ g^{X_1}_{t_1 + t'_1}(x),$$
$$= g^{X_m}_{t'_m} \circ \cdots \circ g^{X_1}_{t'_1} \circ g^{X_m}_{t_m} \circ \cdots \circ g^{X_1}_{t_1}(x),$$
$$= g^{X_m}_{t'_m} \circ \cdots \circ g^{X_1}_{t'_1}(x) = x,$$

and

$$g(-t_1, ..., -t_m) = g^{X_m}_{-t_m} \circ \cdots \circ g^{X_1}_{-t_1}(x),$$
$$= g^{X_m}_{-t_m} \circ \cdots \circ g^{X_1}_{-t_1} \circ g^{X_m}_{t_m} \circ \cdots \circ g^{X_1}_{t_1}(x),$$
$$= g^{X_m}_{-t_m} \circ \cdots \circ g^{X_1}_{-t_1} \circ g^{X_1}_{t_1} \circ \cdots \circ g^{X_m}_{t_m}(x),$$
$$= g^{X_m}_{-t_m} \circ \cdots \circ g^{X_2}_{-t_2} \circ g^{X_2}_{t_2} \circ \cdots \circ g^{X_m}_{t_m}(x),$$
$$\vdots$$
$$= g^{X_m}_{-t_m} \circ g^{X_m}_{t_m}(x) = x.$$

Hence $(t_1 + t'_1, ..., t_m + t'_m) \in \Lambda$ and $(-t_1, ..., -t_m) \in \Lambda$. Therefore, Λ is stable for addition, the inverse of $(t_1, ..., t_m)$ is $(-t_1, ..., -t_m)$ and consequently, Λ is a subgroup of $\mathbb{R}^m$. We show that Λ is independent of x. Let

$$\Lambda' = \left\{ (t'_1, ..., t'_m) \in \mathbb{R}^m : g(t'_1, ..., t'_m) = g^{X_m}_{t'_m} \circ \cdots \circ g^{X_1}_{t'_1}(y) = y \right\}.$$

By surjectivity, one can find $(s_1, ..., s_m) \in \mathbb{R}^m$ such that: $g^{X_m}_{s_m} \circ \cdots \circ g^{X_1}_{s_1}(x) = y$. Let $(t'_1, ..., t'_m) \in \Lambda'$. We have

$$g^{X_m}_{t'_m} \circ \cdots \circ g^{X_1}_{t'_1}(y) = y,$$
$$g^{X_m}_{t'_m} \circ \cdots \circ g^{X_1}_{t'_1} \circ g^{X_m}_{s_m} \circ \cdots \circ g^{X_1}_{s_1}(x) = g^{X_m}_{s_m} \circ \cdots \circ g^{X_1}_{s_1}(x),$$
$$g^{X_m}_{-s_m + t'_m + s_m} \circ \cdots \circ g^{X_1}_{-s_1 + t'_1 + s_1}(x) = x,$$
$$g^{X_m}_{t'_m} \circ \cdots \circ g^{X_1}_{t'_1}(x) = x.$$

Therefore, $(t'_1, ..., t'_m) \in \Lambda$ and Λ does not depend on x. To show that Λ is discrete, we consider a sufficient small neighborhood V of the point $(0, ..., 0)$ and a neighborhood W of the point x. From (a), the application g is a local diffeomorphism, so $g : V \longrightarrow W$ is bijective and consequently, no point of $W \setminus \{(0, ..., 0)\}$ is sent on x; the points of the subgroup Λ have no accumulation point in $\mathbb{R}^m$.

d) The variety M is diffeomorphic to a m-dimensional real torus. Indeed, let $T^k \times \mathbb{R}^{m-k} = \{(\varphi_1, ..., \varphi_k; u_1, ..., u_{m-k})\}$, $(\varphi_1, ..., \varphi_k) \bmod 2\pi$ be the direct product of k circles and $m - k$ straight lines and $\pi : \mathbb{R}^m \longrightarrow T^k \times \mathbb{R}^{m-k}$, with $\pi(\varphi_1, ..., \varphi_k; u_1, ..., u_{m-k}) = ((\varphi_1, ..., \varphi_k) \bmod 2\pi; (u_1, ..., u_{m-k}))$. The points $f_1, ..., f_k \in \mathbb{R}^m$, where each f_i has the coordinates $\varphi_i = 2\pi$, $\varphi_j = 0$, $u_1 = \cdots = u_{m-k} = 0$, are sent in 0 by this application. Let us first note that the stationary group Λ (see point c) can be written in the form $\Lambda = \mathbb{Z}e_1 \oplus \cdots \oplus \mathbb{Z}e_k$, $1 \leq k \leq m$, where $e_1, ..., e_m$ are linearly independent vectors. Indeed, to fix ideas, let us take $m = 2$, that is, $\Lambda = \left\{(t_1, t_2) \in \mathbb{R}^2 : g(t_1, t_2) = g_{t_2}^{X_2} \circ g_{t_1}^{X_1}(x) = x\right\}$. Three cases are possible: (i) $\Lambda = \{0\}$, (ii) $\Lambda = \mathbb{Z}e_1$ and (iii) $\Lambda = \mathbb{Z}e_1 \oplus \mathbb{Z}e_2$. The first case is to be rejected because we have a diffeomorphism between a non-compact $\mathbb{R}^2/\{0\}$ and a compact M, which is absurd. The second case $\mathbb{R}^2/\mathbb{Z}e_1$ (a cylinder) is also to be rejected for the same reasons as in the first case. The third case remains valid because $\mathbb{R}^2/\mathbb{Z}e_1 \oplus \mathbb{Z}e_2$ is a two-dimensional torus. In general, for every discrete subgroup of $\mathbb{R}^m$, there exist k linearly independent vectors such that this group is the set of all their integer linear combinations. Let $e_1, ..., e_k \in \Lambda \subset \mathbb{R}^m$ be generators of the stationary group Λ. We now apply the vector space $\mathbb{R}^m = \{(\varphi_1, ..., \varphi_k; u_1, ..., u_{m-k})\}$ in a surjective way over space vector $\mathbb{R}^m = \{(t_1, ..., t_m)\}$, such that the vectors f_i are transformed into e_i. Let $h : \mathbb{R}^m \longrightarrow \mathbb{R}^m$ be such an isomorphism and note that $\mathbb{R}^m = \{(\varphi_1, ..., \varphi_k; u_1, ..., u_{m-k})\}$ (respectively, $\mathbb{R}^m = \{(t_1, ..., t_m)\}$) determines charts of $T^k \times \mathbb{R}^{m-k}$ (respectively, of the variety M). The application h determines a diffeomorphism $\widetilde{h} : T^k \times \mathbb{R}^{m-k} \longrightarrow M$ and since by hypothesis M is compact, then $k = m$ and consequently, M is a m-dimensional torus. Let us check this out in more detail. Since Λ is the kernel of g, there exists a canonical surjection

$$\widetilde{h} : \mathbb{R}^m/\Lambda \to M, \quad [(t_1, ..., t_m)] \mapsto \widetilde{h}\,[(t_1, ..., t_m)] = g_{t_m}^{X_m} \circ \cdots \circ g_{t_1}^{X_1}(x).$$

Indeed, let $(t_1, ..., t_m)$ and $(s_1, ..., s_m)$ such that: $\widetilde{h}\,[(t_1, ..., t_m)] = \widetilde{h}\,[(s_1, ..., s_m)]$. We have $g_{t_m}^{X_m} \circ \cdots \circ g_{t_1}^{X_1}(x) = g_{s_m}^{X_m} \circ \cdots \circ g_{s_1}^{X_1}(x)$, hence

$$g_{-s_1}^{X_1} \circ \cdots \circ g_{-s_m}^{X_m} \circ g_{t_m}^{X_m} \circ \cdots \circ g_{t_1}^{X_1}(x)$$
$$= g_{-s_1}^{X_1} \circ \cdots \circ g_{-s_m}^{X_m} \circ g_{s_m}^{X_m} \circ \cdots \circ g_{s_1}^{X_1}(x),$$
$$= g_{-s_1}^{X_1} \circ \cdots \circ g_{-s_{m-1}}^{X_{m-1}} \circ g_{s_{m-1}}^{X_{m-1}} \circ \cdots \circ g_{s_1}^{X_1}(x),$$
$$\vdots$$
$$= g_{-s_1}^{X_1} \circ g_{s_1}^{X_1}(x) = x.$$

Since $X_1, ..., X_m$ are commutative, then $g_{t_m - s_m}^{X_m} \circ \cdots \circ g_{t_1 - s_1}^{X_1}(x) = x$. Consequently, $[(t_1 - s_1, ..., t_m - s_m)] = 0$, $[(t_1, ..., t_m)] = [(s_1, ..., s_m)]$. So $\widetilde{h}$ is a diffeomorphism and the proof of proposition 3.1 is complete. $\square$

THEOREM 3.1.– (Arnold–Liouville theorem). Let $H_1 = H, H_2, ..., H_n$ be n first integrals on a $2n$-dimensional symplectic manifold M that are functionally independent (i.e. $dH_1 \wedge ... \wedge dH_n \neq 0$), and pairwise in involution (i.e. $\{H_i, H_j\} = 0$, $1 \leq i, j \leq n$). For generic $c = (c_1, ..., c_n)$, the following level set $M_c = \bigcap_{i=1}^n \{x \in M : H_i(x) = c_i, \ c = (c_i) \in \mathbb{R}^n\}$ will be an n-manifold. If M_c is compact and connected, it is diffeomorphic to an n-dimensional torus $T^n = \mathbb{R}^n/lattice = \{(\varphi_1, ..., \varphi_n) \text{ mod. } 2\pi\}$. The flows $g_t^{X_1}(x), ..., g_t^{X_n}(x)$ defined by the vector fields $X_{H_1}, ..., X_{H_n}$ are straight-line motions on T^n and determine on T^n a quasi-periodic motion, that is, in angular coordinates $\varphi_1, ..., \varphi_n$, we have $\dot{\varphi}_i = \omega_i(c)$, $\omega_i(c) = $ constants, and $\varphi_i(t) = \varphi_i(0) + \omega_i t$. The system [3.1] is integrable by quadrature (if M_c is not compact but the flow of each of the vector fields X_{H_k} is complete[1] on M_c, then M_c is diffeomorphic to a cylinder $\mathbb{R}^k \times T^{n-k}$ under which the vector fields X_{H_k} are mapped to linear vector fields).

PROOF.– With M_c being compact and connected, it is enough (proposition 3.1) to show that M_c is differentiable of dimension n and that it is equipped with n commutative vectors fields. The differentiability of this variety arises from the implicit function theorem since the vectors $J\frac{\partial H_1}{\partial x}, ..., J\frac{\partial H_n}{\partial x}$ are assumed to be independent. Since $m = 2n$, then the first integrals $H_i(x_1, ..., x_{2n})$ are functions of the variables $x_1, ..., x_n, x_{n+1}, ..., x_{2n}$. Therefore, $\dim\{x \in M : H_i = c_i\} = 2n - 1$ and $\dim(\{x \in M : H_i = c_i\} \cap \{x \in M : H_j = c_j\}) = 2n - 2$, $i \neq j$, and so $\dim M_c = n$. Let X_i and X_j, $1 \leq i, j \leq n$, be differentiable ($\mathcal{C}^\infty$) vector fields on M, so also on the variety M_c. Let us define the differential operator L_X by $L_X : \mathcal{C}^\infty(M_c) \to \mathcal{C}^\infty(M_c)$, with $F \mapsto L_X F$ such that: $L_X F(x) = \frac{d}{dt} F(g_t^X(x))|_{t=0}$, $x \in M_c$. We have $L_{X_i} F = \{F, H_i\}$, $L_{X_j} L_{X_i} F = \{\{F, H_i\}, H_j\}$, and

$$L_{X_i} L_{X_j} F - L_{X_j} L_{X_i} F = \{\{F, H_j\}, H_i\} - \{\{F, H_i\}, H_j\},$$
$$= -\{\{H_j, F\}, H_i\} - \{\{F, H_i\}, H_j\},$$
$$= \{\{H_i, H_j\}, F\},$$

according to the identity of Jacobi. Since H_i and H_j are in involution, then $[L_{X_i}, L_{X_j}] = 0$. The construction of the angular coordinates $\varphi_1, ..., \varphi_m$ mod. 2π on the variety M is obviously valid on the invariant variety M_c. Note that $(\varphi_1, ..., \varphi_m) = h^{-1}(t_1, ..., t_m)$ and that the angular coordinates $\varphi_1, ..., \varphi_m$ vary uniformly under the action of the Hamiltonian flow H, that is, $\frac{d\varphi_k}{dt_i} = \{H_i, \varphi_k\} = \omega_i(c)$, $\omega_i(c) = $ constants. In other words, the motion is quasi-periodic on the invariant torus M_c. Finally, to show that the equations of the problem are integrable by quadratures, as well as information about the variables called action-angles, one can consult Arnold (1989). The demonstration of theorem 3.1 ends. □

---

[1] A vector field is complete if every one of its flow curves exist for all time.

REMARK 3.3.– If we restrict ourselves to an invariant open set, we can always assume that the fibers of M_c (where c is a regular value) are connected. The tori obtained in the theorem are Lagrangian sub-varieties.

LEMMA 3.1.– The rank of the matrix J is even.

PROOF.– Let λ be the eigenvalue associated with the eigenvector Z. We have $JZ = \lambda Z, Z \neq 0, Z^*JZ = \lambda Z^*Z, Z^* \equiv \overline{Z}^\top$, where $\lambda = \frac{Z^*JZ}{Z^*Z}$. Since $\overline{J} = J$ and $J^\top = -J$, then $\overline{Z^*JZ} = Z^\top \overline{JZ} = Z^\top J\overline{Z} = (Z^\top J\overline{Z})^\top = Z^*J^\top Z = -Z^*JZ$, which implies that Z^*JZ is either zero or imaginary pure. Since Z^*Z is real, it follows that all the eigenvalues of J are either null or imaginary pure. Now $J\overline{Z} = \overline{\lambda}\overline{Z}$, so if λ is an eigenvalue, then $\overline{\lambda}$ is also an eigenvalue. Consequently, the eigenvalues (non-zero) of J come in pairs, hence the result. □

REMARK 3.4.– Since the Hamiltonian systems that we will have to solve concern the affine space $\mathbb{R}^m$ (or $\mathbb{C}^m$), we will restrict ourselves to the case where M is an affine space $\mathbb{R}^m$ (or $\mathbb{C}^m$). We shall distinguish two cases:

a) Case 1: $\det J \neq 0$. The rank of the matrix J is even (lemma 3.1), $m = 2n$. Recall (definition 3.2) that a Hamiltonian system [3.1], $x \in \mathbb{R}^m$, is completely integrable or Liouville integrable if there exist n first integrals $H_1 = H, H_2, \ldots, H_n$ pairwise in involution with linearly independent gradients. For generic $c = (c_1, \ldots, c_n)$, the level set $M_c = \bigcap_{i=1}^{n} \{x \in M : H_i(x) = c_i,\ c_i \in \mathbb{R}\}$ will be an n-manifold. By the Arnold–Liouville theorem, if M_c is compact and connected, it is diffeomorphic to an n-dimensional torus $\mathbb{T}^n = \mathbb{R}^n/\mathbb{Z}^n$ and each vector field will define a linear flow there. In some open neighborhood of the torus, there are coordinates $s_1, \ldots, s_n, \varphi_1, \ldots, \varphi_n$ in which ω takes the form $\omega = \sum_{k=1}^{n} ds_k \wedge d\varphi_k$. Here, the functions s_k (called action-variables) give coordinates in the direction transverse to the torus and can be expressed functionally in terms of the first integrals H_k. The functions φ_k (called angle-variables) give standard angular coordinates on the torus, and every vector field X_{H_k} can be written in the form $\dot{\varphi}_k = h_k(s_1, \ldots, s_n)$, that is, its integral trajectories define a conditionally periodic motion on the torus. In a neighborhood of the torus, the Hamiltonian vector field X_{H_k} takes the following form $\dot{s}_k = 0, \dot{\varphi}_k = h_k(s_1, \ldots, s_n)$ and can be solved by using quadratures.

b) Case 2: $\det J = 0$. We reduce the problem to $m = 2n + k$ and we look for k Casimir functions (or trivial invariants) $H_{n+1}, \ldots, H_{n+k}$, leading to identically zero Hamiltonian vector fields $J\frac{\partial H_{n+i}}{\partial x} = 0, 1 \leq i \leq k$. The system is Hamiltonian on a generic symplectic manifold $\bigcap_{i=n+1}^{n+k} \{x \in \mathbb{R}^m : H_i(x) = c_i\}$ of dimension $m - k = 2n$. If, for most values of $c_i \in \mathbb{R}$, the invariant manifolds $\bigcap_{i=1}^{n+k} \{x \in \mathbb{R}^m : H_i(x) = c_i\}$ are compact and connected, then they are n-dimensional tori $\mathbb{T}^n = \mathbb{R}^n/\mathbb{Z}^n$ by the Arnold–Liouville theorem and the Hamiltonian flow is linear in angular coordinates of the torus.

EXAMPLE 3.1.– Let $T^*\mathbb{R}^n$ with coordinates $q_1,...,q_n,p_1,...,p_n$. The system corresponding to the Hamiltonian $H = \frac{1}{2}\sum_{j=1}^{n}(p_j^2 + \lambda_j q_j^2)$ of the harmonic oscillator is integrable. The Hamiltonian structure is defined by the Poisson bracket $\{F,H\} = \sum_{j=1}^{n}\left(\frac{\partial F}{\partial q_j}\frac{\partial H}{\partial p_j} - \frac{\partial F}{\partial p_j}\frac{\partial H}{\partial q_j}\right)$. The Hamiltonian field corresponding to H is written explicitly $\dot{q}_j = p_j, \dot{p}_j = -2\lambda_j q_j$, where $j = 1,...,n$ and admits the following first n integral: $H_j = \frac{1}{2}p_j^2 + \lambda_j q_j^2$, $1 \leq j \leq n$. The latter are independent in involution and the system in question is integrable.

3.2. Rotation of a rigid body about a fixed point

The differential equations of this problem are written in the form

$$\dot{M} = M \wedge \Omega + \mu g\,\Gamma \wedge L, \qquad \dot{\Gamma} = \Gamma \wedge \Omega, \qquad [3.2]$$

where $\wedge$ is the vector product in $\mathbb{R}^3$, $M = (m_1, m_2, m_3)$ is the angular momentum of the solid, $\Omega = (\frac{m_1}{I_1}, \frac{m_2}{I_2}, \frac{m_3}{I_3})$ is the angular velocity, I_1, I_2 and I_3 are moments of inertia, $\Gamma = (\gamma_1, \gamma_2, \gamma_3)$ is the unitary vertical vector, μ is the mass of the solid, g is the acceleration of gravity, and finally, $L = (l_1, l_2, l_3)$ is the unit vector originating from the fixed point and directed toward the center of gravity; all of these vectors are considered in a mobile system whose coordinates are fixed to the main axes of inertia. The configuration space of a solid with a fixed point is the group of rotations $SO(3)$. This is generated by the one-parameter subgroup of rotations,

$$A_1 = \begin{pmatrix} 1 & 0 & 0 \\ 0 & \cos t & -\sin t \\ 0 & \sin t & \cos t \end{pmatrix}, \quad A_2 = \begin{pmatrix} \cos t & 0 & \sin t \\ 0 & 1 & 0 \\ -\sin t & 0 & \cos t \end{pmatrix},$$

$$A_3 = \begin{pmatrix} \cos t & -\sin t & 0 \\ \sin t & \cos t & 0 \\ 0 & 0 & 1 \end{pmatrix}.$$

This is the group of 3×3 orthogonal matrices A and the motion of this solid is described by a curve on this group. The angular velocity space of all rotations (the set of derivatives $\dot{A}(t)\big|_{t=0}$ of the differentiable curves in $SO(3)$ passing through the identity in $t = 0$: $A(0) = I$) is the Lie algebra of the group $SO(3)$; it is the algebra $so(3)$ of the 3×3 antisymmetric matrices and is generated as a vector space by the matrices,

$$e_1 = \dot{A}_1(t)\big|_{t=0} = \begin{pmatrix} 0 & 0 & 0 \\ 0 & 0 & -1 \\ 0 & 1 & 0 \end{pmatrix}, \quad e_2 = \dot{A}_2(t)\big|_{t=0} = \begin{pmatrix} 0 & 0 & 1 \\ 0 & 0 & 0 \\ -1 & 0 & 0 \end{pmatrix},$$

$$e_3 = \dot{A}_3(t)\Big|_{t=0} = \begin{pmatrix} 0 & -1 & 0 \\ 1 & 0 & 0 \\ 0 & 0 & 0 \end{pmatrix},$$

which verify the following commutation relations: $[e_1, e_2] = e_3$, $[e_2, e_3] = e_1$, $[e_3, e_1] = e_2$. We will use the fact (see lemma 1.2) that if we identify $so(3)$ to $\mathbb{R}^3$ by sending (e_1, e_2, e_3) on the canonical basis of $\mathbb{R}^3$, the bracket of $so(3)$ corresponds to the vector product. In other words, consider the application

$$\mathbb{R}^3 \longrightarrow so(3), \quad a = (a_1, a_2, a_3) \longmapsto A = \begin{pmatrix} 0 & -a_3 & a_2 \\ a_3 & 0 & -a_1 \\ -a_2 & a_1 & 0 \end{pmatrix},$$

which defines an isomorphism between Lie algebras $(\mathbb{R}^3, \wedge)$ and $(so(3), [,])$ where $a \wedge b \longmapsto [A, B] = AB - BA$. By using this isomorphism, the system [3.2] can be rewritten in the form

$$\dot{M} = [M, \Omega] + \mu g\, [\Gamma, L], \qquad \dot{\Gamma} = [\Gamma, \Omega], \qquad [3.3]$$

where

$$M = (M_{ij})_{1 \leq i,j \leq 3} \equiv \sum_{i=1}^{3} m_i e_i \equiv \begin{pmatrix} 0 & -m_3 & m_2 \\ m_3 & 0 & -m_1 \\ -m_2 & m_1 & 0 \end{pmatrix} \in so(3),$$

$$\Omega = (\Omega_{ij})_{1 \leq i,j \leq 3} \equiv \sum_{i=1}^{3} \omega_i e_i \equiv \begin{pmatrix} 0 & -\omega_3 & \omega_2 \\ \omega_3 & 0 & -\omega_1 \\ -\omega_2 & \omega_1 & 0 \end{pmatrix} \in so(3),$$

$$\Gamma = (\gamma_{ij})_{1 \leq i,j \leq 3} \equiv \sum_{i=1}^{3} \gamma_i e_i \equiv \begin{pmatrix} 0 & -\gamma_3 & \gamma_2 \\ \gamma_3 & 0 & -\gamma_1 \\ -\gamma_2 & \gamma_1 & 0 \end{pmatrix} \in so(3),$$

and $L = \begin{pmatrix} 0 & -l_3 & l_2 \\ l_3 & 0 & -l_1 \\ -l_2 & l_1 & 0 \end{pmatrix} \in so(3)$. Taking into account that $M = I\Omega$, then the above equations become

$$\dot{M} = [M, \Lambda M] + \mu g\, [\Gamma, L], \qquad \dot{\Gamma} = [\Gamma, \Lambda M], \qquad [3.4]$$

where

$$\Lambda M = \sum_{i=1}^{3} \lambda_i m_i e_i \equiv \begin{pmatrix} 0 & -\lambda_3 m_3 & \lambda_2 m_2 \\ \lambda_3 m_3 & 0 & -\lambda_1 m_1 \\ -\lambda_2 m_2 & \lambda_1 m_1 & 0 \end{pmatrix} \in so(3), \quad \lambda_i \equiv \frac{1}{I_i}.$$

The resolution of this problem was first analyzed by Euler (1765) and in 1758, he published the equations (case $\mu = 0$) that carry his name. Euler's equations were integrated by Jacobi (1850) in terms of elliptic functions and, Poinsot (1851) gave them a remarkable geometric interpretation. Lagrange (1888) found another case ($I_1 = I_2, l_1 = l_2 = 0$) of integrability that Poisson examined at length thereafter. The problem continued to attract mathematicians, but for a long time no new results could be obtained. It was then around 1888–1989 that a memoir (Kowalewski 1889) of the highest interest then appeared, containing a newly discovered case ($I_1 = I_2 = 2I_3$, $l_3 = 0$) of integrability. For this remarkable work, Kowalewski was awarded the Bordin Prize of the Paris Academy of Sciences. In fact, although Kowalewski's work is quite important, it is not at all clear why there would be no other new cases of integrability. This was to be the starting point of a series of fierce research on the question of the existence of new cases of integrability. Moreover, among the remarkable results obtained by Poincaré (2003) with the aid of the periodic solutions of the equations of dynamics, we find the following (around 1891): in order to exist in the motion of a solid body around of a fixed point, with an algebraic first integral not being reduced to a combination of the classical integrals, it is necessary that the ellipsoid of inertia relative to the point of suspension is of revolution. Liouville (not to be confused with Joseph Liouville, well known in complex analysis) also competed for the Bordin prize, and presented a paper (Liouville 1896), indicating the necessary and sufficient conditions ($I_3 = 0$, $2I_3/I_1 =$ integer) of existence of a fourth algebraic integral. These conditions have been reproduced in most conventional treatises (e.g. Whittaker (1988)) and in scientific journals. It was not until the year 1906, when Husson (1906), working under the direction of Appell and Painlevé, discovered an erroneous demonstration in the work of Liouville. Indeed, paragraphs I and III of Liouville's dissertation devoted to the search for the necessary conditions seem at first satisfactory, but a more careful study shows that the demonstrations are at least insufficient and that it is impossible to accept conclusions. In fact, although the conditions found by Liouville are necessary, they cannot be deduced from the calculations indicated and these conditions are not sufficient. It was Husson who first solved completely the question of looking for new cases of integrability. Inspired by Poincaré's research on the problem of the three bodies and Painlevé on the generalization of Bruns's theorem, Husson demonstrated that any algebraic integral is a combination of classical integrals, except in the cases of Euler, Lagrange and Kowalewski. Moreover, the question of the existence of analytic integrals has been studied rigorously by Ziglin (1983a, 1983b) and Holmes and Marsden (1983). In the last section, we will mention some special cases: the cases of Hess–Appel'rot (Hess 1890; Appel'rot 1894), Goryachev–Chaplygin (Goryachev 1900; Chaplygin 1948) and Bobylev–Steklov (Bobylev 1896; Steklov 1896).

3.2.1. *The Euler problem of a rigid body*

In this case, we have $l_1 = l_2 = l_3 = 0$, that is, the fixed point is its center of gravity. The Euler rigid body motion (Euler 1765) (also called Euler–Poinsot Poinsot (1851) motion of the solid) expresses the free motion of a rigid body around a fixed point. Euler obtained a solution of this problem when the center of mass coincides with the fixed point.

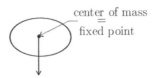

Figure 3.2. *Euler top*

Then the motion of the body is governed by

$$\dot{M} = [M, \Lambda M]. \tag{3.5}$$

Equation [3.5] is explicitly given by

$$\begin{aligned} \dot{m}_1 &= (\lambda_3 - \lambda_2) m_2 m_3, \\ \dot{m}_2 &= (\lambda_1 - \lambda_3) m_1 m_3, \\ \dot{m}_3 &= (\lambda_2 - \lambda_1) m_1 m_2, \end{aligned} \tag{3.6}$$

and can be written as a Hamiltonian vector field $\dot{x} = J\frac{\partial H}{\partial x}$, $x = (m_1, m_2, m_3)^\top$, with the Hamiltonian $H = \frac{1}{2}\left(\lambda_1 m_1^2 + \lambda_2 m_2^2 + \lambda_3 m_3^2\right)$, and $J = \begin{pmatrix} 0 & -m_3 & m_2 \\ m_3 & 0 & -m_1 \\ -m_2 & m_1 & 0 \end{pmatrix}$.

We have $\det J = 0$, so $m = 2n + k$ and $m - k = rk\ J$. Here, $m = 3$ and $rk\ J = 2$, then $n = k = 1$. The system [3.6] has, besides the energy $H_1 = H$, a trivial invariant H_2, that is, such that:

$$J\frac{\partial H_2}{\partial x} = \begin{pmatrix} 0 & -m_3 & m_2 \\ m_3 & 0 & -m_1 \\ -m_2 & m_1 & 0 \end{pmatrix} \begin{pmatrix} \frac{\partial H_2}{\partial m_1} \\ \frac{\partial H_2}{\partial m_2} \\ \frac{\partial H_2}{\partial m_3} \end{pmatrix} = \begin{pmatrix} 0 \\ 0 \\ 0 \end{pmatrix},$$

implying $\frac{\partial H_2}{\partial m_1} = m_1$, $\frac{\partial H_2}{\partial m_2} = m_2$, $\frac{\partial H_2}{\partial m_3} = m_3$, and consequently $H_2 = \frac{1}{2}\left(m_1^2 + m_2^2 + m_3^2\right)$. The system evolves on the intersection of the sphere $H_1 = c_1$ and the ellipsoid $H_2 = c_2$. In $\mathbb{R}^3$, this intersection will be isomorphic to two circles $\left(\text{with } \frac{c_2}{\lambda_3} < c_1 < \frac{c_2}{\lambda_1}\right)$. Since the conditions of the Arnold–Liouville theorem are satisfied, we have the following result.

THEOREM 3.2.– The system [3.6] is completely integrable and the vector $J\frac{\partial H}{\partial x}$ gives a flow on a variety $\bigcap_{i=1}^{2}\{x \in \mathbb{R}^3 : H_i(x) = c_i\}$, diffeomorphic to a real torus of dimension 1, that is to say a circle.

Let us now turn to explicit resolution. We shall show that the problem can be integrated in terms of elliptic functions, as Euler discovered using his then newly invented theory of elliptic integrals. We have just seen that the system in question admits two first quadratic integrals:

$$H_1 = \frac{1}{2}\left(\lambda_1 m_1^2 + \lambda_2 m_2^2 + \lambda_3 m_3^2\right), \qquad H_2 = \frac{1}{2}\left(m_1^2 + m_2^2 + m_3^2\right).$$

We assume that $\lambda_1, \lambda_2, \lambda_3$ are all different from zero (that is, the solid is not reduced to a point and is not focused on a straight line either), then $H_1 = 0$ implies $m_1 = m_2 = m_3 = 0$ and so $H_2 = 0$; the solid is at rest. We dismiss this trivial case and now assume that $H_1 \neq 0$ and $H_2 \neq 0$. When $\lambda_1 = \lambda_2 = \lambda_3$, equations [3.6] obviously show that m_1, m_2 and m_3 are constants. Suppose, for example, that $\lambda_1 = \lambda_2$, equations [3.6] are then written $\dot{m}_1 = (\lambda_3 - \lambda_1)\, m_2 m_3$, $\dot{m}_2 = (\lambda_1 - \lambda_3)\, m_1 m_3$, $\dot{m}_3 = 0$. We then deduce that $m_3 = $ constant $\equiv A$ and $\dot{m}_1 = A(\lambda_3 - \lambda_1)\, m_2$, $\dot{m}_2 = A(\lambda_1 - \lambda_3)\, m_1$. Note that $(m_1 + im_2)^{\cdot} = iA(\lambda_1 - \lambda_3)(m_1 + im_2)$, we obtain $m_1 + im_2 = Ce^{iA(\lambda_1 - \lambda_3)t}$, where C is a constant and so $m_1 = C\cos A(\lambda_1 - \lambda_3)t$, $m_2 = C\sin A(\lambda_1 - \lambda_3)t$. The integration of Euler's equations is delicate in the general case where λ_1, λ_2 and λ_3 are all different; the solutions are expressed in this case using elliptic functions. In the following, we will suppose that λ_1, λ_2 and λ_3 are all different and discard the other trivial cases that pose no difficulty for solving the equations in question. To fix the ideas, we will assume in the following that : $\lambda_1 > \lambda_2 > \lambda_3$. Geometrically, the equations

$$\lambda_1 m_1^2 + \lambda_2 m_2^2 + \lambda_3 m_3^2 = 2H_1, \qquad [3.7]$$

$$m_1^2 + m_2^2 + m_3^2 = 2H_2 \equiv r^2, \qquad [3.8]$$

represent the equations of the surface of a half axis ellipsoid: $\sqrt{\frac{2H_1}{\lambda_1}}$ (half big axis), $\sqrt{\frac{2H_1}{\lambda_2}}$ (middle half axis), $\sqrt{\frac{2H_1}{\lambda_3}}$ (half small axis) and a sphere of radius r. So the movement of the solid takes place on the intersection of an ellipsoid with a sphere.

This intersection makes sense because by comparing [3.7] to [3.8], we see that $\frac{2H_1}{\lambda_1} < r^2 < \frac{2H_1}{\lambda_3}$, which means geometrically that the radius of the sphere [3.8] is between the smallest and largest of the half-axes of the ellipsoid [3.7]. To study the shape of the intersection curves of the ellipsoid [3.7] with the sphere [3.8], set $H_1 > 0$ and let the radius r vary. Like $\lambda_1 > \lambda_2 > \lambda_3$, the semi-axes of the ellipsoid will be $\frac{2H_1}{\lambda_1} > \frac{2H_1}{\lambda_2} > \frac{2H_1}{\lambda_3}$. If the radius r of the sphere is less than the half-axis $\frac{2H_1}{\lambda_3}$ or greater than the half-axis $\frac{2H_1}{\lambda_1}$, then the intersection in question is empty (and no real movement corresponds to these values of H_1 and r). When the radius r equals $\frac{2H_1}{\lambda_3}$, then the intersection is composed of two points. When the radius r increases $\left(\frac{2H_1}{\lambda_3} < r < \frac{2H_1}{\lambda_2}\right)$, we obtain two curves around the ends of the half minor axis. Likewise if $r = \frac{2H_1}{\lambda_1}$, we get both ends of the semi-major axis, and if r is slightly smaller than $\frac{2H_1}{\lambda_1}$, we get two closed curves near these ends. Finally, if $r = \frac{2H_1}{\lambda_2}$, then the intersection in question consists of two circles.

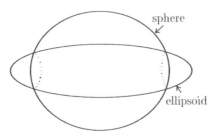

Figure 3.3. *Intersection of an ellipsoid with a sphere*

THEOREM 3.3.– Euler's differential equations [3.6] are integrated by means of Jacobi's elliptic functions.

PROOF.– From the first integrals [3.7] and [3.8], we express m_1 and m_3 as a function of m_2. These expressions are then introduced into the second equation of the system [3.7] to obtain a differential equation in m_2 and $\frac{dm_2}{dt}$ only. In more detail, the following relationships are easily obtained from [3.7] and [3.8]:

$$m_1^2 = \frac{2H_1 - r^2\lambda_3 - (\lambda_2 - \lambda_3)\,m_2^2}{\lambda_1 - \lambda_3}, \qquad [3.9]$$

$$m_3^2 = \frac{r^2\lambda_1 - 2H_1 - (\lambda_1 - \lambda_2)\,m_2^2}{\lambda_1 - \lambda_3}. \qquad [3.10]$$

By substituting these expressions in the second equation of the system [3.6], we obtain

$$\dot{m}_2 = \sqrt{(2H_1 - r^2\lambda_3 - (\lambda_2 - \lambda_3)\,m_2^2)(r^2\lambda_1 - 2H_1 - (\lambda_1 - \lambda_2)\,m_2^2)}.$$

By integrating this equation, we obtain a function $t(m_2)$ in the form of an elliptic integral. To reduce this to the standard form, we can assume that $r^2 > \frac{2H_1}{\lambda_2}$ (otherwise, it is enough to invert the indices 1 and 3 in all the previous formulas). We rewrite the previous equation in the form

$$\frac{dm_2}{\sqrt{(2H_1 - r^2\lambda_3)(r^2\lambda_1 - 2H_1)}\,dt} = \sqrt{(1 - \frac{\lambda_2 - \lambda_3}{2H_1 - r^2\lambda_3}m_2^2)(1 - \frac{\lambda_1 - \lambda_2}{r^2\lambda_1 - 2H_1}m_2^2)}.$$

By setting $\tau = t\sqrt{(\lambda_2 - \lambda_3)(r^2\lambda_1 - 2H_1)}$, $s = m_2\sqrt{\frac{\lambda_2 - \lambda_3}{2H_1 - r^2\lambda_3}}$, we obtain

$$\frac{ds}{d\tau} = \sqrt{(1 - s^2)\left(1 - \frac{(\lambda_1 - \lambda_2)(2H_1 - r^2\lambda_3)}{(\lambda_2 - \lambda_3)(r^2\lambda_1 - 2H_1)}s^2\right)},$$

which suggests choosing elliptic functions as a module $k^2 = \frac{(\lambda_1-\lambda_2)(2H_1-r^2\lambda_3)}{(\lambda_2-\lambda_3)(r^2\lambda_1-2H_1)}$. Inequalities $\lambda_1 > \lambda_2 > \lambda_3$, $\frac{2H_1}{\lambda_1} < r^2 < \frac{2H_1}{\lambda_3}$ and $r^2 > \frac{2H_1}{\lambda_2}$ show that $0 < k^2 < 1$. So we get $\frac{ds}{d\tau} = \sqrt{(1 - s^2)(1 - k^2 s^2)}$. This equation admits the solution (we choose the origin of the times such that $m_2 = 0$ for $t = 0$):

$$\tau = \int_0^s \frac{ds}{\sqrt{(1 - s^2)(1 - k^2 s^2)}}.$$

It is the integral of a holomorphic differential on an elliptic curve

$$\mathcal{E} : w^2 = (1 - s^2)(1 - k^2 s^2). \qquad [3.11]$$

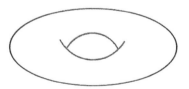

Figure 3.4. *Elliptic curve*

The inverse function $s(\tau)$ is one of Jacobi's elliptic functions: $s = \text{sn}\tau$, which also determines m_2 as a function of time, that is, $m_2 = \sqrt{\frac{2H_1 - r^2\lambda_3}{\lambda_2 - \lambda_3}} \cdot \text{sn}\tau$. According to the equalities [3.9] and [3.10], we know that the functions m_1 and m_3 are expressed algebraically as a function of m_2, so

$$m_1 = \sqrt{\frac{2H_1 - r^2\lambda_3}{\lambda_1 - \lambda_3}} \cdot \sqrt{1 - \text{sn}^2\tau}, \quad m_3 = \sqrt{\frac{r^2\lambda_1 - 2H_1}{\lambda_1 - \lambda_3}} \cdot \sqrt{1 - k^2\text{sn}^2\tau}.$$

Using the other two elliptical functions (see, for example, Lesfari (2008b, 2015b)): $\text{cn}\tau = \sqrt{1 - \text{sn}^2\tau}$, $\text{dn}\tau = \sqrt{1 - k^2\text{sn}^2\tau}$, and the fact that $\tau = t\sqrt{(\lambda_2 - \lambda_3)(r^2\lambda_1 - 2H_1)}$, we finally get the following explicit formulas:

$$m_1 = \sqrt{\frac{2H_1 - r^2\lambda_3}{\lambda_1 - \lambda_3}} \operatorname{cn}(t\sqrt{(\lambda_2 - \lambda_3)(r^2\lambda_1 - 2H_1)}),$$

$$m_2 = \sqrt{\frac{2H_1 - r^2\lambda_3}{\lambda_2 - \lambda_3}} \operatorname{sn}(t\sqrt{(\lambda_2 - \lambda_3)(r^2\lambda_1 - 2H_1)}),$$

$$m_3 = \sqrt{\frac{r^2\lambda_1 - 2H_1}{\lambda_1 - \lambda_3}} \operatorname{dn}(t\sqrt{(\lambda_2 - \lambda_3)(r^2\lambda_1 - 2H_1)}).$$

In other words, the integration of the Euler equations is done by means of elliptic Jacobi functions and the proof is complete. □

REMARK 3.5.– Note that for $\lambda_1 = \lambda_2$, we have $k^2 = 0$. In this case, the elliptic functions $\operatorname{sn}\tau, \operatorname{cn}\tau, \operatorname{dn}\tau$ are reduced, respectively, to functions $\sin\tau, \cos\tau, 1$. From the system above, we easily obtain the expressions

$$m_1 = \sqrt{\frac{2H_1 - r^2\lambda_3}{\lambda_1 - \lambda_3}} \cos\sqrt{(\lambda_1 - \lambda_3)(r^2\lambda_1 - 2H_1)}\,t,$$

$$m_2 = \sqrt{\frac{2H_1 - r^2\lambda_3}{\lambda_1 - \lambda_3}} \sin\sqrt{(\lambda_1 - \lambda_3)(r^2\lambda_1 - 2H_1)}\,t,$$

$$m_3 = \sqrt{\frac{r^2\lambda_1 - 2H_1}{\lambda_1 - \lambda_3}}.$$

We find the solutions established previously with $A = \sqrt{\frac{r^2\lambda_1 - 2H_1}{\lambda_1 - \lambda_3}}$, $C = \sqrt{\frac{2H_1 - r^2\lambda_3}{\lambda_1 - \lambda_3}}$.

3.2.2. *The Lagrange top*

In this case, we have $I_1 = I_2$, $l_1 = l_2 = 0$, that is, the Lagrange top (Lagrange 1888) is a rigid body, in which two moments of inertia are the same and the center of gravity lies on the symmetry axis. In other words, the Lagrange top is a symmetric top with a constant vertical gravitational force acting on its center of mass and leaving the base point of its body symmetry axis fixed.

As in the case of Euler, we also show that in this case the problem is solved by elliptic integrals. Or what amounts to the same, the integration is done using elliptic functions. This problem will be explored later using various methods.

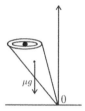

Figure 3.5. *Lagrange top*

3.2.3. *The Kowalewski spinning top*

The Kowalewski spinning top (Kowalewski 1889) is a special symmetric top with a unique ratio of the moments of inertia satisfing the relation: $I_1 = I_2 = 2I_3$, $l_3 = 0$, in which two moments of inertia are equal, the third is half as large and the center of gravity is located in the plane perpendicular to the symmetry axis (parallel to the plane of the two equal points).

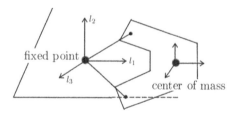

Figure 3.6. *Kowalewski top*

Moreover, we may choose $l_2 = 0$, $\mu g l_1 = l$ and $I_3 = 1$. After the substitution $t \to 2t$, the system [3.4] is written explicitly in the form

$$\dot{m}_1 = m_2 m_3, \qquad \dot{\gamma}_1 = 2m_3\gamma_2 - m_2\gamma_3,$$
$$\dot{m}_2 = -m_1 m_3 + 2\gamma_3, \qquad \dot{\gamma}_2 = m_1\gamma_3 - 2m_3\gamma_1, \qquad [3.12]$$
$$\dot{m}_3 = -2\gamma_2, \qquad \dot{\gamma}_3 = m_2\gamma_1 - m_1\gamma_2.$$

These equations are written in the form of a Hamiltonian vector field $\dot{x} = J\frac{\partial H}{\partial x}$, where $x = (m_1, m_2, m_3, \gamma_1, \gamma_2, \gamma_3)^\mathsf{T}$, $H = \frac{1}{2}\left(m_1^2 + m_2^2\right) + m_3^2 + 2\gamma_1$, is the Hamiltonian and

$$J = \begin{pmatrix} 0 & -m_3 & m_2 & 0 & -\gamma_3 & \gamma_2 \\ m_3 & 0 & -m_1 & \gamma_3 & 0 & -\gamma_1 \\ -m_2 & m_1 & 0 & -\gamma_2 & \gamma_1 & 0 \\ 0 & -\gamma_3 & \gamma_2 & 0 & 0 & 0 \\ \gamma_3 & 0 & -\gamma_1 & 0 & 0 & 0 \\ -\gamma_2 & \gamma_1 & 0 & 0 & 0 & 0 \end{pmatrix}.$$

The above system admits four first integrals:

$$H_1 \equiv H, \qquad H_2 = m_1\gamma_1 + m_2\gamma_2 + m_3\gamma_3,$$
$$H_3 = \gamma_1^2 + \gamma_2^2 + \gamma_3^2, \qquad [3.13]$$
$$H_4 = \left(\left(\frac{m_1 + im_2}{2}\right)^2 - (\gamma_1 + i\gamma_2)\right)\left(\left(\frac{m_1 - im_2}{2}\right)^2 - (\gamma_1 - i\gamma_2)\right).$$

The fourth first integral was obtained by Kowalewski (1889). A second flow commuting with the first is regulated by the equations: $\dot{x} = J\frac{\partial H_4}{\partial x}$, $x = (m_1, m_2, m_3, \gamma_1, \gamma_2, \gamma_3)^\mathsf{T}$, and is written explicitly,

$$\dot{m}_1 = -\frac{1}{4}m_2m_3\left(m_1^2 + m_2^2\right) + m_3\left(m_2\gamma_1 - m_1\gamma_2\right) + \gamma_3\left(2\gamma_2 - m_2m_1\right),$$

$$\dot{m}_2 = \frac{1}{4}m_1m_3\left(m_1^2 + m_2^2\right) - m_3\left(m_1\gamma_1 + m_2\gamma_2\right) + \frac{1}{2}\gamma_3\left(-m_1^2 + m_2^2 + 4\gamma_1\right),$$

$$\dot{m}_3 = m_2m_1\gamma_1 + \frac{1}{2}\gamma_2\left(m_2^2 - m_1^2\right), \qquad [3.14]$$

$$\dot{\gamma}_1 = -\frac{1}{4}\gamma_3\left(m_2m_1^2 - 4m_1\gamma_2 + m_2^3 + 4m_2\gamma_1\right),$$

$$\dot{\gamma}_2 = -\frac{1}{4}\gamma_3\left(-m_1^3 - m_2^2m_1 + 4m_1\gamma_1 + 4m_2\gamma_2\right),$$

$$\dot{\gamma}_2 = \frac{1}{4}\left(\gamma_1 m_2^3 - \gamma_2 m_1^3\right) + \frac{1}{4}m_2m_1\left(m_1\gamma_1 - m_2\gamma_2\right) + m_2\gamma_1^2 + m_2\gamma_2^2.$$

The first integrals H_1 and H_4 are in involution, $\{H_1, H_4\} = \left\langle \frac{\partial H_1}{\partial x}, J\frac{\partial H_4}{\partial x}\right\rangle = 0$, while H_2 and H_3 are trivial, $J\frac{\partial H_2}{\partial x} = J\frac{\partial H_3}{\partial x} = 0$. Let M_c be the affine variety defined by the intersection of the constants of the motion (where $c = (c_1, c_2, c_3 = 1, c_4)$ is not a critical value),

$$M_c = \bigcap_{k=1}^{4} \{x : H_k(x) = c_k\}. \qquad [3.15]$$

We will explain how the affine variety M_c and vector-fields behave after the quotient by some natural involution on M_c, and how these vector-fields become well defined when we take Kowalewski's variables. We show that these variables are naturally related to the so-called Euler's differential equations and can be seen as the addition-formula for the Weierstrass elliptic function. In the theorem below, we will use with Kowalewski the following notations: $c_1 = 6h_1$, $c_2 = 2h_2$ et $c_4 = k^2$.

THEOREM 3.4.– a) Let $(m_1, m_2, m_3, \gamma_1, \gamma_2, \gamma_3) \longmapsto (x_1, x_2, m_3, y_1, y_2, \gamma_3)$, be a birationally map on the affine variety M_c where x_1, x_2, y_1, y_2 are defined as

$$x_1 = \frac{1}{2}(m_1 + im_2), \qquad y_1 = x_1^2 - (\gamma_1 + i\gamma_2), \qquad [3.16]$$

$$x_2 = \frac{1}{2}(m_1 - im_2), \qquad y_2 = x_2^2 - (\gamma_1 - i\gamma_2).$$

Then, the quotient $K \equiv M_c/\sigma$ by the involution

$$\sigma : M_c \longrightarrow M_c \ (x_1, x_2, m_3, y_1, y_2, \gamma_3) \longmapsto (x_1, x_2, -m_3, y_1, y_2, -\gamma_3),$$
$$[3.17]$$

is a Kummer surface

$$K : \begin{cases} y_1 y_2 = k^2, \\ y_1 R(x_2) + y_2 R(x_1) + R_1(x_1, x_2) + k^2(x_1 - x_2)^2 = 0, \end{cases} \qquad [3.18]$$

where

$$R(x) = -x^4 + 6h_1 x^2 - 4h_2 x + 1 - k^2, \qquad [3.19]$$

is a polynomial of degree 4 in x and

$$R_1(x_1, x_2) = -6h_1 x_1^2 x_2^2 + 4h_2 x_1 x_2 (x_1 + x_2) \qquad [3.20]$$
$$- (1 - k^2)(x_1 + x_2)^2 + 6h_1(1 - k^2) - 4h_2^2,$$

is another polynomial of degree 2 in x_1, x_2. The ramification points of M_c on K are given by the eight fixed points of the involution σ.

b) The surface K is a double cover of plane (x_1, x_2), ramified along two elliptic curves intersecting exactly each other at the eight fixed points of the involution σ. These curves give rise to the Euler differential equation

$$\frac{\dot{x}_1}{\sqrt{R(x_1)}} \pm \frac{\dot{x}_2}{\sqrt{R(x_2)}} = 0,$$

to which the variables of Kowalewski are connected

$$s_1 = \frac{R(x_1, x_2) - \sqrt{R(x_1)}\sqrt{R(x_2)}}{(x_1 - x_2)^2} + 3h_1,$$

$$s_2 = \frac{R(x_1, x_2) + \sqrt{R(x_1)}\sqrt{R(x_2)}}{(x_1 - x_2)^2} + 3h_1,$$

where

$$R(x_1, x_2) \equiv -x_1^2 x_2^2 + 6h_1 x_1 x_2 - 2h_2(x_1 + x_2) + 1 - k^2, \qquad [3.21]$$

and can be seen as addition formulas for the Weierstrass elliptic function.

c) In terms of the variables s_1 and s_2, the system of differential equations [3.12] is reduced to the system

$$\frac{ds_1}{\sqrt{P_5(s_1)}} \pm \frac{ds_2}{\sqrt{P_5(s_2)}} = 0, \qquad \frac{s_1 ds_1}{\sqrt{P_5(s_1)}} \pm \frac{s_2 ds_2}{\sqrt{P_5(s_2)}} = dt,$$

where $P_5(s)$ is a fifth-degree polynomial and the problem can be integrated in terms of genus two hyperelliptic functions.

PROOF.– a) Using the change of variables [3.16], with $t \to it$, equations [3.12] and [3.13] become

$$\dot{x}_1 = m_3 x_1 - \gamma_3, \qquad \dot{y}_1 = 2m_3 y_1,$$
$$\dot{x}_2 = -m_3 x_2 + \gamma_3, \qquad \dot{y}_2 = -2m_3 y_1, \qquad [3.22]$$
$$\dot{m}_3 = -x_1^2 + y_1 + x_2^2 - y_2, \qquad \dot{\gamma}_3 = x_1(x_2^2 - y_2) - x_2(x_1^2 - y_1),$$

and

$$y_1 y_2 = k^2, \qquad m_3^2 = 6h_1 + y_1 + y_2 - (x_1 + x_2)^2,$$
$$m_3 \gamma_3 = 2h_2 + x_1 y_2 + x_2 y_1 - x_1 x_2 (x_1 + x_2), \qquad [3.23]$$
$$\gamma_3^2 = 1 - k^2 + x_1^2 y_2 + x_2^2 y_1 - x_1^2 x_2^2.$$

Note that $\sigma : M_c \longrightarrow M_c$ $(x_1, x_2, m_3, y_1, y_2, \gamma_3) \longmapsto (x_1, x_2, -m_3, y_1, y_2, -\gamma_3)$ is an automorphism of M_c of order two. The quotient M_c/σ by the involution σ is a Kummer surface K defined by [3.18]. The variety M_c is a double cover of the surface K branched over the fixed points of the involution σ. To find them, we substitute $m_3 = \gamma_3 = 0$ in the system [3.13] to get

$$y_1 y_2 = k^2, \qquad [3.24]$$
$$y_1 + y_2 = (x_1 + x_2)^2 - 6h_1, \qquad [3.25]$$
$$x_2 y_1 + x_1 y_2 = x_1 x_2 (x_1 + x_2) - 2h_2, \qquad [3.26]$$
$$x_2^2 y_1 + x_1^2 y_2 = x_1^2 x_2^2 + k^2 - 1. \qquad [3.27]$$

Away from the $x_1^2 = x_2^2$, we may solve [3.25] and [3.27] in y_1 and y_2 and substitute into equations [3.24] and [3.26]; we then find two curves in x_1 and x_2 whose equations are

$$R(x_1, x_2) \equiv -x_1^2 x_2^2 + 6h_1 x_1 x_2 - 2h_2(x_1 + x_2) + 1 - k^2 = 0,$$
$$S(x_1, x_2) \equiv \left(x_1^4 + 2x_1^3 x_2 - 6h_1 x_1^2 + 1 - k^2\right)\left(x_2^4 + 2x_1 x_2^3 - 6h_1 x_2^2 + 1 - k^2\right)$$
$$+ k^2(x_1^2 - x_2^2)^2 = 0.$$

These curves intersect at the zeroes of the resultant $\mathrm{Res}(R, S)$ of R, S:

$$\mathrm{Res}(R, S)_{x_2} = x_1^2 \left(x_1^4 + 6h_1 x_1^2 + k^2 - 1\right)^2 P_8(x_1), \qquad [3.28]$$

where $P_8(x_1)$ is a monic polynomial of degree 8. Since the root x_1 must be excluded (it indeed implies that the leading terms of R and S vanish), the possible intersections of the curve R and S will be (i) at the roots of $x_1^4 + 6h_1 x_1^2 + k^2 - 1 = 0$; this is unacceptable, because we then check that the common roots of R and S would have the property that $x_1^2 = x_2^2$, which was excluded; (ii) at the roots of $P_8(x_1) = 0$, for generic k and h, $x_1^2 \neq x_2^2$. Finally, we must analyze the case $x_1^2 = x_2^2$, for which we check that equations [3.24]–[3.27] have no common roots. Consequently, the involution σ has eight fixed points on the affine variety M_c. Clearly the vector field [3.30] vanishes at the fixed points of σ.

b) From equations [3.18], we deduce

$$y_1 = \frac{-1}{2R(x_2)}\left(R_1(x_1, x_2) + k^2(x_1 - x_2)^2 + \Delta\right),$$

$$y_2 = \frac{-1}{2R(x_1)}\left(R_1(x_1, x_2) + k^2(x_1 - x_2)^2 - \Delta\right),$$

where $\Delta^2 = \left(R_1(x_1, x_2) + k^2(x_1 - x_2)^2\right)^2 - 4k^2 R(x_1) R(x_2) \equiv P(x_1, x_2)$. Therefore, the surface K is a double cover of $\mathbb{C}^2$, ramified along the curve $\mathcal{C} : P(x_1, x_2) = 0$. This equation is reducible and $P(x_1, x_2)$ can be written as the product, $P(x_1, x_2) = P_1(x_1, x_2) \cdot P_2(x_1, x_2)$ of two symmetric polynomials (in x_1, x_2) of degree two in each one of the variables x_1, x_2, that is,

$$P_1(x_1, x_2) = a(x_1)x_2^2 + 2b(x_1)x_2 - c(x_1) = a(x_2)x_1^2 + 2b(x_2)x_1 - c(x_2),$$

where $a(x) = -2(k + 3h_1)x^2 + 4h_2 x - 1$, $b(x) = 2h_2 x^2 + (2k(k + 3h_1) - 1)x - 2h_2 k$, $c(x) = x^2 + 4h_2 k x + 2(k^2 - 1)(k + 3h_1) + 4h_2^2$, while the polynomial

$P_2(x_1, x_2)$ is obtained from $P_1(x_1, x_2)$ after replacing k with $-k$. Note that the curve $\mathcal{C}_1 : P_1(x_1, x_2) = 0$ is elliptic:

$$x_1 = \frac{-b(x_2) \pm \sqrt{2(k+3h_1) - 4h_2^2\sqrt{R(x_2)}}}{a(x_2)},$$

$$x_2 = \frac{-b(x_1) \pm \sqrt{2(k+3h_1) - 4h_2^2\sqrt{R(x_1)}}}{a(x_1)},$$

where $R(x)$ is given by [3.19]. Similarly, the curve $\mathcal{C}_2 : P_2(x_1, x_2) = 0$ is elliptic and we note that the two curves $\mathcal{C}_1$ and $\mathcal{C}_2$ intersect exactly at the eight fixed points of involution σ, because (see [3.28]) $\text{Res}(P_1, P_2)_{x_2} = 16k^2 P_8(x_1)$. Differentiating the symmetric equation $P_1(x_1, x_2) = 0$ (or $P_2(x_1, x_2) = 0$) with regard to t, one finds $\frac{\partial P_1}{\partial x_1} \dot{x}_1 + \frac{\partial P_1}{\partial x_2} \dot{x}_2 = 0$, where $\frac{\partial P_1}{\partial x_1} = \pm 2\sqrt{2(k+3h_1) - 4h_2^2\sqrt{R(x_2)}}$, $\frac{\partial P_1}{\partial x_2} = \pm 2\sqrt{2(k+3h_1) - 4h_2^2\sqrt{R(x_1)}}$. Hence

$$\frac{\dot{x}_1}{\sqrt{R(x_1)}} \pm \frac{\dot{x}_2}{\sqrt{R(x_2)}} = 0. \qquad [3.29]$$

Since $R(x_1)$ and $R(x_2)$ are two polynomials of the fourth degree in x_1 and x_2, respectively, with the same coefficients, then [3.29] is the so-called Euler's equation. The reader is referred to Halphen (1888) and Weil (1983) for this theory, which we will summarize later.

Let $F(x) = a_0 x^4 + 4a_1 x^3 + 6a_2 x^2 + 4a_3 x + a_4$ be a polynomial of the fourth degree. The general integral of Euler's equation $\frac{\dot{x}}{\sqrt{F(x)}} \pm \frac{\dot{y}}{\sqrt{F(y)}} = 0$ can be written in two different ways:

i) $F_1(x, y) + 2sF(x, y) - s^2(x-y)^2 = 0$, where $F(x, y) = a_0 x^2 y^2 + 2a_1 xy(x+y) + 3a_2(x^2 + y^2) + 2a_3(x+y) + a_4$ and $F_1(x, y) = \frac{F(x)F(y) - F^2(x,y)}{(x-y)^2}$.

ii) In an irrational form $\frac{F(x,y) \mp \sqrt{F(x)}\sqrt{F(y)}}{(x-y)^2} = s$, which can be seen as the addition-formula for the Weierstrass elliptic function

$$2\wp(u+v) = \frac{(\wp(u) + \wp(v))(2\wp(u)\wp(v) - \frac{1}{2}g_2) - g_3 - \wp'(u)\wp'(v)}{(\wp(u) + \wp(v))^2},$$

$$\wp'^2(u) = \left(\frac{d\wp}{du}\right)^2 = 4\wp^3 - g_2\wp - g_3,$$

with $\wp(u) = x$, $\wp(v) = y$, $\wp'^2(u) = F(x)$, $\wp'^2(v) = F(y)$, $F(x) = 4x^3 - g_2 x - g_3$, $2\wp(u+v) = s$, and g_2, g_3 are constants.

Let us apply these facts to Kowalewski's problem with $F(x) = R(x)$, $F(x_1, x_2) = R(x_1, x_2) + 3h_1(x_1 - x_2)^2$, $a_0 = -1$, $a_1 = 0$, $a_2 = h_1$, $a_3 = -h_2$, $a_4 = 1 - k^2$ and $s = k + 3h_1$. So the polynomial $P_1(x_1, x_2)$, which can also be regarded as a solution of [3.29], can also be written as $R_1(x_1, x_2) + 2sR(x_1, x_2) - s^2(x_1 - x_2)^2 = 0$, where $R_1(x_1, x_2)$ is given by [3.20] and has the form $R_1(x_1, x_2) = \frac{R(x_1)R(x_2) - R^2(x_1, x_2)}{(x_1 - x_2)^2}$. Remember that $R(x_1, x_2)$ is given by [3.21]. The solution of [3.29] can also be expressed

$$\frac{R(x_1, x_2) \mp \sqrt{R(x_1)}\sqrt{R(x_2)}}{(x_1 - x_2)^2} + 3h_1 = s. \qquad [3.30]$$

c) Let us carry out the calculations, assuming the polynomial $R(x)$ is reduced to the form $4x^3 - g_2 x - g_3$ and call s_1 (respectively, s_2) the relation [3.30] with the sign $-$ (respectively, $+$). Now, outside the branch locus of K [3.18] over $\mathbb{C}^2$, equation [3.29] is not identically zero and may be written in the form

$$\begin{cases} \frac{\dot{x}_1}{\sqrt{R(x_1)}} + \frac{\dot{x}_2}{\sqrt{R(x_2)}} = \frac{\dot{s}_1}{\sqrt{4s_1^3 - g_2 s_1 - g_3}} \neq 0, \\ \frac{\dot{x}_1}{\sqrt{R(x_1)}} - \frac{\dot{x}_2}{\sqrt{R(x_2)}} = \frac{\dot{s}_2}{\sqrt{4s_2^3 - g_2 s_2 - g_3}} \neq 0 \end{cases} \qquad [3.31]$$

where $g_2 = k^2 - 1 + 3h_1^2$, $g_3 = h_1(k^2 - 1 - h_1^2) + h_2^2$. After some algebraic manipulation, we deduce from [3.23],

$$(m_3 x_1 - \gamma_3)^2 = R(x_1) + (x_1 - x_2)^2 y_1,$$

$$(m_3 x_2 - \gamma_3)^2 = R(x_2) + (x_1 - x_2)^2 y_2,$$

$$(m_3 x_1 - \gamma_3)(m_3 x_2 - \gamma_3) = R(x_1, x_2),$$

and from [3.22], $\dot{x}_1^2 = R(x_1) + (x_1 - x_2)^2 y_1$, $\dot{x}_2^2 = R(x_2) + (x_1 - x_2)^2 y_2$. This, together with [3.18] and [3.31], implies that

$$\frac{\dot{s}_1^2}{4s_1^3 - g_2 s_1 - g_3} = \left(\frac{\dot{x}_1}{\sqrt{R(x_1)}} + \frac{\dot{x}_2}{\sqrt{R(x_2)}} \right)^2,$$

$$= \frac{(x_1 - x_2)^4}{R(x_1)R(x_2)} \left[\left(\frac{R(x_1, x_2) - \sqrt{R(x_1)}\sqrt{R(x_2)}}{(x_1 - x_2)^2} \right) - k^2 \right],$$

$$= 4 \frac{(s_1 - 3h_1)^2 - k^2}{(s_1 - s_2)}.$$

In the same way, we find

$$\frac{\dot{s}_2^2}{4s_2^3 - g_2 s_2 - g_3} = \left(\frac{\dot{x}_1}{\sqrt{R(x_1)}} - \frac{\dot{x}_2}{\sqrt{R(x_2)}} \right)^2 = 4 \frac{(s_2 - 3h_1)^2 - k^2}{(s_2 - s_1)}.$$

In terms of the variables s_1 and s_2, the system [3.12] becomes

$$\frac{\dot{s}_1}{\sqrt{P(s_1)}} + \frac{\dot{s}_2}{\sqrt{P(s_2)}} = 0, \qquad \frac{s_1 \dot{s}_1}{\sqrt{P(s_1)}} + \frac{s_2 \dot{s}_2}{\sqrt{P(s_2)}} = i,$$

where $P_5(s) = ((s - 3h_1)^2 - k^2)(4s^3 - g_2 s - g_3)$. These equations are integrable by the Abel transformation $\mathcal{H} \longrightarrow Jac(\mathcal{H}) = \mathbb{C}^2/\Lambda$, $p \longmapsto \left(\int_{p_0}^p \theta_1, \int_{p_0}^p \theta_2\right)$, where $\mathcal{H}$ is the hyperelliptic curve of genus 2 with equation: $w^2 = P_5(s)$.

Figure 3.7. *Genus 2 hyperelliptic curve*

Λ is the lattice generated by the vectors $n_1 + \Omega_{\mathcal{H}} n_2$, $(n_1, n_2) \in \mathbb{Z}^2$, $\Omega_{\mathcal{H}}$ is the matrix of periods of $\mathcal{H}$, (θ_1, θ_2) is a basis of holomorphic differentials on $\mathcal{H}$, that is, $\theta_1 = \frac{ds}{\sqrt{P_5(s)}}$, $\theta_2 = \frac{sds}{\sqrt{P_5(s)}}$, and p_0 is a fixed point on $\mathcal{H}$. The theorem is thus proved.
□

3.2.4. *Special cases*

In the introduction of this section we mentioned that any algebraic integral of equations [3.3] is a combination of classical integrals except in the cases of Euler, Lagrange and Kowalewski, and that there could therefore be no first algebraic integral other than those highlighted in these three cases. In addition, there are a few special cases as follows:

– the case of Hess (1890); Appel'rot (1894): $l_2 = 0$, $l_1\sqrt{I_1(I_2 - I_3)} + l_3\sqrt{I_3(I_1 - I_2)} = 0$. In this case, equation $l_1 m_1 + l_3 m_3 = 0$ represents a particular first integral obtained by Hesse and the integration is carried out using elliptic functions;

– the case of Goryachev (1900); Chaplygin (1948): $I_1 = I_2 = 4I_3$, $l_2 = l_3 = 0$. The system [3.3] admits the first integral $\lambda_3 m_3(\lambda_1^2 + m_1^2 + \lambda_2^2 m_2^2) + \mu g l_1 \lambda_1 \lambda_3 m_1 \gamma_3 = g$, $\lambda_i = \frac{1}{I_i}$, $i = 1, 2, 3$, and integration is carried out using hyperelliptic functions of genus 2;

– the case of Bobylev–Steklov Bobylev (1896); Steklov (1896): $I_2 = 2I_1$, $l_1 = l_3 = 0$. The integration of the equations in this case is easy, using elliptic functions.

3.3. Motion of a solid through ideal fluid

The equations of motion of a solid in an ideal fluid have the form (Kirchoff 1876):

$$\dot{p}_1 = p_2\frac{\partial H}{\partial l_3} - p_3\frac{\partial H}{\partial l_2}, \qquad \dot{l}_1 = p_2\frac{\partial H}{\partial p_3} - p_3\frac{\partial H}{\partial p_2} + l_2\frac{\partial H}{\partial l_3} - l_3\frac{\partial H}{\partial l_2},$$
$$\dot{p}_2 = p_3\frac{\partial H}{\partial l_1} - p_1\frac{\partial H}{\partial l_3}, \qquad \dot{l}_2 = p_3\frac{\partial H}{\partial p_1} - p_1\frac{\partial H}{\partial p_3} + l_3\frac{\partial H}{\partial l_1} - l_1\frac{\partial H}{\partial l_3}, \qquad [3.32]$$
$$\dot{p}_3 = p_1\frac{\partial H}{\partial l_2} - p_2\frac{\partial H}{\partial l_1}, \qquad \dot{l}_3 = p_1\frac{\partial H}{\partial p_2} - p_2\frac{\partial H}{\partial p_1} + l_1\frac{\partial H}{\partial l_2} - l_2\frac{\partial H}{\partial l_1},$$

where (p_1, p_2, p_3) is the velocity of a point fixed relatively to the solid, (l_1, l_2, l_3) the angular velocity of the body expressed with regard to a frame of reference also fixed relatively to the solid and H is the Hamiltonian. Equations [3.32] can be regarded as the equations of the geodesics (section 3.5) of the right-invariant metric on the group $E(3) = SO(3) \times \mathbb{R}^3$ of motions of three-dimensional Euclidean space $\mathbb{R}^3$, generated by rotations and translations. Equations [3.31] have the trivial first integrals (or invariants):

$$H_1 = H, \quad H_2 = \sum_{k=1}^{3} p_k^2, \quad H_3 = \sum_{k=1}^{3} p_k l_k. \qquad [3.33]$$

We distinguish two integrable cases: the case of Clebsch (1871) and the case of Lyapunov (1893); Steklov (1893).

3.3.1. *Clebsch's case*

In Clebsch's case, we have

$$H_1 = H = \frac{1}{2}\sum_{k=1}^{3}\left(a_k p_k^2 + b_k l_k^2\right), \qquad [3.34]$$

with $\frac{a_2-a_3}{b_1} + \frac{a_3-a_1}{b_2} + \frac{a_1-a_2}{b_3} = 0$. An additional integral is

$$H_4 = \frac{1}{2}\sum_{k=1}^{3}\left(b_k p_k^2 + \varrho l_k^2\right), \qquad [3.35]$$

where $\varrho = \frac{b_1(b_2-b_3)}{a_2-a_3} = \frac{b_2(b_3-b_1)}{a_3-a_1} = \frac{b_3(b_1-b_2)}{a_1-a_2}$. We shall briefly study the Kötter (1892) solution by quadratures of equations [3.31], in terms of genus 2 hyperelliptic integrals. In fact, the transformation to the separating coordinates s_1 and s_2 that leads to the quadratures in terms of hyperelliptic integrals is quite involved. We find that this transformation requires a great deal of luck and ingenuity. After the substitution $b_k \to \varrho b_k$, $1 \leq k \leq 3$, and an appropriate linear combination of H_1 and H_2, equations [3.32]–[3.34] can be written in the form

$$p_1^2 + p_2^2 + p_3^2 = A,$$

$$b_1 p_1^2 + b_2 p_2^2 + b_3 p_3^2 + l_1^2 + l_2^2 + l_3^2 = B,$$

$$b_1 l_1^2 + b_2 l_2^2 + b_3 l_3^2 - b_2 b_3 p_1^2 - b_1 b_3 p_2^2 - b_1 b_2 p_3^2 = C,$$

$$p_1 l_1 + p_2 l_2 + p_3 l_3 = D,$$

where A, B, C, D are constants. Following (Kötter 1892; Dubrovin 1981), we introduce coordinates φ_k, ψ_k, $1 \leq k \leq 3$ by setting

$$\varphi_k = p_k \left(\frac{\sqrt{\prod_{j=1}^3 (z_1 - b_j)}}{\sqrt{z_1 - b_k}\sqrt{\partial R/\partial z_1}} + \sqrt{-1}\frac{\sqrt{\prod_{j=1}^3 (z_2 - b_j)}}{\sqrt{z_2 - b_k}\sqrt{\partial R/\partial z_2}} \right)$$

$$+ l_k \left(\frac{\sqrt{z_1 - b_k}}{\sqrt{\partial R/\partial z_1}} + \sqrt{-1}\frac{\sqrt{z_2 - b_k}}{\sqrt{\partial R/\partial z_2}} \right),$$

$$\psi_k = p_k \left(\frac{\sqrt{\prod_{j=1}^3 (z_1 - b_j)}}{\sqrt{z_1 - b_k}\sqrt{\partial R/\partial z_1}} - \sqrt{-1}\frac{\sqrt{\prod_{j=1}^3 (z_2 - b_j)}}{\sqrt{z_2 - b_k}\sqrt{\partial R/\partial z_2}} \right)$$

$$+ l_k \left(\frac{\sqrt{z_1 - b_k}}{\sqrt{\partial R/\partial z_1}} - \sqrt{-1}\frac{\sqrt{z_2 - b_k}}{\sqrt{\partial R/\partial z_2}} \right),$$

where $R(z) = \prod_{i=1}^4 (z - z_i)$, and z_1, z_2, z_3, z_4 are the roots of the equation: $A^2 \left(z^2 - z \sum_{k=1}^3 b_k \right) + Bz - C + 2D\sqrt{\prod_{k=1}^3 (z - b_k)} = 0$. Let s_1 and s_2 be the roots of the equation: $\frac{\psi_1^2}{\nu_1^2 - s} + \frac{\psi_2^2}{\nu_2^2 - s} + \frac{\psi_3^2}{\nu_3^2 - s} = 0$, where

$$\nu_k = \frac{\frac{\sqrt{z_3 - b_k}}{\sqrt{\partial R/\partial z_3}} + \sqrt{-1}\frac{\sqrt{z_4 - b_k}}{\sqrt{\partial R/\partial z_4}}}{\frac{\sqrt{z_1 - b_k}}{\sqrt{\partial R/\partial z_1}} + \sqrt{-1}\frac{\sqrt{z_2 - b_k}}{\sqrt{\partial R/\partial z_2}}}, \quad 1 \leq k \leq 3.$$

An expression of the original variables $p_1, p_2, p_3, l_1, l_2, l_3$ in terms of s_1 and s_2 can be found in Kötter (1892). After some algebraic manipulations, we obtain the following equations for s_1 and s_2:

$$\frac{ds_1}{dt} = \frac{(as_1 + b)\sqrt{P_5(s_1)}}{s_2 - s_1}, \qquad \frac{ds_2}{dt} = \frac{(as_2 + b)\sqrt{P_5(s_2)}}{s_1 - s_2},$$

where a, b are constants and $P_5(s)$ is a polynomial of degree 5 of the form $P_5(s) = s(s - \nu_1^2)(s - \nu_2^2)(s - \nu_3^2)(s - \nu_1^2\nu_2^2\nu_3^2)$. These equations can be integrated by the Abelian mapping $\mathcal{H} \to Jac(\mathcal{H}) = \mathbb{C}^2/\Lambda$, $P \longmapsto \left(\int_{P_0}^{P} \theta_1, \int_{P_0}^{P} \theta_2\right)$, where the hyperelliptic curve $\mathcal{H}$ of genus 2 is given by the equation $w^2 = P_5(s)$, Λ is the lattice generated by the vectors $n_1 + Mn_2$, $(n_1, n_2) \in \mathbb{Z}^2$, M is the matrix of period of the curve $\mathcal{H}$, (θ_1, θ_2) is a canonical basis of holomorphic differentials on $\mathcal{H}$, that is, $\theta_1 = \frac{ds}{\sqrt{P_5(s)}}$, $\theta_2 = \frac{sds}{\sqrt{P_5(s)}}$, P_0 is a fixed point. So, we have:

THEOREM 3.5.– The system of differential equations [3.31] in the Clebsch's case can be integrated in terms of genus 2 hyperelliptic functions of time.

An expression of the Kötter's variables in terms of the theta functions on the torus $Jac(\mathcal{H})$ can be found in (Kötter 1892).

3.3.2. *Lyapunov–Steklov's case*

In Lyapunov–Steklov's case, we have $H_1 = H = \frac{1}{2}\sum_{k=1}^{3}\left(a_k p_k^2 + b_k l_k^2\right) + \sum_{k=1}^{3} c_k p_k l_k$, with $a_1 = A^2 b_1 (b_2 - b_3)^2 + B$, $a_2 = A^2 b_2 (b_3 - b_1)^2 + B$, $a_3 = A^2 b_3 (b_1 - b_2)^2 + B$, $c_1 = Ab_2 b_3 + C$, $c_2 = Ab_1 b_3 + C$, $c_3 = Ab_1 b_2 + C$, where A, B, C are constants. A fourth first integral is given by $H_4 = \frac{1}{2}\sum_{k=1}^{3}\left(d_k p_k^2 + l_k^2\right) - A\sum_{k=1}^{3} b_k p_k l_k$, where $d_1 = A^2 (b_2 - b_3)^2$, $d_2 = A^2 (b_3 - b_1)^2$, $d_3 = A^2 (b_1 - b_2)^2$. A long calculation (Kötter 1900) shows that in this case, the integration is also done using hyperelliptic functions of genus two and the solutions can be expressed in terms of theta functions.

3.4. Yang–Mills field with gauge group $SU(2)$

We begin by briefly introducing some notions related to the Yang–Mills field with $SU(2)$ as gauge group. For general considerations and the details of certain notions, one can consult, for example, Dubrovin *et al.* (1990). Our aim is to show the link between the Yang–Mills equations and the Hamiltonian systems and to solve them (here and in other chapters) using different methods. Consider the special unitary group $SU(2)$ of degree 2, that is, the set of 2×2 unitary matrices with determinant 1. This is a real Lie group of dimension three. It is compact, simply connected, simple and semi-simple. The group $SU(2)$ is isomorphic to the group of quaternions of norm one and is diffeomorphic to the 3-sphere S^3. It is well known that the quaternions represent the rotations in three-dimensional space and hence there exists a surjective homomorphism of $SU(2)$ on the rotation group $SO(3)$, whose kernel is $\{+I, -I\}$ (the identical application and its opposite). Also recall that $SU(2)$ is identical to one of the symmetry spinor groups, $Spin(3)$, that enables a spinor presentation of rotations. The Lie algebra $su(2)$ corresponding to $SU(2)$ consists of the 2×2 antihermitian complex matrices with null trace, the standard commutator serving as a Lie bracket. It is a real algebra. The algebra $su(2)$ is isomorphic to the Lie algebra $so(3)$. In this section we consider the Yang–Mills field F_{kl} as a vector field with values in the algebra $su(2)$. It is a local expression of the gauge field or connection defining the covariant derivative of F_{kl} in the adjoint representation of $su(2)$. To determine this expression, note that each Lorentz component of the Yang–Mills field develops on a basis $(\sigma_1, \sigma_2, \sigma_3)$ of $su(2)$, $A_k = A_k^\alpha \sigma_\alpha$, $\alpha = 1, 2, 3$ where the σ_α are the Pauli matrices $\sigma_1 = \begin{pmatrix} 0 & -i \\ i & 0 \end{pmatrix}$, $\sigma_2 = \begin{pmatrix} 0 & 1 \\ 1 & 0 \end{pmatrix}$, $\sigma_3 = \begin{pmatrix} 1 & 0 \\ 0 & -1 \end{pmatrix}$. These matrices are often used in quantum mechanics to represent the spin of particles. The group $SU(2)$ is associated with gauge symmetry in the description of the weak or weak force interaction (one of the four fundamental forces of nature) and is of particular importance in the physics of particles. The dynamics of the Yang–Mills theory is determined by the Lagrangian density $\mathcal{L} = -\frac{1}{2} Tr\{F_{kl} F^{kl}\}$, $1 \leq k, l \leq 4$, where $F_{kl} = \frac{\partial A_l}{\partial \tau_k} - \frac{\partial A_k}{\partial \tau_l} + [A_k, A_l]$ is the expression of the anti-symmetric Faraday tensors with values in $su(2)$. These tensors are not invariant under gauge transformations. On the other hand, we verify that $Tr\{F_{kl} F^{kl}\}$ is actually gauge invariant. The trace relates to the internal space $su(2)$. The equations of the motion are given by $D_k F^{kl} = \frac{\partial F^{kl}}{\partial \tau_k} + [A_k, F^{kl}] = 0$, $F_{kl}, A_k \in su(2)$, $1 \leq k, l \leq 4$, with D_k as the covariant derivative in the adjoint representation of the algebra $su(2)$, and in which $[A_k, F^{kl}]$ is the crochet of the two fields in $su(2)$. The Yang–Mills theory extends the principle of gauge invariance of electromagnetism to other groups of continuous Lie transformations. Thus, the F_{kl} tensor generalizes the electromagnetic field and the Yang–Mills equations are the non-commutative generalization of the Maxwell equations. The self-dual Yang–Mills (SDYM) equations is an universal system for which some reductions include all classical tops from Euler to Kowalewski (0+1-dimensions), KdV, Nonlinear Schrödinger, Sine-Gordon, Toda lattice and N-waves equations (1+1-dimensions), KP and D-S equations (2+1-dimensions). In the case of homogeneous double-component field, we have $\frac{\partial A_l}{\partial \tau_k} = 0, (k \neq 1)$, $A_1 = A_2 = 0, A_3 = n_1 U_1 \in su(2), A_4 = n_2 U_2 \in su(2)$, where n_i are $su(2)$-generators (i.e. they satisfy commutation relations: $n_1 = [n_2, [n_1, n_2]], n_2 = [n_1, [n_2, n_1]]$). The system becomes $\frac{\partial^2 U_1}{\partial t^2} + U_1 U_2^2 = 0$, $\frac{\partial^2 U_2}{\partial t^2} + U_2 U_1^2 = 0$, with $t = \tau_1$. By setting $U_1 = q_1, U_2 = q_2$, $\frac{\partial U_1}{\partial t} = p_1, \frac{\partial U_2}{\partial t} = p_2$, the Yang–Mills equations are reduced to Hamiltonian system $\dot{x} = J \frac{\partial H}{\partial x}, x = (q_1, q_2, p_1, p_2)^\mathsf{T}, J = \begin{pmatrix} O & -I \\ I & O \end{pmatrix}$, where $H = \frac{1}{2} \left(p_1^2 + p_2^2 + q_1^2 q_2^2 \right)$ is the Hamiltonian. The symplectic transformation $p_1 = \frac{\sqrt{2}}{2} (x_1 + x_2)$, $p_2 = \frac{\sqrt{2}}{2} (x_1 - x_2)$, $q_1 = \frac{1}{2} \left(\sqrt[4]{2} \right)^3 (y_1 + iy_2)$ and $q_2 = \frac{1}{2} \left(\sqrt[4]{2} \right)^3 (y_1 - iy_2)$, takes this Hamiltonian into $H = \frac{1}{2}\left(x_1^2 + x_2^2\right) + \frac{1}{4}\left(y_1^2 + y_2^2\right)^2$. The Hamiltonian dynamical system associated with H is written as

$$\dot{y}_1 = x_1, \qquad \dot{x}_1 = -\left(y_1^2 + y_2^2\right)y_1, \qquad [3.36]$$
$$\dot{y}_2 = x_2, \qquad \dot{x}_2 = -\left(y_1^2 + y_2^2\right)y_2.$$

These equations give a vector field on $\mathbb{R}^4$. The existence of a second independent first integral in involution with $H_1 \equiv H$ is enough for the system to be completely integrable. The system [3.36] implies $\ddot{y}_1 + \left(y_1^2 + y_2^2\right)y_1 = 0$, $\ddot{y}_2 + \left(y_1^2 + y_2^2\right)y_2 = 0$, so that, of course, the function (the moment) $H_2 = x_1 y_2 - x_2 y_1$ is a first integral. The functions H_1 and H_2 are in involution $\{H_1, H_2\} = 0$. Obviously, this second first integral determines with H_1 a completely integrable system. Let $M_c = \{x \equiv (y_1, y_2, x_1, x_2) \in \mathbb{R}^4 : H_1(x) = c_1, H_2(x) = c_2\}$ be the invariant surface (where $c = (c_1, c_2)$ is not a critical value). Substituting $y_1 = r\cos\theta$, $y_2 = r\sin\theta$, in equations $H_1 = \frac{1}{2}\left(x_1^2 + x_2^2\right) + \frac{1}{4}\left(y_1^2 + y_2^2\right)^2 = c_1$, $H_2 = x_1 y_2 - x_2 y_1 = c_2$, we obtain $\frac{1}{2}\left(\dot{r}^2 + (r\dot{\theta})^2\right) + \frac{1}{4}r^2 = c_1$, $r^2\dot{\theta} = -c_2$. Hence $(r\dot{r})^2 + \frac{1}{2}r^4 - 2c_1 r^2 + c_2^2 = 0$, and $w^2 + P(z) = 0$, where $w \equiv r\dot{r}$, $z \equiv r^2$, and $P(z) = \frac{1}{2}z^3 - 2c_1 z + c_2^2$. The polynomial $P(z)$ is of degree 3, the genus of the Riemann surface $\mathcal{C}$,

$$\mathcal{C} = \overline{\{(w, z) : w^2 + P(z) = 0\}}, \qquad [3.37]$$

is $g = 1$ (elliptic curve). We have a holomorphic differential $\omega = \dfrac{dz}{\sqrt{P(z)}}$, and the linearization occurs on $\mathcal{C}$. Although the variety M_c has dimension 2, here we have a reduction of dimension 1 and we get the result:

THEOREM 3.6.– The differential system [3.35] is completely linearized on the Jacobian variety of $\mathcal{C}$, that is, on the elliptic curve $\mathcal{C}$ [3.36].

REMARK 3.6.– Further information and methods for resolving the Yang–Mills system will be provided in section 4.10 and exercise 8.8.

3.5. Appendix (geodesic flow and Euler–Arnold equations)

The second-order differential equation for geodesic motion (i.e. the Euler–Lagrange equation) on a Lie group G equipped with a left or right invariant Riemannian metric can be reduced to a first-order differential equation on the Lie algebra $\mathcal{G}$, called the Euler–Arnold equation. The aim of this appendix is to give a brief presentation of some aspects of geodesic flow on groups and the Euler–Arnold equations. Arnold (1989) showed that Euler's equation for the motions of a rigid body is the geodesic equation on the group of rotations of three-dimensional

Euclidean space endowed with a left-invariant metric. He also proved that Euler's equation for an ideal fluid is the geodesic equation on the group of volume preserving diffeomorphisms with respect to the right invariant metric. In general, the geodesic flow of a left-invariant metrics on a Lie group G after G-reduction reduces to the Euler equations on $\mathcal{G}^*$ (the dual space of the Lie algebra $\mathcal{G}$), which are Hamiltonian equations with respect to the Lie–Poisson bracket on $\mathcal{G}^*$. A multidimensional generalization of the Euler case has been suggested by Manakov (1976). Using his idea, Mishchenko and Fomenko (1978) proposed the argument shift method and constructed integrable examples of Euler equations for all compact groups and proved the integrability of the original geodesic flows. Let (M, g) be a Riemannian manifold with metric $g = (g_{ij})$. The geodesic flow on M can be understood as a Hamilton flow on the cotangent space T^*M (or the tangent bundle TM, the two are naturally isomorphic via the Riemann metric). When studying the geometry of the geodesic flow, there are various ways to place the geodesic equations of M in the context of Hamiltonian dynamics. One way is to consider the cotangent bundle T^*M that is a symplectic manifold. Let $q_1, ..., q_n$ be a local coordinate system on M. The Riemannian metric is given by the formula $ds^2 = \sum g_{ij}dq_idq_j$, where dq_i are the coordinates of a tangent vector. The geodesic flow of M with the above metric is the Lagrangian system in the tangent bundle TM with the Lagrangian $\mathcal{L} = \sum g_{ij}\dot{q}_i\dot{q}_j$. Identifying TM with T^*M by using the metric-induced diffeomorphism $TM \cong T^*M$ to pull-back the energy function on TM to T^*M, we may assume that the geodesic flow acts on T^*M. The corresponding system is Hamiltonian and the Hamiltonian is the Legendre transform of the Lagrangian. In local coordinates, we define the Hamilton function H on T^*M by $H : T^*M \longrightarrow \mathbb{R}$, $H(p,q) = \frac{1}{2}\sum g_{ij}dq_idq_j$. An equivalent approach is to put a symplectic structure directly on the tangent bundle TM. The Hamiltonian equation of motion becomes exactly the geodesic equation (more precisely, the geodesic flow is the one-parameter group of diffeomorphisms defined by this Hamiltonian equation). The geodesic flow can be seen from the Lagrangian point of view, or geometrically via local length minimization. Let G be a Lie group and $\mathcal{G} = T_eG$ be the Lie algebra, where $e \in G$ is the identity element with the Lie bracket $[.,.]$. Let $\langle .,. \rangle$ be the inner product on $\mathcal{G}$ and g an element of the group G. In section 1.8, we saw how the Lie group G operates on itself by left translation: $L_g : G \longrightarrow G$, $h \longmapsto gh$. A Riemannian metric on the group G is left-invariant if it is preserved by all left translations L_g, that is, if the derivative of left translation carries every vector to a vector of the same length. It suffices to define a left-invariant Riemannian metric at one point (often we take the identity); for the other points it is defined by left translations. Let A be a symmetric positive linear operator on the Lie algebra $\mathcal{G}$, $A : \mathcal{G} \longrightarrow \mathcal{G}^*$, $(A\xi, \eta) = (A\eta, \xi)$, $\forall \xi, \eta \in \mathcal{G}$, and define a symmetric operator A_g by left translation, $A_g : TG_g \longrightarrow T^*G_g$, $\xi \longmapsto A_g\xi = L^*_{g^{-1}}AL_{g^{-1}*}\xi$. The scalar product defined by A_g will be denoted, $\langle \xi, \eta \rangle_g = (A_g\xi, \eta) = (A_g\eta, \xi) = \langle \eta, \xi \rangle_g$. On the group G, this scalar product defines a Riemannian metric invariant by left translations. On a group there are as many metrics as there are linear symmetric operators A. Define geodesics as extremals of the Lagrangian $\mathcal{L}(g) = \int E(g(t).\dot{g}(t))dt$, where $E(X) = \frac{1}{2}\langle X_g, X_g\rangle = \frac{1}{2}(A_g X_g, X_g)_g$ is the kinetic energy or energy functional. If $g(t)$ is a geodesic, the velocity $\dot{g}(t)$ can be translated to the identity. This can be done in two ways via left or right translation, and as a result we obtain two different elements of the Lie algebra $\mathcal{G}$: (i) by left translation, we obtain $\omega_L = L_{g^{-1}}\dot{g} \in \mathcal{G}$, called the left angular velocity; (ii) by right translation, we obtain $\omega_R = R_{g^{-1}}\dot{g} \in \mathcal{G}$, called the right angular velocity. We will define $A_g \dot{g} = M \in TG_g^*$, as the angular momentum. The vector M lies in the cotangent space to the group at the point g, and it can be carried to the cotangent space to the group at the identity by both left and right translations. We obtain two vectors $M_L = L_g^* M \in \mathcal{G}^*$, $M_R = R_g^* M \in \mathcal{G}^*$ (we did use g instead of g^{-1} for L_g^* and R_g^*. By definition, L_g^* is a function of TG_g^* to $TG_g^* = \mathcal{G}^*$. These statements come from expressions of kinetic energy in terms of angular momentum). We have the following relations: $\omega_R = Ad_g \omega_L$, $M_L = AM_L$, $M_R = Ad_g^* M_L$, where (see section 1.8) the adjoint representation of the group G is defined as the derivative of $R_g^{-1} L_g$ in the unit e, that is, the induced application of tangent spaces as follows, $Ad_g : \mathcal{G} \longrightarrow \mathcal{G}, \xi \longmapsto \frac{d}{dt} R_g^{-1} L_g(e^{t\xi})\big|_{t=0}$, and the transpose operator (coadjoint representation of the Lie group G), $Ad^* : \mathcal{G}^* \longrightarrow \mathcal{G}^*$, $g \longmapsto Ad^*(g) \equiv Ad_g^*$, where g runs through the Lie group G is defined by $\langle Ad_g^*(\zeta), \eta\rangle = \langle \zeta, Ad_g \eta\rangle$, $\zeta \in \mathcal{G}^*$, $\eta \in \mathcal{G}$. Using these notations, we can rewrite the kinetic energy above as follows: $E = \frac{1}{2}\langle \dot{g}, \dot{g}\rangle_g = \frac{1}{2}\langle \omega_L, \omega_L\rangle = \frac{1}{2}(M_L, \omega_L) = \frac{1}{2}(A_g \dot{g}, \dot{g})_g$.

EXAMPLE 3.2.– We consider the group $G = SO(n)$ and its Lie algebra $\mathcal{G} = so(n)$ paired with itself (for definitions, see section 1.9). The kinetic energy of an n-dimensional rigid body (S) is defined by $E(t) = \frac{1}{2} \int_S \|\dot{g}r\|^2 d\nu = -\frac{1}{2} tr(\Omega J \Omega)$, where $S = supp(\nu)$ is the reference state, ν is a positive measure (the mass distribution of (S)) on $\mathbb{R}^3$ with compact support, Ω lies in the Lie algebra $so(n)$ and J is the symmetric matrix with entries $J_{ij} = \int_S x_i x_j d\nu$. Here, we have $\Omega = \omega_L$ and $A(\Omega) = M_L \in so(n)^*$. Consider with a little more detail the classical example of the geodesic flow of a left-invariant metric on the Lie group $G = SO(3)$. The geodesic flow of such a metric describes the motion of a rigid body about a fixed point under its own inertia as well. For more details on the constructions to follow, see Arnold (1989). A motion of the body is then described by a curve $g = g(t)$ in the group. The Lie algebra $\mathcal{G} = so(3)$ is the three-dimensional space of angular velocities of all possible rotations and the commutator in this algebra is the usual vector product (for more information on the group $SO(3)$ and its algebra $so(3)$, see sections 1.9 and 3.2). The rotation velocity $\dot{g}$ of the body is a tangent vector to the group G at the point g and as we did above, by transporting this vector into the tangent space to the identity of the group $SO(3)$, that is, into the algebra $so(3)$, we get the angular velocity and this can be done in two ways: by left translation, $\omega_L = L_{g^{-1}} \dot{g} \in so(3)$ (the left angular velocity), and by right translation, $\omega_R = R_{g^{-1}} \dot{g} \in so(3)$ (the right angular velocity). Indeed, let us take a reference in space, let $g \in SO(3)$ be the position of the body at a given instant and $\omega \in so(3)$, such that $e^{\omega t}$ is a one-parameter of rotations with angular velocity ω. Consider the displacement $g(t+h) = e^{\omega h}g(t)$, where $g(t) \in SO(3)$, $\omega \in so(3)$ and $h \ll 1$. This displacement is obtained from the displacement g by a rotation with angular velocity ω after a small time h. If $\dot{g}(t) = \frac{d}{dh}g(t+h)|_{h=0} = \frac{d}{dh}e^{\omega h}g(t)|_{h=0} = \omega g(t)$, then $\omega = \dot{g}(t)g^{-1}(t) = R_{g^{-1}*}\dot{g}(t)$. Since we are in the space reference, we call ω the angular velocity relative to space and denote it ω_R. So ω_R is deduced from $\dot{g}$ by a right translation. Let us consider a small displacement in the reference related to the solid, we have, in an analogous way, $g(t+h) = g(t)e^{\omega h}$, $h \ll 1$. If $\dot{g}(t) = \frac{d}{dh}g(t+h)|_{h=0} = \frac{d}{dh}g(t)e^{\omega h}|_{h=0} = g(t)\omega$, then $\omega = g^{-1}(t)\dot{g}(t) = L_{g^{-1}*}\dot{g}(t)$, and ω is called the angular velocity in the body and is denoted by ω_L. Thus, ω_L is deduced from $\dot{g}$ by a left translation. The kinetic energy of the body is defined by the vector of angular velocity in the solid and does not depend on the position of the body in space. So the kinetic energy of the body defines a left-invariant Riemannian metric on the group $G = SO(3)$. The symmetric positive definite operator $A_g : TG_g \longrightarrow TG_g^*$ given by this metric is called the moment of inertia operator or tensor. The latter is related to kinetic energy as follows: $E = \frac{1}{2}\langle \dot{g}, \dot{g}\rangle_g = \frac{1}{2}\langle L_{g^{-1}*}\dot{g}, L_{g^{-1}*}\dot{g}\rangle_e$, ($e \equiv$ identity); hence, $E = \frac{1}{2}(AL_{g^{-1}*}\dot{g}, L_{g^{-1}*}\dot{g})\frac{1}{2}(A\omega_L, \omega_L) = \frac{1}{2}(L_{g^{-1}}^* AL_{g^{-1}*}\dot{g}, \dot{g}) = \frac{1}{2}(A_g\dot{g}, \dot{g})$, where $A : so(3) \longrightarrow so(3)^*$ is the value of A_g for $g = e$. We will define $A_g\dot{g} = M \in TG_g^*$ (= cotangent space to the group $SO(3)$ at the point g) as the angular momentum. The space $\mathcal{G}^* = so(3)^*$ is here the space of angular moments. The vector M can be carried to $so(3)^*$ in two different ways, either by a left translation or by a right translation. We obtain $M_L = L_g^*M \in so(3)^*$, that is, the angular momentum relative to the body, and $M_R = R_g^*M \in so(3)^*$, that is, the angular momentum relative to space. Recall that we did use g instead of g^{-1} for L_g^* and R_g^*. By definition, L_g^* is a function of $TSO(3)_g^*$ to $TSO(3)_e^* = so(3)^*$. These statements come from expressions of kinetic energy in terms of angular momentum. We have $E = \frac{1}{2}(A_g\dot{g}, \dot{g}) = \frac{1}{2}(A_g\dot{g}, L_{g^*}L_{g^{-1}*}\dot{g}) = \frac{1}{2}(L_g^*A_g\dot{g}, L_{g^{-1}*}\dot{g}) = \frac{1}{2}(L_g^*M, \omega_L)$, hence $E = \frac{1}{2}(M_L, \omega_L)$. Similarly, we have $E = \frac{1}{2}(A_g\dot{g}, \dot{g}) = \frac{1}{2}(A_g\dot{g}, R_{g^*}R_{g^{-1}*}\dot{g}) = \frac{1}{2}(R_g^*A_g\dot{g}, R_{g^{-1}*}\dot{g}) = \frac{1}{2}(R_g^*M, \omega_R)$, hence $E = \frac{1}{2}(M_R, \omega_R)$. By virtue of the principle of least action, the motion of a rigid body under inertia (with no external forces) is a geodesic in the group of rotations with the left invariant metric described above.

PROPOSITION 3.2.– The vector of angular momentum relative to space is preserved under rotation: $\frac{dM_R}{dt} = 0$.

PROOF.– The Lagrangian of the system is equal to the kinetic energy. However, the latter is constant and invariant by left translation. Indeed, let $\sigma : TG \longrightarrow \mathcal{G}^*$ be the momentum map defined by $\sigma((g, \dot{g}))(\xi) = \frac{\partial K}{\partial \dot{g}}z_\xi = \langle \dot{g}, R_g\xi\rangle_g = (M, R_g\xi) = (R_g^*M, \xi) = M_R(\xi)$, where z_ξ is the left-invariant vector field generated by $\xi \in \mathcal{G}$. The existence of the map σ is due to the invariance of the energy with respect to left translations. According to the Noether's theorem, this map is constant along a geodesic. So, we have $\frac{dM_R}{dt} = 0$. □

PROPOSITION 3.3.– The vector of angular momentum relative to the body satisfies the so-called Euler–Arnold equation: $\frac{dM_L}{dt} = \{\omega_L, M_L\} = [M_L, \omega_L]$.

PROOF.– Consider the map Ad^* above and its derivative (see section 1.8) in the unity of the group $ad^* \equiv (Ad^*)_{*e} : \mathcal{G} \longrightarrow \text{End}(\mathcal{G}^*)$, $\xi \longmapsto ad^*_\xi$. Using the relation $M_R = Ad^*_g M_L$ (see example 3.2) and computing the derivative, we find that $\dot{M}_L = \frac{dM_L}{dt} = ad^*_{\omega_L} M_L$. In other words, according to the previous proposition, we have $\frac{dM_R}{dt} = 0$. Since $M_R = R^*_g M = R^*_g L^*_{g^{-1}} M_L = gM_L g^{-1}$, then $M_R = gM_L g^{-1}$. Indeed, we have $(L^*_g \xi, \eta) = (\xi g, \eta)$ and $(L^*_g \xi, \eta) = (\xi, L_{g*} \eta) = (\xi, g\eta) = -\frac{1}{2} tr(\xi g \eta) = (\xi g, \eta)$, so $L^*_g \xi = \xi g$. Similarly, we have $(R^*_g \xi, \eta) = (g\xi, \eta)$ and $(R^*_g \xi, \eta) = (\xi, R_{g*} \eta) = (\xi, \eta g) = -\frac{1}{2} tr(\xi \eta g) = -\frac{1}{2} tr(g\xi \eta) = (g\xi, \eta)$, so $R^*_g \xi = g\xi$. Thus, $R^*_g L^*_{g^{-1}} M_L = gM_L g^{-1}$. Since $(g^{-1})\cdot = g^{-1} \dot{g} g^{-1}$, then,

$$0 = \frac{dM_R}{dt} = \dot{g} M_L g^{-1} + g\dot{M}_L g^{-1} - gM_L g^{-1} \dot{g} g^{-1},$$
$$= g(g^{-1} \dot{g} M_L + \dot{M}_L - M_L g^{-1} \dot{g}) g^{-1} = g(\omega_L M_L + \dot{M}_L - M_L \omega_L) g^{-1},$$

which implies $\omega_L M_L + \dot{M}_L - M_L \omega_L = 0$. So, $\dot{M}_L = \frac{dM_L}{dt} = [M_L, \omega_L]$. □

REMARK 3.7.– We have shown that geodesic flow on G equipped with a left or right invariant Riemannian metric is governed by the Euler–Arnold equation: $\dot{M}_L = [M_L, \omega_L]$. This equation defines a vector field on $\mathcal{G}$ and should be solved first for M_L, which also gives ω_L. This is the one that has been studied in some problems in various sections. Also, we have seen another equation above (which we will not need), namely $\dot{g} = g\omega$, which becomes a linear non-autonomous and can be solved for g, giving the geodesic. The Euler–Lagrange equations of the problem study in example 3.2, are given by $\dot{M}_L = [M_L, \omega_L]$ and $\dot{g} = g\omega$, so if the metric is bi-invariant (i.e. is both left and right-invariant), then ω_L is constant and geodesics are one-parameter subgroups.

We are interested here in the case of the geodesic flow on the group of rotations because this is related to several questions treated in several sections. But let us point out that some other conservative systems of hydrodynamical type can be written as geodesic equations on the group of diffeomorphisms, or the group of volume preserving diffeomorphisms of a Riemannian manifold, as well as on extensions of these groups. Examples of such Euler–Arnold equations include the Burgers's equation, Korteweg–de Vries, the Camassa–Holm shallow water equation, the equations of motion of an ideal fluid in the Yang–Mills field, the superconductivity equation, the equations of ideal magneto-hydrodynamics, the template matching equation, the superconductivity equation, etc. Another interesting example (mentioned in section 4.6) concerns the geodesic flow of the ellipsoid. This problem was solved by Jacobi (1969) by the separation of variables in elliptic coordinates. Moser (1980) found a Lax pair representation and commuting integrals in Euclidian coordinates by using the geometry of quadrics and the geometric interpretation of the integrability is described by Chasles; the tangent line of a geodesic on the ellipsoid is also tangent to a fixed set of confocal quadrics (see Knörrer 1980). The Neumann problem (section 4.6) was shown by Moser (1980) to contain, in particular, the geodesic motion on an ellipsoid, which was found to be integrable by Jacobi and as mentioned in section 4.6, there is a remarkable correspondence between the Neumann system and the geodesic flow on the ellipsoid via the Gauss mapping (Knörrer 1982).

3.6. Exercises

EXERCISE 3.1.– Prove (as in the Euler problem of a rigid body) that in the Lagrange top, the integration is done using elliptic functions.

EXERCISE 3.2.– Study, in detail, the system of n-independent harmonic oscillators whose evolution is governed by the Hamiltonian $H = \sum_{i=1}^{n} \frac{1}{2} \left(p_i^2 + \omega_i^2 q_i^2 \right)$, with Poisson bracket $\{q_i, q_j\} = \{p_i, p_j\} = 0, \{p_i, q_j\} = \delta_{ij}$.

EXERCISE 3.3.– Complete the proof of theorem 3.7 by showing that the equations of the problem are integrable by quadratures and construct action-angle variables for systems with phase space $\mathbb{R}^{2n}$, and also for a system on an arbitrary symplectic manifold (hint: see Arnold (1989)).

EXERCISE 3.4.– Consider the (two-body) Kepler problem (see section 2.6) of planetary motion whose equations of motions (in Cartesian coordinates) are given by $\frac{\partial^2 x_i}{\partial t^2} + \frac{\partial V(r)}{\partial x_i} = 0, r = \|x\| = \sqrt{x_1^2 + x_2^2 + x_3^2}$ (as in the traditional Kepler problem, we can consider the corresponding scalar potential $V(r) = \frac{k}{r}$, where k is a constant). These equations can be expressed in Hamiltonian form with $H = \frac{1}{2}\left(p_1^2 + p_2^2 + p_3^2\right) + V(r)$, the Hamiltonian and with respect to Poisson brackets $\{p_i, x_i\} = \delta_{ij}$. The phase space is the cotangent bundle T^*S^2. Determine two additional first integrals commuting with H and express them, as well as the Hamiltonian in terms of spherical coordinates: $x_1 = r\sin\theta\cos\varphi$, $x_2 = r\sin\theta\sin\varphi$, $x_3 = r\cos\varphi$. Explain how this Hamiltonian system is integrable and give a geometric description of the results.

EXERCISE 3.5.– A spherical pendulum is a simple pendulum in $3D$ that oscillates on a sphere $S^2 \subset \mathbb{R}^3$ of radius centered on the axis of the pendulum. Let $T^*\mathbb{R}^3 \cong T\mathbb{R}^3 \cong \mathbb{R}^3 \times \mathbb{R}^3$ be endowed with the standard symplectic structure. Let us see S^2 as a submanifold of $\mathbb{R}^3$ with the canonical basis (e_1, e_2, e_3). The spherical pendulum has the phase space $TS^2 = \{(q,p) \in T\mathbb{R}^3 : \|q\| = 1, \langle q, p \rangle = 0\}$, and we treat the spherical pendulum as a Hamiltonian system, where $H(q,p) = \frac{1}{2}\|p\|^2 + \langle q, e_3\rangle$ is the Hamiltonian. Show that the spherical pendulum is a Liouville integrable Hamiltonian system.

EXERCISE 3.6.– Prove that the general integral of Euler's equation $\frac{\dot{x}}{\sqrt{F(x)}} \pm \frac{\dot{y}}{\sqrt{F(y)}} = 0$, $F(x) = a_0 x^4 + 4a_1 x^3 + 6a_2 x^2 + 4a_3 x + a_4$ can be written in two different ways: $F_1(x,y) + 2sF(x,y) - s^2(x-y)^2 = 0$, where $F(x,y) = a_0 x^2 y^2 + 2a_1 xy(x+y) + 3a_2(x^2+y^2) + 2a_3(x+y) + a_4$, and $F_1(x,y) = \frac{F(x)F(y) - F^2(x,y)}{(x-y)^2}$, or in an irrational form $\frac{F(x,y) \mp \sqrt{F(x)}\sqrt{F(y)}}{(x-y)^2} = s$ (hint: see Halphen (1888)).

EXERCISE 3.7.– Consider the following equations (Dubrovin 1981),

$$\ddot{q}_1 - q_2 - \frac{5}{2}q_1^2 = 0, \qquad \ddot{q}_2 - 5q_1 q_2 - \frac{5}{2}q_1^3 + aq_1 + b = 0,$$

corresponding to the Hamiltonian $H_1 = p_1 p_2 - \frac{5}{8}q_1^4 - \frac{5}{2}q_1^2 q_2 + \frac{a}{2}q_1^2 + bq_1 - \frac{1}{2}q_2^2$, where a and b are constants. Show that this system is integrable in the sense of Liouville, the second first integral being $H_2 = q_1^5 + aq_1^3 - 4q_1 q_2^2 + 2q_1 p_1 p_2 + (2q_2 - a)p_2^2 + 2aq_1 q_2 + p_1^2 + 2bq_2$, and the problem can be integrated in terms of genus two hyperelliptic functions.

EXERCISE 3.8.– Let X_H be the vector field on $\mathbb{R}^6$ defined by (see example 1.13 and section 3.3 for further information):

$$\dot{p} = p \wedge \frac{\partial H}{\partial l}, \qquad \dot{l} = p \wedge \frac{\partial H}{\partial p} + l \wedge \frac{\partial H}{\partial l}, \qquad [3.38]$$

where $p = (p_1, p_2, p_3) \in \mathbb{R}^3$, $l = (l_1, l_2, l_3) \in \mathbb{R}^3$, $\wedge$ is the vector product in $\mathbb{R}^3$, H is the Hamiltonian, $\frac{\partial H}{\partial p} = \left(\frac{\partial H}{\partial p_1}, \frac{\partial H}{\partial p_2}, \frac{\partial H}{\partial p_3}\right)$ and $\frac{\partial H}{\partial l} = \left(\frac{\partial H}{\partial l_1}, \frac{\partial H}{\partial l_2}, \frac{\partial H}{\partial l_3}\right)$.

a) Show that the vector field X_H is tangent to the affine variety $\mathcal{A}$ defined by $H_1 \equiv p_1^2 + p_2^2 + p_3^2 = c_1$, $H_2 \equiv p_1 l_1 + p_2 l_2 + p_3 l_3 = c_2$, where c_1, c_2 are generic constants.

b) Show that X_H is a Hamiltonian vector field on the variety $\mathcal{A}$ for the symplectic 2-form: $\omega = \frac{1}{p_3}(dp_1 \wedge dl_2 - dp_2 \wedge dl_1) - \frac{l_3}{p_3^2} dp_1 \wedge dp_2$.

c) Show that the Hamiltonian vector fields X_{H_1}, X_{H_2} generated by H_1 and H_2 with respect to the symplectic structure ω are identically zero.

d) Let $H = H_3 \equiv \alpha_1 p_1^2 + \alpha_2 p_2^2 + \alpha_3 p_3^2 + l_1^2 + l_2^2 + l_3^2$, and $H_4 \equiv \alpha_1 l_1^2 + \alpha_2 l_2^2 + \alpha_3 l_3^2 - \alpha_2 \alpha_3 p_1^2 - \alpha_1 \alpha_3 p_2^2 - \alpha_1 \alpha_2 p_3^2$, where $\alpha_1, \alpha_2, \alpha_3$ denote arbitrary numbers. Show that the Hamiltonian vector fields X_{H_3}, X_{H_4} generated by H_3 and H_4 with respect to the symplectic structure ω, commute. Deduce that for generic constants of c_1, c_2, c_3, c_4 the variety: $M_c = \{(p, l) : H_1 = c_1, H_2 = c_2, H_3 = c_3, H_4 = c_4\}$, $c = (c_1, c_2, c_3, c_4)$ is a torus of dimension 2.

EXERCISE 3.9.– Prove the assertion of section 3.3.2 that the Lyapunov–Steklov's case of the problem of motion of a solid in an ideal fluid can be integrated in terms of genus two hyperelliptic functions.

EXERCISE 3.10.– Let $H = \frac{1}{2}\left(y_1^2 + y_2^2\right) + \frac{1}{2}x_1 x_2^2 + x_1^3 + \frac{\alpha}{8x_2^3}$ be a Hamiltonian where α is a free constant parameter, and x_1, x_2, y_1, y_2 are canonical coordinates and moments, respectively, satisfying the usual canonical Poisson bracket. Study the integrability of the system corresponding to this Hamiltonian, while knowing that a second first integral is $F = x_1 y_2^2 - x_2 y_1 y_2 - \frac{1}{2}x_1^2 x_2^2 - \frac{1}{8}x_2^4 + \frac{\alpha x_1}{4x_2^2}$.

4

Spectral Methods for Solving Integrable Systems

The aim of this chapter is to present an overview of the active area via the spectral linearization methods for solving integrable systems. New examples of completely integrable Hamiltonian systems, which have been discovered, are based on the so-called Lax representation (Lax pairs) of the equations of motion. We will explain how these systems can be realized as straight line motions on a Jacobi variety of a spectral curve. These methods are exemplified by several problems of integrable systems of relevance in mathematics and mathematical physics: we study the geodesic flow on the group $SO(n)$ and more particularly, on $SO(4)$. We are also interested in the study of Euler, Lagrange, Kowalewski and Goryachev–Chaplygin spinning tops, Jacobi geodesic flow on an ellipsoid and the Neumann problem. Other important examples include a family of integrable systems, the coupled nonlinear Schrödinger equations, the Yang–Mills equations and the periodic infinite band matrix.

4.1. Lax equations and spectral curves

A Lax equation is given by a differential equation of the form

$$\dot{A}(h,t) = [A(h,t), B(h,t)] \text{ or } [B(h,t), A(h,t)], \quad \left(\cdot \equiv \frac{d}{dt}\right) \qquad [4.1]$$

where $A(h,t) = \sum_{k=l}^{m} A_k(t)h^k$, $B(h,t) = \sum_{k=l}^{m} B_k(t)h^k$ are functions depending on a parameter h (spectral parameter), whose coefficients A_k and B_k are matrices in Lie algebras. The pair (A, B) is called a Lax pair. Thereafter, the notations $A(h,t)$, $A(h)$, A_h, $A(t)$ or simply A will be used indifferently. The bracket $[,]$ is the usual Lie bracket of matrices. We will also have to consider Lax equations where A and B are differential operators. Equation [4.1] establishes a link between the theoretical Lie group and the algebraic geometric approaches to complete integrability.

THEOREM 4.1.– The polynomial $P(h,z)$ is independent of t. Moreover, the functions $tr\,(A^n)$ are first integrals for [4.1].

PROOF.– We have $\dot{P} = \det C.tr\left(C^{-1}\dot{C}\right) = \det C.tr\left(C^{-1}BC - B\right) = 0$, where $C \equiv A - zI$ and because $trC^{-1}BC = trB$. On the other hand

$$\begin{aligned}\dot{A}^n &= \dot{A}A^{n-1} + A\dot{A}A^{n-2} + \cdots + A^{n-1}\dot{A},\\ &= [A,B]A^{n-1} + A[A,B]A^{n-2} + \cdots + A^{n-1}[A,B],\\ &= (AB - BA)A^{n-1} + \cdots + A^{n-1}(AB - BA),\\ &= ABA^{n-1} - BA^n + \cdots + A^nB - A^{n-1}BA,\\ &= A\left(BA^{n-1}\right) - \left(BA^{n-1}\right)A + \cdots + A\left(A^{n-1}B\right) - \left(A^{n-1}B\right)A.\end{aligned}$$

Since $tr(X + Y) = trX + trY$, $trXY = trYX$, $X, Y \in \mathcal{M}_n(\mathbb{C})$, we obtain $\frac{d}{dt}tr\,(A^n) = tr\frac{d}{dt}(A^n) = 0$, and $tr\,(A^n)$ are first integrals of motion. □

Another proof can be obtained as follows. we show that if $A(t)$ is a solution of the Cauchy problem for equation [4.1] with initial condition $A(0)$, there exists a smooth curve $g(t) \in GL(n,\mathbb{C})$ such that: $g(0) = I$ and the solution to [4.1] has the form $A(t) = g(t)A(0)g(t)^{-1}$. The evolution of $g(t)$ is determined by solving the initial value problem for the equation: $\dot{g}(t) = -A(t)g(t)$. The characteristic polynomials of $A(0)$ and $g(t)A(0)g(t)^{-1}$ are the same, which shows that the eigenvalues are the same. Thus, the constants of the movement are easily deduced.

We form the polynomial $P(h,z) = \det(A - zI)$ or $\det(zI - A)$, where z is another variable and I is the $n \times n$ identity matrix. We define the curve (spectral curve) to be the normalization of the complete algebraic curve whose affine equation is

$$P(h,z) = 0. \qquad [4.2]$$

4.2. Integrable systems and Kac–Moody Lie algebras

We have shown that a Hamiltonian flow of the type [4.1] preserves the spectrum of A and its characteristic polynomial. The curve $\mathcal{C} : P(z,h) = \det(A(h) - zI) = 0$ is time independent, that is, its coefficients $tr\,(A^n)$ are integrals of the motion (equivalently, $A(t)$ undergoes an isospectral deformation). Some Hamiltonian flows on Kostant–Kirillov coadjoint orbits in subalgebras of infinite dimensional Lie algebras (Kac–Moody Lie algebras) yield large classes of extended Lax pairs. A general statement leading to such situations is given by the Adler–Kostant–Symes theorem, which we will study in the following (for more information, one can consult, for example, (Adler 1979; Kostant 1979; Symes 1980; Adler and van Moerbeke 2004; Lesfari 2010)).

THEOREM 4.2.– Let $\mathcal{L}$ be a Lie algebra paired with itself via a non-degenerate, ad-invariant bilinear form $\langle , \rangle$, $\mathcal{L}$ having a vector space decomposition $\mathcal{L} = \mathcal{K} + \mathcal{N}$ with $\mathcal{K}$ and $\mathcal{N}$ Lie subalgebras. Then, with respect to $\langle , \rangle$, we have the splitting $\mathcal{L} = \mathcal{L}^* = \mathcal{K}^\perp + \mathcal{N}^\perp$ and $\mathcal{K}^\perp \approx \mathcal{N}^* (\equiv$ the dual of $\mathcal{N}$) paired with $\mathcal{N}$ via an induced form $\langle\langle , \rangle\rangle$ that inherits the coadjoint symplectic structure of Kostant and Kirillov; its Poisson bracket between functions H_1 and H_2 on $\mathcal{N}^*$ reads $\{H_1, H_2\}(a) = \langle\langle a, [\nabla_{\mathcal{N}^*} H_1, \nabla_{\mathcal{N}^*} H_2]\rangle\rangle$, $a \in \mathcal{N}^*$. Let $V \subset \mathcal{N}^*$ be an invariant manifold under the above coadjoint action of $\mathcal{N}$ on $\mathcal{N}^*$ and let $\mathcal{A}(V)$ be the algebra of functions defined on a neighborhood of V, invariant under the coadjoint action of $\mathcal{L}$ (which is distinct from the $\mathcal{N} - \mathcal{N}^*$ action). Then the functions H in $\mathcal{A}(V)$ lead to commuting Hamiltonian vector fields of the Lax isospectral form $\dot{a} = [a, pr_{\mathcal{K}}(\nabla H)]$, $pr_\mathcal{K}$ projection onto $\mathcal{K}$.

PROOF.– The gradient ∇H of a function H on a vector space E is defined by $dH = (\nabla H, dv)_V$, $v \in E$, $\nabla H \in E^*$, where E^* is the dual of E and $(.,.)_V$ is the pairing between E, E^*. As mentioned above, $\nabla H \in (\mathcal{L}^*)^* = \mathcal{L}$ is the gradient of H when viewed as a function of $\mathcal{L}^* (\approx \mathcal{L})$, while, in general, we have that $pr_\mathcal{K}$, $pr_\mathcal{N}$, $pr_{\mathcal{K}^\perp}$, $pr_{\mathcal{N}^\perp}$ are, respectively, the projections onto $\mathcal{K}$, $\mathcal{N}$, $\mathcal{K}^\perp$, $\mathcal{N}^\perp$ along $\mathcal{K}^\perp$, $\mathcal{K}^\perp$, $\mathcal{N}^\perp$, $\mathcal{K}^\perp$. Note that if $H \in \mathcal{L}^* \approx \mathcal{L}$, then $\nabla_{\mathcal{K}^\perp} H = pr_\mathcal{N}(\nabla H)$, $\nabla_{\mathcal{N}^\perp} H = pr_\mathcal{K}(\nabla H)$. Let $V \subset \mathcal{K}^\perp$ be an invariant manifold under the coadjoint action of $\mathcal{N}$ on $\mathcal{K}^\perp \approx \mathcal{N}^*$. From the identity $\frac{d}{dt} H\left(Ad_{g(t)}(a)\right)\big|_{t=0} = 0$, where $g(t) = 1 + bt + o(t)$, $b \in \mathcal{L}$, $a \in V$, we deduce the relation $[\nabla H(a), a] = 0$, $a \in V$, or what amounts to the same

$$[a, \nabla_{\mathcal{K}^\perp} H] = -[a, pr_\mathcal{K}(\nabla H)]. \qquad [4.3]$$

The Poisson bracket between two functions H_1 and H_2 on $\mathcal{N}^*$ is written as $\{H_1, H_2\}(a) = \langle\langle a, [\nabla_{\mathcal{N}^*} H_1, \nabla_{\mathcal{N}^*} H_2]\rangle\rangle$, $a \in \mathcal{N}^*$, where $\langle\langle .,. \rangle\rangle$ is the natural pairing between $\mathcal{N}$ and $\mathcal{N}^*$, and where $\nabla_{\mathcal{N}^*} H_1 \in \mathcal{N}$ is the natural gradient of H defined by $dH_1(X) = \langle\langle dX, \nabla_{\mathcal{N}^*} H_1, \rangle\rangle$. Since $\mathcal{K}^\perp \approx \mathcal{N}^*$ and $\langle\langle .,. \rangle\rangle = \langle .,. \rangle|_{\mathcal{K}^\perp \times \mathcal{N}}$,

$$\{H_1, H_2\}(a) = \langle a, [\nabla_{\mathcal{K}^\perp} H_1, \nabla_{\mathcal{K}^\perp} H_2] \rangle. \qquad [4.4]$$

Suppose that $H_1, H_2 \in \mathcal{A}(V)$ satisfy the relation [4.3]. Then, by virtue of the relations [4.3] and [4.4] and the fact that $\langle .,. \rangle$ is ad-invariant, we obtain

$$\{H_1, H_2\} = \langle [a, \nabla_{\mathcal{K}^\perp} H_1], \nabla_{\mathcal{K}^\perp} H_2 \rangle = -\langle [a, pr_\mathcal{K} H_1], \nabla_{\mathcal{K}^\perp} H_2 \rangle,$$
$$= -\langle a, [pr_\mathcal{K} H_1, \nabla_{\mathcal{K}^\perp} H_2] \rangle.$$

Using a similar reasoning for H_2, we get $\{H_1, H_2\} = \langle a, [pr_\mathcal{K} H_1, \nabla_{\mathcal{K}^\perp} H_2] \rangle$. Since $\mathcal{K}$ is a Lie algebra and $a \in \mathcal{K}^\perp$, we obtain $\{H_1, H_2\} = 0$. The Hamiltonian vector field is written as $X_{H_1}(H_2) = \{H_1, H_2\} = \langle [\nabla_{\mathcal{K}^\perp} H_1, a], \nabla_{\mathcal{K}^\perp} H_2 \rangle$, and we have $X_{H_1}(a) = pr_{\mathcal{K}^\perp}[\nabla_{\mathcal{K}^\perp} H_1, a]$. Consequently, the corresponding Hamiltonian flow is $\dot{a} = pr_{\mathcal{K}^\perp}[\nabla_{\mathcal{K}^\perp} H_1, a]$, $H_1 \in \mathcal{A}(V)$, and from [4.3], we have $\dot{a} = pr_{\mathcal{K}^\perp}[a, \nabla_{\mathcal{K}^\perp} H_1]$. Now $[\mathcal{K}^\perp, \mathcal{K}] \subset \mathcal{K}^\perp$, so we have $\dot{a} = [a, \nabla_{\mathcal{K}^\perp} H_1]$. □

This theorem produces Hamiltonian systems with many commuting integrals; some precise results are known for interesting classes of orbits in both the case of finite and infinite dimensional Lie algebras. Paradoxically, the finite-dimensional Lie algebras usually lead to non-compact systems, and the infinite-dimensional ones to compact systems. Any finite dimensional Lie algebra $\mathcal{L}$ with bracket $[,]$ and Killing form $\langle,\rangle$ leads to an infinite dimensional formal Laurent series extension $\mathcal{L} = \sum_{-\infty}^{m} A_i h^i : A_i \in \mathcal{L}, m \in \mathbb{Z}$ free, with bracket $\left[\sum A_i h^i, \sum B_j h^j\right] = \sum_{i,j} [A_i, B_j] h^{i+j}$, and ad-invariant, symmetric forms $\left\langle \sum A_i h^i, \sum B_j h^j \right\rangle_k = \sum_{i+j=-k} \langle A_i, B_j \rangle$, depending on $k \in \mathbb{Z}$. The forms $\langle,\rangle_k$ are non-degenerate if $\langle,\rangle$ is so. Let $\mathcal{L}_{p,q}$ ($p \leq q$) be the vector subspace of $\mathcal{L}$, corresponding to powers of h between p and q. A first interesting class of problems is obtained by taking $\mathcal{L} = \mathcal{G}l(n, \mathbb{R})$ and by putting the form $\langle,\rangle_1$ on the Kac–Moody extension. Then, we have the decomposition into Lie subalgebras $\mathcal{L} = \mathcal{L}_{0,\infty} + \mathcal{L}_{-\infty,-1} = \mathcal{K} + \mathcal{N}$, with $\mathcal{K} = \mathcal{K}^\perp, \mathcal{N} = \mathcal{N}^\perp$ and $\mathcal{K} = \mathcal{N}^*$. Consider the invariant manifold V_m, $m \geq 1$ in $\mathcal{K} = \mathcal{N}^*$, defined as $V_m = \left\{ A = \sum_{i=1}^{m-1} A_i h^i + \alpha h^m, \ \alpha = diag(\alpha_1, \cdots, \alpha_n)\right.$ fixed$\}$, with $diag(A_{m-1}) = 0$. We state the following theorem (Adler and van Moerbeke 1980):

THEOREM 4.3.– The manifold V_m has a natural symplectic structure, and the functions $H = \left\langle f(Ah^{-j}), h^k \right\rangle_1$ on V_m for good functions f lead to complete integrable commuting Hamiltonian systems of the form $\dot{A} = [A, pr_\mathcal{K}(f'(Ah^{-j})h^{k-j})], A = \sum_{i=0}^{m-1} A_i h^i + \alpha h^m$, and their trajectories are straight line motions on the Jacobian of the curve $\mathcal{C}$ of genus $(n-1)(nm-2)/2$, defined by $P(z,h) = \det(A - zI) = 0$. The coefficients of this polynomial provide the orbit invariants of V_m and an independent set of integrals of the motion (the flows where $j = m, k = m + 1$ are of particular interest, and have the following form

$$\dot{A} = [A, \, ad_\beta \, ad_\alpha^{-1} A_{m-1} + \beta h], \ \beta_i = f'(\alpha_i), \qquad [4.5]$$

the flow only depends on f through the relation $\beta_i = f'(\alpha_i)$).

Another class is obtained by choosing any semi-simple Lie algebra L. Then, the Kac–Moody extension $\mathcal{L}$ equipped with the form $\langle,\rangle = \langle,\rangle_0$ has the natural level decomposition $\mathcal{L} = \sum_{i \in \mathbb{Z}} L_i, [L_i, L_j] \subset L_{i+j}, [L_0, L_0] = 0, L_i^* = L_{-i}$. Let $B^+ = \sum_{i \geq 0} L_i, B^- = \sum_{i < 0} L_i$. Then the product Lie algebra $\mathcal{L} \times \mathcal{L}$ has the following bracket $\left[(l_1, l_2), (l'_1, l'_2)\right] = \left([l_1, l'_1], -[l_2, l'_2]\right)$, and pairing $\left\langle (l_1, l_2), (l'_1, l'_2) \right\rangle = \langle l_1, l'_1 \rangle - \langle l_2, l'_2 \rangle$. It admits the decomposition into $\mathcal{K} + \mathcal{N}$ with $\mathcal{K} = \{(l, -l) : l \in \mathcal{L}\}, \mathcal{K}^\perp = \{(l,l) : l \in \mathcal{L}\}, \mathcal{N} = \{(l_-, l_+) : l_- \in B^-, l_+ \in B^+, pr_0(l_-) = pr_0(l_+)\}, \quad \mathcal{N}^\perp = \{(l_-, l_+) : l_- \in B^-, \ l_+ \in B^+, pr_0(l_+ + l_-) = 0\}$, where pr_0 denotes projection onto L_0. Then from the last theorem, the orbits in $\mathcal{N}^* = \mathcal{K}^\perp$ possess a lot of commuting Hamiltonian vector fields of Lax form.

The following result (the van Moerbeke–Mumford (1979) linearization method) can be thought of as an example of theorem 4.3.

THEOREM 4.4.– The N-invariant manifold $V_{-j,k} = \sum_{-j \leq i \leq k} L_i \subseteq \mathcal{L} \simeq \mathcal{K}^\perp$ has a symplectic structure and the functions $H(l_1, l_2) = f(l_1)$ on $V_{-j,k}$ lead to commuting vector fields of the Lax form $\dot{l} = \left[l, (pr^+ - \frac{1}{2}pr_0)\nabla H(l)\right]$, pr^+ projection onto B^+, and their trajectories are straight line motions on the Jacobian of a curve defined by the characteristic polynomial of elements in $V_{-j,k}$, which are thought of as functions of h (where $\nabla H(l) \in \mathcal{N}$ is the gradient of H, which is thought of as a function on $\mathcal{L}$).

Using the van Moerbeke–Mumford (1976) linearization method Adler and van Moerbeke (1980b) showed that the linearized flow could be realized on the Jacobian variety $Jac(\mathcal{C})$ (or some subabelian variety of it) of the algebraic curve (spectral curve) $\mathcal{C}$ associated with [4.1]. We then construct an algebraic map from the complex invariant manifolds of these Hamiltonian systems to the Jacobian variety $Jac(\mathcal{C})$ of the curve $\mathcal{C}$. Therefore, all of the complex flows generated by the constants of the motion are straight line motions on these Jacobian varieties, that is the linearizing equations are given by $\int_{s_1(0)}^{s_1(t)} \omega_k + \int_{s_2(0)}^{s_2(t)} \omega_k + \cdots + \int_{s_g(0)}^{s_g(t)} \omega_k = c_k t$, $0 \leq k \leq g$, where $\omega_1, \ldots, \omega_g$ span the g-dimensional space of holomorphic differentials on the curve $\mathcal{C}$ of genus g. In an unifying approach, (Griffiths 1985) found necessary and sufficient conditions on B for the Lax flow [4.1] to be linearizable on the Jacobi variety of its spectral curve, without reference to Kac–Moody Lie algebras (for further details, see Chapter 6).

4.3. Geodesic flow on $SO(n)$

In the particular case $m = 1$, that is, for V_1 (theorem 4.3), we choose $A = X + \alpha h$, $X \in so(n)$. In this case, the Hamiltonian flow described by equation [4.5] (where α_i and β_i can be taken arbitrarily) is reduced to the study of the Euler–Arnold equation for the geodesic flow on $SO(n)$, $\dot{X} = [X, \lambda X]$, $(\lambda X)_{ij} = \lambda_{ij} X_{ij}$, $\lambda_{ij} = \frac{\beta_i - \beta_j}{\alpha_i - \alpha_j}$, for a left-invariant diagonal metric $\sum \lambda_{ij} X_{ij}$. The natural phase space for this motion is an orbit defined in $SO(n)$ by $[n/2]$ orbit invariants. And according to theorem 2.3, the problem is completely integrable and the trajectories are straight lines on $Jac(\mathcal{C})$ of dimension $(n-2)(n-1)/2$, and more specifically, on the Prym variety $Prym(\mathcal{C}/\mathcal{C}_0) \subset Jac(\mathcal{C})$ of dimension $(n(n-1)/2 - [n/2])/2$ induced by the natural involution $\mathcal{C} \longrightarrow \mathcal{C}$, $(z, h) \longmapsto (-z, -h)$, on $\mathcal{C}$ as a result of $X \in so(n)$; $\mathcal{C}_0$ is the curve obtained by identifying (z, h) with $(-z, -h)$. This problem is studied in greater detail for $n = 3$ and $n = 4$ in several sections through various methods.

4.4. The Euler problem of a rigid body

We shall use the Lax representation of the equations of motion to show that the linearized Euler flow can be realized on an elliptic curve isomorphic to the original elliptic curve [3.11]. The solution to equation [3.5], that is,

$$\dot{M} = [M, \Lambda M], \text{ or } \dot{M} = [M, \Omega], \qquad [4.6]$$

has the form $M(t) = O(t)M(t)M^\top(t)$, where $O(t)$ is one parameter sub-group of $SO(3)$. The Hamiltonian flow [4.6] preserves the spectrum of X, and therefore its characteristic polynomial $\det(M - zI) = -z\left(z^2 + m_1^2 + m_2^2 + m_3^2\right)$. Unfortunately, the spectrum of a 3×3 skew-symmetric matrix only provides one piece of information; the conservation of energy does not appear as part of the spectral information. The basic observation, due to Manakov (1976), is that equation [4.6] is equivalent to the Lax equation $\dot{A} = [A, B]$, where $A = M + \alpha h$, $B = \Lambda M + \beta h$, with a formal indeterminate h and

$$\alpha = \begin{pmatrix} \alpha_1 & 0 & 0 \\ 0 & \alpha_2 & 0 \\ 0 & 0 & \alpha_3 \end{pmatrix}, \quad \beta = \begin{pmatrix} \beta_1 & 0 & 0 \\ 0 & \beta_2 & 0 \\ 0 & 0 & \beta_3 \end{pmatrix},$$

$$\lambda_1 = \frac{\beta_3 - \beta_2}{\alpha_3 - \alpha_2}, \quad \lambda_2 = \frac{\beta_1 - \beta_3}{\alpha_1 - \alpha_3}, \quad \lambda_3 = \frac{\beta_2 - \beta_1}{\alpha_2 - \alpha_1},$$

and all α_i distinct. The characteristic polynomial of A is

$$P(h, z) = \det(A - zI) = \det(M + \alpha h - zI),$$

$$= \prod_{j=1}^{3}(\alpha_j h - z) + \left(\sum_{j=1}^{3} \alpha_j m_j^2\right) h - \left(\sum_{j=1}^{3} m_j^2\right) z. \qquad [4.7]$$

The spectrum of the matrix $A = M + \alpha h$ as a function of $h \in \mathbb{C}$ is time independent, and it is given by the zeroes of the polynomial $P(h, z)$ [4.7], thus defining an algebraic curve. By setting $w = h/z$, we obtain the following elliptic curve: $z^2 \prod_{j=1}^{3}(\alpha_j w - 1) + 2H_1 w - 2H_2 = 0$, which is shown to be isomorphic to the original elliptic curve [3.11]. Finally, we have the following theorem:

THEOREM 4.5.– *The Euler rigid body motion is a completely integrable system and the linearized flow can be realized on an elliptic curve.*

4.5. The Manakov geodesic flow on the group $SO(4)$

We consider the group $SO(4)$ and its Lie algebra $so(4)$ paired with itself, via the customary inner product $\langle X, Y \rangle = -\frac{1}{2} tr\,(X.Y)$, where

$$X = (X_{ij})_{1 \leq i,j \leq 4} = \sum_{i=1}^{6} x_i e_i = \begin{pmatrix} 0 & -x_3 & x_2 & -x_4 \\ x_3 & 0 & -x_1 & -x_5 \\ -x_2 & x_1 & 0 & -x_6 \\ x_4 & x_5 & x_6 & 0 \end{pmatrix} \in so(4).$$

A left invariant metric on $SO(4)$ is defined by a non-singular symmetric linear map $\Lambda : so(4) \longrightarrow so(4)$, $X \longmapsto \Lambda.X$, and by the following inner product; given two vectors gX and gY in the tangent space $SO(4)$ at the point $g \in SO(4)$, $\langle gX, gY \rangle = \langle X, \Lambda^{-1}.Y \rangle$, regardless of g. Then the geodesic flow for this metric takes the following commutator form (Euler–Arnold equation):

$$\dot{X} = [X, \Lambda.X], \quad \cdot \equiv \frac{d}{dt} \quad [4.8]$$

where

$$\Lambda.X = (\lambda_{ij} X_{ij})_{1 \leq i,j \leq 4} = \sum_{i=1}^{6} \lambda_i x_i e_i = \begin{pmatrix} 0 & -\lambda_3 x_3 & \lambda_2 x_2 & -\lambda_4 x_4 \\ \lambda_3 x_3 & 0 & -\lambda_1 x_1 & -\lambda_5 x_5 \\ -\lambda_2 x_2 & \lambda_1 x_1 & 0 & -\lambda_6 x_6 \\ \lambda_4 x_4 & \lambda_5 x_5 & \lambda_6 x_6 & 0 \end{pmatrix}.$$

In view of the isomorphism (lemma 1.2) between $(\mathbb{R}^6, \wedge)$ and $(so(4), [,])$, we write the system [4.8] as

$$\begin{aligned} \dot{x}_1 &= (\lambda_3 - \lambda_2)\, x_2 x_3 + (\lambda_6 - \lambda_5)\, x_5 x_6, \\ \dot{x}_2 &= (\lambda_1 - \lambda_3)\, x_1 x_3 + (\lambda_4 - \lambda_4)\, x_4 x_6, \\ \dot{x}_3 &= (\lambda_2 - \lambda_1)\, x_1 x_2 + (\lambda_5 - \lambda_4)\, x_4 x_5, \qquad [4.9] \\ \dot{x}_4 &= (\lambda_3 - \lambda_5)\, x_3 x_5 + (\lambda_6 - \lambda_2)\, x_2 x_6, \\ \dot{x}_5 &= (\lambda_4 - \lambda_3)\, x_3 x_4 + (\lambda_1 - \lambda_6)\, x_1 x_6, \\ \dot{x}_6 &= (\lambda_2 - \lambda_4)\, x_2 x_4 + (\lambda_5 - \lambda_1)\, x_1 x_5. \end{aligned}$$

This flow is Hamiltonian with regard to the usual Kostant–Kirillov symplectic structure induced on the orbit $\mathcal{O}^*_{SO(4)} = \{Ad^*_g(X) = g^{-1} X g : g \in SO(4)\}$, formed by the coadjoint action $Ad^*_g(X)$ of the group $SO(4)$ on the dual Lie algebra $so(4)^* \approx so(4)$. Let $z_1, z_2 \in so(4)$ and $\xi_1 = [X, z_1], \xi_2 = [X, z_2]$ be two tangent vectors to the orbit at the point $X \in so(4)$. Then the symplectic structure is defined by $\omega(X)(\xi_1, \xi_2) = \langle X, [z_1, z_2] \rangle$. This orbit is four-dimensional and is defined by setting two trivial quadratic invariants H_1 and H_2 equal to generic constants c_1 and c_2:

$$H_1 = \sqrt{\det X} = x_1 x_4 + x_2 x_5 + x_3 x_6 = c_1, \qquad [4.10]$$

$$H_2 = -\frac{1}{2} tr\left(X^2\right) = x_1^2 + x_2^2 + \cdots + x_6^2 = c_2. \qquad [4.11]$$

Functions H defined on the orbit lead to the Hamiltonian vector fields $\dot{X} = [X, \nabla H]$. In particular

$$H = \frac{1}{2} \langle X, \Lambda X \rangle = \frac{1}{2}\left(\lambda_1 x_1^2 + \lambda_2 x_2^2 + \cdots + \lambda_6 x_6^2\right), \qquad [4.12]$$

induces geodesic motion [4.8]. The constants of the motion are given by the Hamiltonian H[4.12] and the two quadratic invariants H_1[4.10], H_2[4.11]. Since the system in question is Hamiltonian on a four-dimensional symplectic manifold $\{H_1 = c_1\} \cap \{H_2 = c_2\}$, to make it completely integrable, one needs one independent invariant. Note that equations [4.9] can be written as a Hamiltonian vector field in another form

$$\dot{x}(t) = J\frac{\partial H}{\partial x}, \quad x \in \mathbb{R}^6, \quad J = \begin{pmatrix} 0 & -x_3 & x_2 & 0 & -x_6 & x_5 \\ x_3 & 0 & -x_1 & x_6 & 0 & -x_4 \\ -x_2 & x_1 & 0 & -x_5 & x_4 & 0 \\ 0 & -x_6 & x_5 & 0 & -x_3 & x_2 \\ x_6 & 0 & -x_4 & x_3 & 0 & -x_1 \\ -x_5 & x_4 & 0 & -x_2 & x_1 & 0 \end{pmatrix}.$$

Since $\det J = 0$, then (with the notations of the remark 3.4, case 2), we have $m = 2n + k$ and $m - k = rg J$. Here, $m = 6$ and $rg J = 4$, then $n = k = 2$. The system [4.9] has, beside the energy H [4.12], two trivial constants of motion H_1[4.10] and H_2[4.11] because: $J\frac{\partial H_2}{\partial x} = J\frac{\partial H_3}{\partial x} = 0$, and in order for the Hamiltonian system [4.9], to be completely integrable, it suffices to have one more integral, which we take as $H_4 = \frac{1}{2}\left(\mu_1 x_1^2 + \mu_2 x_2^2 + \cdots + \mu_6 x_6^2\right)$. The four invariants must be functionally independent and in involution, so in particular $\{H_4, H_3\} = \left\langle \frac{\partial H_4}{\partial x}, J\frac{\partial H_3}{\partial x} \right\rangle = 0$, that is,

$$((\lambda_3 - \lambda_2)\mu_1 + (\lambda_1 - \lambda_3)\mu_2 + (\lambda_2 - \lambda_1)\mu_3) x_1 x_2 x_3$$
$$+ ((\lambda_6 - \lambda_5)\mu_1 + (\lambda_1 - \lambda_6)\mu_5 + (\lambda_5 - \lambda_1)\mu_6) x_1 x_5 x_6$$
$$+ ((\lambda_4 - \lambda_6)\mu_2 + (\lambda_6 - \lambda_2)\mu_4 + (\lambda_2 - \lambda_4)\mu_6) x_2 x_4 x_6$$
$$+ ((\lambda_5 - \lambda_4)\mu_3 + (\lambda_3 - \lambda_5)\mu_4 + (\lambda_4 - \lambda_3)\mu_5) x_3 x_4 x_5 = 0.$$

Then

$$(\lambda_3 - \lambda_2)\mu_1 + (\lambda_1 - \lambda_3)\mu_2 + (\lambda_2 - \lambda_1)\mu_3 = 0,$$

$$(\lambda_6 - \lambda_5)\mu_1 + (\lambda_1 - \lambda_6)\mu_5 + (\lambda_5 - \lambda_1)\mu_6 = 0,$$

$$(\lambda_4 - \lambda_6)\mu_2 + (\lambda_6 - \lambda_2)\mu_4 + (\lambda_2 - \lambda_4)\mu_6 = 0,$$

$$(\lambda_5 - \lambda_4)\mu_3 + (\lambda_3 - \lambda_5)\mu_4 + (\lambda_4 - \lambda_3)\mu_5 = 0.$$

Put

$$\mathcal{A} = \begin{pmatrix} \lambda_3 - \lambda_2 & \lambda_1 - \lambda_3 & \lambda_2 - \lambda_1 & 0 & 0 & 0 \\ \lambda_6 - \lambda_5 & 0 & 0 & 0 & \lambda_1 - \lambda_6 & \lambda_5 - \lambda_1 \\ 0 & \lambda_4 - \lambda_6 & 0 & \lambda_6 - \lambda_2 & 0 & \lambda_2 - \lambda_4 \\ 0 & 0 & \lambda_5 - \lambda_4 & \lambda_3 - \lambda_5 & \lambda_4 - \lambda_3 & 0 \end{pmatrix}.$$

The number of solutions of this system is equal to the number of columns of the matrix $\mathcal{A}$ minus the rank of $\mathcal{A}$. If $rk\mathcal{A} = 4$, we have two solutions: $\mu_i = 1$ leads to the invariant H_2 and $\mu_i = \lambda_i$ leads to the invariant H_3. This is unacceptable. If $rk\mathcal{A} = 3$, each four-order minor of $\mathcal{A}$ is singular. Now

$$\begin{pmatrix} \lambda_3 - \lambda_2 & \lambda_1 - \lambda_3 & \lambda_2 - \lambda_1 & 0 \\ \lambda_6 - \lambda_5 & 0 & 0 & 0 \\ 0 & \lambda_4 - \lambda_6 & 0 & \lambda_6 - \lambda_2 \\ 0 & 0 & \lambda_5 - \lambda_4 & \lambda_3 - \lambda_5 \end{pmatrix} = -(\lambda_6 - \lambda_5)C,$$

$$\begin{pmatrix} \lambda_1 - \lambda_3 & \lambda_2 - \lambda_1 & 0 & 0 \\ 0 & 0 & 0 & \lambda_1 - \lambda_6 \\ \lambda_4 - \lambda_6 & 0 & \lambda_6 - \lambda_2 & 0 \\ 0 & \lambda_5 - \lambda_4 & \lambda_3 - \lambda_5 & \lambda_4 - \lambda_3 \end{pmatrix} = (\lambda_1 - \lambda_6)C,$$

$$\begin{pmatrix} \lambda_2 - \lambda_1 & 0 & 0 & 0 \\ 0 & 0 & \lambda_1 - \lambda_6 & \lambda_5 - \lambda_1 \\ 0 & \lambda_6 - \lambda_2 & 0 & \lambda_2 - \lambda_4 \\ \lambda_5 - \lambda_4 & \lambda_3 - \lambda_5 & \lambda_4 - \lambda_3 & 0 \end{pmatrix} = -(\lambda_2 - \lambda_1)C,$$

where $C \equiv \lambda_1\lambda_6\lambda_4 + \lambda_1\lambda_2\lambda_5 - \lambda_1\lambda_2\lambda_4 + \lambda_3\lambda_6\lambda_5 - \lambda_3\lambda_6\lambda_4 - \lambda_3\lambda_2\lambda_5 + \lambda_4\lambda_2\lambda_5 + \lambda_4\lambda_1\lambda_3 - \lambda_4\lambda_1\lambda_5 + \lambda_6\lambda_2\lambda_3 - \lambda_6\lambda_2\lambda_5 - \lambda_1\lambda_6\lambda_3$, and it follows that the condition for which these minors are zero is $C = 0$. Note that this relation holds by cycling the indices: $1 \frown 4, 2 \frown 5, \frown 3 \frown 6$. Under Manakov (1976) conditions,

$$\lambda_1 = \frac{\beta_2 - \beta_3}{\alpha_2 - \alpha_3}, \quad \lambda_2 = \frac{\beta_1 - \beta_3}{\alpha_1 - \alpha_3}, \quad \lambda_3 = \frac{\beta_1 - \beta_2}{\alpha_1 - \alpha_2},$$

$$\lambda_4 = \frac{\beta_1 - \beta_4}{\alpha_1 - \alpha_4}, \quad \lambda_5 = \frac{\beta_2 - \beta_4}{\alpha_2 - \alpha_4}, \quad \lambda_6 = \frac{\beta_3 - \beta_4}{\alpha_3 - \alpha_4},$$

[4.13]

where $\alpha_i, \beta_i \in \mathbb{C}$, $\prod_{i<j}(\alpha_i - \beta_j) \neq 0$, equations [4.9] admit a Lax equation with an indeterminate h:

$$\dot{\widetilde{(X + \alpha h)}} = [X + \alpha h, \Lambda X + \beta h], \qquad [4.14]$$

$$\alpha = \begin{pmatrix} \alpha_1 & 0 & 0 & 0 \\ 0 & \alpha_2 & 0 & 0 \\ 0 & 0 & \alpha_3 & 0 \\ 0 & 0 & 0 & \alpha_4 \end{pmatrix}, \quad \beta = \begin{pmatrix} \beta_1 & 0 & 0 & 0 \\ 0 & \beta_2 & 0 & 0 \\ 0 & 0 & \beta_3 & 0 \\ 0 & 0 & 0 & \beta_4 \end{pmatrix}$$

which is equivalent to $\dot{X} = [X, \Lambda.X] \iff [4.8]$, $[X, \beta] + [\alpha, \Lambda.X] = 0 \iff [4.13]$, and $[\alpha, \beta] = 0$, trivially satisfied for diagonal matrices. The parameters $\mu_1, \ldots, \mu_6$ can be parameterized (like $\lambda_1, \ldots, \lambda_6$) by

$$\mu_1 = \frac{\gamma_2 - \gamma_3}{\alpha_2 - \alpha_3}, \quad \mu_2 = \frac{\gamma_1 - \gamma_3}{\alpha_1 - \alpha_3}, \quad \mu_3 = \frac{\gamma_1 - \gamma_2}{\alpha_1 - \alpha_2},$$

$$\mu_4 = \frac{\gamma_1 - \gamma_4}{\alpha_1 - \alpha_4}, \quad \mu_5 = \frac{\gamma_2 - \gamma_4}{\alpha_2 - \alpha_4}, \quad \mu_6 = \frac{\gamma_3 - \gamma_4}{\alpha_3 - \alpha_4}.$$

To use the method of isospectral deformations, consider the Kac–Moody extension ($n = 4$): $\mathcal{L} = \widetilde{gl(n, \mathbb{R})} = \{\sum_{-\infty}^{N} A_i h^i : N \text{ arbitrary} \in \mathbb{Z}, A_i \in gl(n, \mathbb{R})\}$, of $gl(n, \mathbb{R})$ with the bracket: $[A(h), B(h)] = [\sum A_i h^i, \sum B_j h^j] = \sum_k \left(\sum_{i+j=k} [A_i, B_j]\right) h^k$, and the non-degenerate, invariant inner product $\langle A(h), B(h) \rangle = \langle \sum A_i h^i, \sum B_j h^j \rangle = \sum_{i+j=-1} \langle A_i, B_j \rangle$, where $\langle , \rangle$ is the usual form defined on $gl(n, \mathbb{R})$. This Lie algebra has a natural decomposition $\mathcal{L} = \mathcal{L}_{-\infty,-1} + \mathcal{L}_{0,\infty}$, $\mathcal{L}_{ij} = \{\sum_{i \geq 0} A_k h^k\}$, where $\mathcal{L}_{0,\infty}$ and $\mathcal{L}_{-\infty,-1}$ are, respectively, the ≥ 0 and < 0 powers of h in $\mathcal{L}$. Observe that $\mathcal{L}_{-\infty,-1}^\perp = \mathcal{L}_{-\infty,-1}$ and $\mathcal{L}_{0,\infty}^\perp = \mathcal{L}_{0,\infty}$ where $\perp$ is taken with respect the form above. The infinite-dimensional Lie group underlying $\mathcal{L}_{-\infty,-1}$ acts conjointly on the dual Kac–Moody Lie algebra $\mathcal{L}_{-\infty,-1}^* \approx \mathcal{L}_{0,\infty}^\perp = \mathcal{L}_{0,\infty}$, according to the rule of customary conjugation followed by registering the non-negative powers of h only. The orbits described in this way come equipped with a symplectic structure with Poisson bracket $\{H_1, H_2\}(\alpha) = \left\langle \alpha, \left[\nabla_{\mathcal{L}_{-\infty,-1}^*} H_1, \nabla_{\mathcal{L}_{-\infty,-1}^*} H_2\right] \right\rangle$, where $\alpha \in \mathcal{L}_{-\infty,-1}^*$ and $\nabla_{\mathcal{L}_{-\infty,-1}^*} H \in \mathcal{L}_{-\infty,-1}$. The functions defined on this orbit are all in involution and the flow [4.14] evolves on the coadjoint orbit through the point $X + ah \in \mathcal{L}_{0,\infty}$, $X \in so(4)$. According to the Adler–Kostant–Symes theorem where $\mathcal{K} = \mathcal{L}_{0,\infty}$ and $\mathcal{N} = \mathcal{L}_{-\infty,-1}$, the flow [4.14] is Hamiltonian on an orbit through the point $X + ah$, $X \in so(4)$ formed by the coadjoint action of the subgroup $G_\mathcal{N} \subset SL(n)$ of lower triangular matrices on the dual Kac–Moody algebra

$\mathcal{L}^*_{-\infty,-1} \approx \mathcal{L}^\perp_{0,\infty} = \mathcal{L}_{0,\infty}$. As a result, the coefficients of $z^i h^i$ appearing in the curve:

$$\mathcal{C}: \{(z,h) \in \mathbb{C}^2 : \det(X + ah - zI) = 0\}, \qquad [4.15]$$

associated with [4.14] are invariant of the system in involution for the symplectic structure of this orbit. Note that $\det(gXg^{-1}) = \det X = (x_1x_4 + x_2x_5 + x_3x_6)^2$ and $tr(gXg^{-1})^2 = tr(gX^2g^{-1}) = tr(X^2) = -2(x_1^2 + x_2^2 + \cdots + x_6^2)$. The complex flows generated by these invariants can be realized as straight lines on the Abelian variety defined by the periods of the curve $\mathcal{C}$. Explicitly, equation [4.15] is presented as follows:

$$\mathcal{C}: \quad \prod_{i=1}^{4}(\alpha_i h - z) + 2H_4 h^2 - 2H_1 zh + 2H_2 z^2 + H_3^2 = 0, \qquad [4.16]$$

where $H_1(X) = c_1$, $H_2(X) = c_2$, $H_3(X) = 2H = c_3$, $H_4(X) = c_4$, with c_1, c_2, c_3, c_4 generic constants. $\mathcal{C}$ is a curve of genus 3, and it has a natural involution $\sigma : \mathcal{C} \longrightarrow \mathcal{C}$, $(z,h) \longmapsto (-z,-h)$. The Jacobian variety $Jac(\mathcal{C})$ of $\mathcal{C}$ splits up into an even and old part: the even part is an elliptic curve $\mathcal{C}_0 = \mathcal{C}/\sigma$ and the odd part is a two-dimensional Abelian surface $Prym(\mathcal{C}/\mathcal{C}_0)$ (also noted, $Prym_\sigma(\mathcal{C})$) the Prym variety: $Jac(\mathcal{C}) = \mathcal{C}_0 \oplus Prym_\sigma(\mathcal{C})$. The van Moerbeke–Mumford linearization method (1979) provides an algebraic map from the complex affine variety $\bigcap_{i=1}^{4}\{H_i(X) = c_i\} \subset \mathbb{C}^6$ to the Jacobi variety $Jac(\mathcal{C})$. By the antisymmetry of $\mathcal{C}$, this map sends this variety to $Prym_\sigma(\mathcal{C})$, $\bigcap_{i=1}^{4}\{H_i(X) = c_i\} \longrightarrow Prym_\sigma(\mathcal{C})$, $p \longmapsto \sum_{k=1}^{3} s_k$, and the complex flows generated by the constants of the motion are straight lines on $Prym_\sigma(\mathcal{C})$. We have the following theorem:

THEOREM 4.6.– The geodesic flow [4.8] is a Hamiltonian system. The Hamiltonian is given by the function $H \equiv H_1 = \frac{1}{2}(\lambda_1 x_1^2 + \lambda_2 x_2^2 + \cdots + \lambda_6 x_6^2)$ and the system [4.9] has two trivial invariants; $H_2 = \frac{1}{2}(x_1^2 + x_2^2 + \cdots + x_6^2)$ and $H_3 = x_1 x_4 + x_2 x_5 + x_3 x_6$. Moreover, if $\lambda_1 \lambda_6 \lambda_4 + \lambda_1 \lambda_2 \lambda_5 - \lambda_1 \lambda_2 \lambda_4 + \lambda_3 \lambda_6 \lambda_5 - \lambda_3 \lambda_6 \lambda_4 - \lambda_3 \lambda_2 \lambda_5 + \lambda_4 \lambda_2 \lambda_5 + \lambda_4 \lambda_1 \lambda_3 - \lambda_4 \lambda_1 \lambda_5 + \lambda_6 \lambda_2 \lambda_3 - \lambda_6 \lambda_2 \lambda_5 - \lambda_1 \lambda_6 \lambda_3 = 0$, the system has a fourth independent constant of the motion of the form $H_4 = \frac{1}{2}(\mu_1 x_1^2 + \mu_2 x_2^2 + \cdots + \mu_6 x_6^2)$. Then the system [4.9] is completely integrable and can be linearized on the Prym variety $Prym_\alpha(\mathcal{C})$.

REMARK 4.1.– As mentioned in section 3.3, the Kirchhoff's equations of motion of a solid in an ideal fluid can be regarded as the equations of the geodesics of the right-invariant metric on the $E(3) = SO(3) \times \mathbb{R}^3$ group of motions of three-dimensional Euclidean space $\mathbb{R}^3$, generated by rotations and translations. The motion has the trivial coadjoint orbit invariants $\langle p,p \rangle$ and $\langle p,l \rangle$. As it turns out, this is a special case of a more general system of equations written as $\dot{x} = x \wedge \frac{\partial H}{\partial x} + y \wedge \frac{\partial H}{\partial y}$, $\dot{y} = y \wedge \frac{\partial H}{\partial x} + x \wedge \frac{\partial H}{\partial y}$, where $x = (x_1, x_2, x_3) \in \mathbb{R}^3$ and $y = (y_1, y_2, y_3) \in \mathbb{R}^3$. The first set can be obtained from the second by putting $(x,y) = (l, p/\varepsilon)$ and letting $\varepsilon \to 0$. The latter set of equations is the geodesic flow on $SO(4)$ for a left invariant metric defined by the quadratic form H.

4.6. Jacobi geodesic flow on an ellipsoid and Neumann problem

For $m = 2$, that is, V_2 (theorem 4.3), if one chooses $A = \alpha h^2 - hx \wedge y - y \otimes y$, $x, y \in \mathbb{R}^n$, which can also be considered (Adler and van Moerbeke 1980a,b) as a rank 2 perturbation of the diagonal matrix α, then equation [4.5] reduces to $\dot{A} = [A, ad_\beta ad_\alpha^{-1}(y \wedge x) + \beta h]$, $\beta_i = f'(\alpha_i)$. This equation can be reduced to the following Hamiltonian system:

$$\dot{x} = -(ad_\beta ad_\alpha^{-1}(y \wedge x))x - \beta y = -\frac{\partial H_\beta}{\partial y},$$

$$\dot{y} = -(ad_\beta ad_\alpha^{-1}(y \wedge x))y = \frac{\partial H_\beta}{\partial x},$$

$$H_\beta = \tfrac{1}{2}\sum_i \beta_i \left(y_i^2 + \sum_{j \neq i} \frac{(x_i y_j - x_j y_i)^2}{\alpha_i - \alpha_j} \right).$$

i) For $f(z) = \ln z$, that is, $\beta_i = \frac{1}{\alpha_i}$, we obtain the problem of Jacobi geodesic flow on the ellipsoid $\frac{x_1^2}{\alpha_1^2} + \cdots + \frac{x_n^2}{\alpha_n^2} = 1$, expressing the motion of the tangent line $x + sy : s \in \mathbb{R}$ to the ellipsoid in the direction y of the geodesic. (ii) For $f(z) = \tfrac{1}{2}z^2$, that is, $\beta_i = \alpha_i$, we get the Neumann motion of a point on the sphere S^{n-1}, $|x| = 1$, under the influence of the force $-\alpha x$. From theorem 4.3, both motions are straight lines on $Jac(\mathcal{C})$, where $\mathcal{C}$ turns out to be hyperelliptic of genus $n-1$ (much lower than the generic one) ramified at the following $2n$ points: some point at ∞, the n points α_i and $n-1$ other points λ_i of geometrical significance, based on the observation that generically, a line in $\mathbb{R}^n$ touches $n-1$ confocal quadrics. To be precise, the set of all common tangent lines to $n-1$ confocal quadrics $Q_{\lambda_i}(x,x) + l = 0$, $i = 1, ..., n-1$, where $Q_u(x, y) = \langle (u - \alpha)^{-1} x, y \rangle$, can be parameterized by the quotient of the Jacobian of the hyperelliptic curve $\mathcal{C}$ above by an Abelian group G. The group is generated by the discrete action obtained by flipping the signs of x_k and y_k and some trivial one-dimensional action. Letting $h \to 0$ in the matrix A and excising the largest eigenvalue from this matrix leads to a new isospectral symmetric matrix $L = (I - P_y)(\alpha - x \otimes x)(I - P_y)$, and a flow $\dot{L} = [ad_\beta ad_\alpha^{-1} x \wedge y, L]$, where the spectrum of L is given by the $n-1$ branch points λ_i above and zero. From these considerations, it follows that the tangent line $\{x+sy : s \in \mathbb{R}\}$ to the ellipsoid remains tangent to $n-2$ other confocal quadrics, and the corresponding $n-1$ eigenfunctions of L provide the orthogonal set of normals to the $n-1$ quadrics at the points of tangency, hence recovering a theorem of Chasles. The close relationship between Jacobi's and Neumann's problems, which in fact live on the same orbits, was implemented by Knörrer (1982), who showed that the normal vector to the ellipsoid moves according to the Neumann problem, when the point moves according to the geodesic. These facts are also investigated by Knörrer (1980) and others; the set of all $n-1$ dimensional linear subspaces in the intersection of two quadrics $X_1^2+\cdots+X_n^2-Y_1^2-\cdots-Y_{n-1}^2=0$ and $\alpha_1 X_1^2+\cdots+\alpha_n X_n^2-\lambda_1 Y_1^2-\cdots-\lambda_{n-1}Y_{n-1}^2=X_0^2$ in $\mathbb{P}^{2n-1}$ is the Jacobian of the curve $\mathcal{C}$ defined above. This is done by observing that the set of linear subspaces in the above quadrics is the same as the set of $(n-2)$-dimensional linear subspaces tangent to $n-1$ quadrics $(\alpha_1-\lambda_j)X_1^2+\cdots+(\alpha_n-\lambda_j)X_n^2=X_0^2, j=1,2,...,n-1$, which is dual to the set of tangents to the confocal quadrics. The Neumann problem is also strikingly related to the Korteweg–de Vries equation and various other nonlinear partial differential equations (see Deift *et al.* (1980)).

4.7. The Lagrange top

For another example of V_2 in theorem 4.3, we consider the Lagrange top (see section 3.2.2), which evolves on an orbit of type V_2; $n=3$, $A=\Gamma+Mh+ch^2$, $\Gamma\in so(3)\simeq\mathbb{R}^3$ is the unit vector in the direction of gravity and $M\in so(3)\simeq\mathbb{R}^3$ is the angular momentum in body coordinates with regard to the fixed point; moreover, $c=(\lambda+\mu)\Upsilon$, where $\Upsilon\in so(3)\simeq\mathbb{R}^3$ expresses the coordinates of the center of mass, and where $(\lambda+\mu,\lambda+\mu,2\lambda)$ is the inertia tensor in diagonalized form. The situation then leads to a linear flow on an elliptic curve (see Adler and van Moerbeke (2004) and for higher dimensional generalizations, see Ratiu (1982)).

4.8. Quartic potential, Garnier system

We consider the Hamiltonian

$$H=\frac{1}{2}\left(y_1^2+y_2^2+a_1 x_1^2+a_2 x_2^2\right)+\frac{1}{4}x_1^4+\frac{1}{4}a_3 x_2^4+\frac{1}{2}a_4 x_1^2 x_2^2, \qquad [4.17]$$

where a_1,a_2,a_3,a_4 are constants. The corresponding system is given by

$$\ddot{x}_1+\left(a_1+x_1^2+a_4 x_2^2\right)x_1=0, \qquad \ddot{x}_2+\left(a_2+a_3 x_2^2+a_4 x_1^2\right)x_2=0. \quad [4.18]$$

The integrability of this Hamiltonian system has been studied by several authors. It arises in connection with some problems in scalar field theory and in the semi-classical method in quantum field theory. It corresponds to the Garnier system (Garnier 1919) and to the anisotropic harmonic oscillator in a radial quartic potential. Also, this Hamiltonian can be connected to the coupled nonlinear Schrödinger equations (see section 4.9) and to the Yang–Mills system for a field with gauge group $SU(2)$ (see section 4.10). Note that the Hamiltonian [4.17] has the Painlevé property (i.e. the general solutions have no movable singularities other than poles; these concepts will be discussed in more detail in Chapters 7 and 8) only if (i) $a_1=a_2$, $a_3=a_4=1$ and (ii) $a_1=a_2$, $a_3=1$, $a_4=3$. In the case (i), the second integral has the form $H_2=y_2 x_1-y_1 x_2$, whereas in the case (ii) the second integral is

$H_2 = y_1 y_2 + x_1 x_2 \left(a_1 + x_1^2 + x_2^2\right)$. Let us point out that if (iii) $a_2 = 4a_1$, $a_3 = 16$, $a_4 = 6$, the system [4.18] is integrable and the second integral is $H_2 = a_1 x_1^2 x_2 + x_1 x_2 \left(x_1^3 + 2x_1 x_2^2\right) + y_1 \left(y_2 x_1 - y_1 x_2\right)$. This case can be deduced from the paper (Dorizzi et al. 1983), and will be studied in exercise 8.8. Also, it was shown (Ramani 1982; Grammaticos et al. 1983, 1984; Hietarinta 1984) that if (iv) $a_2 = 4a_1$, $a_3 = 8$, $a_4 = 3$, the system [4.18] is integrable and the second integral is $H_2 = y_1^4 + x_1^4 y_1^2 + 6x_1^2 x_2^2 y_1^2 + 2a_1 x_1^2 y_1^2 - 4x_1^3 x_2 y_1 y_2 + x_1^4 y_2^2 + a_1^2 x_1^4 + a_1 x_1^6 + 2a_1 x_1^4 x_2^2 + \frac{1}{4} x_1^8 + x_1^6 x_2^2 + x_1^4 x_2^4$. This section deals with the problem of integrability of the system [4.18] corresponding to the choice: a_1, a_2 arbitrary constants and $a_3 = a_4 = 1$, that is,

$$H = \frac{1}{2}\left(y_1^2 + y_2^2 + a_1 x_1^2 + a_2 x_2^2\right) + \frac{1}{4}\left(x_1^2 + x_2^2\right)^2. \qquad [4.19]$$

The corresponding system is given by

$$\ddot{x}_1 + \left(a_1 + x_1^2 + x_2^2\right) x_1 = 0, \qquad \ddot{x}_2 + \left(a_2 + x_1^2 + x_2^2\right) x_2 = 0. \qquad [4.20]$$

For $a_1 = a_2$, it is easy to show that the problem can be integrated in terms of elliptic functions. We suppose that $a_1 \neq a_2$. Using the results given in Christiansen et al. (1995), Eilbeck et al. (1993) and Eilbeck et al. (1994), we consider the Lax representation in the form

$$\dot{A}_h = [B_h, A_h], \qquad \dot{} \equiv \frac{\partial}{\partial t} \qquad [4.21]$$

with the following ansatz for the Lax operator

$$A_h = \begin{pmatrix} u(h) & v(h) \\ w(h) & -u(h) \end{pmatrix}, \qquad B_h = \begin{pmatrix} 0 & 1 \\ r(h) & 0 \end{pmatrix},$$

where

$$v(h) = -(a_1 + h)(a_2 + h)\left(1 + \frac{1}{2}\left(\frac{x_1^2}{a_1 + h} + \frac{x_2^2}{a_2 + h}\right)\right),$$

$$u(h) = \frac{1}{2}(a_1 + h)(a_2 + h)\left(\frac{y_1 x_1}{a_1 + h} + \frac{y_2 x_2}{a_2 + h}\right), \qquad [4.22]$$

$$w(h) = (a_1 + h)(a_2 + h)\left(\frac{1}{2}\left(\frac{y_1^2}{a_1 + h} + \frac{y_2^2}{a_2 + h}\right) - h + \frac{1}{2}\left(x_1^2 + x_2^2\right)\right),$$

$$r(h) = h - x_1^2 - x_2^2.$$

The system [4.20] admits a Lax representation given by [4.21]. The proof is straightforward and based on direct computation, we have

$$[B_h, A_h] = \begin{pmatrix} w(h) - v(h)r(h) & -2u(h) \\ 2u(h)r(h) & v(h)r(h) - w(h) \end{pmatrix},$$

and it follows from [4.19], [4.20] and [4.22] that

$$\dot{u}(h) = w(h) - v(h)r(h), \qquad \dot{v}(h) = -2u(h), \qquad \dot{w}(h) = 2u(h)r(h).$$

Recall that equation [4.21] means that for $h \in \mathbb{C}$ and under the time evolution of the system, $A_h(t)$ remain similar to $A_h(0)$. So the spectrum of A_h is conserved, that is, it undergoes an isospectral deformation. The eigenvalues of A_h, viewed as functionals, represent the integrals (constants of the motion) of the system. To be precise, a Hamiltonian flow of the type [4.21] preserves the spectrum of A_h, and therefore its characteristic polynomial $\det(A_h - zI)$. We form the Riemann surface in (z, h) space: $\Gamma : \det(A_h - zI) = 0$, whose coefficients are functions of the phase space. This equation is presented as follows

$$\begin{aligned}\Gamma : z^2 &= u^2(h) + v(h)w(h) \equiv P_5(h), \quad [4.23]\\ &= (a_1 + h)(a_2 + h)(h^3 + (a_1 + a_2)h^2 + (a_1 a_2 - H_1)h - H_2),\end{aligned}$$

where $H_1 = H$ is defined by [4.19] with a_1, a_2 arbitrary, $a_3 = a_4 = 1$ and a second quartic integral H_2 of the form :

$$\begin{aligned}H_2 &= \frac{1}{4}(a_2 x_1^4 + a_1 x_2^4 + (a_1 + a_2)x_1^2 x_2^2 + (y_1 x_2 - y_2 x_1)^2) \quad [4.24]\\ &+ \frac{1}{2}(a_2 y_1^2 + a_1 y_2^2 + a_1 a_2 (x_1^2 + x_2^2)).\end{aligned}$$

The curve Γ determined by the fifth-order equation [4.23] is smooth, hyperelliptic and its genus is two. Obviously, Γ is invariant under the involution $(h, z) \mapsto (h, -z)$. The second Hamiltonian vector field is written as

$$\begin{aligned}\dot{x}_1 &= \frac{1}{2}(y_1 x_2 - y_2 x_1)x_2 + a_2 y_1, \qquad \dot{x}_2 = -\frac{1}{2}(y_1 x_2 - y_2 x_1)x_1 + a_1 y_2,\\ \dot{y}_1 &= -a_2 x_1^3 - \frac{1}{2}(a_1 + a_2)x_1 x_2^2 + \frac{1}{2}(y_1 x_2 - y_2 x_1)y_2 - a_1 a_2 x_1,\\ \dot{y}_2 &= -a_1 x_2^3 - \frac{1}{2}(a_1 + a_2)x_1^2 x_2 - \frac{1}{2}(y_1 x_2 - y_2 x_1)y_1 - a_1 a_2 x_2.\end{aligned}$$

These vector fields are in involution with respect to the associated Poisson bracket. For generic $c = (c_1, c_2) \in \mathbb{C}^2$,

$$M_c = \{H_1 = c_1, H_2 = c_2\}, \quad [4.25]$$

is a smooth affine surface. The linearized flow could be realized on the Jacobian variety $Jac(\Gamma)$ of the curve [4.23]. Indeed, the structure of the matrices A_h and B_h (degree of the polynomials in h) is the same as for the Neumann system, so both systems are linearized in the same way. We construct an algebraic map from M_c to the Jacobi variety $Jac(\Gamma)$: $M_c \longrightarrow Jac(\Gamma)$, $p \in M_c \longmapsto (s_1 + s_2) \in Jac(\Gamma)$, and the flows generated by the constants of the motion are straight lines on $Jac(\Gamma)$, that is, the linearizing equations are given by $\sum_{i=1}^{2} \int_{s_i(0)}^{s_i(t)} \omega_k = c_k t$, $0 \leq k \leq 2$, where ω_1, ω_2 span the two-dimensional space of holomorphic differentials on the curve Γ and s_1, s_2, two appropriate variables algebraically related to the originally given ones, for which the Hamilton–Jacobi equation could be solved by the separation of variables. Thus, we have the following theorem:

THEOREM 4.7.– The system [4.18] is completely integrable for all $a_1, a_2, a_3 = a_4 = 1$ (i.e. [4.20]) and admits a Lax representation given by [4.21]. The invariants of A_h are integrals of motion in involution. The first integral is given by the Hamiltonian $H_1 = H$[4.19], whereas the second integral H_2[4.24] is also quartic. The flows generated by H_1 and H_2 are straight line motions on the Jacobian variety $Jac(\Gamma)$ of a smooth genus two hyperelliptic curve Γ[4.23] associated with Lax equation [4.21].

We introduce coordinates s_1 and s_2 on the affine surface M_c[4.25], such that $v(s_1) = v(s_2) = 0$, $a_1 \neq a_2$, $x_1^2 = 2\frac{(a_1+s_1)(a_1+s_2)}{a_1-a_2}$, $x_2^2 = 2\frac{(a_2+s_1)(a_2+s_2)}{a_2-a_1}$, that is, $s_1 + s_2 = \frac{1}{2}(x_1^2 + x_2^2) - a_1 - a_2$, $s_1 s_2 = -\frac{1}{2}(a_2 x_1^2 + a_1 x_2^2) + a_1 a_2$. After some manipulations, we obtain the following equations:

$$\dot{s}_1 = 2\frac{\sqrt{P_5(s_1)}}{s_1 - s_2}, \qquad \dot{s}_2 = 2\frac{\sqrt{P_5(s_2)}}{s_2 - s_1},$$

where $P_5(s)$ is defined by [4.23]. These equations can be integrated by the Abelian mapping $\Gamma \longrightarrow Jac(\Gamma) = \mathbb{C}^2/L$, $p \longmapsto \left(\int_{p_0}^{p} \omega_1, \int_{p_0}^{p} \omega_2\right)$, where the genus two hyperelliptic curve Γ is given by [4.23], L is the lattice generated by the vectors $n_1 + \Omega n_2$, $(n_1, n_2) \in \mathbb{Z}^2$, Ω is the matrix of period of Γ, (ω_1, ω_2) is a basis of holomorphic differentials on the curve Γ, that is, $\omega_1 = \frac{ds}{\sqrt{P_5(s)}}$, $\omega_2 = \frac{sds}{\sqrt{P_5(s)}}$, and p_0 is a fixed point. Thus, we have the following theorem:

THEOREM 4.8.– The system of differential equations [4.20] can be integrated in terms of genus two hyperelliptic functions of time.

4.9. The coupled nonlinear Schrödinger equations

The system of two coupled nonlinear Schrödinger equations is given by

$$i\frac{\partial a}{\partial z} + \frac{\partial^2 a}{\partial t^2} + \Omega_0 a + \frac{2}{3}\left(|a|^2 + |b|^2\right)a + \frac{1}{3}\left(a^2 + b^2\right)\bar{a} = 0, \qquad [4.26]$$

$$i\frac{\partial b}{\partial z} + \frac{\partial^2 b}{\partial t^2} - \Omega_0 b + \frac{2}{3}\left(|a|^2 + |b|^2\right)b + \frac{1}{3}\left(a^2 + b^2\right)\bar{b} = 0,$$

where $a(z,t)$ and $b(z,t)$ are functions of z and t, the bar "−" denotes the complex conjugation, "||" denotes the modulus and Ω_0 is a constant. We seek solutions of [4.30] in the form, $a(z,t) = y_1(t)\exp(i\Omega z)$, $b(z,t) = y_2(t)\exp(i\Omega z)$, where $y_1(t)$ and $y_2(t)$ are two functions and Ω is an arbitrary constant. Then, we obtain the system

$$\ddot{y}_1 + \left(y_1^2 + y_2^2\right)y_1 = (\Omega - \Omega_0)y_1, \qquad \ddot{y}_2 + \left(y_1^2 + y_2^2\right)y_2 = (\Omega + \Omega_0)y_2.$$

The latter coincides with [4.20] for $a_1 = \Omega_0 - \Omega$ and $a_2 = -\Omega_0 - \Omega$ and we can use the results obtained previously to study it.

4.10. The Yang–Mills equations

In section 3.4 we saw that the Yang–Mills equations for a field with gauge group $SU(2)$ are reduced to Hamiltonian system

$$\dot{x} = J\frac{\partial H}{\partial x}, \quad x = (y_1, y_2, x_1, x_2)^\mathsf{T}, \quad J = \begin{pmatrix} O & -I \\ I & O \end{pmatrix},$$

with $H = \frac{1}{2}(x_1^2 + x_2^2) + \frac{1}{4}\left(y_1^2 + y_2^2\right)^2$, the Hamiltonian. The latter coincides with [4.19] for $a_1 = a_2 = 0$ and the results obtained previously can be used to study it.

4.11. The Kowalewski top

Various Lax forms have been proposed for the Kowalewski top, in view of its resolution by the isospectral deformation method, which had been an open problem for several years. A detailed geometrical study of Kowalewski top has revealed rational relationships between various systems. For example, for a long time no one had suspected a rational relationship between the Kowalewski spinning top, the geodesic flow on the group $SO(4)$ for the Manakov metric and the Hénon–Heiles differential system. Such a relation was obtained by Adler and van Moerbeke (1988), and the use of our results obtained in Lesfari (1988) allowed them to provide a pair of Lax for the Kowalewski top. In what follows, we will use the form of Lax (see Adler and van Moerbeke (2004) and the references therein), in order to linearize the Kowalewski problem.

THEOREM 4.9.– The Kowalewski system admits the Lax pair $\dot{A} = i[A, B]$, with

$$A = \begin{pmatrix} 0 & -\Gamma_2 & -\frac{1}{2}x_2 h & -\gamma_3 \\ \Gamma_1 & 0 & \gamma_3 & \frac{1}{2}x_1 h \\ -\frac{1}{2}x_1 h & -\gamma_3 & -m_3 h & \Gamma_1 - h^2 \\ \gamma_3 & \frac{1}{2}x_2 h & -\Gamma_2 + h^2 & m_3 h \end{pmatrix},$$

$$B = \begin{pmatrix} -m_3 & 0 & \frac{1}{2}x_2 & 0 \\ 0 & m_3 & 0 & -\frac{1}{2}x_1 \\ \frac{1}{2}x_1 & 0 & m_3 & h \\ 0 & -\frac{1}{2}x_2 & -h & m_3 \end{pmatrix}$$

where $x_1 = m_1 + im_2$, $x_2 = m_1 - im_2$, $\Gamma_1 = \gamma_1 + i\gamma_2$, $\Gamma_2 = \gamma_1 - i\gamma_2$.

PROOF.– We sketch a proof. The projective complex algebraic curve (or spectral curve), defined by the affine equation $C : P(h,z) = \det(A - zI) = 0$ is written explicitly

$$z^4 + (h^4 - c_1 h^2 + 2c_3)z^2 + c_4 h^4 + (c_2^2 - c - 1c - 3)h^2 + c_3^2 = 0,$$

where the constants (supposed generic) c_1, c_2, c_3, c_4 denote, respectively, the constants H_1, H_2, H_3, H_4 (see Kowalewski top, section 3.2.3) and this equation describes isospectral deformation. The curve C is smooth and is a double cover

$$\varphi : C \longrightarrow \mathcal{H}, \quad (z,h) \longmapsto (w,h), \quad w = \frac{2z^2 + h^4 - c_1 h^2 + 2c_3}{h},$$

of a hyperelliptic curve $\mathcal{H}$ defined by

$$\mathcal{H} : w^2 = \left((h^2 - c_1)^2 + 4(c_3 - c_4)\right) h^2 - 4c_2^2.$$

The projection $\pi : \mathcal{H} \longrightarrow \mathbb{C}$, $(w,h) \longmapsto h$, realizes $\mathcal{H}$ as a double cover of $\mathbb{C}$ branched in six points. Consequently, the compactified $\overline{\mathcal{H}}$ of $\mathcal{H}$ is a hyperelliptic curve of genus 2. The covering φ has four branch points (w,h) on $\mathcal{H}$ for which $z = 0$, that is, $wh = h^4 - c_1 h^2 + 2c_3$. It is shown that the genus of $\overline{C}$ is 5. Moreover, the curve $\mathcal{H}$ is an unramified double cover $\mathcal{H} \longrightarrow \mathcal{E}$, $(w,h) \longmapsto (\zeta, \xi) = (wh, h^2)$, of an elliptic curve $\mathcal{E} : \zeta^2 = \left((\xi - c_1)^2 + 4(c_3 - c_4)\right)\xi^2 - 4c_2\xi$. Moreover, the curve C can be seen as an unramified double cover $C \longrightarrow \mathcal{S}$, $(z,h) \longmapsto (z,\xi) = (z, h^2)$, of a new curve $\mathcal{S}$, defined by

$$\mathcal{S} : z^4 + (h^4 - c_1 h^2 + 2c_3)z^2 + c_4 h^4 + (c_2^2 - c - 1c - 3)h^2 + c_3^2 = 0.$$

The curve $\mathcal{S}$ itself is a double cover of the elliptic curve $\mathcal{E}$,

$$\mathcal{S} \longrightarrow \mathcal{E}, \quad (z,\xi) \longmapsto (\zeta, \xi) = (2z^2 + \xi^2 - c_1\xi + 2c_3, \xi),$$

and the genus of $\overline{\mathcal{S}}$ is equal to 3. Finally, it is easily verified that the linearization of the flow is done on the Jacobian variety of the curve C.

4.12. The Goryachev–Chaplygin top

The differential equations of the Goryachev–Chaplygin top (Goryachev 1900; Chaplygin 1948) correspond to the case $I_1 = I_2 = 4I_3$, $l_2 = l_3 = 0$ (see section 3.2.4) and are explicitly written in the form (without restricting generality, values have been given to the constants in order to not weigh down the notations),

$$\begin{aligned}
\dot{m}_1 &= 3m_2 m_3, & \dot{\gamma}_1 &= 4m_3\gamma_2 - m_2\gamma_3, \\
\dot{m}_2 &= -3m_1 m_3 - 4\gamma_3, & \dot{\gamma}_2 &= m_1\gamma_3 - 4m_3\gamma_1, \\
\dot{m}_3 &= 4\gamma_2, & \dot{\gamma}_3 &= m_2\gamma_1 - m_1\gamma_2.
\end{aligned} \qquad [4.27]$$

This system admits the following four invariants:

$$\begin{aligned}
H_1 &= m_1^2 + m_2^2 + 4m_3^2 - 8\gamma_1 = 6b_1, \\
H_2 &= (m_1^2 + m_2^2)m_3 + 4m_1\gamma_3 = 2b_2, \\
H_3 &= \gamma_1^2 + \gamma_2^2 + \gamma_3^2 = b_3, \\
H_4 &= m_1\gamma_1 + m_2\gamma_2 + m_3\gamma_3 = 0,
\end{aligned}$$

where b_1, b_2, b_3 are generic constants. The system [4.27] is integrable in the sense of Liouville, H_1 (Hamiltonian) and H_4 are in involution, while H_2, H_3 are Casimir invariants. The differential system [4.27] has a Lax pair of the form

$$\dot{A} = [A, B], \qquad [4.28]$$

with

$$A = \begin{pmatrix} 0 & \gamma_3 & \frac{1}{2}x_1 h \\ -\gamma_3 & -m_3 h & \Gamma_1 - h^2 \\ \frac{1}{2}x_2 h & -\Gamma_2 + h^2 & m_3 h \end{pmatrix}, \quad B = \begin{pmatrix} 3im_3 & 0 & -ix_1 \\ 0 & 2im_3 & 2ih \\ -ix_2 & -2ih & -2im_3 \end{pmatrix},$$

$x_1 = m_1 + im_2$, $x_2 = m_1 - im_2$, $\Gamma_1 = \gamma_1 + i\gamma_2$, $\Gamma_1 = \gamma_1 - i\gamma_2$. The transformation $(t, m_1, m_2, m_3, \gamma_1, \gamma_2, \gamma_3) \longmapsto (-t, -m_1, -m_2, -m_3, -\gamma_1, -\gamma_2, -\gamma_3)$ shows that the systems [4.27] and [4.28] are equivalent. The spectral curve, $\mathcal{C} : P(h, z) = \det(A - zI) = 0$, is written explicitly

$$\mathcal{C} : w^3 + \left(h^4 - \frac{3}{2}b_1 h^2 + b_3\right) w + \frac{b_2}{2} h^3 = 0,$$

and one easily checks that the flow linearizes on the Jacobian of this curve.

4.13. Periodic infinite band matrix

Review the example of V_{ij} in theorem 4.4 (see also van Moerbeke (1979) Adler and van Moerbeke (1980a,b)). Consider the periodic infinite band matrix M of period n having $j + h + 1$ diagonals; the spectrum of M is defined by the points $(z, h) \in \mathbb{C}^2$, such that $Mv(h) = zv(h)$, $v(h) = (..., h^{-1}v, v, hv, ...)$, $v \in \mathbb{C}^n$. Let M_h be the square matrix obtained from M and let $\mathcal{C}$ be the curve defined by $\det(M_h - zI) = 0$. Then in the case of the set of infinite band matrices with $j + k + 1$ diagonals, in higher dimensions many partial results seem to lead to rigidity. In fact, it was shown that a discrete two-dimensional Laplacian cannot be deformed, given its periodic spectrum; the proof can be summarized by the observation that the Picard variety of most algebraic surfaces is trivial; the proof that the specific spectral surface defined by the two-dimensional Laplacian has trivial Picard variety is based on the technique of toroidal embedding, which reduces cohomological computations to combinatorial questions. Finally, inspired by the dynamical systems, Mumford (1983a, 1983b) has given a beautiful description of hyperelliptic Jacobians of dimension g. Let $y^2 = R(z)$ be the monic polynomial of degree $2g + l$ defining the curve $\mathcal{C}$ and let θ be the theta divisor. Then $\text{Jac}(\mathcal{C})\backslash\theta$ is a variety of polynomials U, V with $\deg U = g$, $\deg V \leq g - 1$ and U monic, such that $U | B - V^2$.

4.14. Exercises

EXERCISE 4.1.– Show that a Lax pair is not unique and that power A^k of A also satisfies the Lax equation.

EXERCISE 4.2.– Let $H = \frac{1}{2}\left(y_1^2 + y_2^2\right) + (a + bx_2)x_1^2 + 2(2a + bx_2)x_2^2$ be a Hamiltonian where a, b are constant parameters, and y_1, y_2 are the momenta conjugate to x_1, x_2, respectively, satisfying the usual canonical Poisson bracket. This yields the motion,

$$\dot{x}_1 = y_1, \qquad \dot{y}_1 = -2(a + bx_2)x_1,$$
$$\dot{x}_2 = y_2, \qquad \dot{y}_2 = -bx_1^2 - 2(4a + 3bx_2)x_2.$$

Using the method developed in this chapter, show that the second integral of motion is $F = y_1(x_1y_2 - x_2y_1) + (2a + bx_2)x_1^2x_2 + \frac{b}{4}x_1^4$, the system above is integrable in the Liouville sense and determines its solutions, if possible, explicitly.

EXERCISE 4.3.– Let a, b, c, d be constant parameters, x_1, x_2, y_1, y_2 canonical coordinates and momenta, respectively, satisfying the usual canonical Poisson bracket. Study the complete integrability of the following Hamiltonian systems via the methods developed in this chapter.

1) The Sawada–Kotera system, with Hamiltonian

$$H = \frac{1}{2}\left(y_1^2 + y_2^2\right) + 2c\left(x_1^2 + x_2^2\right) + d\left(x_1^2 + \frac{1}{3}x_2^2\right)x_2^2.$$

2) The Kaup–Kuperschdmit system, with Hamiltonian

$$H = \frac{1}{2}\left(y_1^2 + y_2^2\right) + a\left(x_1^2 + 16x_2^2\right) + d\left(x_1^2 + \frac{16}{3}x_2^2\right)x_2^2.$$

3) The Korteweg–de Vries system, with Hamiltonian

$$H = \frac{1}{2}\left(y_1^2 + y_2^2\right) + \frac{1}{2}\left(ax_1^2 + bx_2^2\right) + d\left(x_1^2 + 2x_2^2\right)x_2^2.$$

EXERCISE 4.4.– Consider the Hamiltonian of the elliptic Calogero system Perelomov (1990):

$$H = \frac{1}{2}\sum_{k=1}^{n} p_k^2 + \sum_{k<l}^{n} \wp(q_k - q_l),$$

where $q_1, q_2, ..., q_n$ are coordinates of n particles on the circle, interacting with the integrable potential $\sum_{k<l}^{n} \wp(q_k - q_l)$, and $\wp(q)$ is the Weierstrass elliptic function. This yields the following canonical Hamiltonian system

$$\frac{d^2 q_1}{dt^2} = -\sum_{k\neq 1} \wp'(q_1 - q_k), ..., \frac{d^2 q_n}{dt^2} = -\sum_{k\neq n} \wp'(q_n - q_k).$$

Show that this differential system is integrable, and admits a Lax pair of the form $i\dot{A} = [A, B]$, where $A = (a)_{kl}$ and $B = (b)_{kl}$ are two matrices of order n,

$$a)_{kl} = p_k \delta_{kl} + i(1 - \delta_{kl})\varphi(q_k - q_l, h),$$

$$b)_{kl} = \delta_{kl}\left(\sum_{j\neq k} \wp(q_k - q_j)\right) + (1 - \delta_{kl})\varphi'(q_k - q_l, h),$$

with $\varphi(q, h) \equiv \frac{\sigma(q-h)}{\sigma(q)\sigma(h)} \exp(\zeta(h)q)$, where σ (resp. ζ) is the Weierstrass sigma (respectively, zeta) function,

$$\sigma(q) \equiv q \prod_{m_1, m_2}\left(1 - \frac{q}{\omega_{m_1, m_2}}\right)\exp\left(\frac{q}{\omega_{m_1, m_2}} + \frac{1}{2}\left(\frac{q}{\omega_{m_1, m_2}}\right)^2\right),$$

and $\omega_{m_1, m_2} = m_1\omega_1 + m_2\omega_2$, $\frac{\omega_2}{\omega_1} \notin \mathbb{R}$. Verify that the linearization of the flow is done on the Jacobian variety of a spectral curve.

EXERCISE 4.5.– Consider the Hamiltonian of Calogero–Moser system is of the form:

$$H = \frac{1}{2}\sum_k p_k^2 + g^2 \sum_{k\neq l} \mathcal{U}(q_k - q_l), \quad g > 0$$

where the potential $\mathcal{U}$ can have several forms: the rational case $\frac{1}{x^2}$, the hyperbolic case $\frac{\alpha^2}{4\sinh^2 \frac{\alpha x}{2}}$, $\alpha > 0$, the trigonometric case $\frac{\alpha^2}{4\sin^2 \frac{\alpha x}{2}}$, $\alpha > 0$, and the elliptic case $\wp(x)$ (studied in the above exercise). The trigonometric system is also called the Sutherland system.

1) Show that the rational case, the hyperbolic case and the trigonometric case can be viewed as limiting cases of the elliptic case.

2) Show that Hamilton's equations for the rational case admit a Lax pair of the form $\dot{A} = [A, B]$, where $A = (a)_{kl}$ and $B = (b)_{kl}$ are two matrices of order n,

$$a_{kl} = p_k \delta_{kl} + \frac{(-g)^{1/2}}{q_k - q_l}(1 - \delta_{kl}), \quad 1 \leq k, l \leq n,$$

$$(-g)^{1/2} b_{kl} = \delta_{kl} \sum_{j\neq k,l} \frac{1}{(q_k - q_j)^2} - (1 - \delta_{kl})\frac{1}{(q_k - q_l)^2}.$$

3) Same question for the hyperbolic case with

$$a_{kl} = p_k \delta_{kl} + ig(1 - \delta_{kl})\frac{\alpha}{2\sinh \alpha(q_k - q_l)/2}, \quad p_k = \dot{q}_k,$$

$$b_{kl} = \frac{ig\alpha^2}{4}\left(-\delta_{kl}\sum_{j\neq k}\frac{1}{\sinh^2 \alpha(q_k - q_j)/2} + (1 - \delta_{kl})\frac{\cosh \alpha(q_k - q_l)/2}{\sinh^2 \alpha(q_k - q_l)/2}\right).$$

Also show that the first integrals are given explicitly by

$$H_1 = \operatorname{tr} L = \sum_{l=1}^n p_l, \quad H_2 = \frac{1}{2}\operatorname{tr} L^2 = H, \quad H_l = \frac{1}{l}\sum_{k=1}^n p_k^l + V_l(x, p), l > 2,$$

where V_l has a degree strictly inferior to l in $p_1, ..., p_n$.

4) Show, as above, that the Hamiltonian of Calogero–Moser system in the trigonometric case is completely integrable and give an explicit formula for the first integrals, as well as an explicit solution of the system.

EXERCISE 4.6.– Consider the Hamiltonian

$$H = \frac{1}{2}\left(y_1^2 + y_2^2\right) + \frac{1}{8}(x_1^2 + x_2^2)^2 - \frac{3\alpha + 2\beta}{8}x_1^2 - \frac{2\alpha + 3\beta}{8}x_2^2 + \frac{A}{8x_1^2} + \frac{B}{8x_2^2},$$

where $\alpha, \beta, \alpha \neq \beta, A, B$ are constant, x_1, x_2, y_1, y_2 are canonical coordinates and momenta, respectively, satisfying the usual canonical Poisson bracket. Using the method developed in this chapter, determine a Lax representation of equations of motion. Show that

$$F = \frac{1}{2}(x_1 y_2 - x_2 y_1)^2 - \frac{1}{2}\left(\beta x_1^2 + \alpha x_2^2\right) - \frac{1}{8}(x_1^2 + x_2^2)(\beta x_1^2 + \alpha x_2^2)$$

$$+ \frac{1}{8}\left(\beta(3\alpha + 2\beta)x_1^2 + \alpha(2\alpha + 3\beta)x_2^2\right) - \frac{A\beta}{8x_1^2} - \frac{B\alpha}{8x_2^2} + \frac{Ax_2^2}{8x_1^2} + \frac{Bx_1^2}{8x_2^2}$$

is a second integral of motion and analyzes, in detail, the integrability of the system associated with this Hamiltonian.

EXERCISE 4.7.– The Holt system is defined by the Hamiltonian

$$H = \frac{1}{2}\left(y_1^2 + y_2^2\right) + \frac{3ab}{4}x_1^{4/3} + ax_1^{-2/3}x_2^2 + acx_1^{-2/3},$$

where a and c are arbitrary constants. Show that this system is Liouville integrable for $b \in \{1, 6, 16\}$ and determine the Lax pair for the integrable cases (hint: see Tsiganov (1999)).

EXERCISE 4.8.– Consider the nonlinear Schrödinger equation (NLS):

$$i\frac{\partial \psi}{\partial t} + \frac{\partial^2 \psi}{\partial x^2} + 2|\psi|^2 \psi = 0, \quad \psi \in \mathbb{C}.$$

This equation describes the stationary two-dimensional self-focusing and the associated transverse instability of a plane monochromatic wave. Determine a Lax pair for this equation.

EXERCISE 4.9.– Consider a nonlinear hyperbolic partial differential equation, called sine-Gordon equation (see also exercises 9.6, 9.10),

$$\frac{\partial^2 u}{\partial t^2} - \frac{\partial^2 u}{\partial x^2} + \sin u = 0, \quad u = u(x, t).$$

This equation describes locally the isometric embedding of surfaces with constant negative Gaussian curvature in the Euclidean space $\mathbb{R}^3$. Show that this equation can be written in the form $\frac{\partial^2 u}{\partial \xi \partial \eta} = \sin u$, where $\xi = \frac{1}{2}(x-t)$, $\eta = \frac{1}{2}(x+t)$, (light-cone coordinates). Also, show that the latter equation admits the following Lax representation: $i\dot{A} = [A, B]$, where the operators A, B are given by

$$A = i \begin{pmatrix} \frac{\partial}{\partial \eta} & \frac{1}{2}\frac{\partial}{\partial \eta} \\ \frac{1}{2}\frac{\partial u}{\partial x} & -\frac{\partial u}{\partial \eta} \end{pmatrix}, \qquad B = -\frac{1}{4} \begin{pmatrix} \cos u & \sin u \\ \sin u & -\cos u \end{pmatrix} A^{-1}.$$

EXERCISE 4.10.— Let $H = \sum_{j,\sigma} \varepsilon_j \bar{c}_{j\sigma} c_{j\sigma} + w\bar{b}b + g \sum_j \left(\bar{b} c_{j\downarrow} c_{j\uparrow} + b \bar{c}_{j\uparrow} \bar{c}_{j\downarrow} \right)$ be the Hamiltonian of the boson-fermion condensate, where $\bar{c}_{j\sigma}$, $c_{j\sigma}$ are creation and annihilation operators for fermions in the state $\sigma = \uparrow$ or $\downarrow$, and $\bar{b}$, b are creation annihilation operators of a molecule at zero momentum.

a) Show that this Hamiltonian can be rewritten in terms of pseudo spins $2s_j^z = \sum_\sigma \bar{c}_{j\sigma} c_{j\sigma} - 1$, $s_j^- = c_{j\downarrow} c_{j\uparrow}$, $s_j^+ = \bar{c}_{j\uparrow} \bar{c}_{j\downarrow}$, in the Jaynes–Cummings–Gaudin Hamiltonian,

$$H = \sum_{j=0}^{n-1} 2\varepsilon_j s_j^z + w\bar{b}b + g \sum_{j=0}^{n-1} \left(\bar{b} s_j^- + b s_j^+ \right).$$

b) The Poisson brackets read $\{s_j^a, s_j^b\} = -\varepsilon_{abc} s_j^c$, $\{b, \bar{b}\} = i$, the $\vec{s}_j$ brackets are degenerate and we fix the value of the Casimir function $\vec{s}_j \cdot \vec{s}_j = s^2$. The phase space has dimension $2(n+1)$. Show that the equations of motion read

$$\dot{b} = -iwb - ig\sum_{j=0}^{n-1} s_j^-, \qquad \dot{s}_j^z = ig\left(\bar{b} s_j^- - b s_j^+\right),$$

$$\dot{s}_j^+ = 2i\varepsilon_j s_j^+ - 2ig\bar{b} s_j^z, \qquad \dot{s}_j^- = -2i\varepsilon_j s_j^- + 2igb s_j^z.$$

c) Show that the equations of motion are equivalent to the Lax equation:

$$\dot{A}(h) = [B(h), A(h)], \qquad A(h) = \begin{pmatrix} U(h) & V(h) \\ W(h) & -U(h) \end{pmatrix},$$

where

$$U(h) = \frac{2h}{g^2} - \frac{\omega}{g^2} + \sum_{j=0}^{n-1} \frac{s_j^z}{h - \varepsilon_j}, \quad V(h) = \frac{2b}{g} + \sum_{j=0}^{n-1} \frac{s_j^-}{h - \varepsilon_j},$$

$$W(h) = \frac{2\bar{b}}{g} + \sum_{j=0}^{n-1} \frac{s_j^+}{h - \varepsilon_j}.$$

d) The spectral curve $\det(A(h) - \lambda I) = 0$ is independent of time. Show that it reads

$$\lambda^2 = \frac{1}{g^4}(2h-\omega)^2 + \frac{4}{g^2}H_n + \frac{2}{g^2}\sum_j \frac{H_j}{h-\varepsilon_j} + \sum_j \frac{s^2}{(h-\varepsilon_j)^2},$$

where the $(n+1)$ Hamiltonians read, $H_n = b\bar{b} + \sum_j s_j^z$, and

$$H_j = (2\varepsilon_j - \omega)s_j^z + g\left(bs_j^+ + \bar{b}s_j^-\right) + g^2 \sum_{k \neq j} \frac{s_j \cdot s_k}{\varepsilon_j - \varepsilon_k}, \quad 0 \leq j \leq n-1.$$

Verify that the Hamiltonian H is written as: $H = \omega H_n + \sum_{j=0}^{n-1} H_j$.

e) Examine the exact solution of the Jaynes–Cummings–Gaudin model by interpreting the result (hint: see arXiv: 0703124v1 (2007), for more information).

5

The Spectrum of Jacobi Matrices and Algebraic Curves

In this chapter, we present a study on the spectrum of periodic Jacobi matrices, infinite continued fractions, difference operators, Cauchy–Stieltjes transforms and Abelian integrals.

5.1. Jacobi matrices and algebraic curves

A Jacobi matrix is a doubly infinite matrix (a_{ij}) for $i,j \in \mathbb{Z}$ such that: $a_{ij} = 0$ if $|i - j|$ is large enough. The set of these matrices forms an associative algebra and consequently a Lie algebra by antisymmetrization. Consider the Jacobi matrix

$$\Gamma = \begin{pmatrix} b_1 & a_1 & 0 & \cdots & 0 \\ a_1 & b_2 & a_2 & & \vdots \\ 0 & a_2 & \ddots & \ddots & 0 \\ \vdots & & \ddots & \ddots & \\ 0 & \cdots & 0 & & \end{pmatrix}, a_i \in \mathbb{R}^+, b_i \in \mathbb{R}, 1 \leq i < \infty \qquad [5.1]$$

As an example of $V_{-j,k}$ (theorem 4.4), consider the infinite Jacobi matrix (symmetric, tridiagonal and N-periodic):

$$A = \begin{pmatrix} \ddots & \ddots & & & & & \\ \ddots & b_0 & a_0 & 0 & \cdots & 0 & \\ & a_0 & b_1 & a_1 & & \vdots & \\ & 0 & a_1 & \ddots & \ddots & 0 & \\ & \vdots & & \ddots & \ddots & a_{N-1} & \\ & 0 & \cdots & 0 & a_{N-1} & b_N & \ddots \\ & & & & & \ddots & \ddots \end{pmatrix}, \qquad [5.2]$$

with $a_i, b_i \in \mathbb{C}$. The matrix A is N-periodic when $a_{i+N} = a_i$, $b_{i+N} = b_i$, $\forall i \in \mathbb{Z}$. We denote by $f = (..., f_{-1}, f_0, f_1, ...)$ the (infinite) column vector and by D the operator passage of degree +1, $Df_i = f_{i+1}$. Since the matrix A is N-periodic, we have $AD^N = D^N A$. Reciprocally, this relation of commutation means that N is the period of A. Let

$$A(h) = \begin{pmatrix} b_1 & a_1 & 0 & \cdots & a_N h^{-1} \\ a_1 & b_2 & a_2 & & \vdots \\ 0 & a_2 & \ddots & \ddots & 0 \\ \vdots & & \ddots & \ddots & a_{N-1} \\ a_N h & \cdots & 0 & a_{N-1} & b_N \end{pmatrix}, h \in \mathbb{C}^*$$

be the finite Jacobi matrix (symmetric tridiagonal and N-periodic). The determinant of the matrix

$$A(h) - zI = \begin{pmatrix} b_1 - z & a_1 & 0 & \cdots & a_N h^{-1} \\ a_1 & b_2 - z & a_2 & & \vdots \\ 0 & a_2 & \ddots & \ddots & 0 \\ \vdots & & \ddots & \ddots & a_{N-1} \\ a_N h & \cdots & 0 & a_{N-1} & b_N - z \end{pmatrix},$$

is

$$F(h, h^{-1}, z) \equiv \det(A(h) - zI) = (-1)^{N+1} \left(\alpha \times (h + h^{-1}) - P(z) \right), \quad [5.3]$$

where $(z,h) \in \mathbb{C} \times \mathbb{C}^*$, $\alpha = \prod_{i=1}^{N} a_i$, $P(z) = z^N + \cdots$, is a polynomial of degree N with real coefficients:

$$P(z) = \det \begin{pmatrix} b_1 - z & a_1 & 0 & \cdots & 0 \\ a_1 & b_2 - z & a_2 & & \vdots \\ 0 & a_2 & \ddots & \ddots & 0 \\ \vdots & & \ddots & \ddots & a_{N-1} \\ 0 & \cdots & 0 & a_{N-1} & b_N - z \end{pmatrix}$$

$$- a_0^2 \det \begin{pmatrix} b_2 - z & a_2 & 0 & \cdots & 0 \\ a_2 & b_3 - z & a_3 & & \vdots \\ 0 & a_3 & \ddots & \ddots & 0 \\ \vdots & & \ddots & \ddots & a_{N-2} \\ 0 & \cdots & 0 & a_{N-2} & b_{N-1} - z \end{pmatrix},$$

$$= z^N + \cdots$$

Let $\mathcal{C}$ be the Riemann surface defined by

$$\mathcal{C} = \{(z,h) \in \mathbb{C} \times \mathbb{C}^* : Af = zf, D^N f = hf\},$$
$$= \{(z,h) \in \mathbb{C} \times \mathbb{C}^* : F(h, h^{-1}, z) = 0\}. \qquad [5.4]$$

We suppose that $\alpha \neq 0$. From the equation $F(h, h^{-1}, z) = 0$, we derive the following relation: $h = \frac{P(z) \pm \sqrt{P^2(z) - 4\alpha^2}}{2\alpha}$. Note that $\mathcal{C}$ is a hyperelliptic curve branched in $2N$ points given by the zeros of the polynomial $P^2(z) - 4\alpha^2$ and admits two points at infinity $\mathcal{P}$ and $\mathcal{Q}$; the point $\mathcal{P}$ covering the case $z = \infty, h = \infty$ while the point $\mathcal{Q}$ is relative to the case $z = \infty, h = 0$.

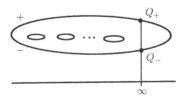

Figure 5.1. *Point relative to $z = \infty, h = 0$*

According to the Riemann–Hurwitz formula, the genus of $\mathcal{C}$ is $N - 1$. The meromorphic function h has neither zero nor poles except in the neighborhood of $z = \infty$. When $z \nearrow \infty$, we have on the sheet +, $h \simeq \frac{P(z) + P(z)}{2\alpha} = \frac{P(z)}{\alpha} = \frac{z^N}{\alpha} + \cdots$, which shows that h has a pole of order N. Similarly, when $z \nearrow \infty$, we have on the sheet -, $h = \frac{P(z)-\sqrt{P^2(z)-4\alpha^2}}{2\alpha} = \frac{2\alpha}{P(z)+\sqrt{P^2(z)-4\alpha^2}} \simeq \frac{\alpha}{z^N} + \cdots$, hence h has a zero of order N. Therefore, the divisor (h) of the function h on $\mathcal{C}$ is $(h) = -N\mathcal{P} + N\mathcal{Q}$, where $\mathcal{P}$ and $\mathcal{Q}$ are the two points covering ∞ on the sheets + and -, respectively. The map $\sim: \mathcal{C} \longrightarrow \mathcal{C}, (z,h) \longmapsto (\bar{z}, \bar{h}^{-1})$, is an antiholomorphic involution. In other words, $\sim: p \longmapsto \tilde{p}$ such that: $\widetilde{\mathcal{P}} = \mathcal{Q}$. When $|h| = 1$, the finite matrix $A(h)$ is self-adjoint and admits a real spectrum. Hence, the set of fixed points of this involution denoted by $\mathcal{C}^\sim$ is determined by $\mathcal{C}^\sim = \{p \in \mathcal{C} : \tilde{p} = p\} = \left\{(z,h) : h = \bar{h}^{-1}, \bar{z} = z\right\} = \{(z,h) : |h| = 1\}$. Note that this set divides $\mathcal{C}$ into two distinct regions $\mathcal{C}_+$ and $\mathcal{C}_-$. More precisely, we have $\mathcal{C}\backslash\mathcal{C}^\sim = \mathcal{C}_+ \cup \mathcal{C}_- = \{p \in \mathcal{C} : |h| > 1\} \cup \{p \in \mathcal{C} : |h| < 1\}$, so $\mathcal{C} = \mathcal{C}_+ \cup \mathcal{C}^\sim \cup \mathcal{C}_-$. The first region $\mathcal{C}_+$ contains the point $\mathcal{P}$ while the second $\mathcal{C}_-$ contains the point $\mathcal{Q}$. In fact, $\mathcal{C}^\sim$ can be seen as the frontier of $\mathcal{C}_+$ and $\mathcal{C}_-$, so $\mathcal{C}^\sim$ is homologous to zero. Moreover, the involution $\sim$ extends to an involution $*$ on the field of meromorphic functions as follows: $\varphi^*(p) = \overline{\varphi(\tilde{p})}$, and on the differential space as follows: $(\varphi d\psi)^* = \varphi^* d\psi^*$. Hence, we have $h^* = h^{-1}$ and $z^* = z$. The condition that the matrices A and D^N have an eigenvector in common is parameterized by the Riemann surface $\mathcal{C}$ [5.4]; let $f = (..., f_{-1}, f_0, f_1, ...)$ be such an eigenvector. In the following, appropriate standardization is used by selecting $f_0 \equiv 1$, from where $F_N = h$. Therefore, let us consider $\bar{f} = (f_1, f_2, ..., f_{N-1})^\top$. Since $\bar{f}$ satisfies $(A(h) - zI)\bar{f} = 0$, we then have $f_k = \frac{\Delta_{1,k}}{\Delta_{1,l}} f_l = \frac{\Delta_{2,k}}{\Delta_{2,l}} f_l = \cdots = \frac{\Delta_{N,k}}{\Delta_{N,l}} f_l$, $1 \leq k, l \leq N$, where $\Delta_{k,l}$ is the (k,l)-cofactor of $(A(h) - zI)$, that is to say,

$$\Delta_{k,l} = (-1)^{k+l} \times (k,l) - \text{minor of } (A(h) - zI). \quad [5.5]$$

The (k,l) − minor of $(A(h) - zI)$ is the determinant of the $N - 1$ submatrix obtained by removing the kth line and the lth column of the matrix $(A(h) - zI)$. In particular, f can be expressed as a rational function in z and h, $f_k = \frac{\Delta_{N,k}}{\Delta_{N,N}} h = \frac{\Delta_{k,k}}{\Delta_{k,N}} h$. According to matrix $A(h) - zI$, we have

$$\Delta_{N,1} = \prod_{j=1}^{N-1} a_j + (-1)^N a_N h^{-1} \left((-z)^{N-2} + \cdots\right),$$

$$\Delta_{1,N} = \prod_{j=1}^{N-1} a_j + (-1)^N a_N h \left((-z)^{N-2} + \cdots\right).$$

Note that $(-z)^{N-2} + \cdots$ is a polynomial of degree $N - 2$, and $\Delta_{N,N} = (-z)^{N-1} + \cdots$ is a polynomial of degree $N - 1$. To determine the divisor structure of f_k, one proceeds as follows: for f_1, we have

$$(f_1)_\infty = (\Delta_{N,1})_\infty + (h) - (\Delta_{N,1})_\infty,$$
$$= -(2N-2)\mathcal{Q} - N\mathcal{P} + N\mathcal{Q} + (N-1)\mathcal{P} + (N-1)\mathcal{Q},$$
$$= \mathcal{Q} - \mathcal{P}.$$

For the other f_k, we first consider the matrix $(A(h) - zI)$ shifted by one, that is,

$$\begin{pmatrix} b_2 - z & a_2 & 0 & \cdots & a_1 h^{-1} \\ a_2 & b_3 - z & a_3 & & \vdots \\ 0 & a_3 & \ddots & \ddots & 0 \\ \vdots & & \ddots & b_N - z & a_N \\ a_1 h & \cdots & 0 & a_N & b_1 - z \end{pmatrix}.$$

Hence,

$$\begin{pmatrix} b_2 - z & a_2 & 0 & \cdots & a_1 h^{-1} \\ a_2 & b_3 - z & a_3 & & \vdots \\ 0 & a_3 & \ddots & \ddots & 0 \\ \vdots & & \ddots & b_N - z & a_N \\ a_1 h & \cdots & 0 & a_N & b_1 - z \end{pmatrix} \begin{pmatrix} \frac{f_2}{f_1} \\ \frac{f_3}{f_1} \\ \vdots \\ \frac{f_N}{f_1} \\ h \end{pmatrix} = 0,$$

and $\left(\frac{f_2}{f_1}\right)_\infty = \mathcal{Q} - \mathcal{P} \implies (f_2)_\infty = \left(\frac{f_2}{f_1}\right)_\infty + (f_1)_\infty = 2\mathcal{Q} - 2\mathcal{P}$. In general, we get $(f_k)_\infty = k\mathcal{Q} - k\mathcal{P}$. Let $\mathcal{D}$ be a minimal positive divisor on $\mathcal{C}$ such that: $(f_k) + \mathcal{D} \geq -k\mathcal{P} + k\mathcal{Q}, \forall k \in \mathbb{Z}$. It is shown that the degree of $\mathcal{D}$ is $\deg \mathcal{D} = g = N - 1$. We show that the divisor $\mathcal{D}$ is regular with respect to $\mathcal{P}$ and $\mathcal{Q}$, that is, such that $\dim \mathcal{L}(\mathcal{D} + k\mathcal{P} - (k+1)\mathcal{Q}) = 0, \forall k \in \mathbb{Z}$. The proof consists of showing, first, that the divisor $\mathcal{D}$ is general. A positive divisor $\mathcal{D}$ of degree g on $\mathcal{C}$ is general if $(\omega_j(p_j)) \neq 0$, $p_j \in \mathcal{C}, 1 \leq j \leq g$ where $(\omega_1, ..., \omega_g)$ is a normalized base of differential forms on $\mathcal{C}$. It is shown that $\mathcal{C}$ is general if and only if $\dim \mathcal{L}(\mathcal{D}) = 1$ (where $\mathcal{L}(\mathcal{D})$ denotes the set of meromorphic functions f on $\mathcal{C}$ such that: $(f) + \mathcal{D} \geq 0$) or if and only if $\dim \Omega(-\mathcal{D}) = 0$ where $\Omega(\mathcal{D})$ denotes the set of meromorphic differential forms ω on $\mathcal{C}$ such that the divisor $(\omega) + \mathcal{D} \geq 0$. Consider an integer $k > g - 2$, we then deduce from the Riemann–Roch theorem, $\dim \mathcal{L}(\mathcal{D} + k\mathcal{P}) = \dim \Omega(-\mathcal{D} - k\mathcal{P}) + g + k - g + 1 = \dim \Omega(-\mathcal{D} - k\mathcal{P}) + k + 1$. Since $\dim \Omega(-\mathcal{D} - k\mathcal{P}) = 0$, because a holomorphic differential can have at most $2g - 2$ zeros, then $\dim \mathcal{L}(\mathcal{D} + k\mathcal{P}) = k + 1$. Moreover, $\mathcal{L}(\mathcal{D} + j\mathcal{P})$ is strictly larger than $\mathcal{L}(\mathcal{D} + (j-1)\mathcal{P})$, because f_j belongs to the first space and not to the second. Therefore, by lowering the index j down to 0, it follows that $\dim \mathcal{L}(\mathcal{D}) = 1$, which shows that $\mathcal{D}$ is general. Let us now show that $\mathcal{D}$ is regular. It suffices to proceed by induction. Since $\dim \mathcal{L}(\mathcal{D}) = 1$ and $\mathcal{L}(\mathcal{D} - \mathcal{Q}) \subsetneq \mathcal{L}(\mathcal{D})$ (i.e. $f_0 = 1 \notin \mathcal{L}(\mathcal{D} - \mathcal{Q})$; the function $f_0 = 1$ belongs to the second space but not the first), then $\dim \mathcal{L}(\mathcal{D} - \mathcal{Q}) = 0$. Assume that $\dim \mathcal{L}(\mathcal{D} + k\mathcal{P} - (k+1)\mathcal{Q}) = 0$, then, by the Riemann–Roch theorem, $\dim \mathcal{L}(\mathcal{D} + (k+1)\mathcal{P} - (k+2)\mathcal{Q}) \leq \dim \mathcal{L}(\mathcal{D} + k\mathcal{P} - (k+1)\mathcal{Q}) + 1 = 1$ implies equality since f_{k+1} belongs to the first space. Since f_{k+1} belongs to $\mathcal{L}(\mathcal{D} + (k+1)\mathcal{P} - (k+1)\mathcal{Q})$ but not to $\mathcal{L}(\mathcal{D} + (k+1)\mathcal{P} - (k+2)\mathcal{Q})$, we have $\dim \mathcal{L}(\mathcal{D} + (k+1)\mathcal{P} - (k+2)\mathcal{Q}) = 0$. Consider the differential of F [5.3] while taking into account that z appears on only the diagonal of the matrix $A(h) - zI$. We have $-\sum_{i=1}^{N} \Delta_{ii} dz + h \frac{\partial F}{\partial h} \frac{dh}{h} = 0$, and either $\omega = \frac{-i \Delta_{NN} dz}{h \frac{\partial F}{\partial h}}$. So

$$\omega = \frac{-i \frac{dh}{h}}{\sum_{i=1}^{N} \frac{\Delta_{ii}}{\Delta_{NN}}} = \frac{-i \frac{dh}{h}}{\sum_{i=1}^{N} \frac{\Delta_{ii}}{\Delta_{iN}} \cdot \frac{\Delta_{iN}}{\Delta_{NN}}} = \frac{-i \frac{dh}{h}}{\sum_{i=1}^{N} \frac{\Delta_{Ni}}{\Delta_{NN}} \cdot \frac{\Delta_{iN}}{\Delta_{NN}}}.$$

Since $\Delta_{iN} = \Delta_{Ni}^*$, $1 \leq i \leq N$, then

$$\omega = \frac{-i \frac{dh}{h}}{\sum_{i=1}^{N} \frac{\Delta_{Ni}}{\Delta_{NN}} \cdot \left(\frac{\Delta_{iN}}{\Delta_{NN}}\right)^*} = \frac{-i \frac{dh}{h}}{\sum_{i=1}^{N} f_i f_i^*} = \pm \frac{\Delta_{NN} dz}{\sqrt{P^2(z) - 4A^2}}.$$

The differential ω belongs to $\Omega(-\mathcal{D} + \mathcal{P} + \mathcal{Q})$ and we deduce that $\omega^* = \omega$. In addition, $\omega \geq 0$ on $\mathcal{C}^\sim$. Indeed, on $\mathcal{C}^\sim$ we have $\sum_{i=1}^{N} f_i f_i^* = \sum_{i=1}^{N} |f_i|^2 \geq 0$. Let $h = \rho e^{i\theta}$. Note that in all finite number points, h is a local parameter on $\mathcal{C}$ while θ is a local parameter on $\mathcal{C}^\sim$. Like $-ih^{-1}dh = d\theta$, $\omega \geq 0$, at these points and by continuity at all points. We also have a relation that shows that the scalar product between f_k and f_l is $\langle f_k, f_l \rangle = \int_{\mathcal{C}^\sim} f_k \cdot f_l^* \omega = 0$ if $k \neq l$ and > 0 if $k = l$. That is, the functions f_k, $k \in \mathbb{Z}$, are orthogonal to $\mathcal{C}^\sim$ with respect to ω. We deduce from these properties that the divisor of ω is $(\omega) = \mathcal{D} + \widetilde{\mathcal{D}} - \mathcal{P} - \mathcal{Q}$, for the involution $\sim$ introduced previously. Given a matrix of the form A [5.2], we have obtained a series of data $\{\mathcal{C}, z, h, \mathcal{D}, \omega\}$. What is remarkable is that the reverse is also true. More specifically, we have the following theorem:

THEOREM 5.1.– There is a one-to-one correspondence between the following sets of data: (a) Let $a_i, b_i \in \mathbb{C}$, $a_i \neq 0$, where $a_{i+N} = a_i$, $b_{i+N} = b_i$, $-\infty < i < +\infty$. An infinite N-periodic matrix

$$\begin{pmatrix} \ddots & \ddots & & & & & \\ \ddots & b_0 & a_0 & 0 & \cdots & & 0 \\ & \bar{a}_0 & b_1 & a_1 & & & \vdots \\ & 0 & \bar{a}_1 & \ddots & \ddots & & 0 \\ & \vdots & & \ddots & \ddots & a_{N-1} & \\ & 0 & \cdots & 0 & \bar{a}_{N-1} & b_N & \ddots \\ & & & & & \ddots & \ddots \end{pmatrix},$$

modulo conjugation by N-periodic diagonal matrices with real entries. (b) A curve $\mathcal{C}$ (possibly singular) of genus $N-1$ with two points $\mathcal{P}$ and $\mathcal{Q}$ on $\mathcal{C}$, a divisor $\mathcal{D}$ of degree $N-1$ on $\mathcal{C}$, two meromorphic functions h and z on $\mathcal{C}$ such that: $(h) = -N\mathcal{P} + N\mathcal{Q}$ and $(z) = -\mathcal{P} - \mathcal{Q} +$ a positive divisor not containing the points $\mathcal{P}$ and $\mathcal{Q}$. The curve $\mathcal{C}$ is equipped with an antiholomorphic involution $\sim: (z, h) \longmapsto (\overline{z}, \overline{h}^{-1})$, such that: $\mathcal{C} = \mathcal{C}_+ \cup \mathcal{C}^\sim \cup \mathcal{C}_-$, where $\mathcal{C}^\sim = \{p \in \mathcal{C} : \widetilde{p} = p\} = \{(z,h) : |h| = 1\}$, $\mathcal{C}_+ = \{p \in \mathcal{C} : |h| > 1\}$, $\mathcal{C}_- = \{p \in \mathcal{C} : |h| < 1\}$, and such that: $\mathcal{P} \in \mathcal{C}_+$ with $\mathcal{Q} \in \mathcal{C}_-$. The involution $\sim$ extends to an involution $*$ on the field of the meromorphic functions with $\varphi^*(p) = \overline{\varphi(\widetilde{p})}$ and to the space of differentials with $(\varphi d\psi)^* = \varphi^* d\psi^*$ and so $h^* = h^{-1}$, $z^* = z$. The divisor of a differential form ω on $\mathcal{C}$ is $(\omega) = \mathcal{D} + \widetilde{\mathcal{D}} - \mathcal{P} - \mathcal{Q}$.

5.2. Difference operators

For any difference operator X, we define

$$\left(X^{[+]}\right)_{ij} = \begin{cases} X_{ij} & \text{if } i < j, \\ \frac{1}{2} X_{ij} & \text{if } i = j, \\ 0 & \text{if } i > j, \end{cases} \qquad X^{[-]} = X - X^{[+]}.$$

Let $\mathcal{M}$ be the vector space of infinite N-periodic matrices A such that for some K, $c_{ij} = 0$ if $|i - j| > K$. On $\mathcal{M}$, we consider the following scalar product: $\langle C, D \rangle = Tr(CD^\top) = \sum_{(i,j) \in \mathbb{Z}^2} c_{ij} d_{ij}$. We say that a functional F is differentiable if there exists a matrix $\frac{\partial F}{\partial C}$ in $\mathcal{M}$ such that for every D, $\lim_{\epsilon \to 0} \frac{F(C+\epsilon D) - F(C)}{\epsilon} = \langle \frac{\partial F}{\partial C}, D \rangle$. Note that $\langle [A, B], C \rangle = \langle [A^\top, C], B \rangle$. Define the following bracket between two differentiable functionals F and G on $\mathcal{M}$,

$$\{F, G\} = \left\langle \left[\left(\frac{\partial F}{\partial X}\right)^{[+]}, \left(\frac{\partial G}{\partial X}\right)^{[+]} \right] - \left[\left(\frac{\partial F}{\partial X}\right)^{[-]}, \left(\frac{\partial G}{\partial X}\right)^{[-]} \right], X \right\rangle.$$

$\{,\}$ satisfies the Jacobi identity. Let $P(A, S, S^{-1})$ be a polynomial in $S + S^{-1}$ and A with real coefficients. Consider the following Lax equation:

$$\dot{A} = \left[P(A, S, S^{-1})^{[+]} - P(A, S, S^{-1})^{[-]}, A \right]. \qquad [5.6]$$

When the matrix $A(t)$ deforms with t, then only the divisor $\mathcal{D}$ varies while $\{\mathcal{C}, z, h, \mathcal{P}, \mathcal{Q}\}$ remain fixed. As we have already shown, the coefficients of $z^i h^j$ in equation [5.3] are invariants of this motion. The divisor $\mathcal{D}(t)$ evolves linearly on the Jacobian manifold $\text{Jac}(\mathcal{C})$. Any linear flow over $\text{Jac}(\mathcal{C})$ is equivalent to equation [5.6] and is a Hamiltonian flow with respect to the above (Poisson) bracket. In particular, the flow $\dot{A} = \left[A, (S^{-k} A^l)^{[+]} \right]$ is written in terms of the (Poisson) bracket as follows: $\dot{a}_{ij} = \{F, a_{ij}\}$, $F = \frac{1}{l+1} Tr\left(S^{-k} A^{l+1}\right)$. The (Poisson) bracket of two functionals of the form $Tr\left(S^{-k}A^{l+1}\right)$ is zero, which means that we have a set of integrals in involution. Let $(\omega_1, ..., \omega_g)$ be a holomorphic differential basis on the hyperelliptic curve $\mathcal{C}$. We have $\omega_k = \frac{z^{k-1}}{\sqrt{P^2(z)-4Q^2}}$, and let $c_k = \operatorname{Res}_p(\omega_k z^j)$, $1 \leq j \leq g$. Since the order of the zeros of ω_k at the points at infinity $\mathcal{P}$, $\mathcal{Q}$ is equal to $g - k$, then $c_k = 0$ for $k < g - j + 1$ and $c_k \neq 0$ for $k = g - j + 1$. Therefore, a complete set of flows is given by the functions $z, z^2, ..., z^g$ and the flow that leaves the spectrum of A and X invariant is given by a polynomial $P(z)$ of degree at most equal to g: $\dot{A} = \frac{1}{2}[A, P(A)^+ - P(A)^-]$, where $P(A)^+$ (respectively, $-P(A)^-$) is the upper (respectively, lower) triangular part of $P(A)$, including the diagonal of $P(A)$. The (Poisson) bracket between two functional F and G can still be written in the form

$$\{F,G\} = \left\langle \begin{pmatrix} \frac{\partial F}{\partial a} \\ \frac{\partial F}{\partial b} \end{pmatrix}^\top, J \begin{pmatrix} \frac{\partial G}{\partial a} \\ \frac{\partial G}{\partial b} \end{pmatrix} \right\rangle,$$

where $\frac{\partial F}{\partial a}$ and $\frac{\partial F}{\partial b}$ are the column vectors whose elements are given by $\frac{\partial F}{\partial a_i}$ and $\frac{\partial F}{\partial b_i}$, respectively, while J is the antisymmetric matrix of order $2n$ defined by

$$J = \begin{pmatrix} O & \mathcal{A} \\ -\mathcal{A}^\top & O \end{pmatrix}, \quad \mathcal{A} = 2\begin{pmatrix} a_1 & 0 & 0 & \cdots & -a_N \\ -a_1 & a_2 & 0 & & \vdots \\ 0 & -a_2 & a_3 & & \vdots \\ \vdots & & & & \vdots \\ 0 & \cdots & \cdots & -a_{N-1} & a_N \end{pmatrix}.$$

The symplectic structure is given by

$$\omega = \sum_{j=2}^{N} db_j \wedge \sum_{j \leq i \leq N} \frac{da_i}{a_i}. \qquad [5.7]$$

Flaschka variables (Flaschka 1974): $a_j = \frac{1}{2}e^{x_j - x_{j+1}}$, $b_j = -\frac{1}{2}y_j$, applied to the form [5.7] with $x_{N+1} = 0$, leads to the symplectic structure $\omega = \frac{1}{2}\sum_{j=2}^{N} dx_j \wedge dy_j$, used by Moser (1980) during the study of the dynamical system of $N - 1$ particles moving freely on the real axis under the influence of the exponential potential. See also the example given in section 6.5 concerning the study of Toda lattice. We have $\det(A_h - zI)|_{h=i} = (-1)^N z^N + \sum_{i=1}^{N} \beta_i z^{N-1}$, where $\beta_2, ..., \beta_N$ are the g invariant, functionally independent and in involution. These invariants are given by the $g = N - 1$ points chosen from the spectrum of A_1 and A_{-1}, that is, by the branch points of the hyperelliptic curve $\mathcal{C}$ or by the quantities $tr A^k$, $2 \leq k \leq N$.

5.3. Continued fraction, orthogonal polynomials and Abelian integrals

Consider again the Jacobi's matrix Γ [5.1] and the associated continued Γ-fraction,

$$\varphi(z) = \cfrac{a_0^2}{z - b_1 - \cfrac{a_1^2}{z - b_2 - \cfrac{a_2^2}{z - b_3 - \cdots}}} \qquad [5.8]$$

where a_0 is a positive real number. By cutting off the Γ-fraction $\varphi(z)$ at the kth term, we obtain the kth Padé approximant $\frac{A_k(z)}{B_k(z)}$ of $\varphi(z)$, that is,

$$\varphi(z) = \lim_{k \to \infty} \frac{A_k(z)}{B_k(z)}. \qquad [5.9]$$

The degree of the polynomial $A_k(z)$ is $k - 1$, while the degree of $B(z)$ is k. Moreover, $\varphi(z)$ admits formal series expansion in a neighborhood of the pole $z = 0$ in the following form: $\varphi(z) = \frac{c_0}{z} + \frac{c_1}{z^2} + \frac{c_2}{z^3} + \cdots = \sum_{j=0}^{\infty} \frac{c_j}{z^{j+1}}$. Note that the characteristic polynomial B_k of the Γ-Jacobi matrix

$$B_k(z) = \det \begin{pmatrix} b_1 - z & a_1 & 0 & \cdots & 0 \\ a_1 & b_2 - z & a_2 & & \vdots \\ 0 & a_2 & \ddots & \ddots & 0 \\ \vdots & & \ddots & \ddots & a_{k-1} \\ 0 & \cdots & 0 & a_{k-1} & b_k - z \end{pmatrix},$$

is the last term of the second-order recursion $B_j(z) - (z - b_k)B_{j-1}(z) + a_{j-1}^2 B_{j-2}(z) = 0$. The polynomials $A_k(z)$, $B_k(z)$ form a pair of linearly independent solutions of a second-order finite difference equation (the eigenvectors of the Jacobi matrix from which we remove the first row and column): $a_j y_j + b_{j+1} y_{j+1} + a_{j+1} y_{j+2} = z y_{j+1}$, $j = 0, 1, \ldots$ with the boundary conditions: $y_0 \neq 0$, $y_1 = 0$, $y_{N+1} = 0$. We have also the relation: $A_{j-1}(z)B_j(z) - A_j(z)B_{j-1}(z) = \frac{1}{a_{j-1}}$, $j = 1, 2, \ldots$ The polynomials B_k form an orthogonal system with respect to a Stieltjes measure $d\sigma(x)$ on the real axis, $\int_{-\infty}^{\infty} B_k(x) B_l(x) d\sigma(x) = \delta_{kl}$. Conversely, if a family of polynomials $P_n(x)$ is orthogonal for $d\sigma(x)$, then $P_n(x)$ satisfies the following recurrence relation: $P_j(x) - (\lambda_j x - \mu_j) P_{j-1}(x) + \gamma_{j-1} P_{j-2}(x) = 0$, where $\lambda_j, \mu_j, \gamma_j > 0$ are constants. Moreover, if we consider the continued fraction,

$$\psi(z) = \cfrac{\gamma_0}{\lambda_1 z - \mu_1 - \cfrac{\gamma_1}{\lambda_2 z - \mu_2 - \cfrac{\gamma_2}{\lambda_3 z - \mu_3 - \cdots}}}$$

and realize an equivalent transformation

$$\psi(z) = \cfrac{\gamma_0}{z - \cfrac{\mu_1}{\lambda_1} - \cfrac{\frac{\gamma_1}{\lambda_1 \lambda_2}}{z - \cfrac{\mu_2}{\lambda_2} - \cfrac{\frac{\gamma_2}{\lambda_2 \lambda_3}}{z - \cfrac{\mu_3}{\lambda_3} - \ddots}}}$$

we reconstruct the Γ-fraction corresponding to $d\sigma(x)$ and put $\frac{\gamma_j}{\lambda_j \lambda_{j+1}} = a_j^2$, $\frac{\mu_j}{\lambda_j} = b_j$. It follows that there is a one-to-one correspondence between the set of Jacobi matrices and all the orthogonal polynomial systems on $\mathbb{R}$. If we choose the orthogonal polynomials $P_n = \frac{\gamma_0}{\prod_{j=1}^{n-1}} B_{n-1}(x)$, as the basis of the vector space consisting of all polynomials, then the Jacobi matrix represents the multiplication by x. With the Jacobi matrix, we associate an operator T on a separable Hilbert space E as follows, $Te_0 = b_0 e_0 + a_0 e_1$, $Te_j = a_{j-1} e_{j-1} + b_j e_j + a_j e_{j+1}$, $j = 1, 2, \ldots$ where $(e_1, \ldots)$ is an orthonormal basis in E. The operator T is symmetric. Indeed, for any two finite vectors u and v, we have $\langle Tu, v \rangle = \langle u, Tv \rangle$, according to the symmetry of the Jacobi matrix. Moreover, if the (Carleman's) condition: $\sum_{j=0}^{\infty} \frac{1}{a_j} = +\infty$, is satisfied, then the operator T is self-adjoint and its spectrum is simple with e_0 a generating element. In this case, the information about the spectrum of T is contained in function,

$$\varphi(z) = \langle (T - zI)^{-1} e_0, e_0 \rangle = \int_{-\infty}^{\infty} \frac{d\sigma(x)}{z - x}, \qquad [5.10]$$

defined at $z \notin \sigma(T)$ where $\sigma(x) = \langle I_x e_0, e_0 \rangle$ and I_x is the resolution of the identity of the operator T. Recall that the infinite continued fraction converges if the limit [5.9] exists. If the operator T is self-adjoint, then the continued fraction $\varphi(z)$ converges uniformly in any closed bounded domain of z without common points with the real axis to the analytic function defined by [5.10]. If the support of $d\sigma(x)$ is bounded, then the sequence $\left(\frac{A_k(z)}{B_k(z)} \right)$ converges uniformly to a holomorphic function near $z = \infty$. Moreover, if a Jacobi matrix is bounded, that is, if there exists $\rho > 0$ such that $\forall j$, $|a_j| \leq \frac{\rho}{3}$, $|b_j| \leq \frac{\rho}{3}$, then the associated Γ-fraction converges uniformly on the following domain $\{z : |z| \geq \rho\}$ and the support of $d\sigma(x)$ is included in $[-\rho, \rho]$. In the case of a periodic Jacobi matrix, this one is obviously bounded and therefore the associated Γ-fraction converges near $z = \infty$. In addition, the function $\varphi(z)$ is written in the form [5.10] (Cauchy–Stieltjes transform of $d\sigma(x)$), which shows that $\varphi(z)$ has zero of first order at $z = \infty$ and, for any point z belonging to the upper-half plane, the imaginary part of $\varphi(z)$ is non-positive.

We will now extend the Jacobi matrix Γ to the infinite symmetric, tridiagonal and N-periodic Jacobi matrix A [5.2] and use the results obtained previously. We consider $\varphi(z)$ [5.8] as being the associated N-periodic Γ-fraction. The latter converges near the infinite point $z = \infty$. After an analytic prolongation, the function $\varphi(z)$ coincides with $a_0 f_1$ where f_1 is a meromorphic function on the genus $N-1$ hyperelliptic curve $\mathcal{C}$ [5.4]. The latter is branched at the $2N$ real zeros $\xi_1, \xi_2,...,\xi_{2N}$ of the polynomial $P^2(z) - 4\alpha^2$. The interval $[\xi_{2j-1}, \xi_{2j}]$, $1 \leq j \leq N$, is called the stable band and the interval $[\xi_{2j}, \xi_{2j+1}]$, $1 \leq j \leq N-1$, is called the unstable band.

THEOREM 5.2.– Each zero $\sigma_1 < \sigma_2 < \cdots < \sigma_{N-1}$ of $\Delta_{k,l}$ [5.5] belongs to the jth finite unstable band $[\lambda_{2j}, \lambda_{2j+1}]$, $1 \leq j \leq N-1$.

The function $\varphi(z)$ can be expressed (see below) by means of Abelian integrals on the hyperelliptic curve $\mathcal{C}$ [5.4]. For $N = 1$, $B_k(x)$ is the kth Tschebyscheff polynomial of the second type. For $N > 1$, a new phenomenon related to discrete measures was discovered by authors including Kato. We have seen that $\varphi(z) = a_0 f_1 = a_0 \frac{\Delta_{N,1}}{\Delta_{N,N}} h$, belonging to $\mathcal{L}(\mathcal{D} + \mathcal{P} - \mathcal{Q})$. We then have the following theorem:

THEOREM 5.3.– The function $\varphi(z)$ can be explicitly written by means of Abelian integrals on the hyperelliptic curve $\mathcal{C}$ [5.4] as follows:

$$\varphi(z) = \sum_{j=1}^{N-1} \frac{\operatorname{Res}(\varphi(z), \sigma_j^-)}{z - \sigma_j} + \sum_{j=1}^{N} \frac{(-1)^{N+1}}{2\pi i} \int_{\xi_{2l-1}}^{\xi_{2l}} \frac{\sqrt{P^2(x) - 4\alpha^2}}{(z-x)\Delta_{N,N}(x)} dx,$$

[5.11]

where $\operatorname{Res}(\varphi(z), \sigma_j^-) = \frac{\alpha h(\sigma_j^-) + (-1)^N a_0^2 \cdot \Lambda}{\prod_{l \neq j}(\sigma_j - \sigma_l)}$, and

$$\Lambda \equiv \det \begin{pmatrix} b_2 - \sigma_j & a_2 & 0 & \cdots & 0 \\ a_2 & b_3 - \sigma_j & a_3 & & \vdots \\ 0 & a_3 & \ddots & \ddots & 0 \\ \vdots & & \ddots & \ddots & a_{N-2} \\ 0 & \cdots & 0 & a_{N-2} & b_{N-1} - \sigma_j \end{pmatrix}.$$

The differentials obtained in the previous section, $a \frac{\Delta_{N,N}(x)}{\sqrt{P^2(x) - 4\alpha^2}} dx$, $b \frac{\sqrt{P^2(x) - 4\alpha^2}}{\Delta_{N,N}(x)} dx$, ($a$ and b are constants), are positive measures on each stable band $[\xi_{2j-1}, \xi_{2j}]$. Therefore, expression [5.11] means that $\varphi(z)$ can be obtained by the Cauchy–Stieltjes transform of

$$d\sigma = \sum_{j=1}^{N-1} \operatorname{Res}(\varphi(z), \sigma_j^-) . \delta(x - \sigma_j) dx + \frac{(-1)^{N+1}}{2\pi i} \cdot \frac{\sqrt{P^2(x) - 4\alpha^2}}{\Delta_{N,N}(x)} dx,$$

$$= \text{discrete measure} + \text{continuous measure},$$

as follows, $\varphi(z) = \int_{-\infty}^{\infty} \frac{d\sigma}{z-x}$. The function $\varphi(z)$ belongs to $\mathcal{L}(\mathcal{D}' + \mathcal{P} - \mathcal{Q})$ where $\mathcal{D}' = \sigma_1^+ + \cdots + \sigma_{N-1}^+$ is contained in $\mathcal{C}_+ = \{p \in \mathcal{C} : |h| > 1\}$ (theorem 5.1). From expression [5.11], we have $\mathcal{D} = \sigma_{j_1}^- + \cdots + \sigma_{j_l}^- + \sigma_{j_{l+1}}^+ + \cdots + \sigma_{j_{N-1}}^+$, where $j_1 < j_2 < \ldots < j_l$ denote the numbers for which $\text{Res}(\varphi(z), \sigma_j^-) > 0$ and $j_{l+1} < j_{l+2} < \ldots < j_{N-1}$ the numbers for which $\text{Res}(\varphi(z), \sigma_j^-) = 0$. Hence,

$$\text{Res}(\varphi(z), \sigma_j^-) = 0 \text{ or } -\frac{\sqrt{P^2(\sigma_j^-) - 4\alpha^2}}{\prod_{l \neq j}(\sigma_j - \sigma_l)}.$$

5.4. Exercises

EXERCISE 5.1.– Prove the statement made in section 5.2 that the symplectic structure is given by formula [5.7].

EXERCISE 5.2.– Prove theorem 5.2.

EXERCISE 5.3.– Let $I_k(z)$ be the infinite part of continued fraction [5.8] starting with nth elements of sequences a_n and b_n. In the notation of section 5.3, express $I_k(z)$ in terms of polynomials A_k and B_k.

EXERCISE 5.4.– We have shown (section 5.3) how, with the Jacobi matrix, we can associate an operator T on a separable Hilbert space. Assume that $\lim_{k \to \infty} |b_k| = \infty$, $\limsup \frac{a_{k-1}^2}{|b_{k-1}b_k|} < \frac{1}{4}$. Show that the operator T has a discrete spectrum.

EXERCISE 5.5.– Let $\lambda_1, \ldots, \lambda_n$, $n > 3$, be n real numbers. Show that there exists a periodic Jacobi matrix:

$$\Gamma = \begin{pmatrix} b_1 & a_1 & 0 & \cdots & a_N \\ a_1 & b_2 & a_2 & & \vdots \\ 0 & a_2 & \ddots & \ddots & 0 \\ \vdots & & \ddots & \ddots & a_{N-1} \\ a_N & \cdots & \cdots & a_{N-1} & b_N \end{pmatrix}, \quad a_i \neq 0,$$

with the λ_i as periodic spectrum if and only if they can be ordered so that either (i) $\lambda_1 > \lambda_2 \geq \lambda_3 > \lambda_4 \geq \lambda_5 > \cdots$, and in this case the matrix Γ can be taken to have all a_i positive or (ii) $\lambda_1 = \lambda_2 > \lambda_3 \geq \lambda_4 > \lambda_5 \geq \cdots$, and in this case the matrix Γ will have some a_i negative. An alternative statement in this case is that there exists a periodic Jacobi matrix Γ with all a_i positive and the λ_i as antiperiodic spectrum (hint: see (Ferguson 1980)).

EXERCISE 5.6.– Prove the assertions made in section 5.3 that if the operator T is self-adjoint, then the continued fraction $\varphi(z)$ [5.8] converges uniformly in any closed bounded domain of z without common points with the real axis, to the analytic function defined by [5.10].

6

Griffiths Linearization Flows on Jacobians

The aim of this chapter is to present the Griffiths linearization method of studying integrable systems, summarizing the situations discussed in Chapter 4. Griffiths has found necessary and sufficient conditions on the matrix B, without reference to the Kac–Moody algebras, so that the flow of the Lax form [4.1] can be linearized on the Jacobian variety $\text{Jac}(\mathcal{C})$ for the spectral curve $\mathcal{C}$ defined by [4.2]. These conditions are cohomological and we will see that the Lax equations turn out to have a very natural cohomological interpretation. These results are exemplified by the Toda lattice, the Lagrange top, Nahm's equations and the n-dimensional rigid body.

6.1. Spectral curves

Suppose that for every $p(h, z)$ belonging to the curve of affine equation [4.1], that is,

$$\mathcal{C} = \{(h, z) : \det(A - zI) = 0\}, \qquad [6.1]$$

with $\dim \ker(A - zI) = 1$ (i.e. the corresponding eigenspace of A is one-dimensional) and generated by a vector $v(t, p) \in V$ where $V \simeq \mathbb{C}^n$ is an n-dimensional vector space. There is then a family of holomorphic mappings that send $(h, z) \in \mathcal{C}$ to $\ker(A - zI)$:

$$f_t : \mathcal{C} \longrightarrow \mathbb{P}V, \quad p \longmapsto \mathbb{C}v(t, p). \qquad [6.2]$$

(We call this the eigenvector map associated with the Lax equation). We set:

$$L_t = f_t^*\left(\mathcal{O}_{\mathbb{P}V}(1)\right) \in \text{Pic}^d(\mathcal{C}) \cong \text{Jac}(\mathcal{C}), \qquad L = L_0 \qquad [6.3]$$

where $d = \deg f_t(\mathcal{C})$; $\mathcal{O}_{\mathbb{P}V}(1)$ is the hyperplane line bundle on $\mathbb{P}V$ and $\text{Pic}^d(\mathcal{C})$ the Picard variety of $\mathcal{C}$. Let us recall that it is the set of straight bundles of degree d on $\mathcal{C}$.

By continuity, the degree of L_t does not vary with time t. Let $\mathbf{H}$ be the hyperplane class of $\mathbb{P}V$. We have $\deg L_t = \int_\mathcal{C} f_t^* \mathbf{H} = \int_{f_t(\mathcal{C})} \mathbf{H}$. This expression is the Poincaré dual of the class $[\mathcal{C}]$ of $\mathcal{C}$ and coincides with the degree of $\mathcal{C}$. Hence, $\deg L_t = \deg(\mathcal{C})$. While t varies, L_t moves in $\mathrm{Pic}^d(\mathcal{C})$. Therefore, if we fix a line bundle $L_0 \in \mathrm{Pic}^d(\mathcal{C})$, the line bundle $L_0^{-1} \otimes L_t$ moves in the Jacobian variety $\mathrm{Jac}(\mathcal{C}) = H^1(\mathcal{C}, \mathcal{O}_\mathcal{C})/H^1(\mathcal{C}, \mathbb{Z}) \simeq H^0(\mathcal{C}, \Omega_\mathcal{C})^*/H_1(\mathcal{C}, \mathbb{Z})$, that is, the mapping $L \longmapsto L_0^{-1} \otimes L$ induces a morphism $\mathrm{Pic}^d(\mathcal{C}) \simeq \mathrm{Jac}(\mathcal{C})$. The motion of the line bundle $L_0^{-1} \otimes L_t$ depends on the choice of the matrix B. A question arises: determine necessary and sufficient conditions on the matrix B so that the flow

$$t \longmapsto L_t \in \mathrm{Jac}(\mathcal{C}), \qquad [6.4]$$

can be linearized on the Jacobian variety $\mathrm{Jac}(\mathcal{C})$.

6.2. Cohomological deformation theory

As we have pointed out, Griffiths has found necessary and sufficient conditions of a cohomological nature on B that the flow $t \longmapsto L_t \in \mathrm{Jac}(\mathcal{C})$ be linear. His method is based on the observation that the tangent space to any deformation lies in a suitable cohomology group and on algebraic curves, and higher cohomology can always be eliminated using duality theory. In fact, by applying more or less standard cohomological techniques from deformation theory (Arbarello *et al.* 1985), we may give necessary and sufficient conditions that the map $t \longmapsto L_t$ be linear. Let

$$f : \mathcal{C} \longrightarrow X, \qquad [6.5]$$

be a non-constant holomorphic map where $\mathcal{C}$ is a given smooth algebraic curve and X is a complex manifold. We define the normal sheaf of $\mathcal{C}$ in X by the exact sequence

$$0 \longrightarrow \Theta_\mathcal{C} \xrightarrow{f_*} f^*\Theta_X \longrightarrow N_f \longrightarrow 0 \qquad [6.6]$$

where $\Theta_\mathcal{C}, \Theta_X$ are the respective tangent sheaves and f_* is the differential of f. Then the Kodaira–Spencer tangent space (Arbarello *et al.* 1985) to the moduli space of the map (6.5) is given by $H^0(\mathcal{C}, N_f)$. If $f_t : \mathcal{C} \longrightarrow X$, $f_0 = f$, is a deformation of [6.5], then $\dot{f} \in H^0(\mathcal{C}, N_f)$ the corresponding infinitesimal deformation at $t = 0$, that is, in local product coordinates (z, t) on $\cup_t \mathcal{C}_t$ and $w = (w^1, w^2, ..., w^n)$ of X, f_t is given by $(t, \xi) \longmapsto w(t, \xi)$, then the section $\dot{f} \in H^0(\mathcal{C}, N_f)$ is locally given by $\left.\frac{\partial w(t,\xi)}{\partial t}\right|_{t=0}$ modulo $\frac{\partial w(0,\xi)}{\partial z}$. The corresponding cohomological sequence of [6.6] is

$H^0(\Theta_\mathcal{C}) \longrightarrow H^0(f^*\Theta_X) \longrightarrow H^0(N_f) \xrightarrow{\bar{\partial}} H^1(\Theta_\mathcal{C})$. Here, $H^1(\Theta_\mathcal{C})$ is the tangent space to the moduli space of $\mathcal{C}$ as an abstract curve and $\bar{\partial}(\dot{f}) \equiv \dot{\mathcal{C}} \in H^1(\Theta_\mathcal{C})$ is the tangent to the family of curves $\{\mathcal{C}_t\}$. Thus, the tangent space to deformations of [6.5] where the curve $\mathcal{C}$ remains fixed is given by $H^0(f^*\Theta_X)/H^0(\Theta_\mathcal{C}) \subset H^0(N_f)$. Since the isospectral curve $\mathcal{C}$ is independent of t, this is the situation that we are interested in.

Here, we take again the vector space V of dimension n and assume that $X = \mathbb{P}V$ (projective space) and consider the Euler sequence

$$0 \longrightarrow \mathcal{O}_{\mathbb{P}V} \overset{i}{\longrightarrow} V \otimes \mathcal{O}_{\mathbb{P}V}(1) \overset{p}{\longrightarrow} \Theta_{\mathbb{P}V} \longrightarrow 0 \qquad [6.7]$$

This is an exact sequence of vector bundles, and it remains exact after pulling back to $\mathcal{C}$ via f^* (combining this with [6.6]). We have a diagram of exact sequences ($L = f^*\mathcal{O}_{\mathbb{P}V}(1)$):

$$\begin{array}{c} 0 \\ \downarrow \\ \mathcal{O}_\mathcal{C} \\ \downarrow v \\ V \otimes L \\ \downarrow \\ 0 \longrightarrow \Theta_\mathcal{C} \overset{f_*}{\longrightarrow} f^*\Theta_{\mathbb{P}V} \longrightarrow N_f \longrightarrow 0 \\ \downarrow \\ 0 \end{array}$$

The associated cohomology diagram contains the following piece:

$$\begin{array}{c} H^0(\mathcal{C}, V \otimes L) \\ \downarrow \tau \\ H^0(\mathcal{C}, \Theta_\mathcal{C}) \longrightarrow H^0(\mathcal{C}, f^*\Theta_{\mathbb{P}V}) \overset{j}{\longrightarrow} H^0(\mathcal{C}, N_f) \overset{\bar{\delta}}{\longrightarrow} H^1(\mathcal{C}, \Theta_\mathcal{C}) \\ \downarrow \delta \\ H^1(\mathcal{C}, \mathcal{O}_\mathcal{C}) \end{array}$$

Consider the family of holomorphic maps $f_t : \mathcal{C} \longrightarrow \mathbb{P}V$. Locally, choose a coordinate ξ on $\mathcal{C}$ and also a position vector mapping $(t, \xi) \longmapsto v(t, \xi) \in V \backslash \{0\}$, that is, a local lift v_t of f_t to $V \backslash \{0\}$, such that $f_t(\xi) = \mathbb{C}.v(t, \xi) \subset V$. Note that v_t is a time-dependent map $\mathcal{C} \longrightarrow V \backslash \{0\}$. This lift is not canonical and exists only locally, but we are going to use it to define an object denoted by $\dot{v}$, which will be independent of the lift and therefore will be globally well defined. Since $\mathcal{O}_{\mathbb{P}V}$ is the tautological bundle of $\mathbb{P}V$, the fiber of $f^*\mathcal{O}_{\mathbb{P}V}(-1)$ at a point $p \in \mathcal{C}$ may be identified with the space $\mathbb{C}v_t(p)$, which defines the maps $f^*\mathcal{O}_{\mathbb{P}V}(-1)V \otimes \mathcal{O}_\mathcal{C}$ and $v_t : \mathcal{O}_\mathcal{C} \longrightarrow V \otimes L_t$, $\phi \longmapsto \phi v_t$, where v_0 coincides with the application v mentioned in the previous diagram. If $\widetilde{v}$ is another lift given by $\widetilde{v}(t, \xi) = \kappa(t, \xi)v(t, \xi)$, $\kappa \neq 0$, then we have $\dot{\widetilde{v}} = \kappa \dot{v} + \dot{\kappa} v$. Let us set, $\dot{v}(\xi) = \left.\frac{\partial v(t,\xi)}{\partial t}\right|_{t=0}$ modulo $v(t, \xi)$. The latter quantity is well defined of the representative position mapping of v, that is, since the inclusion $\mathcal{O}_\mathcal{C} \overset{v}{\hookrightarrow} V \otimes L$, $L = f^*\mathcal{O}_{\mathbb{P}V}(1)$, is locally given by $\mathcal{O}_\mathcal{C} \ni \phi \longmapsto \phi.v$, it follows that $\dot{v} \in H^0(\mathcal{C}, V \otimes L/\mathcal{O}_\mathcal{C}) = H^0(\mathcal{C}, f^*\Theta_{\mathbb{P}V})$ is well defined and independent of the choice of the lift. Then we have

$$j(\dot{v}) = \dot{f} \qquad [6.8]$$

We are interested in the tangent vector $\dot{L} \equiv \frac{dL_t}{dt}\big|_{t=0} \in H^1(\mathcal{C}, \mathcal{O}_\mathcal{C})$.

THEOREM 6.1.– We have $\dot{L} = \delta(\dot{v})$, where $\dot{v}$ is the infinitesimal variation of $f_t : \mathcal{C} \longrightarrow \mathbb{P}V$ and in particular, $\dot{L} = 0$, if and only if $\dot{v} = \tau(w)$ for some $w \in H^0(V \otimes L)$ where τ is the map in the above diagram.

PROOF.– Let $(w_0, ..., w_n)$ be homogeneous coordinates in $\mathbb{P}V$ associated with a fixed basis of V and let U_i be the corresponding open cover of $\mathbb{P}V$. We shall write $w^k o v = v^k$. The quantity $\dot{v}$ as a vector field on $\mathbb{P}V$ may be written as $\sum_{i=0}^{n} \dot{v}^i \frac{\partial}{\partial w^i}$. In order to calculate the action of the connecting morphism δ on $\dot{v}$, we must take a counterimage of $\dot{v}$ under the morphism p of [6.7]; this is the n-ple $(\dot{v}_0, ..., \dot{v}_n)$. Now we must apply the Čech differential associated with the cover $f^{-1}(Ui)$ of $\mathcal{C}$ and invert the map i. Since the latter is $\dot{f} \longmapsto (\dot{f}w_0, ..., \dot{f}w_n)$, we get the cocycle $\frac{\dot{v}_k}{v_k} - \frac{\dot{v}_i}{v_i}$ that describes a class in $H^1(\mathcal{C}, \mathcal{O}_\mathcal{C})$. On the other hand, the transition functions of the line bundle L_t may be written as $g_{ik}(t) = \frac{w_k}{w_i} o f_t = \frac{v_k}{v_i}$, and the cocycle corresponding to $\dot{L}$ is therefore $\varepsilon_{ik} = \dot{g}_{ik} g_{ik}^{-1} = \frac{\dot{v}_k}{v_k} - \frac{\dot{v}_i}{v_i}$. The second assertion results from the exactness of the column in the diagram above. □

We write $B(t,h) = \sum_{k=0}^{N} B_k(t) h^k = \sum_{k=0}^{N} B_k(t) h_0^{N-k} h_1^k$, where we have regarded h as an affine coordinate in the $\mathbb{P}^1$, which is the base of the covering $\pi : \mathcal{C} \longrightarrow \mathbb{P}^1$, while h_0, h_1 are homogeneous coordinates. Recall that $B(t,h) \in H^0(\mathcal{C}, \mathrm{Hom}(V, V(N)))$, where V is the sheaf of sections of the trivial bundle $\mathcal{C} \times V$, $V(D) = V \otimes \mathcal{O}_\mathcal{C}(D)$. Here, $B(t,h)$ is a holomorphic section of the bundle $\mathrm{Hom}(V,V) \otimes \mathcal{O}_\mathcal{C}(N)$, $\mathcal{O}_\mathcal{C}(N) = \pi^*\mathcal{O}_{\mathbb{P}^1}(N)$, that is, we are viewing $h = [h_0, h_1]$ as a homogeneous coordinate on $\mathbb{P}^1$ pulled up to $\mathcal{C}$. Let $D = (h_0^N)$ be the divisor $N.\pi^{-1}(\infty)$ on the curve $\mathcal{C}$. Then $B/h_0^N \in H^0(\mathcal{C}, \mathrm{Hom}(V, V(D)))$ and $v \in H^0(V \otimes L)$, where $V(D) \cong V(N)$ are the sections of $V \otimes \mathcal{O}_\mathcal{C}(D)$ (here B/h_0^N is a matrix in $\mathrm{Hom}(V,V)$ with meromorphic functions in $H^0(\mathcal{C}, \mathcal{O}_\mathcal{C}(D))$ as entries, that is, we are viewing h_1/h_0 as a function in $H^0(\mathcal{C}, \mathcal{O}_\mathcal{C}(D))$). Hence, $\left(\frac{B}{h_0^N}\right).v \in H^0(\mathcal{C}, V \otimes L(D))$ and the cohomological interpretation of the Lax equation is given by the following theorem:

THEOREM 6.2.– We have

$$\dot{v} = \tau\left(\frac{B}{h_0^N}.v\right). \qquad [6.9]$$

In addition, $\dot{L} = 0$ if and only if there is a meromorphic function $\varphi \in H^0(\mathcal{C}, \mathcal{O}_\mathcal{C}(D))$ such that $\frac{B}{h_0^N}.v + \varphi v \in H^0(\mathcal{C}, V \otimes L(D))$ is holomorphic.

PROOF.– We consider a diagram of exact sequence of sheaves on the curve $\mathcal{C}$,

$$\begin{array}{ccccccccc}
& & 0 & & 0 & & 0 & & \\
& & \downarrow & & \downarrow & & \downarrow & & \\
0 & \longrightarrow & \mathcal{O}_\mathcal{C} & \longrightarrow & \mathcal{O}_\mathcal{C}(\mathcal{D}) & \longrightarrow & \mathcal{O}_\mathcal{D}(\mathcal{D}) & \longrightarrow & 0 \\
& & \downarrow v & & \downarrow v & & \downarrow & & \\
0 & \longrightarrow & V \otimes L & \longrightarrow & V \otimes L(\mathcal{D}) & \longrightarrow & V \otimes L \otimes \mathcal{O}_\mathcal{D}(\mathcal{D}) & \longrightarrow & 0 \\
& & \downarrow \tau & & \downarrow \tau & & \downarrow & & \\
0 & \longrightarrow & f^*\Theta_{\mathbb{P}V} & \longrightarrow & f^*\Theta_{\mathbb{P}V}(\mathcal{D}) & \longrightarrow & f^*\Theta_{\mathbb{P}V} \otimes \mathcal{O}_\mathcal{D}(\mathcal{D}) & \longrightarrow & 0 \\
& & \downarrow & & \downarrow & & \downarrow & & \\
& & 0 & & 0 & & 0 & &
\end{array}$$

whence we get the cohomology diagram,

$$\begin{array}{ccccccc}
& & 0 & & 0 & & \\
& & \downarrow & & \downarrow & & \\
& & H^0(\mathcal{C},\mathcal{O}_\mathcal{C}(\mathcal{D})) & \longrightarrow & H^0(\mathcal{C},\mathcal{O}_\mathcal{D}(\mathcal{D})) & \xrightarrow{\delta_1} & \\
& & \downarrow v & & \downarrow \sigma & & \\
0 \longrightarrow & H^0(\mathcal{C}, V \otimes L) \xrightarrow{i} & H^0(\mathcal{C}, V \otimes L(\mathcal{D})) & \xrightarrow{j} & H^0(\mathcal{C}, V \otimes L \otimes \mathcal{O}_\mathcal{D}(\mathcal{D})) & & \\
& \downarrow \tau & \downarrow \tau & & \downarrow \tau & & \\
0 \longrightarrow & H^0(\mathcal{C}, f^*\Theta_{\mathbb{P}V}) \xrightarrow{i} & H^0(\mathcal{C}, f^*\Theta_{\mathbb{P}V}(\mathcal{D})) & \xrightarrow{j} & H^0(\mathcal{C}, f^*\Theta_{\mathbb{P}V} \otimes \mathcal{O}_\mathcal{D}(\mathcal{D})) & & \\
& \downarrow \delta & \downarrow \delta & & & & \\
\xrightarrow{\delta_1} & H^1(\mathcal{C}, \mathcal{O}_\mathcal{C}) \longrightarrow & H^1(\mathcal{C}, \mathcal{O}_\mathcal{C}(\mathcal{D})) & & & &
\end{array}$$

Then the meaning of formula [6.9] is $\frac{B}{h_0^N}.v \in H^0(V \otimes L(\mathcal{D}))$, $\tau\left(\frac{B}{h_0^N}.v\right) = \dot{i}(\dot{v})$. Working in $\mathbb{C}^2$ with coordinates (h, z), $A(t, h)$ and $B(t, h)$ are polynomials in $h \in \mathbb{C}$ whose coefficients are holomorphic functions of t. We write $B(t, h) = \frac{B}{h_0^N}$ where $B \in H^0(V, V(N))$ is considered a homogeneous polynomial in h_0, h_1. Since the tangent space to any algebro-geometric moduli space is computed cohomologically, the answer to the question above is expressed in terms of an H^1. The cohomology H^1 can always be reduced to the cohomology H^0 using the duality. Near the point $p = (h, z) \in \mathcal{C}$, differentiating with regard to t the eigenvalue problem $Av(t, p) = zv(t, p)$ leads to $\dot{A}v + A\dot{v} = z\dot{v}$. Using the Lax equation: $\dot{A} = [B, A]$, we obtain $A(\dot{v} - Bv) = z(\dot{v} - Bv)$. Since generically the eigenvalues have multiplicity 1, we have

$$Bv = \dot{v} + \lambda v, \qquad [6.10]$$

for a some λ. Hence $\tau(Bv) = \dot{v} \in V \otimes L/\mathbb{C}.v$. The existence of φ is equivalent to the existence of $b \in H^0(\mathcal{C}, \mathcal{O}_\mathcal{C}(\mathcal{D}))$ with $i(b) = \frac{B}{h_0^N}.v + \varphi v \in H^0(\mathcal{C}, V \otimes L(\mathcal{D}))$, and then by the above commutative diagram, we get $\dot{L} = \delta \dot{v} = \delta \tau(b) = 0$. □

Since $Bv = \dot{v} + \lambda v$, that is, $Bv = \dot{v} + \lambda_j v$, where λ_j is the principal part of the Laurent series expansion of λ at p, then given the curve $\mathcal{C}$ defined by [4.1] and $p \in \mathcal{C}$, Griffiths defines $[\text{Laurent tail}(B)]_p \equiv \{\text{principal part of the Laurent series expansion of } \lambda \text{ at } p\}$, and shows that the Lax flow can be linearized on the Jacobian variety $\text{Jac}(\mathcal{C})$ if and only if for every $p \in (h)_\infty$ (divisor of the poles of h), we have

$$\frac{d}{dt}[\text{Laurent tail}(B)]_p \in \text{linear combination } \{[\text{Laurent tail}(B)]_p; \text{Laurent}$$

$$\text{tail at } p \text{ of any meromorphic function } f \text{ on } \mathcal{C}$$

$$\text{such that} : (f) \geq n(h)_\infty\}.$$

Equation [4.1] is invariant under the substitution $B \longmapsto B + P(h, A)$, $P(h, g) \in \mathbb{C}[h, g]$, which shows that B is not unique and that its natural place is somewhere in a cohomology group. Let $B(t, h) = \sum_{k=0}^n B_k h^k$ be a polynomial of degree n. Let $\mathcal{D} = h^{-1}(\infty) = \sum_j n_j p_j$, $n_j \geq 0$ (where h is seen as a meromorphic function) be a positive divisor on $\mathcal{C}$ and let z_j be a local coordinate around p_j. B must be interpreted as an element of $H^0(\mathcal{C}, \text{Hom}(V, V(\mathcal{D})))$ where V is the sheaf of sections of the trivial bundle $\mathcal{C} \times V$ and $V(\mathcal{D}) = V \otimes \mathcal{O}_\mathcal{C}(\mathcal{D})$. A section of $\mathcal{O}_\mathcal{D}(\mathcal{D})$ is written as $\varphi = \sum \varphi_j$, $\varphi_j = \sum_{k=-n_j}^{-1} a_k z_j^k$; it is a principal part (Laurent tail) centered on p_j.

6.3. Mittag–Leffler problem

The Mittag–Leffler problem can be formulated as follows: given a principal part φ_j, find conditions for a function $\varphi \in H^0(\mathcal{C}, \mathcal{O}_\mathcal{C}(\mathcal{D}))$ such that $\varphi - \varphi_j$ is holomorphic around p_j. The answer is provided by the following theorem:

THEOREM 6.3.– Let $\mathcal{D} = \sum_j a_j p_j$. Given the Laurent tail $\{\varphi_j\}$, there exists $\varphi \in H^0(\mathcal{C}, \mathcal{O}_\mathcal{C}(\mathcal{D}))$ such that $\varphi - \varphi_j$ is holomorphic near p_j if and only if

$$\sum_j \text{Res}_{p_j}(\varphi_j.\omega) = 0, \qquad [6.11]$$

for every holomorphic differential ω on $\mathcal{C}$.

PROOF.– The exact sheaf sequence of Mittag–Leffler attached to the divisor $\mathcal{D}$ is $0 \longrightarrow \mathcal{O}_\mathcal{C} \longrightarrow \mathcal{O}_\mathcal{C}(\mathcal{D}) \longrightarrow \mathcal{O}_\mathcal{D}(\mathcal{D}) \longrightarrow 0$, and the last term $\mathcal{O}_\mathcal{D}(\mathcal{D})$ is a skyscraper sheaf that may be identified with the collections $\{\varphi_j\}$ of the Laurent tail. We deduce the cohomology sequence of this exact sequence as well as its dual

$$H^0(\mathcal{C}, \mathcal{O}_\mathcal{C}(\mathcal{D})) \xrightarrow{res} H^0(\mathcal{C}, \mathcal{O}_\mathcal{D}(\mathcal{D})) \xrightarrow{\delta_1} H^1(\mathcal{C}, \mathcal{O}_\mathcal{C}) \longrightarrow H^1(\mathcal{C}, \mathcal{O}_\mathcal{C}(\mathcal{D})) \longrightarrow 0,$$

$$H^1(\mathcal{C}, \mathcal{O}_\mathcal{C}(-D)) \longleftarrow H^0(\mathcal{C}, \mathcal{O}_\mathcal{D}(D))^* \xleftarrow{\delta_1^*} H^0(\mathcal{C}, \Omega_\mathcal{C}) \longleftarrow H^0(\mathcal{C}, \mathcal{O}_\mathcal{C}(-D)),$$

where $H^0(\mathcal{C}, \Omega_\mathcal{C})$ is the space of holomorphic 1-forms on $\mathcal{C}$. The problem is therefore equivalent to the resolution of the equation $\delta\varphi = 0$ and because of the duality this amounts to solving the system of linear equations: $\langle \delta_1\varphi, \omega \rangle = 0, \forall \omega \in H^0(\mathcal{C}, \Omega_\mathcal{C})$. Consider an open cover of $\mathcal{C}$ by small discs U_j centered in p_j. We suppose that on every U_j there exists a meromorphic function f_j such that: $\mathrm{Res}_{p_j}(f_j) = \varphi_j$, if U_j contains p_j (otherwise, we will choose the zero function). Then the cocycle $\delta_1(\varphi)_{ik} = \{f_i - f_k\}$ represents an element of $H^1(\mathcal{C}, \mathcal{O}_\mathcal{C})$. The corresponding $(0,1)$-form under the Dolbeault isomorphism $H^1(\mathcal{C}, \mathcal{O}_\mathcal{C}) \simeq H^{0,1}_{\bar\partial}(\mathcal{C})$ is $h = \sum_j \bar\partial(z_j f_j)$, where $\{z_j\}$ is a partition of the unity (with supp $z_j \subset U_j$ such that $\sum_j z_j = 1$ in a neighborhood of p_j) and we define $h(p_j) = 0$. Recall that, by definition, $\langle \delta_1\varphi, \omega \rangle = \int_\mathcal{C} \phi$, where ϕ is a Dolbeault representative of the cup-product $\delta\varphi.\omega$. With the notations introduced above, we therefore have $\langle \delta_1\varphi, \omega \rangle = \int_\mathcal{C} h \wedge \omega = \int_\mathcal{C} \sum_j \bar\partial(z_j f_j \omega) = \int_\mathcal{C} \sum_j d(z_j f_j \omega)$. Let $V_j(\epsilon)$ be a disc of radius ϵ and center p_j. We have

$$\langle \delta_1\varphi, \omega \rangle = \lim_{\epsilon \to 0} \int_{\mathcal{C}\setminus(\cup_j V_j(\epsilon))} \sum_j d(z_j f_j \omega) = -\lim_{\epsilon \to 0} \int_{\partial V_j(\epsilon)} \sum_j d(z_j f_j \omega),$$

so $\langle \delta_1\varphi, \omega \rangle = -2\pi i \sum_j \mathrm{Res}_{p_j}(f_j \omega) = -2\pi i \sum_j \mathrm{Res}_{p_j}(\varphi_j \omega)$, and $\forall \omega \in H^0(\mathcal{C}, \Omega_\mathcal{C})$, $\langle \delta_1\varphi, \omega \rangle = 0$ is equivalent to [6.11]. The number of independent equations of this system of g linear equations with d unknowns is equal to $g - \dim \mathcal{I}(-D)$. Now, two meromorphic functions having the same principal part differ only by a constant, so $\dim H^0(\mathcal{C}, \mathcal{O}_\mathcal{C}(D)) = \deg D - g + 1 + \dim \mathcal{I}(-D)$, which is the Riemann–Roch theorem. The theorem results from [6.11] and the fact that the above sequences are exact. $\square$

6.4. Linearizing flows

The residue of B, denoted by $\rho(B) \in H^0(\mathcal{C}, \mathcal{O}_\mathcal{D}(D))$, is the collection of Laurent tails $\{\lambda_j\}$ given above (recall that λ_j is the principal part of the Laurent series expansion of λ at p). We shall say that the flow L_t is linear if there exists a complex number a such that $\frac{d^2 L_t}{dt^2} = a \frac{dL_t}{dt}$. The Griffiths theorem is as follows:

THEOREM 6.4.– (a) We have $\dot{L} = \frac{dL_t}{dt}\big|_{t=0} = \delta_1(\rho(B))$. (b) Let Im res $\subset H^0(\mathcal{C}, \mathcal{O}_\mathcal{D}(D))$ be the Laurent tails of meromorphic functions in $H^0(\mathcal{C}, \mathcal{O}_\mathcal{D}(D))$. Then the flow $L_t[6.4]$ in $\mathrm{Pic}^d(\mathcal{C})$ is linear if and only if

$$\rho(\dot{B}) = 0 \mod(\rho(B), \mathrm{Im\,res}). \qquad [6.12]$$

PROOF.– (a) By the commutative diagram above, we let $E \in H^0(\mathcal{C}, V \otimes L(D))$ satisfy $\tau(E) = i(w)$ for some $w \in H^0(f^*\mathcal{O}_{\mathbb{P}V})$. In particular, take $E = \frac{B}{h_0^N}.v$ and

$w = \dot{v}$ as in theorem 6.2. By commutativity $\tau j(E) = j\tau(E) = ji(\dot{v}) = 0$ and so there exists $\lambda \in H^0(\mathcal{C}, \mathcal{O}_\mathcal{D}(\mathcal{D}))$ such that $\sigma(\lambda) = j(E)$. Since (from the diagram above) $H^0(\mathcal{C}, \mathcal{O}_\mathcal{D}(\mathcal{D}))$ occurs in the top right corner as well as the bottom left corner, we get $L = \delta(\dot{v}) = \delta_1(\lambda) = \delta(\rho(B))$. (b) Note that the equality $\frac{d^2 L_t}{dt^2} = a\frac{dL_t}{dt}$ holds in the fixed vector space $H^1(\mathcal{C}, \mathcal{O}_\mathcal{C})$ and $\rho(\ddot{B}) = 0$ in the fixed vector space $H^0(\mathcal{C}, \mathcal{O}_\mathcal{D}(\mathcal{D})) \cong \mathbb{C}^k$, where $k = \deg \mathcal{D}$. Let $\phi \in H^0(\mathcal{C}, \mathcal{O}_\mathcal{C}(\mathcal{D}))$. We deduce from $\rho(\dot{B}) = a\rho(B) + \operatorname{Res}\phi$, that $\delta_1(\rho(\dot{B})) = a\delta_1(\rho(B))$, and from (a), $\ddot{L} = a\dot{L}$. Conversely, if $\ddot{L} = a\dot{L}$, then $\delta_1(\rho(\ddot{B})) - a\delta_1(\rho(B)) = 0$, and therefore there exists $\phi \in H^0(\mathcal{C}, \mathcal{O}_\mathcal{C}(\mathcal{D}))$ such that $\rho(\dot{B}) - a\rho(B) = \operatorname{Res}\phi$, which completes the proof. $\square$

The condition [6.12] is equivalent to $\sum_j \operatorname{Res}_{p_j}(\dot{\rho}_j(B))\omega) = t\sum_j \operatorname{Res}_{p_j}(\rho_j(B))\omega)$, $\omega \in H^0(\mathcal{C}, \Omega_\mathcal{C})$. If this is satisfied, then the linear flow on $\operatorname{Jac}(\mathcal{C})$ is given by the bilinear map

$$(t, \omega) \longmapsto t\sum_j \operatorname{Res}_{p_j}(\rho_j(B))\omega) = t\sum_j \operatorname{Res}_{p_j}(\lambda_j \omega). \qquad [6.13]$$

6.5. The Toda lattice

The Toda lattice (Toda 1967) is a system of n particles connected by nonlinear springs with a restoring force depending exponentially on displacement. Such a system has been known for some time as a discrete version of the Korteweg–de Vries equation[1] and is governed by the following Hamiltonian $H = \frac{1}{2}\sum_{j=1}^{N} y_j^2 + \sum_{j=1}^{N} e^{x_j - x_{j+1}}$. The Hamiltonian equations are $\dot{x}_j = y_j$, $\dot{y}_j = -e^{x_j - x_{j+1}} + e^{x_{j-1} - x_j}$. Flaschka variables (Flaschka 1974): $a_j = \frac{1}{2}e^{x_j - x_{j+1}}$, $b_j = -\frac{1}{2}y_j$ can be used to express the symplectic structure ω [5.7] in terms of x_j and y_j as follows: $\frac{da_j}{a_j} = dx_j - dx_{j+1}$, $2db_j = -dy_j$, then $\omega = -\frac{1}{2}\sum_{j=2}^{N} dy_j \sum_{i=j}^{N}(dx_i - dx_{i+1}) = \frac{1}{2}\sum_{j=2}^{N} dx_j^* \wedge dy_j^*$. We will study the integrability of this problem with the Griffiths approach. There are two cases: (i) The non-periodic case, that is, $x_0 = -\infty$, $x_{N+1} = +\infty$, where the masses are arranged on a line.

Figure 6.1. Toda lattice (non-periodic case)

In terms of the Flaschka variables above, Toda's equations take the following form: $\dot{a}_j = a_j(b_{j+1} - b_j)$, $\dot{b}_j = 2(a_j^2 - a_{j+1}^2)$, with $a_{N+1} = a_1$ and $b_{N+1} = b_1$. To

---

[1] In short, the KdV equation: $\frac{\partial u}{\partial t} - 6u\frac{\partial u}{\partial x} + \frac{\partial^3 u}{\partial x^3} = 0$. This is an infinite-dimensional completely integrable system. For more information on this equation, see Chapter 9.

show that this system is completely integrable, one should find N first integrals independent and in involution each other. From the second equation, we have $\left(\sum_{j=1}^{N} b_j\right)^{\cdot} = \sum_{j=1}^{N} \dot{b}_j = 0$, and we normalize the b_j's by requiring that $\sum_{j=1}^{N} b_j = 0$ (applying this fact to [5.7] leads to the original symplectic form $\omega = \frac{1}{2} \sum_{j=2}^{N} dx_j \wedge dy_j$). This is a first integral for the system and to show that it is completely integrable, we must find $N - 1$ other integrals that are functionally independent and in involution. We further define $N \times N$ matrices A and B with

$$A = \begin{pmatrix} b_1 & a_1 & 0 & \cdots & a_N \\ a_1 & b_2 & \vdots & & \vdots \\ 0 & \ddots & \ddots & \ddots & 0 \\ \vdots & & \ddots & b_{N-1} & a_{N-1} \\ a_N & \cdots & 0 & a_{N-1} & b_N \end{pmatrix}, B = \begin{pmatrix} 0 & a_1 & \cdots & & -a_N \\ -a_1 & 0 & \vdots & & \vdots \\ \vdots & & \ddots & \ddots & \vdots \\ \vdots & & \ddots & \ddots & a_{N-1} \\ a_N & \cdots & \cdots & -a_{N-1} & 0 \end{pmatrix}.$$

The proposed system is equivalent to the Lax equation $\dot{A} = [B, A]$. From theorem 4.1, the quantities $I_k = \frac{1}{k} tr A^k$, $1 \leq k \leq N$, are first integrals of motion. To be more precise, $\dot{I}_k = tr(\dot{A} \cdot A^{k-1}) = tr([B, A] \cdot A^{k-1}) = tr(BA^k - ABA^{k-1}) = 0$. Note that I_1 is the first already known integral. Since these N first integrals are shown to be independent and in involution with each other, the system in question is thus completely integrable. (ii) The periodic case, that is, $y_{j+N} = y_j$, $x_{j+N} = x_j$, the connected masses will be arranged on a circle.

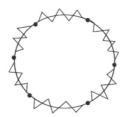

Figure 6.2. *Toda lattice (periodic case)*

We show that in this case, the spectrum of the periodic Jacobi matrix

$$A = \begin{pmatrix} b_1 & a_1 & 0 & \cdots & a_N h^{-1} \\ a_1 & b_2 & \vdots & & \vdots \\ 0 & \ddots & \ddots & \ddots & 0 \\ \vdots & & \ddots & b_{N-1} & a_{N-1} \\ a_N h & \cdots & 0 & a_{N-1} & b_N \end{pmatrix},$$

remains invariant in time. The matrix B depending on the spectral parameter h is

$$B = \begin{pmatrix} 0 & a_1 & \cdots & \cdots & -a_N h^{-1} \\ -a_1 & 0 & \vdots & & \vdots \\ \vdots & \ddots & \ddots & \ddots & \vdots \\ \vdots & & \ddots & \ddots & a_{N-1} \\ a_N h & \cdots & \cdots & -a_{N-1} & 0 \end{pmatrix},$$

and the rest follows from the general theory. Note that if $a_j(0) \neq 0$, then $a_j(t) \neq 0$ for all t. Since $A^\top(h) = A(h^{-1})$, then $P(h,z) = \det(A(h) - zI) = P(h^{-1}, z)$. Therefore, the application

$$\sigma : \mathcal{C} \longrightarrow \mathcal{C}, \quad (h,z) \longmapsto (h^{-1}, z), \qquad [6.14]$$

is an involution on the spectral curve $\mathcal{C}$. We choose

$$A(h) = \begin{pmatrix} 0 & \cdots & a_N \\ \vdots & \ddots & \vdots \\ 0 & \cdots & 0 \end{pmatrix} h^{-1} + \begin{pmatrix} b_1 & a_1 & & & \\ a_1 & b_2 & & & \\ & & \ddots & & \\ & & & b_{N-1} & a_{N-1} \\ & & & b_N & a_N \end{pmatrix}$$

$$+ \begin{pmatrix} 0 & \cdots & 0 \\ \vdots & \ddots & \vdots \\ a_N & \cdots & 0 \end{pmatrix} h.$$

The matrix A is meromorphic (previously we considered it to be a polynomial in h) but we will see that we can also adopt the theory explained in this chapter to this situation. We have $P(h,z) = -\prod_{j=1}^{N-1} a_j \cdot (h + h^{-1}) + z^N + c_1 z^{N-1} + \cdots + c_N$. Let us assume that $\prod_{j=1}^{N-1} a_j \neq 0$ and let

$$Q(h,z) \equiv \frac{P(h,z)}{\prod_{j=1}^{N-1} a_j} = h + h^{-1} + \frac{z^N + c_1 z^{N-1} + \cdots + c_N}{\prod_{j=1}^{N-1} a_j},$$

$$= h + h^{-1} + d_0 z^N + d_1 z^{N-1} + \cdots + d_N.$$

In $\mathbb{P}^2(\mathbb{C})$, the affine algebraic curve of equation $Q(h,z) = 0$ is singular at infinity for $n \geq 4$. We will compute the genus of the normalization $\mathcal{C}$ of this curve. Note that $\mathcal{C}$ is a double covering of $\mathbb{P}^1(\mathbb{C})$ branched into $2N$ points coinciding with the fixed points of involution σ [6.14], that is, points where $h = \pm 1$. According to the Riemann–Hurwitz formula (see Appendix 2), the genus g of the curve $\mathcal{C}$ is

$g = 2\left(g(\mathbb{P}^1(\mathbb{C})) - 1\right) + 1 + \frac{2N}{2} = N - 1$. Consider the covering $\mathcal{C} \longrightarrow \mathbb{P}^1(\mathbb{C})$ below and set $\frac{1}{z}(\infty) = \mathcal{P} + \mathcal{Q}$, where $\mathcal{P}$ and $\mathcal{Q}$ are located on two separate sheets. From the equation $Q(h, z) = 0$, the divisor of h is $(h) = N\mathcal{P} - N\mathcal{Q}$. The divisor $\mathcal{D}$ is written $\mathcal{D} = N\mathcal{P} + N\mathcal{Q}$, hence $B \in H^0(\mathcal{C}, \mathrm{Hom}(V, V(\mathcal{D})))$.

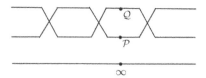

Figure 6.3. *Divisor* $\mathcal{D}$

The residue $\rho(B) \in H^0(\mathcal{C}, \mathcal{O}_\mathcal{D}(\mathcal{D}))$ satisfies the conditions of theorem 6.4 and consequently the linear flow is given by the application [6.13]. To compute the residue $\rho(B)$ of B, we will determine a set of holomorphic eigenvectors, using the van Moerbeke–Mumford method described above. Let us calculate the residue in $\mathcal{Q}$ and the result will be similarly deduced in $\mathcal{P}$. Let $\mathcal{E} = \sum_{j=1}^{g} r_j$ be a general divisor of degree g such that: $\forall k, \dim \mathcal{L}(\mathcal{E} + (k - 1)\mathcal{P} - k\mathcal{Q}) = 0$. According to the Riemann–Roch theorem, $\dim \mathcal{L}(\mathcal{E} + k\mathcal{P} - k\mathcal{Q}) \geq 1$, hence $\dim \mathcal{L}(\mathcal{E} + k\mathcal{P} - k\mathcal{Q}) = 1$, for all k. Let $(f_k) \in \mathcal{L}(\mathcal{E} + k\mathcal{P} - k\mathcal{Q}) = H^0(\mathcal{C}, \mathcal{O}_\mathcal{C}(\mathcal{E} + k\mathcal{P} - k\mathcal{Q}))$, $1 \leq k \leq N$, be a base with $f_N = h$. We can choose a vector v of the following form $v = (f_1, ..., f_N)^\top$, such that v is an eigenvector of A, that is, $Av = zv$, $(h, z) \in \mathcal{C}$. Hence, $V = h^{-1}v$ is a holomorphic eigenvector. Without restricting generality, we take $N = 3$. The system $Av = zv$ is explicitly written $b_1 f_1 + a_2 f_2 + a_3 = z f_1$, $a_1 f_1 + b_2 f_2 + a_2 h = z f_2$, $a_3 h f_1 + a_2 f_2 + b_3 h = z h$. By multiplying each equation of this system by h^{-1}, everything becomes holomorphic except the last equation, that is, $a_3 f_1 = z + $ Taylor. Recall that the residue $\rho(B)$ of B is the section of $\mathcal{O}_\mathcal{D}(\mathcal{D})$ induced by λ in the equation $Bv = \dot{v} + \lambda v$. In other words, $Bv = \varrho(B)v + $ Taylor, and therefore

$$\begin{pmatrix} \frac{a_1 f_2}{h} - \frac{a_3}{h} \\ -\frac{a_1 f_1}{h} + a_2 \\ a_3 f_1 - \frac{a_2 f_2}{h} \end{pmatrix} = \begin{pmatrix} 0 \\ 0 \\ z \end{pmatrix} + \text{Taylor}.$$

Hence, $\rho(B) = h^{-1}z$ and $\rho(\dot{B}) = 0$. The same conclusion holds for the residue in $\mathcal{P}$. Then, the flow in question linearizes on the Jacobian variety of $\mathcal{C}$.

6.6. The Lagrange top

We consider the Lagrange top discussed in sections 3.2.2 and 4.7. It is proved in Ratiu and van Moerbeke (1982) that equations [3.3] or [3.4] (corresponding to the

Lagrange case) may be written in the particular form $\overbrace{(\Gamma + Mh + ch^2)}^{\cdot} = [\Gamma + Mh + ch^2, \Omega + Lh]$ of a Lax equation with a parameter if, and only if, $\lambda_1 = \lambda_2 \equiv \alpha$, $l_1 = l_2 = 0$. We set

$$A(h) = \Gamma + Mh + ch^2 \in so(3). \qquad [6.15]$$

We have $P(h, z) = \det(zI - A) = z(z + |A|^2)$, where $|A|^2$ is the sum of the squares of the entries of A. By [6.15], we have

$$|A|^2 = \gamma_0 + \gamma_1 h + \gamma_2 h^2 + \gamma_3 h^3 + \gamma_4 h^4, \qquad \gamma_0 \equiv |\Gamma|^2, \gamma_4 \equiv |c|^2. \qquad [6.16]$$

The spectral curve is reducible with one component ($z = 0$) corresponding to the zero eigenvalue ($z = 0$) of any matrix in $so(3)$. The other component $z^2 + |A(h)|^2 = 0$ is by [6.16] an elliptic curve, which is smooth and can be realized by $(h, z) \longrightarrow h$, as two-sheeted branched covering of $\mathbb{P}^1$ with sheet interchange given by $(h, z) \longmapsto (h, -z)$. The spectral curve is a two-sheeted covering $\varphi : \mathcal{C} \longrightarrow \mathbb{P} = \mathbb{P}^1(h)$, branched over four points h_j with all $h_j \neq \infty$. Since $B(h) = \Omega + Lh = \Lambda M + Lh$, we have $D = \varphi^{-1}(\infty) = P + Q$, where $z = \frac{\alpha+\beta}{h} + \cdots$ near P, $z = -\frac{\alpha+\beta}{h} + \cdots$ near Q, ($\alpha \equiv \lambda_1 + \lambda_2, \beta \equiv \lambda_3$). Then the residue $\rho(B) \in H^0(\mathcal{O}_D(D))$ is given by $\rho(B) = \frac{1}{h} + \cdots$ near P, $\rho(B) = -\frac{1}{h} + \cdots$ near Q. We deduce that $\rho(\dot{B}) = 0$ and [6.13] is satisfied. Therefore, the flow is linearized on the Jacobian variety of $\mathcal{C}$.

6.7. Nahm's equations

Nahm's equations (Nahm 1981) involve three functions $(T_j(t))_{j=1,2,3}$ with values in the algebra $u(n)$ formed by the complex antihermitian matrices of order n. These equations, which arise in the study of monopoles, are $\dot{T}_1 = [T_2, T_3]$, $\dot{T}_2 = [T_3, T_1]$, $\dot{T}_3 = [T_1, T_2]$. We have $\dot{T}_j = \frac{1}{2}\sum_{k,l} \epsilon_{jkl}[T_k, T_l]$, where ϵ_{jkl} is the Levi–Civita symbol,

$$\epsilon_{jkl} = \begin{cases} +1 \text{ if } (j, k, l) = (1, 2, 3), (3, 1, 2) \text{ where } (2, 3, 1) \\ -1 \text{ if } (j, k, l) = (3, 2, 1), (1, 3, 2) \text{ where } (2, 1, 3) \\ 0 \text{ if } j = k, k = l \text{ where } l = j \end{cases}$$

Nahm's equations are equivalent (Hitchin 1983) to the Lax equation $\dot{A} = [B, A]$, where

$$A = (T_1 + iT_2) - 2iT_3 h + (T_1 - iT_2)h^2, \qquad [6.17]$$

$$B = -\frac{1}{2}\frac{dA}{dh} = iT_3 - (T_1 - iT_2)h.$$

Let $\mathcal{C}$ be the spectral curve (in the twistor space $T\mathbb{P}^1$, the tangent bundle to $\mathbb{P}^1$ and geometrically may be thought of as the space of oriented straight lines in $\mathbb{R}^3$) associated with these equations. It is a smooth curve given by the equation

$$P(h,z) = \det((T_1 + iT_2) - 2iT_3 h + (T_1 - iT_2)h^2 - zI) = 0,$$

and its genus is $(n-1)^2$. The coefficients of the polynomial $P(h,z)$ are independent of t and are invariants of Nahm's equations. We have $\mathcal{D} = h^{-1}(\infty) = \sum_{j=1}^n \mathcal{P}_j$. Let $z_j = \frac{\lambda_j}{\xi_j} + \frac{1}{\xi_j} +$ Taylor, where $\xi_j = \frac{1}{h}$ is a local coordinate around p_j. According to [6.17], we have around p_j, $A = (T_1+iT_2) - 2i\frac{T_3}{\xi_j} + \frac{(T_1-iT_2)}{\xi_j^2}$, $B = iT_3 - \frac{(T_1-iT_2)}{\xi_j}$. The eigenvectors v_j satisfy $Av_j = z_j v_j$, whence $(T_1 + iT_2 - 2iT_3 h + (T_1 - iT_2)h^2)v_j = (\lambda_j h^2 + h + \text{Taylor})v_j$, and $Bv_j = iT_3 - (T_1 - iT_2)hv_j$. Hence, $(T_1 - iT_2)h^2 v_j = z_j v_j + o(h)$ and $Bv_j = -(T_1 - iT_2)hv_j + o(1)$. We thus have $(T_1 - iT_2)v_j(p_j) = \lambda_j v_j(p_j)$, while the residue is given by $\rho(B) = \sum_j \frac{\lambda_j}{z_j}$. Therefore, $\rho\left(\dot{B}\right) = 0$, and [6.13] linearizes the flow in question on the Jacobian variety of $\mathcal{C}$.

6.8. The n-dimensional rigid body

Let $J = \text{diag}(\lambda_1, ..., \lambda_n)$, $\lambda_j > 0$, be the matrix representing the tensor of inertia of a rigid body in a principal axis system, and let $\Omega(t) \in so(n)$ be the skew-symmetric matrix associated with the angular velocity vector of the rigid body in the usual way. Define $M = \Omega J + J\Omega \in so(n)$, the equations of motion of the rigid body can be written as $\dot{M} = [M, \Omega]$. These equations are Hamiltonian on each adjoint orbit of $so(n)$ defined by initial conditions with Hamiltonian $H(M) = \frac{1}{2}(M, \Omega) = -\frac{1}{4}Tr(M\Omega)$ (the cases $n = 3$ and $n = 4$ have and will be studied in several places). With Manakov's trick (Manakov 1976), these equations are equivalent to a Lax equation with parameter $\overbrace{(M + J^2 h)} = [M + J^2 h, \Omega + Jh]$. Hence, $\mathcal{D} = h^{-1}(\infty) = \sum_i p_i$ is the divisor with p_i being the n distinct points lying over $h = \infty$. If z_i is a local coordinate on $\mathcal{C}$ near p_i (say $z_i = h^{-1}$), then from equation [6.10] and taking $B = \Omega + Jh$, we obtain $\rho(B) = \sum_i \frac{\lambda_i}{z_i}$. Since λ_i are constant, one clearly has $\rho(\dot{B}) = 0$ so the flow is linear on $Jac(\mathcal{C})$. We note that since $A = M + J^2 h$ where $M + M^\top = 0$, $J^2 - J^{2\top} = 0$, we have $P(h,z) = (-1)^n P(-h,-z)$. Thus, there is an involution of the spectral curve $\sigma : \mathcal{C} \longrightarrow \mathcal{C}$, $(h,z) \longmapsto (-h,-z)$. We note that Ω moves on an adjoint orbit $\mathcal{O}_\nu \subset so(n)$ and to linearize the flow in question we need $\frac{1}{2} \dim \mathcal{O}_\nu$ integrals of motion that are in involution where for general ν,

$$\dim \mathcal{O}_\nu = \frac{n(n-1)}{2} - \left[\frac{n}{2}\right], \qquad [6.18]$$

(see Mironov (2010) for more information). Let $g(\mathcal{C})$ be the genus of the spectral curve $\mathcal{C}$ and $g(\mathcal{C}_0)$ the genus of the quotient $\mathcal{C}_0 = \mathcal{C}/\sigma$ of $\mathcal{C}$ by the involution σ. Since $g(\mathcal{C}) = \frac{(n-1)(n-2)}{2}$, then by the Riemann–Hurwitz formula,

$$g(\mathcal{C}_0) = g(\mathcal{C}) - \frac{1}{2}\left(\frac{n(n-1)}{2} - \begin{bmatrix} n \\ 2 \end{bmatrix}\right) = \begin{cases} \frac{(n-2)^2}{4} & n \equiv 0 \bmod 2 \\ \frac{(n-1)(n-3)}{4} & n \equiv 1 \bmod 2 \end{cases} \quad [6.19]$$

Associated with the double covering $\mathcal{C} \longrightarrow \mathcal{C}_0$ is the Prym variety $Prym(\mathcal{C}/\mathcal{C}_0)$ and since $\sigma(\rho(B)) = -\rho(B)$, the flow in question actually occurs on this complex torus. From [6.19], it follows that

$$\dim Prym(\mathcal{C}/\mathcal{C}_0) = \frac{1}{2}\left(\frac{n(n-1)}{2} - \begin{bmatrix} n \\ 2 \end{bmatrix}\right) = \begin{cases} \frac{n(n-2)}{4} & n \equiv 0 \bmod 2 \\ \frac{(n-1)^2}{4} & n \equiv 1 \bmod 2 \end{cases}$$

In comparison with [6.18], we obtain $\dim Prym(\mathcal{C}/\mathcal{C}_0) = \frac{1}{2}\dim \mathcal{O}_\nu$, and the motion linearizes on a torus $Prym(\mathcal{C}/\mathcal{C}_0)$ of exactly the right dimension.

6.9. Exercises

EXERCISE 6.1.– Use the Griffiths linearization method to study the Jacobi geodesic flow on an ellipsoid and Neumann problem (section 4.6).

EXERCISE 6.2.– Consider the system $\dot{x}_1 = y_1$, $\dot{y}_1 = 2x_1 x_2$, $\dot{x}_2 = y_2$, $\dot{y}_2 = x_1^2 + 6x_2^2$, corresponding to a particular case of Hénon–Heiles Hamiltonian: $H = \frac{1}{2}(y_1^2 + y_2^2) - x_1^2 x_2 - 2x_2^3$, where x_1, x_2, y_1, y_2 are canonical coordinates and momenta, respectively (for detailed information on the Hénon–Heiles system in general, see section 7.6. For ease of calculation, we chose here a particular case of this system with a slight change of sign).

a) Show that this system admits a Lax pair of the form $\dot{A} = [A, B]$, where

$$A(h) = \begin{pmatrix} y_1 x_1 - y_2 h & -x_1^2 + 2hx_2 + h^2 \\ y_1^2 - \left(\frac{x_1^2}{2} + 2x_2^2\right)h + x_2^2 h^2 - \frac{h^3}{2} & -y_1 x_1 + y_2 h \end{pmatrix}, \quad h \in \mathbb{C}^*,$$

and $B(h) = \begin{pmatrix} 0 & 1 \\ 2x_2 - \frac{h}{2} & 0 \end{pmatrix}$, $h \in \mathbb{C}^*$. Determine a second first integral F of this system, such that H and F are functionally independent and in involution.

b) Using the Griffiths linearization method, show that the Hamiltonian flows corresponding to H and F are linear.

EXERCISE 6.3.– Let $gl(n, \mathbb{C})$ be a Lie algebra of all $n \times n$ matrices and $\mathcal{G}$ the graded Lie algebra of all Laurent polynomials in complex variable h with coefficients in $gl(n, \mathbb{C})$ and under the bracket operation: $[Az^m, Bz^n] = [A, B]z^{m+n}$. Let us denote $A(h, t) \equiv A_h$ and assume that A_h is a generic solution of Lax equation [4.1] (i.e. A_h has simple eigenvalues for all but a finite number of values h). Let X be a compact Riemann surface corresponding to the spectral curve $\mathcal{C}$ of A_h.

a) Prove that there exists a holomorphic line bundle $E(t)$ (i.e. an eigenbundle) over X with the property that if $(h, z) \in \mathcal{C}$ and z is a simple eigenvalue for A_h, then the eigenspace $E(h, z, t) = \{x \in \mathbb{C}^n : A_h x = zx\}$ of A_h equals the fiber $E(t)|_{(h,z)}$ of $E(t)$ over (h, z).

b) Let $Q(h, z)$ be a polynomial and $R(h, z)$ the polynomial part of the Laurent polynomial $Q(h, A_h)$. Show that the Jacobian flow of the Lax equation $\dot{A}_h = [A_h, R(h, A_h)]$ is linear on the Jacobian torus.

EXERCISE 6.4.– Using the method developed in this chapter, study again the Kowalewski spinning top and its Lax representation.

EXERCISE 6.5.– We use the notation from section 6.4, with $\mathcal{D} = np_1 + np_2$. Show that there exist local coordinates z_1 and z_2 near p_1 and p_2 such that the residue of B at $\mathcal{D}$ is given by $\rho(B) = \frac{1}{z_1^2} - \frac{1}{z_2^2}$.

EXERCISE 6.6.– Consider the Hamiltonian system: $\dot{x}_j = x_j(x_{j+1} - x_{j-1})$, $x_{n+j} = x_j$, $1 \leq j \leq n$. This is a special case of the Lotka–Volterra system: $\dot{x}_j = \alpha_j x_j + \frac{1}{\beta_j} \sum_{k=1}^n a_{jk} x_j x_k$, $1 \leq j \leq n$. The system that interests us here corresponds to the case $\alpha_j = \beta_j = 0$ and (a_{ij}) skew-symmetric. The Hamiltonian function is $H = x_1 + x_2 + \cdots + x_n$, the Poisson bracket is defined by $\{x_j, x_{j+1}\} = x_j x_{j+1}$ with all other brackets equal to zero and the system is written in the Hamiltonian form: $\dot{x}_j = \{x_j, H\}, 1 \leq j \leq n$.

a) Show that $F = (x_1 - \log x_1) + (x_2 - \log x_2) + \cdots + (x_n - \log x_n)$ is a first integral of this differential system and the latter admits the Lax pair $\dot{A} = [A, B]$, where

$$A(h) = \begin{pmatrix} 0 & \sqrt{x_1} & \cdots & & \sqrt{x_n}h^{-1} \\ \sqrt{x_1} & 0 & \sqrt{x_2} & & \vdots \\ \vdots & \sqrt{x_2} & \ddots & & \vdots \\ \vdots & & \ddots & & \sqrt{x_{n-1}} \\ \sqrt{x_n}h & \cdots & & \sqrt{x_{n-1}} & 0 \end{pmatrix}, \quad h \in \mathbb{C}^*,$$

$$B(h) = \begin{pmatrix} 0 & 0 & \sqrt{x_1 x_2} & \cdots & -\sqrt{x_{n-1}x_n}h^{-1} & 0 \\ 0 & 0 & 0 & & & -\sqrt{x_1 x_n}h^{-1} \\ -\sqrt{x_1 x_2} & 0 & 0 & \ddots & & \vdots \\ \vdots & & \ddots & \ddots & \ddots & \\ & & & & & \sqrt{x_{n-2}x_{n-1}} \\ \sqrt{x_{n-1}x_n}h & \ddots & 0 & 0 & & 0 \\ 0 & \sqrt{x_1 x_n}h & \cdots & -\sqrt{x_{n-2}x_{n-1}} & 0 & 0 \end{pmatrix}$$

b) Using the Griffiths linearization method with $\mathcal{D} = np_1 + np_2$, show that the system above linearizes on the Jacobi variety $\text{Jac}(\mathcal{C})$ where $\mathcal{C}$ is the spectral curve associated with the system. Show that the system in question linearizes also on a Prym variety associated with spectral curve $\mathcal{C}$ of dimension $\left[\frac{n}{2}\right]$.

7

Algebraically Integrable Systems

This chapter presents an excellent introduction to the problems, techniques and results of algebraic complete integrability. We will mainly focus on algebraic integrability in the sense of Adler–van Moerbeke, where the fibers of the momentum map are affine parts of complex algebraic tori (Abelian varieties). It is well known that most of the problems of classical mechanics are of this form. Many important problems will be studied: Euler and Kowalewski tops, the Hénon–Heiles system, geodesic flow on $SO(n)$, the Kac–van Moerbeke lattice, generalized periodic Toda systems, the Gross–Neveu system, the Kolossof potential, as well as other systems.

7.1. Meromorphic solutions

Consider the system of nonlinear differential equations

$$\frac{dz_1}{dt} = f_1(t, z_1, ..., z_n), ..., \frac{dz_n}{dt} = f_n(t, z_1, ..., z_n), \qquad [7.1]$$

where $f_1, ..., f_n$ are functions of $n+1$ complex variables $t, z_1, ..., z_n$ and which apply a domain of $\mathbb{C}^{n+1}$ into $\mathbb{C}$. The Cauchy problem is the search for a solution $(z_1(t), ..., z_n(t))$ in a neighborhood of a point t_0, passing through the given point $(t_0, z_1^0, ..., z_n^0)$, that is, satisfying the initial conditions $z_1(t_0) = z_1^0, ..., z_n(t_0) = z_n^0$. The system [7.1] can be written in a vector form in $\mathbb{C}^n$, $\frac{dz}{dt} = f(t, z(t))$, by putting $z = (z_1, ..., z_n)$ and $f = (f_1, ..., f_n)$. In this case, the Cauchy problem will be to determine the solution $z(t)$, such that $z(t_0) = z_0 = (z_1^0, ..., z_n^0)$. We know that when the functions $f_1, ..., f_n$ are holomorphic in the neighborhood of the $(t_0, z_1^0, ..., z_n^0)$, then the Cauchy problem admits a unique holomorphic solution. A question arises: can the Cauchy problem admit some non-holomorphic solution in the neighborhood of point $(t_0, z_1^0, ..., z_n^0)$? When the functions $f_1, ..., f_n$ are holomorphic, the answer is negative. Other circumstances may arise for the Cauchy problem concerning the system of differential equations [7.1], when the holomorphic hypothesis relative to the functions $f_1, ..., f_n$ is no longer satisfied in the neighborhood of a point. In such a case, it can be seen that the behavior of the solutions can take on the most diverse aspects. In general, the singularities of the solutions are of two types: mobile or fixed, depending on whether or not they depend on the initial conditions. Important results have been obtained by Painlevé (1975). Suppose, for example, that the system [7.1] is written in the form

$$\frac{dz_1}{dt} = \frac{P_1(t, z_1, ..., z_n)}{Q_1(t, z_1, ..., z_n)}, ..., \frac{dz_n}{dt} = \frac{P_n(t, z_1, ..., z_n)}{Q_n(t, z_1, ..., z_n)},$$

where

$$P_k(t, z_1, ..., z_n) = \sum_{0 \le i_1, ..., i_n \le p} A^{(k)}_{i_1, ..., i_n}(t) z_1^{i_1} ... z_n^{i_n}, \ 1 \le k \le n,$$

$$Q_k(t, z_1, ..., z_n) = \sum_{0 \le j_1, ..., j_n \le q} B^{(k)}_{j_1, ..., j_n}(t) z_1^{j_1} ... z_n^{j_n}, \ 1 \le k \le n,$$

are polynomials with several indeterminate $z_1, ..., z_n$ and algebraic coefficients in t. We know (i) that the fixed singularities are constituted by four sets of points. The first is the set of singular points of the coefficients $A^{(k)}_{i_1, ..., i_n}(t)$, $B^{(k)}_{j_1, ..., j_n}(t)$ intervening in the polynomials $P_k(t, z_1, ..., z_n)$ and $Q_k(t, z_1, ..., z_n)$. In general, this set contains $t = \infty$. The second set consists of the points α_j such that $Q_k(t, z_1, ..., z_n) = 0$, which occurs if all of the coefficients $B^{(k)}_{j_1, ..., j_n}(t)$ vanish for $t = \alpha_j$. The third is the set of points β_l such that for some values $(z_{1'}, ..., z_{n'})$ of $(z_1, ..., z_n)$, we have $P_k(\beta_l, z_{1'}, ..., z_{n'}) = Q_k(\beta_l, z_{1'}, ..., z_{n'}) = 0$. Then the second members of the above system are presented in the indeterminate form $\frac{0}{0}$ at the points $(\beta_l, z_{1'}, ..., z_{n'})$. Finally, there is the set of points γ_n such that there exist $u_1, ..., u_n$, for which $R_k(\gamma_n, u_1, ..., u_n) = S_k(\gamma_n, u_1, ..., u_n) = 0$, where R_k and S_k are polynomials in $u_1, ..., u_n$ obtained from P_k and Q_k by setting $z_1 = \frac{1}{u_1}, ..., z_n = \frac{1}{u_n}$. Each of these sets contains only a finite number of elements. The system in question has a finite number of fixed singularities. (ii) The mobile singularities of solutions of this system are algebraic mobile singularities: poles and (or) algebraic critical points. There are no essential singular points for the solution $(z_1, ..., z_n)$.

Considering the system of differential equations [7.1], can we find sufficient conditions for the existence and uniqueness of meromorphic solutions? We will establish a theorem of existence and uniqueness for the solution of the Cauchy problem concerning the system of differential equations [7.1] using the method of indeterminate coefficients. The solution will be explained in the form of a Laurent series. The problem of convergence will therefore arise. This will be solved by the method of major functions. There are many theoretical and practical problems, or differential equations, of which the second member is not holomorphic. In the following, we will consider the Cauchy problem concerning the normal system [7.1], where $f_1, ..., f_n$ do not depend explicitly on t, that is,

$$\frac{dz_1}{dt} = f_1(z_1, ..., z_n), ..., \frac{dz_n}{dt} = f_n(z_1, ..., z_n). \qquad [7.2]$$

We suppose that $f_1, ..., f_n$ are rational functions in $z_1, ..., z_n$ and that the system [7.2] is weight-homogeneous. That is, there exist positive integers $s_1, ..., s_n$ such that $f_i(\alpha^{s_1} z_1, ..., \alpha^{s_n} z_n) = \alpha^{s_i+1} f_i(z_1, ..., z_n)$, $1 \leq i \leq n$, for each non-zero constant α. In other words, the system [7.2] is invariant under the transformation $t \to \alpha^{-1} t$, $z_1 \to \alpha^{s_1} z_1, ..., z_n \to \alpha^{s_n} z_n$. Note that if the determinant

$$\Delta \equiv \det\left(z_j \frac{\partial f_i}{\partial z_j} - \delta_{ij} f_i\right)_{1 \leq i,j \leq n}, \qquad [7.3]$$

is not identically zero, then the choice of the numbers $s_1, ..., s_n$ is unique. In what follows, we will assume that $t_0 = z_0 = 0$, which does not affect the generality of the results.

THEOREM 7.1.– Suppose that

$$z_i = \frac{1}{t^{s_i}} \sum_{k=0}^{\infty} z_i^{(k)} t^k, \quad 1 \leq i \leq n, \quad z^{(0)} \neq 0 \qquad [7.4]$$

($s_i \in \mathbb{Z}$, some $s_i > 0$) is the formal solution (Laurent series), obtained by the method of undetermined coefficients of the weight-homogeneous system [7.2]. Then the coefficients $z_i^{(0)}$ satisfy the nonlinear equation

$$s_i z_i^{(0)} + f_i(z_1^{(0)}, ..., z_n^{(0)}) = 0, \qquad [7.5]$$

where $1 \leq i \leq n$, while $z_i^{(1)}, z_i^{(2)}, ...$ each satisfy a system of linear equations of the form

$$(L - k\mathcal{I})z^{(k)} = \text{some polynomial in the } z^{(j)}, \quad 0 \leq j \leq k, \qquad [7.6]$$

where $z^{(k)} = (z_1^{(k)}, ..., z_n^{(k)})^\top$ and $L \equiv \left(\frac{\partial f_i}{\partial z_j}(z^{(0)}) + \delta_{ij} s_i\right)_{1 \leq i,j \leq n}$ is the Jacobian matrix (Kowalewski matrix) of [7.5]. Moreover, the formal series [7.4] are convergent.

PROOF.– By substituting [7.4] into [7.2], taking into account the weight-homogeneity of the system, we obtain

$$\sum_{k=0}^{\infty}(k-s_i)z_i^{(k)} t^{k-s_i-1} = f_i\left(\sum_{k=0}^{\infty} z_1^{(k)} t^{k-s_1}, ..., \sum_{k=0}^{\infty} z_n^{(k)} t^{k-s_n}\right),$$

$$= f_i\left(t^{-s_1}(z_1^{(0)} + \sum_{k=1}^{\infty} z_1^{(k)}t^k), ..., t^{-s_n}(z_n^{(0)} + \sum_{k=1}^{\infty} z_n^{(k)}t^k)\right),$$

$$= t^{-s_i-1} f_i\left(z_1^{(0)} + \sum_{k=1}^{\infty} z_1^{(k)}t^k, ..., z_n^{(0)} + \sum_{k=1}^{\infty} z_n^{(k)}t^k\right).$$

Then, the second member is developed as follows:

$$\sum_{k=0}^{\infty}(k-s_i)z_i^{(k)}t^k = f_i(z_1^{(0)}, ..., z_n^{(0)}) + \sum_{j=1}^{n} \frac{\partial f_i}{\partial z_j}(z_1^{(0)}, ..., z_n^{(0)}) \sum_{k=1}^{\infty} z_j^{(k)}t^k$$

$$+ \sum_{k=2}^{\infty} t^k \sum_{(\alpha,\tau)\in D_k} \frac{1}{\alpha!} \frac{\partial^\alpha f_i}{\partial z^\alpha}(z_1^{(0)}, ..., z_n^{(0)}) \prod_{j=1}^{n} (z_j^{(\tau_j)})^{\alpha_j},$$

where $\alpha = (\alpha_1, ..., \alpha_n)$, $\tau = (\tau_1, ..., \tau_n)$, $|\alpha| = \sum_{j=1}^{n} \alpha_j$, $\alpha! = \prod_{j=1}^{n} \alpha_j!$ and $D_k = \{(\alpha, \tau) : \tau_j > 0, \forall, |\alpha| > 2, \sum_{j=1}^{n} \alpha_j \tau_j = k\}$. By identifying the terms that have the same power at the first and the second member, we obtain, successively for $k = 0$, the expression [7.5], for $k = 1$, $(L - \mathcal{I})c^{(1)} = 0$, and for $k \geq 2$,

$$\left((L - k\mathcal{I})z^{(k)}\right)_i = -\sum_{(\alpha,\tau)\in D_k} \frac{1}{\alpha!} \frac{\partial^\alpha f_i}{\partial z^\alpha}(z_1^{(0)}, ..., z_n^{(0)}) \prod_{j=1}^{n} (z_j^{(\tau_j)})^{\alpha_j}, \quad [7.7]$$

where $\tau_j > 0$, $\sum_{j=1}^{n} \alpha_j \tau_j = k$, which leads to [7.6]. The solution obtained by the method of indeterminate coefficients is formal because we obtain it by performing on various series, which we assume *a priori* convergent, various operations whose validity remains to be justified. The theorem will therefore be established as soon as we have verified that these series are convergent. This will be done using the majorant method (Adler and van Moerbeke 1989). Note that free parameters either appear in the system [7.5] of n equations with n unknown, when the latter admits a continuous set of solutions, or because of the fact that $\lambda_i \equiv k \in \mathbb{N}^*$, $1 \leq i \leq n$, is an eigenvalue of the matrix L. The coefficients can be seen as rational functions on an affine variety V, fibered over the indicial locus: $\bigcap_{i=1}^{n}\left\{s_i z_i^{(0)} + f_i\left(z_1^{(0)}, ..., z_n^{(0)}\right) = 0\right\}$. Let $n_0 \in V$ and fix a compact subset K of V, containing an open neighborhood of n_0. Note that K can be equipped with the topology of the complex plan. Let $A = 1 + \max\left\{\left|z_1^{(\tau_1)}(n_0)\right|, \left|z_2^{(\tau_2)}(n_0)\right|, ..., \left|z_n^{(\tau_n)}\right|(n_0)\right\}$, where $1 \leq \tau_i \leq \lambda_n$, $1 \leq i \leq n$ and λ_n denotes the largest eigenvalue of the matrix L. Let B and C be two constants with $C > A$, such that in the compact K we have $\left|\frac{\partial^\alpha f_i}{\partial z^\alpha}(n_0)\right| \leq \alpha! B^{|\alpha|}$, $\left|(L(n_0) - k\mathcal{I}_n)^{-1}\right| \leq C$, $k \geq \lambda_n + 1$. From [7.7], we deduce that $\left|z_i^{(k)}(n_0)\right| \leq C \sum_{(\alpha,\tau)\in D} B^{|\alpha|} \prod_{j=1}^{n} \left|z_j^{(\tau_j)}\right|^{\alpha_j}$, $k \geq \lambda_n + 1$. Consider the series

$\Phi(t) = At + \sum_{k=2}^{\infty} \beta_k t^k$, where β_k are real numbers inductively defined by $\beta_1 \equiv A$ and $\beta_k \equiv C \sum_{(\alpha,\tau) \in D} B^{|\alpha|} \prod_{j=1}^{n} \beta_{\tau_j}^{\alpha_j}$, $k \geq 2$. Note that the series $\Phi(t)$ is an upper bound for $\sum_{k=1}^{\infty} z_i^{(k)} t^k$, $1 \leq i \leq n$. Indeed, we have $\left|z_i^{(1)}\right| \leq A$. Suppose that $\left|z_i^{(j)}\right| \leq \beta_j, j < k, \forall i$. Then, for $k \geq \lambda_n + 1$,

$$\left|z_i^{(k)}(n_0)\right| \leq C \sum_{(\alpha,\tau) \in D} B^{|\alpha|} \prod_{j=1}^{n} \left|z_j^{(\tau_j)}\right|^{\alpha_j} \leq C \sum_{(\alpha,\tau) \in D} B^{|\alpha|} \prod_{j=1}^{n} \left|\beta_{\tau_j}^{\alpha_j}\right| = \beta_k.$$

It results from the definition of the numbers β_k that $\Phi(t) = At + CB^2 \frac{(n\Phi(t))^2}{1 - Bn\Phi(t)}$. By writing the formal series [7.4] in the form, $z_i(t) = \frac{1}{t^{s_i}}\left(z_i^{(0)} + g_i(t)\right)$, we see that the root

$$\Phi(t) = \frac{1 + nABtz - \sqrt{(1 - 2nAB(1 + 2nBC)t + n^2 A^2 B^2 t^2)}}{2nB(1 + nBC)},$$

provides the required majorant for the functions $g_i(t)$, which therefore converge for sufficiently small t. $\square$

REMARK 7.1.– The series [7.4] is the only meromorphic solution in the sense that this solution results from the fact that the coefficients $z_i^{(k)}$ are unequivocally determined using the adopted method of calculation.

REMARK 7.2.– The result of the previous theorem applies to the following *quasi-homogeneous* differential equation of order n:

$$\frac{d^n z}{dt^n} = f\left(z, \frac{dz}{dt}, \ldots, \frac{d^{n-1} z}{dt^{n-1}}\right), \qquad [7.8]$$

with f being a rational function in $z, \frac{dz}{dt}, \ldots, \frac{d^{n-1}z}{dt^{n-1}}$ and $z(t_0) = z_1^0, \frac{dz}{dt}(t_0) = z_2^0, \ldots, \frac{d^{n-1}z}{dt^{n-1}}(t_0) = z_n^0$. Indeed, equation [7.8] reduces to a system of n first-order differential equations by setting $z(t) = z_1(t), \frac{dz}{dt}(t) = z_2(t), \ldots, \frac{d^{n-1}z}{dt^{n-1}}(t) = z_n(t)$. We thus obtain

$$\frac{dz_1}{dt} = z_2, \quad \frac{dz_2}{dt} = z_3, \ldots, \frac{dz_{n-1}}{dt} = z_n, \quad \frac{dz_n}{dt} = f(z_1, z_2, \ldots, z_n).$$

Such a system constitutes a particular case of the normal system [7.2].

7.2. Algebraic complete integrability

We will work with complexes instead of real ones. Concepts such as Liouville integrability, involution, commutativity of vector fields and so on can be defined as in the real case. On the other hand, difficulties arise: we know that there are no compact holomorphic submanifolds in the complex space $\mathbb{C}^m$ (maximum principle), therefore the complex tori that we can get in Arnold–Liouville's theorem are not compact. So the problem of compactification of invariant varieties arises. In addition, the solutions of the system in question are not uniform (single-valued). First, we will recall some results, then we will define and explain the concept of algebraic complete integrability of Hamiltonian systems in more detail. The definition of the algebraic complete integrability of a Hamiltonian system varies according to the literature and is usually found (with some minor variants) in any modern text on integrable systems. For integrable systems treated in several chapters (which are important and are often encountered elsewhere), we will have to consider the affine space $\mathbb{C}^m$. The integrable systems that we will deal with here are complex integrable systems where M is an affine space $\mathbb{C}^m$, the algebra that we consider is just that of the polynomial functions and we focus on algebraic complete integrability in the sense of Adler–van Moerbeke. Consider a Hamiltonian completely integrable system

$$X_H : \dot{z} = J \frac{\partial H}{\partial z} \equiv f(z), \ z \in \mathbb{R}^m, \quad m = 2n + k, \qquad [7.9]$$

($J(z)$ polynomial in z), with $n + k$ functionally independent invariants $H_1, ..., H_{n+k}$ of which k invariants (Casimir functions) lead to zero vector fields $J\frac{\partial H_{n+j}}{\partial z}(z) = 0$, $1 \leq j \leq k$, the $n = (m-k)/2$ remaining ones are in involution (i.e. $\{H_i, H_j\} = 0$), which give rise to n commuting vector fields. According to the Arnold–Liouville theorem 3.1, if the invariant manifolds $\bigcap_{i=1}^{n+k} \{z \in \mathbb{R}^m : H_i(z) = c_i\}$, are compact, then for most values of $c_i \in \mathbb{R}$, their connected components are diffeomorphic to real tori $\mathbb{R}^n/Lattice$, and the flows $g_t^{X_1}(x),...,g_t^{X_n}(x)$ defined by the vector fields $X_{H_1},...,X_{H_n}$ are straight-line motions on these tori. Now consider $z \in \mathbb{C}^m$ and $t \in \mathbb{C}$. Let $\Delta \subset \mathbb{C}^m$ be a non-empty Zariski open set. By the functional independence of the first integrals, the map (momentum mapping) $\varphi \equiv (H_1,...,H_{n+k}) : \mathbb{C}^m \longrightarrow \mathbb{C}^{n+k}$, is a generic submersion (i.e. $dH_1(z),...,dH_{n+k}(z)$ are linearly independent) on Δ. Let $\Omega = \varphi(\mathbb{C}^m \backslash \Delta)$, that is, $\Omega = \{c = (c_i) \in \mathbb{C}^{n+k} : \exists z \in \varphi^{-1}(c) \text{ with } dH_1(z) \wedge ... \wedge dH_{n+k}(z) = 0\}$, be the set of critical values of the map φ and denote by $\overline{\Omega}$ the Zariski closure of Ω in $\mathbb{C}^{n+k}$.

PROPOSITION 7.1.– The set defined by $\Gamma = \{z \in \mathbb{C}^m : \varphi(z) \in \mathbb{C}^{n+k} \backslash \overline{\Omega}\}$ is everywhere dense in $\mathbb{C}^m$ for the usual topology.

PROOF.– Indeed, it suffices to show that the set $\Gamma = \varphi^{-1}(\mathbb{C}^{n+k} \backslash \overline{\Omega})$, is a non-empty Zariski open set in $\mathbb{C}^m$. Since a polynomial mapping between affine algebraic sets is continuous for the Zariski topology, then the above set is indeed a Zariski open set in $\mathbb{C}^m$ and it is non-empty. Suppose this one is empty, that is $\varphi(\mathbb{C}^m) \subset \overline{\Omega}$. Since the map φ is submersive on a non-empty open set of Zariski $\Delta \subset \mathbb{C}^m$, then $\varphi(\Delta)$ is open in $\mathbb{C}^{n+k}$. According to Sard's theorem for varieties, $\mathbb{C}^{n+k}\backslash\overline{\Omega}$ is a non-empty Zariski open set, and therefore everywhere dense for the usual topology in $\mathbb{C}^{n+k}$. So $\varphi(\Delta) \cap (\mathbb{C}^{n+k}\backslash\overline{\Omega}) \neq \emptyset$, which is absurd. This completes the proof. □

Let M_c be the complex affine variety defined by

$$M_c \equiv \varphi^{-1}(c) = \bigcap_{i=1}^{n+k} \{z \in \mathbb{C}^m : H_i(z) = c_i\}. \quad [7.10]$$

For all $c \equiv (c_1, ..., c_{n+k}) \in \mathbb{C}^{n+k}\backslash\overline{\Omega}$, the fiber M_c is smooth.

DEFINITION 7.1.– *The system [7.9] will be called algebraic complete integrable (a.c.i.) in the sense of Adler–van Moerbeke with Abelian functions z_i when, for every $c \in \mathbb{C}^{n+k}\backslash\overline{\Omega}$, the fiber M_c [7.10] is the affine part of an Abelian variety (complex algebraic torus) $\widetilde{M_c} = T^n \simeq \mathbb{C}^n/L_c$, ($L_c$ a lattice in $\mathbb{C}^n$), and moreover, the flows $g_{X_i}^t(z)$, $z \in M_c$, $t \in \mathbb{C}$, defined by the vector fields X_{H_i}, $1 \leq i \leq n$, are straight line on T^n, that is, $\left[g_{X_i}^t(z)\right]_j = f_j\left(p + t(k_1^i, ..., k_n^i)\right)$, where $f_j(t_1, ..., t_n)$ are Abelian functions on T^n, $f_j(p) = z_j$, $1 \leq j \leq m$.*

We will be concerned with algebraically completely integrable systems that are irreducible, that is, when the generic Abelian variety is irreducible (it does not contain a subtorus). The following remark is intended to present several interrelated definitions, all involving algebraically completely integrable systems. For more information and comments on some definitions, see Vanhaecke (2001).

REMARK 7.3.– 1) Let H be a smooth function on a $2n$-dimensional symplectic manifold (M, ω). The Hamiltonian system defined by the vector field X_H is algebraically completely integrable if there exists a smooth algebraic variety $\mathcal{M}$, a co-symplectic structure $\widetilde{\omega}$ that restricts to ω along M, that is, $\widetilde{\omega} \in \Lambda^2 T_{\mathcal{M}}$ and a morphism $h : \mathcal{M} \longrightarrow \mathcal{U}$, where $\mathcal{U}$ is a Zariski open subset of $\mathbb{C}^n$, all defined over the real field such that (i) h is a proper submersive whose components are in involution (i.e. $\{X_i oh, X_j oh\} \equiv \widetilde{\omega}(d(X_i oh), d(X_j oh)) = 0$, X_i being coordinates on $\mathbb{R}^n$); (ii) M is a component of $\mathcal{M}_{\mathbb{R}}$, the $\widetilde{\omega}$ on M is the $\widetilde{\omega}$ on $\mathcal{M}$ along M, and H is a C^∞-function of $X_j.oh|_M$ (in such a situation, the fibers of the map h are Abelian varieties or extensions of these by $\mathbb{C}^{*k}$. This definition also includes the non-compact case and allows fibers that are more general than Abelian varieties).

2) Let $(M, \{.,.\}, \varphi)$ be a complex integrable system where M is a non-singular affine variety and $\varphi = (H_1, ..., H_s)$ is given by regular algebraic functions H_i. This system is algebraically completely integrable if, for generic $c \in \mathbb{C}^s$, the fiber of $\varphi^{-1}(c)$ is an affine part of an Abelian variety and if the Hamiltonian vector fields X_{H_i} are translation invariant when restricted to these fibers.

3) Let $(M, \{.,.\})$ be a smooth Poisson variety. An algebraically completely integrable Hamiltonian system consists of a proper flat morphism $H : M \longrightarrow B$, where B is a smooth variety, such that over the complement $B\backslash\Delta$ of some proper closed subvariety $\Lambda \subset B$, the morphism H is a Lagrangian fibration whose fibers are isomorphic to Abelian varieties.

4) Let M be a $2n$-dimensional complex manifold with a holomorphic symplectic structure ω, a holomorphic function $H : M \longrightarrow \mathbb{C}$ and n holomorphic functions which are pairwise in involution and also with H, that is, $\{H_i, H_j\} = \{H_i, H\} = 0$, $1 \leq i, j \leq n$. Let B be an open dense subset of $\mathbb{C}^n$ and $F : F^{-1}(B) \subset M \longrightarrow B$, a submersive map. The Hamiltonian system defined by the vector field X_H is an algebraic completely integrable system if there exists a bundle $\pi : A \longrightarrow B$ of Abelian varieties, a divisor $D \subset A$, an isomorphism $\sigma : F^{-1}(B) \longrightarrow A\backslash D$ and a vector field Y on $A\backslash D$ that restricts to a linear vector field on the fibers of π, so that the following diagram:

$$\begin{array}{ccc} F^{-1}(B) & \xrightarrow{\sigma} & A\backslash D \\ {}_F\searrow & & \swarrow_\pi \\ & B & \end{array}$$

is commutative and such that the vector field X_H is σ-related to Y.

5) Note that Hitchin (1994) gave a large class of integrable systems that are almost by construction, algebraic completely integrable and showed that the cotangent bundle of moduli spaces of stable vector bundles on Riemann surfaces carry the structure of integrable systems and are indeed algebraic completely integrable.

REMARK 7.4.– a) The complete algebraic integrability in the sense of Adler–van Moerbeke in the case where $M = \mathbb{C}^m$ means that (i) the system [7.9] with polynomial right hand possesses $n + k$ independent polynomial invariants $H_1 \equiv H, H_2, ..., H_{n+k}$ of which k invariants lead to zero vector fields, the $n = (m - k)/2$ remaining ones are in involution, which give rise to n commuting vector fields. For generic c_i, the invariant manifolds $\bigcap_{i=1}^{n+k} \{z \in \mathbb{R}^m : H_i = c_i\}$ are assumed compact, connected and therefore real tori by the Arnold–Liouville's theorem. (ii) Moreover, the invariant manifolds, thought of as affine varieties in $\mathbb{C}^m$ (non-compact), can be completed into complex algebraic tori, that is,

$$\bigcap_{i=1}^{n+k}\{z \in \mathbb{C}^m : H_i(z) = c_i\} = T^n \backslash \left\{ \mathcal{D} \equiv \begin{pmatrix} \text{one or several} \\ \text{codimension one} \\ \text{subvarieties} \end{pmatrix} \right\},$$

where the tori $T^n = \mathbb{C}^n/Lattice =$ complex algebraic torus (Abelian variety) depends on the c's. In the natural coordinates $(t_1, ..., t_n)$ of these tori, the

Hamiltonian flows (run with complex time) defined by the vector fields generated by the constants of the motion are straight-line motions, and the coordinates $z_i = z_i(t_1, ..., t_n)$ are meromorphic in $(t_1, ..., t_n)$.

b) The existence of polynomial first integrals for a Hamiltonian system does not necessarily imply the complete integrability of this system. For example, the Hamiltonian system where $H(x,y) = \frac{x^2}{2} + P(y)$, ($P(y)$ being a polynomial in y), will be algebraically completely integrable with Abelian (here elliptic) functions if and only if $P(y)$ is a polynomial of degree 3 or 4. Following Mumford (1983a, 1983b), the commuting vector fields $X_{H_1}, ..., X_{H_n}$ define on the real torus $M_c \subset \mathbb{R}^{2n}$ defined by the intersection of the constants of the motion $H_1 = c_1, ..., H_n = c_n$, an addition law $\oplus : M_c \times M_c \longrightarrow M_c$, $(x, y) \longmapsto x \oplus y = g_{t+s}(p)$, $p \in M_c$, with $x = g_t(p)$, $y = g_s(p)$, $g_t(p) = g_{t_1}^{X_1} ... g_{t_n}^{X_n}(p)$, where $g_{t_i}^{X_i}(p)$ denote the flows generated by X_{H_i}. Algebraic complete integrability means that this addition law is rational, that is: $(x \oplus y)_j = R_j(x_i, y_i, c)$, where $R_j(x_i, y_i, c)$ is a rational function of all coordinates $x_i, y_i, 1 \leq i \leq n$. Putting $x = p$, $y = g_t^{X_i}(p)$, in this formula, we note that on the real torus M_c, the flows $g_t^{X_i}(p)$ depend rationally on the initial condition p. Moreover, a Weierstrass theorem on functions admitting an addition theorem states that the coordinates x_i are restricted to the real torus: $\mathbb{R}^n/Lattice \longrightarrow M_c$, $(t_1, ..., t_n) \longmapsto x_i(t_1, ..., t_n)$, are Abelian functions. Geometrically, this means that the real torus $M_c \simeq \mathbb{R}^n/Lattice$ is the affine part of an algebraic complex torus (Abelian variety) $\mathbb{C}^n/Lattice$ and that the real functions $x_i(t_1, ..., t_n)$, $(t_i \in \mathbb{R})$, are the restrictions on this real torus of the meromorphic functions $x_i(t_1, ..., t_n)$, $(t_i \in \mathbb{C})$, of n complex variables, with $2n$ periods (n periods + n imaginary periods). In degenerate situations, some of these periods may be infinite, as in the case of a harmonic oscillator.

c) If the Hamiltonian flow [7.9] is a.c.i., it means that the variables z_i are meromorphic on the torus T^n and by compactness they must blow up along a codimension one subvariety (a divisor) $\mathcal{D} \subset T^n$. By the a.c.i. definition, the flow [7.9] is a straight line motion in T^n and thus, it must hit the divisor $\mathcal{D}$ in at least one place.

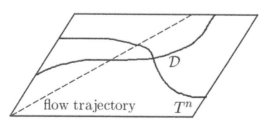

Figure 7.1. *Divisor and flow trajectory*

Through every point of $\mathcal{D}$, there is a straight line motion and a Laurent expansion around that point of intersection. The differential equations must admit Laurent expansions that depend on the $n-1$ parameters defining $\mathcal{D}$ and the $n+k$ constants c_i defining the torus T^n, the total count is $m - 1 = dim\ (phase\ space) - 1$ parameters. The fact that algebraic complete integrable systems possess $(m-1)$-dimensional families of Laurent solutions was implicitly used by Kowalewski (1889) in her classification of integrable rigid body motions. Such a necessary condition (see Adler and van Moerbeke (1989)) for algebraic complete integrability can be formulated as follows.

THEOREM 7.2.– If the Hamiltonian system [7.9] (with invariant tori not containing elliptic curves) is algebraic complete integrable, then each z_i blows up after a finite (complex) time, and for every z_i, there is a family of solutions

$$z_i = \sum_{j=0}^{\infty} z_i^{(j)} t^{j-s_i}, \quad s_i \in \mathbb{Z}, \quad \text{some } s_i > 0, \qquad [7.11]$$

depending on $\dim(\text{phase space}) - 1 = m - 1$, free parameters. The system [7.9] possesses families of Laurent solutions depending on $m-2, m-3, ..., m-n$ parameters. The coefficients of each one of these solutions are rational functions on affine algebraic varieties of dimensions $m-1, m-2, m-3, ..., m-n$.

The question raised is whether this criterion is also sufficient. The main problem will be to complete the affine variety $M_c = \bigcap_{i=1}^{n+k} \{z \in \mathbb{C}^m, H_i(z) = c_i\}$, into an Abelian variety. A naive guess would be to take the natural compactification $\overline{M}_c$ of M_c by projectivizing the equations. Indeed, this can never work for a general reason: an Abelian variety $\widetilde{M}_c$ of dimension bigger or equal than two is never a complete intersection, that is, it can never be described in some projective space $\mathbb{P}^n$ by n-dim $\widetilde{M}_c$ global polynomial homogeneous equations. In other words, if M_c is to be the affine part of an Abelian variety, $\overline{M}_c$ must have a singularity somewhere along the locus at infinity. The trajectories of the vector fields [7.9] hit every point of the singular locus at infinity and ignore the smooth locus at infinity. In fact, the existence of meromorphic solutions to the differential equations [7.9] depending on some free parameters can be used to manufacture the tori, without ever going through the delicate procedure of blowing up and down. Information about the tori can then be gathered from the divisor. More precisely, around the hitting points, the system of differential equations [7.9] admit a Laurent expansion solution depending on $m-1$ free parameters, and in order to regularize the flow at infinity, we use these parameters to blow up the variety $\overline{M}_c$ along the singular locus at infinity. The new complex variety obtained in this fashion is compact, smooth and has commuting vector fields on it; it is therefore an Abelian variety. The system [7.9] with $k+n$ polynomial invariants has a coherent tree of Laurent solutions, when it has families of Laurent solutions in t, depending on $n-1, n-2, ..., m-n$ free parameters. Adler and van Moerbeke (1989) have shown that if the system possesses several families of $(n-1)$-dimensional Laurent solutions (principal Painlevé solutions) they must fit together in a coherent way, and as we mentioned above, the system must possess $(n-2)$-, $(n-3)$-,... dimensional Laurent solutions (lower Painlevé solutions), which are the gluing agents of the $(n-1)$-dimensional family. The gluing occurs via a rational change of coordinates, in which the lower parameter solutions are seen to be genuine limits of the higher parameter solutions, and which in turn appears due to a remarkable propriety of algebraic complete integrable systems; they can be put into quadratic form in both the original variables and their ratios (to see further). As a whole, the full set of Painlevé solutions glue together to form a fiber bundle with a singular base. A partial converse to theorem 7.8 can be formulated as follows (Abenda and Fedorov 2000):

THEOREM 7.3.– If the Hamiltonian system [7.9] satisfies the condition (a)(i) in the remark 7.7 of algebraic complete integrability and if it possesses a coherent tree of Laurent solutions, then the system is algebraic complete integrable and there are no other $m-1$-dimensional Laurent solutions than those provided by the coherent set.

We assume that the divisor is very ample and in addition, projectively normal. Consider a point $p \in \mathcal{D}$, a chart U_j around p on the torus and a function y_j in $\mathcal{L}(\mathcal{D})$ having a pole of maximal order at p. Then the vector $(1/y_j, y_1/y_j, \ldots, y_N/y_j)$ provides a good system of coordinates in U_j. Then, taking the derivative with regard to one of the flows $\left(\frac{y_i}{y_j}\right)^{\cdot} = \frac{\dot{y}_i y_j - y_i \dot{y}_j}{y_j^2}$, $1 \leq j \leq N$, are finite on U_j as well. Therefore, since y_j^2 has a double pole along $\mathcal{D}$, the numerator must also have a double pole (at worst), that is, $\dot{y}_i y_j - y_i \dot{y}_j \in \mathcal{L}(2\mathcal{D})$. Hence, when $\mathcal{D}$ is projectively normal, we have $\left(\frac{y_i}{y_j}\right)^{\cdot} = \sum_{k,l} a_{k,l} \left(\frac{y_k}{y_j}\right)\left(\frac{y_l}{y_j}\right)$, that is, the ratios y_i/y_j form a closed system of coordinates under differentiation. At the bad points, the concept of projective normality plays an important role: this enables us to show that y_i/y_j is a bona fide Taylor series starting from every point in a neighborhood of the point in question. Moreover, the Laurent solutions provide an effective tool to find the constants of the motion. To do this, just search polynomials H_i of z, having the property that when evaluated along all of the Laurent solutions $z(t)$, they have no polar part. Indeed, since an invariant function of the flow does not blow up along a Laurent solution, the series obtained by substituting the formal solutions [7.11] into the invariants should, in particular, have no polar part. The polynomial functions $H_i(z(t))$ being holomorphic and bounded in every direction of a compact space (i.e. bounded along all principle solutions) are thus constant by a Liouville type of argument. In this argument, it is thus important to use all of the generic solutions. To make these informal arguments rigorous is an outstanding question of the subject. As for the system [7.2], we also assume here that Hamiltonian flows are weight-homogeneous with a weight $s_i \in \mathbb{N}$, going with each variable z_i, i.e. $f_i(\alpha^{s_1} z_1, \ldots, \alpha^{s_m} z_m) = \alpha^{s_i+1} f_i(z_1, \ldots, z_m), \forall \alpha \in \mathbb{C}$. Observe that the constants of the motion H can then be chosen to be weight-homogeneous: $H(\alpha^{s_1} z_1, ..., \alpha^{s_m} z_m) = \alpha^k H(z_1, ..., z_m)$, $k \in \mathbb{Z}$. The study of the algebraic complete integrability of Hamiltonian systems includes several passages to prove rigorously. Here, we mention the main passages, leaving out the detail when studying the different problems in the following sections. We saw that if the flow is algebraically completely integrable, the differential equations [7.9] must admit Laurent series solutions [7.11] depending on $m - 1$ free parameters. We must have $k_i = s_i$ and coefficients in the series must satisfy nonlinear equations at the 0th step,

$$f_i\left(z_1^{(0)}, ..., z_m^{(0)}\right) + g_i z_i^{(0)} = 0, 1 \leq i \leq m, \qquad [7.12]$$

and linear systems of equations at the kth step :

$$(L - kI)z^{(k)} = \begin{cases} 0 \text{ for } k = 1 \\ \text{some polynomial in } z^{(1)}, ..., z^{(k-1)} \text{ for } k > 1, \end{cases} \qquad [7.13]$$

where $L =$ Jacobian map of [7.12] $= \frac{\partial f}{\partial z} + gI \mid_{z=z^{(0)}}$. If $m - 1$ free parameters are to appear in the Laurent series, they must either come from the nonlinear equations [7.12] or from the eigenvalue problem [7.13], that is, L must have at least $m - 1$ integer eigenvalues. These conditions are much less than expected, because of the fact that the homogeneity k of the constant H must be an eigenvalue of L. Moreover, the formal series solutions are convergent as a consequence of the majorant method (theorem 7.1). Thus, the first step is to show the existence of the Laurent solutions, which requires an argument precisely every time k is an integer eigenvalue of L and therefore $L - kI$ is not invertible. One shows the existence of the remaining constants of the motion in involution so as to reach the number $n + k$. Then we have to prove that for given $c_1, ..., c_m$, the set

$$\mathcal{D} \equiv \left\{ \begin{array}{l} z_i(t) = t^{-\nu_i}\left(z_i^{(0)} + z_i^{(1)}t + z_i^{(2)}t^2 + \cdots\right), 1 \leq i \leq m \\ \text{Laurent solutions such that : } H_j(z_i(t)) = c_j + \text{Taylor part} \end{array} \right\}$$

defines one or several $n - 1$ dimensional algebraic varieties ("Painlevé" divisor) with the property that $\bigcap_{i=1}^{n+k} \{z \in \mathbb{C}^m : H_i(z) = c_i\} \cup \mathcal{D}$, is a smooth compact, connected variety with n commuting vector fields independent at every point, that is, a complex algebraic torus $\mathbb{C}^n/Lattice$. The flows $J\frac{\partial H_{k+i}}{\partial z}, ..., J\frac{\partial H_{k+n}}{\partial z}$ are straight line motions on this torus. Later, we will see in more detail that having computed the space of functions $\mathcal{L}(\mathcal{D})$ with simple poles at worst along the expansions, it is often important to compute the space $\mathcal{L}(k\mathcal{D})$ of functions having k-fold poles at worst along the expansions. These functions play a crucial role for embedding the invariant tori into projective space. From the divisor $\mathcal{D}$, a lot of information can be obtained with regard to the periods and the action-angle variables. Some others integrable systems appear as coverings of algebraic completely integrable systems. The manifolds invariant by the complex flows are coverings of Abelian varieties and these systems are called algebraic completely integrable in the generalized sense. These systems will be studied in detail in the following chapter.

7.3. The Liouville–Arnold–Adler–van Moerbeke theorem

The idea of the Adler–van Moerbeke's proof (Adler and van Moerbeke 1985) we shall give here is closely related to the geometric spirit of the (real) Arnold–Liouville theorem (Arnold 1989). Namely, a compact complex n-dimensional variety, on which there exist n holomorphic commuting vector fields, which are independent at every point that is analytically isomorphic to a n-dimensional complex torus and the complex flows generated by the vector fields are straight lines on this complex torus.

THEOREM 7.4.– Let $\overline{\mathcal{A}}$ be an irreducible variety defined by an intersection

$$\overline{\mathcal{A}} = \bigcap_i \{Z = (Z_0, Z_1, ..., Z_n) \in \mathbb{P}^N(\mathbb{C}) : P_i(Z) = 0\},$$

involving a large number of homogeneous polynomials P_i with a smooth and irreducible affine part $\mathcal{A} = \overline{\mathcal{A}} \cap \{Z_0 \neq 0\}$. Put $\overline{\mathcal{A}} \equiv \mathcal{A} \cup \mathcal{D}$, that is, $\mathcal{D} = \overline{\mathcal{A}} \cap \{Z_0 = 0\}$ and consider the map $f : \overline{\mathcal{A}} \longrightarrow \mathbb{P}^N(\mathbb{C})$, $Z \longmapsto f(Z)$. Let $\widetilde{\mathcal{A}} = f(\overline{\mathcal{A}}) = \overline{f(\mathcal{A})}$, $\mathcal{D} = \mathcal{D}_1 \cup ... \cup \mathcal{D}_r$, where $\mathcal{D}_i$ are codimension 1 subvarieties and also consider $\mathcal{S} \equiv f(\mathcal{D}) = f(\mathcal{D}_1) \cup ... \cup f(\mathcal{D}_r) \equiv \mathcal{S}_1 \cup ... \cup \mathcal{S}_r$. Assume that (i) f maps $\mathcal{A}$ smoothly and 1-1 onto $f(\mathcal{A})$. (ii) There exist n holomorphic vector fields $X_1, ..., X_n$ on $\mathcal{A}$, which commute and are independent at every point. One vector field, say X_k (where $1 \leq k \leq n$), extends holomorphically to a neighborhood of $\mathcal{S}_k$ in the projective space $\mathbb{P}^N(\mathbb{C})$. (iii) For all $p \in \mathcal{S}_k$, the integral curve $f(t) \in \mathbb{P}^N(\mathbb{C})$ of the vector field X_k through $f(0) = p \in \mathcal{S}_k$ has the property that $\{f(t) : 0 <| t |< \varepsilon, t \in \mathbb{C}\} \subset f(\mathcal{A})$. This condition means that the orbits of X_k through $\mathcal{S}_k$ go immediately into the affine part and in particular, the vector field X_k does not vanish on any point of $\mathcal{S}_k$. Then (a) $\widetilde{M}$ is compact, connected and admits an embedding into $\mathbb{P}^N(\mathbb{C})$. (b) $\widetilde{\mathcal{A}}$ is diffeomorphic to a n-dimensional complex torus. The vector fields $X_1, ..., X_n$ extend holomorphically and remain independent on $\widetilde{\mathcal{A}}$. (c) $\widetilde{\mathcal{A}}$ is a Kähler variety. (d) $\widetilde{M}$ is a Hodge variety. In particular, $\mathcal{A}$ is the affine part of an Abelian variety $\widetilde{\mathcal{A}}$.

PROOF.– a) We will show that the orbits running through $\mathcal{S}_k$ form a smooth variety Σ_p, $p \in \mathcal{S}_k$, such that $\Sigma_p \backslash \mathcal{S}_k \subseteq \mathcal{A}$. Let $p \in \mathcal{S}_k$, $\varepsilon > 0$ small enough, $g^t_{X_k}$ the flow generated by X_k on $\mathcal{A}$ and $\{g^t_{X_k} : t \in \mathbb{C}, 0 <| t |< \varepsilon\}$, the orbit going through the point p. The vector field X_k is holomorphic in the neighborhood of any point $p \in \mathcal{S}_k$ and non-vanishing, by (ii) and (iii). Then, the flow $g^t_{X_k}$ can be straightened out after a holomorphic change of coordinates. Let $\mathcal{H} \subset \mathbb{P}^N(\mathbb{C})$ be a hyperplane transversal to the direction of the flow at p and let Σ_p be the surface element formed by the divisor

$\mathcal{S}_k$ and the orbits going through p. Consider the segment of $\mathcal{S}' \equiv \mathcal{H} \cap \Sigma_p$, and so locally, we have $\Sigma_p = \mathcal{S}' \times \mathbb{C}$.

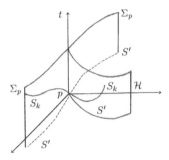

Figure 7.2. *Surface element* Σ_p

We shall show that Σ_p is smooth. Note that $\mathcal{S}'$ is smooth. Indeed, suppose that $\mathcal{S}'$ is singular at 0, then Σ_p would be singular along the trajectory (t-axis), which immediately goes into the affine $f(\mathcal{A})$, by condition (iii). Hence, the affine part would be singular, which is impossible by condition (i). So, $\mathcal{S}'$ is smooth and by the implicit function theorem, Σ_p is smooth too. Consider the map $\overline{\mathcal{A}} \subset \mathbb{P}^m(\mathbb{C}) \longrightarrow \mathbb{P}^N(\mathbb{C})$, $Z \longmapsto f(Z)$, where $Z = (Z_0, Z_1, ..., Z_n) \in \mathbb{P}^m(\mathbb{C})$ and $\widetilde{\mathcal{A}} = f(\overline{\mathcal{A}}) = \overline{f(\mathcal{A})}$. Recall that the flow exists in a full neighborhood of p in $\mathbb{P}^N(\mathbb{C})$ and it has been straightened out. Therefore, near $p \in \mathcal{S}_k$, we have $\Sigma_p = \widetilde{\mathcal{A}}$ and $\Sigma_p \backslash \mathcal{S}_k \subseteq \mathcal{A}$. Otherwise, there would exist an element $\Sigma_p' \subset \widetilde{\mathcal{A}}$, such that $\{g_{X_k}^t : t \in \mathbb{C}, 0 <| t |< \varepsilon\} = (\Sigma_p \cap \Sigma_p') \backslash p \subset \mathcal{A}$, by condition (iii). In other words, $\Sigma_p \cap \Sigma_p' = t$-axis and hence, $\mathcal{A}$ would be singular along the t-axis, which is impossible. Since the variety $\mathcal{A}$ is irreducible and since the generic hyperplane section $\mathcal{H}_{gen.}$ of $\widetilde{\mathcal{A}}$ is also irreducible, all hyperplane sections are connected and hence $\mathcal{D}$ is also connected. Now consider the graph $G_f \subset \mathbb{P}^m(\mathbb{C}) \times \mathbb{P}^N(\mathbb{C})$ of the map f, which is irreducible together with $\widetilde{\mathcal{A}}$. It follows from the irreducibility of G_f that a generic hyperplane section $G_f \cap (\mathcal{H}_{gen.} \times \mathbb{P}^N(\mathbb{C}))$ is irreducible, hence the special hyperplane section $G_f \cap (\{Z_0 = 0\} \times \mathbb{P}^N(\mathbb{C}))$ is connected and therefore the projection map $Proj_{\mathbb{P}^N(\mathbb{C})}[G_f \cap (\{Z_0 = 0\} \times \mathbb{P}^N(\mathbb{C}))] = f(\mathcal{D}) \equiv \mathcal{S}$ is connected. Hence, the variety $\widetilde{\mathcal{A}} = \mathcal{A} \cup \bigcup_{p \in \mathcal{S}_k} \Sigma_p = \mathcal{A} \cup \mathcal{S}_k \subseteq \mathbb{P}^N(\mathbb{C})$ is compact, connected and embeds smoothly into $\mathbb{P}^N(\mathbb{C})$ via f.

b) Let g^{t_i} be the flow generated by X_i on $\mathcal{A}$ and let $p_1 \in \widetilde{\mathcal{A}} \setminus \mathcal{A}$. For small $\varepsilon > 0$ and for all $t_1 \in \mathbb{C}$, such that $0 < |t_1| < \varepsilon$, $q \equiv g^{t_1}(p_1)$ is well defined and $g^{t_1}(p_1) \in f(\mathcal{A})$, using condition (iii). Let $U(q) \subseteq \mathcal{A}$ be a neighborhood of q and let $g^{t_2}(p_2) = g^{-t_1} \circ g^{t_2} \circ g^{t_1}(p_2)$, $\forall p_2 \in U(p_1) \equiv g^{-t_1}(U(q))$, which is well defined since by commutativity, we can see that the right hand side is independent of t_1: $g^{-(t_1+\varepsilon)} \circ g^{t_2} \circ g^{t_1+\varepsilon}(p_2) = g^{-t_1} \circ g^{t_2} \circ g^{t_1}(p_2)$. Note that $g^{t_2}(p_2)$ is a holomorphic function of p_2 and t_2, because in $U(p_1)$, the function g^{t_1} is holomorphic and its image is away from $\mathcal{S}$, that is, in the affine, g^{t_2} is holomorphic. The same argument applies to $g^{t_3}(p_3), ..., g^{t_n}(p_n)$, where $g^{t_n}(p_n) = g^{-t_{n-1}} \circ g^{t_n} \circ g^{t_{n-1}}(p_n)$, $\forall p_n \in U(p_{n-1}) \equiv g^{-t_{n-1}}(U(q))$. Thus, $X_1, ..., X_n$ have been holomorphically extended, remaining independent and commuting on $\widetilde{\mathcal{A}}$. Therefore, we can show along the same lines as in the Arnold–Liouville theorem that $\widetilde{\mathcal{A}}$ is a complex torus $\mathbb{C}^n/lattice$. It suffices to consider the local diffeomorphism $\mathbb{C}^n \longrightarrow \widetilde{\mathcal{A}}$, $t = (t_1, ..., t_n) \longmapsto g^t p = g^{t_1} \circ ... \circ g^{t_n}(p)$, for a fixed origin $p \in f(\mathcal{A})$. The additive subgroup $L = \{t \in \mathbb{C}^n : g^t p = p\}$ is a lattice of $\mathbb{C}^n$ (spanned by $2n$ vectors in $\mathbb{C}^n$, independent over $\mathbb{R}$), hence $\mathbb{C}^n/L \longrightarrow \widetilde{\mathcal{A}}$ is a biholomorphic diffeomorphism.

c) Let $ds^2 = \sum_{k=1}^{n} dt_k \otimes d\bar{t}_k$ be a Hermitian metric on $\widetilde{\mathcal{A}}$ and let ω be its fundamental $(1, 1)$-form. We have $\omega = -\frac{1}{2} \operatorname{Im} ds^2 = \frac{\sqrt{-1}}{2} \sum_{k=1}^{n} dt_k \wedge d\bar{t}_k$. So we see that ω is closed and the metric ds^2 is Kähler, and consequently, $\widetilde{\mathcal{A}}$ is a Kähler variety.

d) On the Kähler variety $\widetilde{\mathcal{A}}$ are defined periods of ω. If these periods are integers (possibly after multiplication by a number), we obtain a variety of Hodge. More specifically, integrals $\int_{\gamma_k} \omega$ of the form ω (where γ_k are cycles in $H_2(\widetilde{\mathcal{A}}, \mathbb{Z})$) determine the periods ω. As they are integers, $\widetilde{\mathcal{A}}$ is a Hodge variety. The variety $\widetilde{\mathcal{A}}$ is equipped with n holomorphic vectors fields, independent and commuting. From (a) and (b), $\widetilde{\mathcal{A}}$ is both a projective variety and a complex torus, and hence an Abelian variety as a consequence of the Chow theorem (Griffiths and Harris 1978). Another proof is to use the result that we have just shown, since every Hodge torus is Abelian, the converse is also true. Also, by Moishezon's theorem (Moishezon 1967), a compact complex Kähler variety having as many independent meromorphic functions as its dimension is an Abelian variety. □

Much of the material used in the following problems can be obtained as corollaries of theorem 7.4. However, I believe that working through it will help deepen the reader's understanding of the processes of analysis.

7.4. The Euler problem of a rigid body

Let us examine the algebraic complete integrability of the Euler problem of a rigid body (see sections 3.2.1 and 4.4). We have seen (section 3.2.1) that [3.6] is Hamiltonian on the two-dimensional sphere $m_1^2 + m_2^2 + m_3^2$ equal to a constant and for appropriate values of the constants, the invariant energy surface (ellipsoid) intersects the sphere according to cirles. As expected from the Arnold–Liouville theorem, the system [3.6] is completely integrable (theorem 3.2) and the vector $J\frac{\partial H}{\partial x}$ gives a flow on a variety $\bigcap_{i=1}^{2} \{x \in \mathbb{R}^3 : H_i(x) = c_i\}$, $c_i \in \mathbb{R}$, diffeomorphic to a real torus of dimension 1, that is to say a circle. But there is more to it (theorem 3.3);

the problem can be integrated in terms of elliptic functions. Moreover, the two circles of the intersection above (with $\frac{c_1}{\lambda_3} < c_2 < \frac{c_1}{\lambda_1}$, otherwise it is empty) form the real part of the one-dimensional complex torus, defined by the elliptic curve $\mathcal{E}$ [3.11]. The complex intersection $(\subset \mathbb{C}^3)$ is the affine part of an elliptic curve

$$\overline{M}_c = \{X \in \mathbb{P}^3(\mathbb{C}) : H_1(X) = c_1 X_0^2\} \cap \{X \in \mathbb{P}^3(\mathbb{C}) : H_2(X) = c_2 X_0^2\}.$$

$\overline{M}_c$ is isomorphic to the elliptic curve $\mathcal{E}$. If $p(t) = (m_1(t), m_2(t), m_3(t))$ is a solution of the system [3.6], the law connecting $p(t_1 + t_2)$ to $p(t_1)$ and $p(t_2)$ is the addition law on the elliptic curve $\mathcal{E}$. From equations [3.6], the unique holomorphic differential on $\overline{M}_c$ is given by

$$\omega = \frac{dm_1}{(\lambda_3 - \lambda_2) m_2 m_3} = \frac{dm_2}{(\lambda_1 - \lambda_3) m_1 m_3} = \frac{dm_3}{(\lambda_2 - \lambda_1) m_1 m_2},$$

so $t = \int_{p(0)}^{p(t)} \omega$, $p(0) \in \overline{M}_c$. For fixed values of the constants of the motion, the coordinates m_i are elliptic functions on the complex torus $\mathbb{C}/lattice$. In fact, the classical way of solving Euler's equations in terms of elliptic functions can be understood as a characterization of this complex torus. In other words, when considering the time t as a complex parameter rather than a real one, then $p(t)$ for $t \in \mathbb{C}$ is an elliptical function, that is, a doubly periodic meromorphic function of t with periods τ_1 and τ_2, since otherwise, $p(t)$ would be bounded and therefore constant.

Figure 7.3. Periods τ_1 and τ_2

After a finite (complex) time t_∞, $p(t)$ must blow up and have a Laurent expansion around t_∞. To mention another feature, the system [3.6] is invariant by the transformations $t \to \alpha^{-1} t$, $m_1 \to \alpha m_1$, $m_2 \to \alpha m_2$, $m_3 \to \alpha m_3$. These are unique since the determinant [7.3] is

$$\Delta = \begin{vmatrix} -(\lambda_3 - \lambda_2) m_2 m_3 & (\lambda_3 - \lambda_2) m_2 m_3 & (\lambda_3 - \lambda_2) m_2 m_3 \\ (\lambda_1 - \lambda_3) m_1 m_3 & -(\lambda_1 - \lambda_3) m_1 m_3 & (\lambda_1 - \lambda_3) m_1 m_3 \\ (\lambda_2 - \lambda_1) m_1 m_2 & (\lambda_2 - \lambda_1) m_1 m_2 & -(\lambda_2 - \lambda_1) m_1 m_2 \end{vmatrix},$$

$$= 4(\lambda_3 - \lambda_2)(\lambda_1 - \lambda_3)(\lambda_2 - \lambda_1) m_1^2 m_2^2 m_3^2 \neq 0.$$

We look for solutions to the system [3.6] or to equation [3.5] in the form of Laurent series

$$M(t) = t^{-1} \left(M^{(0)} + M^{(1)} t + M^{(2)} t^2 + \cdots \right) = \sum_{j=0}^{\infty} M^{(j)} t^{j-1}, \qquad [7.14]$$

depending on $\dim(phase\ space) - 1 = 2$ free parameters. By substituting [7.14] into equation [3.5], one obtains $\sum_{j=0}^{\infty}(j-1)M^{(j)}t^{j-2} = \sum_{j=0}^{\infty}\left(\sum_{i=0}^{j}\left[M^{(i)}, \Lambda M^{(j-i)}\right]\right)t^{j-2}$. Therefore, $(j-i)M^{(j)} = \sum_{i=0}^{j}\left[M^{(i)}, \Lambda M^{(j-i)}\right]$, and we see that the coefficients $M^{(0)}, M^{(1)}, ...$, satisfy the equations

$$M^{(0)} + \left[M^{(0)}, \Lambda M^{(0)}\right] = 0, \qquad [7.15]$$

$$(L - kI)M^{(k)} = -\sum_{i=1}^{k-1}\left[M^{(i)}, \Lambda C^{(k-i)}\right], \ k \geq 1,$$

where L is the linear operator $L : so(3) \longrightarrow so(3)$ defined by

$$L(Y) = Y + \left[M^{(0)}, \Lambda Y\right] + [Y, \Lambda M^{0}] = \text{Jacobian of [7.15]}.$$

The matrix $M^{(0)}$ appearing in L is a solution of the nonlinear equation [7.15]. A simple calculation shows that the matrix $(L - kI)$ is always invertible except for $k = 2$, and therefore, its rank is equal to 1. This shows that the coefficient $M^{(2)}$ contains two free parameters and can be assimilated to c_1 and c_2.

7.5. The Kowalewski top

It is well known that the various procedures for solving the equation of motion for the Kowalewski top always lead to a linear flow on an Abelian surface. In her celebrated paper Kowalewski (1889), integrates the problem in terms of hyperelliptic integrals, using a very beautiful change of variables (see section 3.2.3), which shows that this Abelian surface is the Jacobian of a hyperelliptic curve associated with the integrals of motion. Here, we sketch the integration of the problem using the Laurent solutions, as carried out for the first time Lesfari (1988). The result is that the invariant surfaces could be completed via the flow into complex algebraic tori (Abelian surfaces) $\mathbb{C}^2/lattice$, where the lattice is spanned by the columns of the period matrix $\begin{pmatrix} 1 & 0 & a & c \\ 0 & 2 & c & b \end{pmatrix}$, $\text{Im}\begin{pmatrix} a & c \\ c & b \end{pmatrix} > 0$, that is, the problem is not expressed in terms of hyperelliptic integrals, but rather in terms of Abelian integrals associated with the period matrix. As we have seen in the previous section, such Abelian surfaces come up naturally as Prym varieties of double covers of elliptic curves ramified over four points. We look for weight homogeneous Laurent solutions to the system [3.4] in the case of Kowalewski (i.e. [3.12]). The latter is weight

176 Integrable Systems

homogeneous, with M having weight 1 and Γ weight 2. Let M and Γ have the following asymptotic expansions:

$$M = \sum_{k=0}^{\infty} M^{(k)} t^{k-1}, \quad \Gamma = \sum_{k=0}^{\infty} \Gamma^{(k)} t^{k-2}. \quad [7.16]$$

By substituting [7.16] in the differential equations [3.4] (case of Kowalewski), at the 0th step, the coefficients of t^{-2} for M and t^{-3} for Γ yield a nonlinear system

$$M^{(0)} + \left[M^{(0)}, \Lambda M^{(0)}\right] + \left[\Gamma^{(0)}, L\right] = 0, \quad [7.17]$$

$$2\Gamma^{(0)} + \left[\Gamma^{(0)}, \Lambda M^{(0)}\right] = 0,$$

and at the kth step ($k \geq 1$), the coefficients of t^{k-2} for M and t^{k-3} for Γ lead to a system of linear equations in M^k and Γ^k:

$$(\Psi - kI)\begin{pmatrix} M^{(k)} \\ \Gamma^{(k)} \end{pmatrix} = \begin{cases} 0 \text{ for } k=1 \\ \begin{bmatrix} -\sum_{i=1}^{k-1}\left(M^{(i)}, \Lambda M^{(k-i)}\right) \\ -\sum_{i=1}^{k-1}\left(\Gamma^{(i)}, \Lambda M^{(k-i)}\right) \end{bmatrix} \text{ for } k \geq 2 \end{cases} \quad [7.18]$$

where Ψ denotes the linear operator

$$\Psi\begin{pmatrix} X \\ Y \end{pmatrix} = \begin{pmatrix} [M^{(0)}, \Lambda X] + [X, \Lambda M^{(0)}] + [Y, L] + X \\ [\Gamma^{(0)}, \Lambda X] + [Y, \Lambda M^{(0)}] + 2Y \end{pmatrix}.$$

In the basis $(e_1, ..., e_6)$, Ψ is given by the following matrix

$$\Psi = \begin{pmatrix} 1 & m_3^{(0)} & m_2^{(0)} & 0 & 0 & 0 \\ -m_3^{(0)} & 1 & -m_1^{(0)} & 0 & 0 & 2 \\ 0 & 0 & 1 & 0 & -2 & 0 \\ 0 & -\gamma_3^{(0)} & 2\gamma_2^{(0)} & 2 & 2m_3^{(0)} & -m_2^{(0)} \\ \gamma_3^{(0)} & 0 & -2\gamma_1^{(0)} & -2m_3^{(0)} & 2 & m_1^{(0)} \\ -\gamma_2^{(0)} & \gamma_1^{(0)} & 0 & m_2^{(0)} & -m_1^{(0)} & 2 \end{pmatrix},$$

which is the Jacobian of [7.17].

THEOREM 7.5.– The nonlinear system [7.17] defines two lines and two points. For the case corresponding to the points, the system [7.18] has two degrees of freedom for $k = 2$ and one degree of freedom for $k = 3$ and 4. For the case corresponding to the lines, the system [7.18] has one degree of freedom for $k = 1, 2, 3$ and 4. The solutions to the system [7.18] are obtained explicitly in a direct way.

PROOF.– The proof is a linear algebra problem (a straight forward computation). □

The generic solution blows up after a finite time according to a Laurent series within a five-parameter family of Laurent solutions. When using the majorant method (theorem 7.1), any formal Laurent series solution of a system of differential equations with quadratic right-hand side automatically converges. Now it is easily checked that $(\Psi + I)\begin{pmatrix} M^{(0)} \\ \Gamma^{(0)} \end{pmatrix} = 0$, and it follows from theorem 7.24 that $\det(\Psi - kI) = (k+1)k(k-1)(k-2)(k-3)(k-4)$. Consequently, we have the following theorem:

THEOREM 7.6.– The system [3.4] in the case of Kowalewski (i.e. [3.12]) presents two distinct families of Laurent series solutions: $M = \sum_{k=0}^{\infty} M^{(k)} t^{k-1}$, $\Gamma = \sum_{k=0}^{\infty} \Gamma^{(k)} t^{k-2}$, depending on five free parameters such that the coefficients $M^{(0)}$ and $\Gamma^{(0)}$ satisfy the nonlinear equations,

$$M^{(0)} + \left[M^{(0)}, \Lambda M^{(0)}\right] + \left[\Gamma^{(0)}, L\right] = 0, \quad 2\Gamma^{(0)} + \left[\Gamma^{(0)}, \Lambda M^{(0)}\right] = 0,$$

and depend on a free variable α, while $M^{(k)}$ and $\Gamma^{(k)}$ satisfy linear systems

$$(\mathcal{M} - kI)\begin{pmatrix} M^{(k)} \\ \Gamma^{(k)} \end{pmatrix} = \begin{cases} 0 & \text{if } k = 1 \\ \begin{bmatrix} -\sum_{i=1}^{k-1}\left(M^{(i)}, \Lambda M^{(k-i)}\right) \\ -\sum_{i=1}^{k-1}\left(\Gamma^{(i)}, \Lambda M^{(k-i)}\right) \end{bmatrix} & \text{if } k \geq 2 \end{cases}$$

where $\mathcal{M}$ is the Jacobian matrix of [7.17]. These systems provide a free variable at each level $k = 1, 2, 3, 4$. The first meromorphic solution with the Laurent expansion is

$$m_1(t) = \frac{\alpha}{t} + i(\alpha^2 - 2)\beta + o(t), \qquad \gamma_1(t) = \frac{1}{2t^2} + o(t),$$

$$m_2(t) = \frac{i\alpha}{t} - \alpha^2\beta + o(t), \qquad \gamma_2(t) = \frac{i}{2t^2} + o(t),$$

$$m_3(t) = \frac{i}{t} + \alpha\beta + o(t), \qquad \gamma_3(t) = \frac{\beta}{t} + o(t)$$

and the second meromorphic solution with the Laurent expansion is

$$m_1(t) = \frac{\alpha}{t} - i(\alpha^2 - 2)\beta + o(t), \qquad \gamma_1(t) = \frac{1}{2t^2} + o(t),$$

$$m_2(t) = -\frac{i\alpha}{t} - \alpha^2\beta + o(t), \qquad \gamma_2(t) = -\frac{i}{2t^2} + o(t),$$

$$m_3(t) = -\frac{i}{t} + \alpha\beta + o(t), \qquad \gamma_3(t) = \frac{\beta}{t} + o(t).$$

The study of convergence of these Laurent series solutions will be carried out via the majorant method (theorem 7.1) around all points where $\alpha, \beta \neq 0$ and the techniques of projective normality around bad points. Let $M_c = \bigcap_{k=1}^{4} \{x \in \mathbb{C}^6 : H_k(x) = c_k\}$, $c_3 = 1$, be the affine variety defined by the four constants of motion. The invariant variety M_c is a smooth affine surface for generic values of c_1, c_2 and c_3. Also in this problem, how do we find the compactification of M_c into an Abelian surface? As has been done in the previous two sections, following the methods in Adler and van Moerbeke (1984), the idea of the direct proof as we have seen in theorem 7.4 is closely related to the geometric spirit of the (real) Arnold–Liouville theorem (Arnold 1989). Namely, a compact complex n-dimensional variety on which there exist n holomorphic commuting vector fields that are independent at every point is analytically isomorphic to a n-dimensional complex torus $\mathbb{C}^n/lattice$, and the complex flows generated by the vector fields are straight lines on this complex torus.

Now, the main problem will be to complete M_c into a non-singular compact complex algebraic variety $\widetilde{M}_c = M_c \cup \mathcal{D}$ in such a way that the vector fields X_{H_1} and X_{H_4} generated, respectively, by H_1 and H_4, extend holomorphically along a divisor $\mathcal{D}$ and remain independent there. If this is possible, $\widetilde{M}_c$ is an algebraic complex torus (an Abelian variety) and the coordinates restricted to M_c are Abelian functions. As we have already pointed out, a naive guess would be to take the natural compactification $\overline{M}_c$ of M_c by projectivizing the equations in $\mathbb{P}^6(\mathbb{C})$. Indeed, this can never work for a general reason: an Abelian variety $\widetilde{M}_c$ of dimension bigger or equal than two is never a complete smooth intersection, that is it can never be described in some projective space $\mathbb{P}^n(\mathbb{C})$ by n-dim $\widetilde{M}_c$ global polynomial homogeneous equations. If M_c is to be the affine part of an Abelian surface, $\overline{M}_c$ must have a singularity somewhere along the locus at infinity $\overline{M}_c \cap \{Z = 0\}$. In fact, what happens in this specific case? Let $Z, M_j = \frac{m_j}{Z}, \Gamma_j = \frac{\gamma_j}{Z}$, be the projective coordinates and consider the transformation

$$\overline{M}_c \longrightarrow \overline{M}_c, \quad (Z, M_1, M_2, M_3, \Gamma_1, \Gamma_2, \Gamma_3) \longmapsto (Z, X_1, X_2, M_3, U_1, U_2, \Gamma_3),$$

where $2X_1 = M_1 + iM_2$, $2X_2 = M_1 - iM_2$, $U_1 = \Gamma_1 + i\Gamma_2$, $U_2 = \Gamma_1 - i\Gamma_2$. In these new variables, the equations defining $\overline{M}_c$ are written as

$$F_1 \equiv 2X_1X_2 + M_3^2 + (U_1 + U_2)Z - c_1Z^2 = 0,$$
$$F_2 \equiv X_1U_2 + X_2U_1 + M_3\Gamma_3 - c_2Z^2 = 0,$$
$$F_3 \equiv U_1U_2 + \Gamma_3^2 - Z^2 = 0,$$
$$F_4 \equiv (X_1^2 - U_1Z)(X_2^2 - U_2Z) - c_4Z^4.$$

THEOREM 7.7.– *The projective variety $\overline{M}_c \subset \mathbb{P}^6(\mathbb{C})$ defined by the above equations is not an Abelian surface. In addition, $\overline{M}_c$ intersects the hyperplane at infinity $Z = 0$*

according to the two straight lines $d_1 = (0, X_1, 0, 0, U_1, 0, 0)$, $d_2 = (0, 0, X_2, 0, 0, U_2, 0)$ and the circle $S^1 = (0, 0, 0, 0, U_1, U_2, \Gamma_3)$, where $U_1 U_2 + \Gamma_3^2 = 0$. Moreover, $\overline{M}_c$ is singular along the lines d_1, d_2; singularity of type: $Y^4 + U^3 Z^3 = 0$, (Y, Z) small, and along the circle S^1; singularity of type: $\Gamma_3^4 X^2 = (\Gamma_3^4 + \Gamma_3^2 - 1) Z^2$, (X, Z) small.

PROOF.– Suppose that $\overline{M}_c$ is an Abelian surface. Thus, this surface is isomorphic to a two-dimensional complex algebraic torus $\mathbb{C}^2 / lattice$, where the lattice is generated by four independent vectors in $\mathbb{C}^2$. Let (φ_1, φ_2) be the natural coordinates of this torus. The canonical divisor of this Abelian surface is empty as, up to scalars, the only 2-form is given by $d\varphi_1 \wedge d\varphi_2$, which clearly does not vanish anywhere on $\overline{M}_c$. The canonical bundle $K_{\overline{M}_c}$ of $\overline{M}_c$ is given in terms of the canonical bundle $K_{\mathbb{P}^6(\mathbb{C})}$ of $\mathbb{P}^6(\mathbb{C})$ by the adjunction formula $K_{\overline{M}_c} = \left(K_{\mathbb{P}^6(\mathbb{C})} \otimes [\overline{M}_c] \right)_{\overline{M}_c}$, where $[\overline{M}_c]$ denotes the line bundle associated with the divisor $\overline{M}_c$ in $\mathbb{P}^6(\mathbb{C})$. Let $x_1 = \frac{M_1}{Z}$, $x_2 = \frac{M_2}{Z}$, $x_3 = \frac{M_3}{Z}$, $y_1 = \frac{\Gamma_1}{Z}$, $y_2 = \frac{\Gamma_2}{Z}$, $y_3 = \frac{\Gamma_3}{Z}$, be affine coordinates on $\mathbb{P}^6(\mathbb{C}) \setminus \{Z = 0\}$ and consider the form $\omega = dx_1 \wedge dx_2 \wedge dx_3 \wedge dy_1 \wedge dy_2 \wedge dy_3$. This form has no zeroes or poles in $\{Z \neq 0\}$. Let $u = \frac{Z}{\Gamma_3}$, $u_1 = \frac{M_1}{\Gamma_3}$, $u_2 = \frac{M_2}{\Gamma_3}$, $u_3 = \frac{M_3}{\Gamma_3}$, $v_1 = \frac{\Gamma_1}{\Gamma_3}$, $v_2 = \frac{\Gamma_2}{\Gamma_3}$, be affine coordinates on $\mathbb{P}^6(\mathbb{C}) \setminus \{Z = 0\}$. Then $u = \frac{1}{y_3}$, $u_1 = \frac{x_1}{y_3}$, $u_2 = \frac{x_2}{y_3}$, $u_3 = \frac{x_3}{y_3}$, $v_1 = \frac{y_1}{y_3}$, $v_2 = \frac{y_2}{y_3}$, and so $\omega = \frac{1}{u^7} du \wedge du_1 \wedge du_2 \wedge du_3 \wedge dv_1 \wedge dv_2$. Therefore, the divisor of ω is $(\omega) = -7\mathbf{H}$ where $\mathbf{H}$ is a hyperplane in $\mathbb{P}^6(\mathbb{C})$ and consequently $K_{\mathbb{P}^6(\mathbb{C})} = [-7\mathbf{H}]$. Since every line bundle on projective space is a multiple of $[\mathbf{H}]$, we have $K_{\overline{M}_c} = [\mathbf{H}]_{\overline{M}_c}$ and this implies that every holomorphic 2-form on $\overline{M}_c$ would have a zero on $\overline{M}_c$, a contradiction. The second part of the theorem is straightforward. We will describe the singularities of $\overline{M}_c$ at infinity. First let us analyze the locus at infinity $C = \overline{M}_c \cap \{Z = 0\}$. It is shown that $C = d_1 \cup d_2 \cup S^1$, where $d_1 = (0, X_1, 0, 0, U_1, 0, 0)$ and $d_2 = (0, 0, X_2, 0, 0, U_2, 0)$ are two straight lines tangent to the following circle $S^1 = (0, 0, 0, 0, U_1, U_2, \Gamma_3)$, $U_1 U_2 + \Gamma_3^2 = 0$. In the neighborhoods of these lines and of this circle, we have used $\left. \frac{\partial(F_1, F_2, F_3, F_4)}{\partial(Z, X_1, X_2, M_3, U_1, U_2, \Gamma_3)} \right|_{Z=0} = 3$, and the variety $\overline{M}_c$ is singular along the straight lines d_1, d_2 (singularity of type $Y^4 + U^3 Z^3 = 0$, Y, Z small) and along the circle S^1 (singularity of type $\Gamma_3^4 X^2 = (\Gamma_3^4 + \Gamma_3^2 - 1) Z^2$, X, Z small). $\square$

The solutions to the system [3.4] in the case of Kowalewski (i.e. [3.12]) intersect d_1 and d_2, however the solutions ignore S^1 and the same facts holds for the other vector field [3.14] commuting with the first.

To regularize (i.e. make the trajectories parallel) the flow at infinity, it is necessary to solve the singularities of the projective variety $\overline{M}_c$. We have just seen that these singularities are complicated and we must (probably) blow up $\overline{M}_c$ many times along the lines and blow down $\overline{M}_c$ along the circle. Generally, the method of blowing up and blowing down sub-varieties is considered very delicate and its use in the present case is difficult. In fact, we shall show that the existence of meromorphic solutions to the differential equations in question, depending on four free parameters, can be used to manufacture the tori, without ever going through the delicate procedure of blowing up and down. Information about the tori can then be gathered from the divisor.

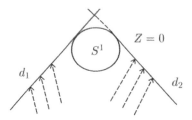

Figure 7.4. Behavior of solutions around d_1, d_2 and S^1

Consider the set of Laurent solutions which remain confined to a fixed affine invariant surface (where $c_3 = 1$ and c_1, c_2, c_4 are generic constants), that is,

$$\mathcal{D}_\varepsilon = \left\{ \begin{array}{c} \text{the Laurent solutions } m_j(t), \gamma_j(t), 1 \leq j \leq 3, \text{ such that :} \\ H_k(m_j(t), \gamma_j(t)) = c_k + \text{Taylor part}, 1 \leq k \leq 4 \end{array} \right\},$$

= 4 polynomial relations between α, β, γ and μ;

$$c_1 = (\alpha^2 - 4)\beta^2 + 18\lambda$$
$$\varepsilon c_2 = -(\alpha^2 + 2)\beta^3 + 6\beta\lambda - 12\theta$$
$$8 = (5\alpha^2 - 2)\beta^4 - 6\beta^2\lambda + 84\beta\theta - 240\mu$$
$$8c_4 = (\alpha^2 - 1)\left((11\alpha^2 + 10)\beta^4 + 54\beta^2\lambda + 108\beta\theta + 240\mu\right)$$

= algebraic curve

$$= \left\{ \begin{array}{c} (\alpha, \beta, \varepsilon) \text{ such that :} \\ P(\alpha, \beta, \varepsilon) \equiv (\alpha^2 - 1)((\alpha^2 - 1)\beta^4 - P(\beta) + c_4 = 0, \\ P(\beta) \equiv c_1\beta^2 - 2\varepsilon c_2\beta - 1, \quad \varepsilon \equiv \pm i \end{array} \right\}. \quad [7.19]$$

The quotient $\mathcal{D}_\varepsilon^0 = \mathcal{D}_\varepsilon / \sigma_\varepsilon$ by the involution,

$$\sigma_\varepsilon : \mathcal{D}_\varepsilon \longrightarrow \mathcal{D}_\varepsilon, \quad (\alpha, \beta, \varepsilon) \longmapsto (-\alpha, \beta, \varepsilon), \qquad [7.20]$$

is an elliptic curve defined by

$$u^2 = P^2(\beta) - 4c_4\beta^4. \qquad [7.21]$$

The curve $\mathcal{D}_\varepsilon$ is a two-sheeted ramified covering of $\mathcal{D}_\varepsilon^0$,

$$\varphi_\varepsilon : \mathcal{D}_\varepsilon \longrightarrow \mathcal{D}_\varepsilon^0, \quad (\alpha, u, \beta, \varepsilon) \longmapsto (u, \beta, \varepsilon), \qquad [7.22]$$

$$\mathcal{D}_\varepsilon : \begin{cases} \alpha^2 = \frac{2\beta^4 + P(\beta) + u}{2\beta^4} \\ u^2 = P^2(\beta) - 4c_4\beta^4. \end{cases} \qquad [7.23]$$

Let us look more closely at certain points of interest on the non-singular version of the curve $\mathcal{D}_\varepsilon$. For β sufficiently small, $\alpha^2 = \frac{2\beta^4 + P(\beta) + \sqrt{P^2(\beta) - 4c_4\beta^4}}{2\beta^4} = 1 + c_4 + o(\beta)$ and $\alpha^2 = \frac{2\beta^4 + P(\beta) - \sqrt{P^2(\beta) - 4c_4\beta^4}}{2\beta^4} = \frac{1}{\beta^4}(-1 + o(\beta))$. At $\beta = \infty$, the curve $\mathcal{D}_\varepsilon$ behaves as follows: $2(\alpha^2 - 1)\beta^2 = c_1 \pm \sqrt{c_1^2 - 4c_4} + o(\beta)$. The curve $\mathcal{D}_\varepsilon$ has four points at infinity p_j ($1 \leq j \leq 4$) and four branch points $q_j \equiv (\alpha = 0, u = -2\beta^4 - P(\beta), \beta^4 + P(\beta) + c_4 = 0)$ ($1 \leq j \leq 4$) on the elliptic curve $\mathcal{D}_\varepsilon^0$. The divisor structure of α and β on $\mathcal{D}_\varepsilon$ is $(\alpha) = \sum_{j=1}^4 q_j - \sum_{j=1}^4 p_j$, $(\beta) = 4$ zeroes $- \sum_{j=1}^4 p_j$. Let $g(\mathcal{D}_\varepsilon) =$ genus of $\mathcal{D}_\varepsilon$, $g(\mathcal{D}_\varepsilon^0) =$ genus of $\mathcal{D}_\varepsilon^0$, $n =$ number of sheets and $v =$ number of branch points. According to the Riemann–Hurwitz formula, $g(\mathcal{D}_\varepsilon) = n\left(g(\mathcal{D}_\varepsilon^0) - 1\right) + 1 + \frac{v}{2} = 3$. The map $(\alpha, u, \beta, \varepsilon) \longmapsto (\alpha, u, -\beta, -\varepsilon)$ is an isomorphism between $\mathcal{D}_{\varepsilon=i}$ and $\mathcal{D}_{\varepsilon=-i}$ and so we have the following commutative diagram

$$\begin{array}{ccc} \mathcal{D}_{\varepsilon=i} & \xrightarrow{\sim} & \mathcal{D}_{\varepsilon=-i} \\ \downarrow\varphi_{\varepsilon=i} & & \downarrow\varphi_{\varepsilon=-i} \\ \mathcal{D}_{\varepsilon=i}^0 & \xrightarrow{\sim} & \mathcal{D}_{\varepsilon=-i}^0 \end{array}$$

Thus, we have proved the following theorem.

THEOREM 7.8.– The divisor on which the solutions of [3.12] blow up consists of two isomorphic irreducible components $\mathcal{D}_{\varepsilon=i}$ and $\mathcal{D}_{\varepsilon=-i}$, which are both non-singular curves of genus 3. Each of these curves $\mathcal{D}_\varepsilon$[7.19], [7.23] is a double cover of an elliptic curve $\mathcal{D}_\varepsilon^0$[7.21] ramified at four points.

Let $\mathcal{D} \equiv \mathcal{D}_{\varepsilon=i} + \mathcal{D}_{\varepsilon=-i}$. For positive integers k, define the space $\mathcal{L}(k\mathcal{D})$ of meromorphic functions having a k-fold pole at worst along $\mathcal{D}$.

THEOREM 7.9.– The space $\mathcal{L}(\mathcal{D})$ is spanned by the following functions: $f_0 = 1$, $f_1 = m_1$, $f_2 = m_2$, $f_3 = m_3$, $f_4 = \gamma_3$, $f_5 = f_1^2 + f_2^2$, $f_6 = 4f_1f_4 - f_3f_5$, $f_7 = (f_2\gamma_1 - f_1\gamma_2)f_3 + 2f_4\gamma_2$. In addition, $\mathcal{D}$ is embedded into $\mathbb{P}^7(\mathbb{C})$ according to $p = (\alpha, u, \beta) \longmapsto \lim_{t \to 0} t(1, f_1(p), ..., f_7(p)) = \left(0, f_1^{(0)}(p), ..., f_7^{(0)}(p)\right)$, in a way that $\mathcal{D}_{\varepsilon=i}$ and $\mathcal{D}_{\varepsilon=-i}$ intersect each other transversally in four points at infinity $\left(\alpha = \pm 1, u = \pm\beta^2\sqrt{c_1^2 - 4c_4}, \beta = \infty\right)$ and that the geometric genus of $\mathcal{D}$ is 9.

PROOF.– Let $L^{(r)}$ be the set of polynomial functions $f = f(x(t))$ of degree less than or equal to r, modulo the constants $H_k = c_k, 1 \leq k \leq 4$ and such that: $f(x(t)) = t^{-1}(x^{(0)} + x^{(1)}t + \cdots)$, $x^{(0)} \neq 0$ on $\mathcal{D}$, with $x(t) = (m_1, m_2, m_3, \gamma_1, \gamma_2, \gamma_3)$ given explicitly above. We look for r such that the geometric genus of $\mathcal{D}^{(r)} = N_r + 2$,

$\mathcal{D}^{(r)} \subset \mathbb{P}^{N_r}(\mathbb{C})$, and we shall show that it is unnecessary to go beyond $r = 4$. Indeed, using the Laurent series obtained previously, one obtains that the spaces $L^{(r)}$, nested according to weighted degree, are generated as follows:

$$L^{(1)} = \{f_0 = 1, f_1 = m_1, f_2 = m_2, f_3 = m_3\},$$

$$L^{(2)} = L^{(1)} \oplus \left\{f_4 = \gamma_3, f_5 = -\frac{1}{4}(f_1^2 + f_2^2)\right\},$$

$$L^{(3)} = L^{(2)} \oplus \{f_6 = f_3 f_5 + f_1 f_4\},$$

$$L^{(4)} = L^{(3)} \oplus \{f_7 = (f_2 \gamma_1 - f_1 \gamma_2)f_3 + 2f_4 \gamma_2\}.$$

Note that the spaces $L^{(1)}$, $L^{(2)}$ and $L^{(3)}$ do not provide the embedding because $g(\mathcal{D}^{(r)}) \neq \dim L^{(r)} + 1, r = 1, 2, 3$. More precisely, since the embedding in $\mathbb{P}^3(\mathbb{C})$ via $L^{(1)}$ does not separate the sheets, we then pass to a $L^{(2)}$ and we find that the embedding in $\mathbb{P}^6(\mathbb{C})$ is invalid because $g\left(\mathcal{D}^{(2)} \text{ embedded in } \mathbb{P}^5(\mathbb{C})\right) - 2 > 5$, which contradicts the fact that $N_r + 1 = g(\mathcal{D}^{(2)}) - 1$. In the same way, using space $L^{(3)}$ we obtain $g\left(\mathcal{D}^{(3)} \text{ embedded in } \mathbb{P}^6(\mathbb{C})\right) - 2 > 6$, and the contradiction persists. We must therefore consider $L^{(4)}$ and study the embedding of $\mathcal{D}$ into $\mathbb{P}^7(\mathbb{C})$ using the functions of $L^{(4)}$. Let $\left(\frac{f_0}{f_4}, \ldots, \frac{f_7}{f_4}\right) \equiv F_k = F_k^{(0)} + F_k^{(1)} t + \cdots, 0 \leq k \leq 7$, and $p_j \equiv \left(\alpha = \pm 1, u = \pm \beta^2 \sqrt{c_1^2 - 4c_4}, \beta = \infty\right), 1 \leq j \leq 4$, be the four points at ∞ of $\mathcal{D}_\varepsilon$. Therefore

$$F_k^{(0)}(p_j) = \left(0, \frac{\alpha}{\beta}, \frac{\varepsilon \alpha}{\beta}, \frac{\varepsilon}{\beta}, 1, \varepsilon \alpha, \varepsilon(\alpha^2 - 1)\beta, -\frac{c_2}{\beta} + \frac{\varepsilon(1-u)}{\beta^2}\right)(p_j),$$

$$= \left(0, 0, 0, 0, 1, \pm \varepsilon, 0, \mp \varepsilon \sqrt{c_1^2 - 4c_4}\right) = 4 \text{ distinct points}.$$

The four points at ∞ are separated on each curve $\mathcal{D}_\varepsilon$, but using the transformation $\mathcal{D}_{\varepsilon=i} \longrightarrow \mathcal{D}_{\varepsilon=-i}$, $(\alpha, u, \beta) \longmapsto (-\alpha, -u, \beta)$, one shows $F_k^{(0)}(p_j)\big|_{\mathcal{D}_{\varepsilon=i}} = F_k^{(0)}(p_j)\big|_{\mathcal{D}_{\varepsilon=-i}}$, that is, the four points at ∞ are identified pairwise with the four points at ∞ on the other curve. Let $s = \frac{1}{\beta}$ be a local parameter for p_j. Since

$$\frac{\partial F_k^{(0)}}{\partial s}(p_j) = (0, \pm 1, \pm \varepsilon, \varepsilon, 0, -c_2) \implies \frac{\partial F_k^{(0)}}{\partial s}(p_j)\bigg|_{\mathcal{D}_{\varepsilon=i}} \neq \frac{\partial F_k^{(0)}}{\partial s}(p_j)\bigg|_{\mathcal{D}_{\varepsilon=-i}},$$

then the curve $\mathcal{D}_{\varepsilon=i}$ intersects the curve $\mathcal{D}_{\varepsilon=-i}$ transversely in four points at infinity

$$b_1, \ldots, b_4 = (\alpha = \infty, \beta = 0), p_1, \ldots, p_4 = (\alpha = \pm 1, u = \pm \beta^2 \sqrt{c_1^2 - 4c_4}, \beta = \infty).$$

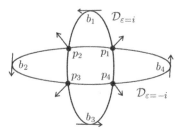

Figure 7.5. *Curves* $\mathcal{D}_{\varepsilon=\pm i}$

In a neighborhood of the points $b_j = (\alpha = \infty, \beta = 0)$, $1 \leq j \leq 4$, one divides by f_1. By setting $\left(\frac{f_0}{f_1}, ..., \frac{f_7}{f_1}\right) \equiv G_k = G_k^{(0)} + G_k^{(1)}t + \cdots$, $0 \leq k \leq 7$, one obtains $G_k^{(0)}(p_j) = (0, 1, \varepsilon, 0, 0, 0, \mp 1, 0)$, which are the coordinates of four different points in $\mathbb{P}^7(\mathbb{C})$. Therefore, $g\left(\mathcal{D}^{(4)} \text{ embedded in } \mathbb{P}^7(\mathbb{C})\right) - 2 = 7$, that is, $g\left(\mathcal{D}^{(4)}\right) = 9$. □

To avoid overburdening the text, we will use the notation $\mathcal{D}$ instead of $\mathcal{D}^{(4)}$ when embedding in $\mathbb{P}^7(\mathbb{C})$. At all points where $\alpha, \beta \neq 0, \infty$, the Laurent solutions are nicely convergent (theorem 7.1). Therefore, at most points of $\mathcal{D}$ there is a transversal fiber to the curve. Hence, this defines a smooth surface strip around $\mathcal{D}$ except at the bad points. Now we need to construct a surface strip around $\mathcal{D}$ at the bad points as well. To do so, we must use the concept of a normally generated line bundle or projective normality. Ultimately, we wish to prove that in the various charts

$$\left(\frac{f_j}{f_k}\right)^{\cdot} = \text{polynomial}\left(\frac{f_j}{f_k}\right), \quad 1 \leq j \leq 7, k \text{ fixed} \qquad [7.24]$$

This enables us to show that $\left(\frac{f_j}{f_k}\right)$ is a bona fide Taylor series starting from every point in a neighborhood of the point in question $\subseteq \mathbb{P}^7(\mathbb{C})$. As mentioned above, let $\mathcal{L} \equiv L^{(4)}$ and $\mathcal{D} \equiv \mathcal{D}^{(4)}$.

THEOREM 7.10.– The orbits of the vector field [3.12] going through the curve $\mathcal{D}$ form a smooth surface Σ near $\mathcal{D}$ such that $\Sigma \backslash \mathcal{D} \subseteq M_c$. Moreover, the variety $\widetilde{M_c} = M_c \cup \Sigma$ is smooth, compact and connected.

PROOF.– Let $\phi(t, p) = \{z(t) = (m_1(t, p), m_2(t, p), m_3(t, p), \gamma_1(t, p), \gamma_2(t, p), \gamma_3(t, p),)\}$, $t \in \mathbb{C}$, $0 < |t| < \varepsilon$, be the orbit of the vector field [3.12] going through the point $p \in \mathcal{D}$. Consider the surface element $\Sigma_p \subset \mathbb{P}^7(\mathbb{C})$ formed by the divisor $\mathcal{D}$ and the orbits going through p, and set $\Sigma \equiv \cup_{p \in \mathcal{D}} \Sigma_p$. Let $\mathcal{D}' = \mathcal{H} \cap \Sigma$ be the curve where $\mathcal{H} \subset \mathbb{P}^7(\mathbb{C})$ is a hyperplane transversal to the direction of the flow. If $\mathcal{D}'$ is smooth, then using the implicit function theorem the surface Σ is smooth. But if $\mathcal{D}'$ is singular at 0, then Σ would be singular along the trajectory (t–axis), which immediately goes into the affine part M_c. Hence, M_c would be singular, which is a contradiction because M_c is the fiber of a morphism from $\mathbb{C}^6$ to $\mathbb{C}^4$ and so smooth for almost all of the three constants of the motion c_k. Let $\overline{M}_c$ be the projective closure of M_c into $\mathbb{P}^6(\mathbb{C})$, let $Z = [Z_0, M_1, M_2, M_3, \Gamma_1, \Gamma_2, \Gamma_3] \in \mathbb{P}^6(\mathbb{C})$ where Z_0, $M_j = \frac{m_j}{Z_0}$, $\Gamma_j = \frac{\gamma_j}{Z_0}$, are the projective coordinates and let $C = \overline{M}_c \cap \{Z_0 = 0\}$ be the locus at infinity. By theorem 7.7, we have $C = d_1 \cup d_2 \cup S^1$ where d_1, d_2 are straight lines and S^1 is a circle with $d_1 \cap d_2 = \emptyset$ and $d_j \cap S^1 =$ point, $j = 1, 2$. Consider the application $\overline{M}_c \subseteq \mathbb{P}^6(\mathbb{C}) \longrightarrow \mathbb{P}^7(\mathbb{C})$, $Z \longmapsto f(Z)$, where $f = (f_0, f_1, ..., f_7) \in L(\mathcal{D})$ (see theorem 7.9) and let $\widetilde{M}_c = f(\overline{M}_c)$. In a neighborhood $V(p) \subseteq \mathbb{P}^7(\mathbb{C})$ of p, we have $\Sigma_p = \widetilde{M}_c$ and $\Sigma_p \backslash \mathcal{S} \subseteq M_c$. Otherwise there would exist an element of surface $\Sigma'_p \subseteq \widetilde{M}_c$ such that: $\Sigma_p \cap \Sigma'_p = (t - $ axis$)$, orbit $\phi(t, p) = (t - $ axis$) \backslash p \subseteq M_c$, and hence M_c would be singular along the t-axis, which is impossible. Since the variety $\overline{M}_c \cap \{Z_0 \neq 0\}$ is irreducible and the generic hyperplane section $\mathcal{H}_{gen.}$ of $\overline{M}_c$ is also irreducible, all hyperplane sections are connected and hence C is also connected. Now, consider the graph $\Gamma_f \subseteq \mathbb{P}^6(\mathbb{C}) \times \mathbb{P}^7(\mathbb{C})$ of the map f, which is irreducible together with $\overline{M}_c$. It follows from the irreducibility of C that a generic hyperplane section $\Gamma_f \cap \{\mathcal{H}_{gen.} \times \mathbb{P}^7(\mathbb{C})\}$ is irreducible, hence the special hyperplane section $\Gamma_f \cap \{\{Z_0 = 0\} \times \mathbb{P}^7(\mathbb{C})\}$ is connected and the projection map $proj_{\mathbb{P}^7(\mathbb{C})}\{\Gamma_f \cap \{\{Z_0 = 0\} \times \mathbb{P}^7\}\} = f(C) \equiv \mathcal{D}$ is connected. Hence, the variety $M_c \cup \Sigma = \widetilde{M}_c$ is compact, connected and embeds smoothly into $\mathbb{P}^7(\mathbb{C})$ via f. $\square$

We shall prove a somewhat stronger statement than [7.24], namely that [7.24] is satisfied with quadratic polynomials. By inspection, we see that the $f_0, ..., f_7$ do not satisfy that property. For example, $\left(\frac{f_0}{f_4}\right)' = \frac{f_1 \gamma_2 - f_2 \gamma_1}{f_4^2} \neq$ polynomial $\left(\frac{f_j}{f_4}\right)$, $1 \leq j \leq 7$. The problem is that $\mathcal{D}$, although very ample, will not be projectively normal (the line bundle $[\mathcal{D}]$ is not normally generated), which forces us to look at the divisor $2\mathcal{D}$ or maybe simply $2\mathcal{D}_{\varepsilon=i} + \mathcal{D}_{\varepsilon=-i}$. Hence, we must take functions with higher order poles. For instance, let us consider the space $\mathcal{L}(2\mathcal{D}_{\varepsilon=i} + \mathcal{D}_{\varepsilon=-i})$ of functions, which have double poles at worst along $\mathcal{D}_{\varepsilon=i}$ and simple ones at worst along $\mathcal{D}_{\varepsilon=-i}$. When using the Riemann–Roch theorem, if $2\mathcal{D}_{\varepsilon=i} + \mathcal{D}_{\varepsilon=-i}$ is a divisor on an Abelian variety, this space must be 18-dimensional. Indeed, let $\mathcal{L}(2\mathcal{D}_{\varepsilon=i} + \mathcal{D}_{\varepsilon=-i}) = L^{(4)} \oplus \{g_8, ..., g_{17}\} \equiv \{g_0, ..., g_{17}\}$, where $g_0 = 1$, $g_1 = \frac{1}{2}(m_1 + im_2)$, $g_2 = \frac{1}{2}(m_1 - im_2)$, $g_3 = f_3$, $g_4 = f_4$, $g_5 = f_5$, $g_6 = f_6$, $g_7 = f_7$, $g_8 = g_2^2$, $g_9 = g_2 g_3$, $g_{10} = g_2 g_4$, $g_{11} = g_2^2 - \gamma_1 - i\gamma_2$, $g_{12} = g_1 g_{11}$, $g_{13} = g_1^2 g_{11}$, $g_{14} = g_1^3 g_{11}$, $g_{15} = g_2 g_5$, $g_{16} = g_2 g_6$, $g_{17} = g_2 g_7$, where one reads off the behavior of the g_k along $\mathcal{D}_\varepsilon$ from the Laurent series solutions; here (g_k) denote the divisor restricted to $\mathcal{D}_\varepsilon$: $(g_1) = -\mathcal{D}_{\varepsilon=-i}$, $(g_2) = -\mathcal{D}_{\varepsilon=i}$, $(g_k) = -\mathcal{D}_{\varepsilon=i} - \mathcal{D}_{\varepsilon=-i}$ where $k = 3, ..., 7$, $(g_8) = -2\mathcal{D}_{\varepsilon=i}$, $(g_k) = -2\mathcal{D}_{\varepsilon=i} - \mathcal{D}_{\varepsilon=-i}$, where $k = 9, 10, 15, 16, 17$, $(g_{11}) = -2\mathcal{D}_{\varepsilon=i} + 2\mathcal{D}_{\varepsilon=-i}$, $(g_{12}) = -2\mathcal{D}_{\varepsilon=i} + \mathcal{D}_{\varepsilon=-i}$, $(g_k) = -2\mathcal{D}_{\varepsilon=-i}$, where $k = 13, 14$. Note that all g_k have the property $(g_k) \geq -2\mathcal{D}_{\varepsilon=i} - \mathcal{D}_{\varepsilon=-i}$. Now consider the embedding of $2\mathcal{D}_{\varepsilon=i} + \mathcal{D}_{\varepsilon=-i}$ via these functions, $2\mathcal{D}_{\varepsilon=i} + \mathcal{D}_{\varepsilon=-i} \longrightarrow \mathbb{P}^{17}(\mathbb{C})$,

$p \longmapsto \lim_{t\to 0} t\left(g_0(p), ..., g_{17}(p)\right) = \left(0, g_1^{(0)}(p), ..., g_{17}^{(0)}(p)\right)$. About the points $b_j = (\alpha = \infty, \beta = 0)$, it is appropriate to divide by g_8, which implies that $\left(\frac{g_k}{g_8}\right)(b_j)$, $0 \leq j \leq 17$, is finite. In a neighborhood of the points at infinity $p_j = \left(\alpha = \pm 1, u = \pm\beta^2\sqrt{c_1^2 - 4c_4}, \beta = \infty\right)$, $1 \leq j \leq 4$, one divides $g_0, ..., g_{17}$ by g_{10}, which makes $\left(\frac{g_k}{g_{10}}\right)(p_j)$, $0 \leq j \leq 17$, finite. Using [3.12], we show that in a neighborhood of the points p_j and modulo linear combination of the constants of motion $\left(\frac{g_k}{g_{10}}\right)^{\cdot} = \frac{\dot{g}_k g_{10} - g_k \dot{g}_{10}}{g_{10}^2}$ = quadratic polynomial of $\left(\frac{g_0}{g_{10}}, ..., \frac{g_{17}}{g_{10}}\right)$. Also, in a neighborhood of the points b_j we have $\left(\frac{g_k}{g_8}\right)^{\cdot}$ = quadratic polynomial of $\left(\frac{g_0}{g_8}, ..., \frac{g_{17}}{g_8}\right)$. It follows that at the bad points p_j, b_j the series $\left(\frac{g_k}{g_{10}}\right)(p_j), \left(\frac{g_k}{g_8}\right)(b_j)$, where $0 \leq k \leq 17$, converge as a consequence of Picard's theorem applied to the system of ordinary differential equations $\left(\frac{g_k}{g_l}\right)^{\cdot}$, $l = 8, 10$. Therefore, we have the following theorem:

THEOREM 7.11.– The divisor $2\mathcal{D}_{\varepsilon=i} + \mathcal{D}_{\varepsilon=-i}$ is projectively normal and has a smooth embedding in $\mathbb{P}^{17}(\mathbb{C})$, which shows that in particular the Laurent series solutions converge everywhere.

THEOREM 7.12.– The two commuting vector fields [3.12] and [3.14] extend holomorphically and remain independent on $\widetilde{M_c}$.

PROOF.– Let $\varphi^{t_1}, \varphi^{t_2}$ be the flows generated, respectively, by vector fields [3.12] and [3.14]. Consider a point $p \in \widetilde{M_c} \setminus M_c = \mathcal{D}$. For δ sufficiently small, $\varphi^{\tau_1}, -\delta < \tau_1 < \delta$, is well defined and $\varphi^{\tau_1} \in M_c$. We may define φ^{t_2} on $\widetilde{M_c}$ by $g^{t_2}(q) = g^{-t_1} g^{t_2} g^{t_1}(q)$, $q \in U(p) = g^{-t_1}(U(g^{t_1}(p)))$, where $U(p)$ is a neighborhood of p. By commutativity one can see that g^{t_2} is independent of t_1; $g^{-t_1-\varepsilon_1} g^{t_2} g^{t_1+\varepsilon_1}(q) = g^{-t_1} g^{-\varepsilon_1} g^{t_2} g^{t_1} g^{\varepsilon_1}(q) = g^{-t_1} g^{t_2} g^{t_1}(q)$. Note that $g^{t_2}(q)$ is holomorphic away from $\mathcal{D}$. This is because $g^{t_2} g^{t_1}(q)$ is holomorphic away from $\mathcal{D}$ and g^{t_1} is holomorphic in $U(p)$ and maps bi-holomorphically $U(p)$ onto $U(g^{t_1}(p))$. □

The flows φ^{t_1} and φ^{t_2} are holomorphic, independent on $\mathcal{D}$ and we can show along the same lines as in the theorem 7.10 that $\widetilde{M_c}$ is a torus. And that will be done by considering the holomorphic map $\zeta : \mathbb{C}^2 \longrightarrow \widetilde{M_c}$, $(t_1, t_2) \longmapsto \zeta(t_1, t_2) = \varphi^{t_1} \varphi^{t_2}(p)$, for a base point $p \in M_c$. Then $L = \{(t_1, t_2) \in \mathbb{C}^2 : \zeta(t_1, t_2) = p\}$, is a lattice of $\mathbb{C}^2$, hence $\zeta : \mathbb{C}^2/L \longrightarrow \widetilde{M_c}$ is a biholomorphic diffeomorphism. Thus, $\widetilde{M_c} \subseteq \mathbb{P}^7$ is conformal to a complex torus $\mathbb{C}^2/L$ and an Abelian surface as a consequence of Chow. We have the following theorem:

THEOREM 7.13.– $\widetilde{M_c}$ is an Abelian surface on which the Hamiltonian flows [3.12] and [3.14] are straight line motions.

THEOREM 7.14.– The three holomorphic differentials on $\mathcal{D}_\varepsilon$ are

$$\omega_0 = \frac{d\beta}{u}, \qquad \omega_1 = \frac{k_1\left(\alpha^2 - 1\right)\beta^2 d\beta}{\alpha u}, \qquad \omega_2 = \frac{k_2 d\beta}{\alpha u}, \qquad [7.25]$$

where u is given by [7.21] and $k_1, k_2 \in \mathbb{C}$. In addition, on $\widetilde{M_c}$ there are two holomorphic differentials dt_1 and dt_2, such that: $dt_1|_{\mathcal{D}_\varepsilon} = \omega_1$, $dt_2|_{\mathcal{D}_\varepsilon} = \omega_2$, where ω_1 and ω_2 are the two holomorphic differentials [7.25] on $\mathcal{D}_\varepsilon$.

PROOF.– From the Poincaré residue formula, we know that the three holomorphic differentials on $\mathcal{D}_\varepsilon$ are of the form $\frac{g(\alpha,\beta,\varepsilon)d\beta}{\frac{\partial P}{\partial \alpha}(\alpha,\beta,\varepsilon)} = \frac{g(\alpha,\beta,\varepsilon)d\beta}{\alpha u}$, where $g(\alpha,\beta,\varepsilon)$ is a polynomial of at most degree five in α and β and $P(\alpha,\beta,\varepsilon)$ is given by [7.19]. It is easy to verify that $\omega_0, \omega_1, \omega_2$ [7.25] effectively form a basis of holomorphic differentials on $\mathcal{D}_\varepsilon$. Let $p \in \mathcal{D}_\varepsilon \cap \{\alpha u \neq 0\}$. Around the point p, we consider two coordinates on $\widetilde{M_c}$,

$$\tau = \frac{1}{m_3} = -\varepsilon t + o(t^2), \qquad x = \begin{cases} m_1 + im_2 = -i\beta + o(t) \text{ along } \mathcal{D}_{\varepsilon=i} \\ m_1 - im_2 = i\beta + o(t) \text{ along } \mathcal{D}_{\varepsilon=-i} \end{cases}$$

We denote by $\frac{\partial}{\partial t_1}$ (resp. $\frac{\partial}{\partial t_2}$) the derivative according to the vector field [3.12] (respectively, [3.14]). Obviously, we have

$$dt_1 = \frac{1}{\Delta(\tau,x)}\left(\frac{\partial x}{\partial t_1}d\tau - \frac{\partial \tau}{\partial t_1}dx\right), \qquad dt_2 = \frac{1}{\Delta(\tau,x)}\left(-\frac{\partial x}{\partial t_1}d\tau + \frac{\partial \tau}{\partial t_1}dx\right),$$

where $\Delta(\tau,x) = \frac{\partial \tau}{\partial t_1}\cdot\frac{\partial x}{\partial t_2} - \frac{\partial \tau}{\partial t_2}\cdot\frac{\partial x}{\partial t_1}$. By direct computation using the asymptotic expansions, we find that

$$\frac{\partial \tau}{\partial t_1} = -\varepsilon + o(t), \qquad \frac{\partial x}{\partial t_1} = -2\alpha\beta^2 + o(t),$$

$$\frac{\partial \tau}{\partial t_2} = -4\varepsilon(\alpha^2 - 1)\beta^2 + o(t), \qquad \frac{\partial x}{\partial t_2} = 8\alpha(\alpha^2 - 1)\beta^4 - P(\beta) + o(t),$$

where $P(\beta) \equiv c_1\beta^2 - 2\varepsilon c_2\beta - 1$, from which one can deduce the two differentials dt_1 and dt_2. The restrictions of dt_1 and dt_2 to the curve $\mathcal{D}_\varepsilon$ are given by

$$dt_1|_{\mathcal{D}_\varepsilon} = \frac{k_1\left(\alpha^2 - 1\right)\beta^2 d\beta}{\alpha u}, \qquad dt_2|_{\mathcal{D}_\varepsilon} = \frac{k_2 d\beta}{\alpha u}, \qquad (k_1, k_2 \in \mathbb{C}),$$

and are the two holomorphic differentials ω_1, ω_2 [7.25] on $\mathcal{D}_\varepsilon$. $\square$

THEOREM 7.15.– The vector field [3.12] (respectively, [3.14]) is regular along $\mathcal{D}$, transversal to $\mathcal{D}$ at every point $\beta \neq 0$ (respectively, $\beta \neq \infty$) and (doubly) tangent at $\beta = 0$ (respectively, $\beta = \infty$).

PROOF.– Using the same notation as in the proof of theorem 7.9, one can see that there exist F_k^0 at F_l^1 such that $\det \begin{pmatrix} \frac{\partial}{\partial s} F_k^0(p_j) & \frac{\partial}{\partial s} F_l^0(p_j) \\ F_k^1(p_j) & F_l^1(p_j) \end{pmatrix} \neq 0$, and consequently the vector field [3.12] is transversal to $\mathcal{D}$ at the four points p_j of $\mathcal{D}_{\varepsilon=i} \cap \mathcal{D}_{\varepsilon=-i}$. From theorem 7.14, the function $\frac{\omega_1}{\omega_2} = \frac{k_1}{k_2}(\alpha^2 - 1)\beta^2 \sim \frac{1}{\beta^2}$ is meromorphic along a neighborhood of $b_j = (\alpha = \infty, \beta = 0)$, $(1 \leq j \leq 4)$ and provides the tangent to the curve $\mathcal{D}$ in the coordinates t_1 and t_2. The function $\frac{\omega_1}{\omega_2}$ vanishes whenever the vector field [3.14] is tangent to $\mathcal{D}$ and has a pole whenever [3.12] is tangent to $\mathcal{D}$. Hence, the zeroes b_j of ω_2 provide the four points of tangency of the vector field [3.12] to $\mathcal{D}$. We find that for all G_k^0, G_l^1, we have $\det \begin{pmatrix} \frac{\partial}{\partial \beta} G_k^0(b_j) & \frac{\partial}{\partial \beta} G_l^0(b_j) \\ G_k^1(b_j) & G_l^1(b_j) \end{pmatrix} = 0$, where the notation G_k^0, G_l^1 has been introduced in the proof of theorem 7.9. Consequently, [3.12] is (doubly) tangent to $\mathcal{D}$ at four points b_j, which concludes the proof of the theorem. □

THEOREM 7.16.– The space of holomorphic differentials on the divisor $\mathcal{D}$ is $\left\{ f_1^{(0)} \omega_2, f_2^{(0)} \omega_2, ..., f_7^{(0)} \omega_2 \right\} \oplus \{\omega_1, \omega_2\}$, where $f_1^{(0)}, f_2^{(0)}, ..., f_7^{(0)}$ are the first coefficients (the residues) of the functions $f_1, f_2, ..., f_7 \in \mathcal{L}(\mathcal{D})$ (theorem 7.9) and the embedding of $\mathcal{D}$ into $\mathbb{P}^7(\mathbb{C})$ is the canonical embedding

$$p = (\alpha, u, \beta) \in \mathcal{D} \longmapsto \left\{ \omega_2, f_1^{(0)} \omega_2, f_2^{(0)} \omega_2, ..., f_7^{(0)} \omega_2 \right\} \in \mathbb{P}^7(\mathbb{C}).$$

PROOF.– The adjunction formula gives us a map, the Poincaré residue map, between meromorphic 2-forms on $\widetilde{M_c}$, with a pole along $\mathcal{D}$ and holomorphic 1-forms on $\mathcal{D}$. Applied to the 2-form $\omega = f_j dt_1 \wedge dt_2$ with $f_j \in \mathcal{L}(\mathcal{D})$,

$$\omega = \frac{dt_1 \wedge dt_2}{\frac{1}{f_j}} \longrightarrow \text{Res } \omega|_\mathcal{D} = -\frac{dt_1}{\frac{\partial}{\partial t_2}\left(\frac{1}{f_j}\right)}\bigg|_\mathcal{D} = \frac{dt_2}{\frac{\partial}{\partial t_1}\left(\frac{1}{f_j}\right)}\bigg|_\mathcal{D},$$

hence,

$$\text{Res } \omega|_\mathcal{D} = \frac{dt_2}{\frac{\partial}{\partial t_1}\left(\frac{t_1}{f_j^{(0)}} + o(t_1^2)\right)}\bigg|_\mathcal{D} = f_j^{(0)} dt_2\bigg|_\mathcal{D} = f_j^{(0)} \omega_2,$$

where $\frac{\partial}{\partial t_1}$ is the derivative according to the vector field [3.12]. Using the second vector field [3.14], we find $\omega = \widetilde{f}_j^{(0)} \omega_1$, with $\widetilde{f}_j^{(0)}$ the residue. The differentials ω_1, ω_2, $f_1^{(0)} \omega_2, ..., f_7^{(0)} \omega_2$, form a basis for the space of holomorphic differential forms on $\mathcal{D}$, that is, $H^0(\mathcal{D}, \Omega_\mathcal{D}^1) = \{\omega_1, \omega_2\} \oplus \left\{f_j^{(0)} \omega_2, 1 \leq j \leq 7\right\} = \{\omega_1, \omega_2\} \oplus W\omega_2$, where the space W consists of all the residues of the functions of $\mathcal{L}(\mathcal{D})$. The embedding into $\mathbb{P}^7(\mathbb{C})$ via the residues of f_j is the canonical embedding of the curve $\mathcal{D}$ via its holomorphic differentials, except for the two differentials ω_1 and ω_2, $p = (\alpha, u, \beta) \in \mathcal{D} \longmapsto \left\{\omega_2, f_1^{(0)} \omega_2, f_2^{(0)} \omega_2, ..., f_7^{(0)} \omega_2\right\} \in \mathbb{P}^7(\mathbb{C})$. This proves theorem 7.16. □

As we have seen (theorem 3.4 (a)), the involution σ [3.17] has eight fixed points on the affine variety M_c. In fact (which corresponds to involution σ_ε [7.20] on $\mathcal{D}_\varepsilon$), it has eight other fixed points at infinity given by the branch points of $\mathcal{D}_\varepsilon$ on $\mathcal{D}_\varepsilon^0$. So the involution in question has 16 fixed points on $\widetilde{M}_c$, thus confirming the number of fixed points that it should have on an Abelian variety.

THEOREM 7.17.– The involution σ [3.17] on the Abelian surface $\widetilde{M}_c$ coming from the one defined on the affine variety M_c has eight fixed points at infinity.

PROOF.– From the asymptotic expansions (see theorem 7.6), the functions $m_1, m_2, \gamma_1, \gamma_2$ remain invariable by the transformation $(t, \alpha, \beta) \longmapsto (-t, -\alpha, \beta)$, whereas m_3, γ_3 change into $-m_3, -\gamma_3$. Then the involution σ [3.17] is transformed at infinity into an involution σ_ε [7.20] on $\mathcal{D}_\varepsilon$. Now the fixed points of σ_ε are given by the branch points of $\mathcal{D}_\varepsilon$ on $\mathcal{D}_\varepsilon^0$. □

THEOREM 7.18.– The Abelian surface $\widetilde{M}_c$ that completes the affine surface M_c is the dual Prym variety $Prym^*\left(\mathcal{D}_\varepsilon / \mathcal{D}_\varepsilon^0\right)$ of the genus 3 curve $\mathcal{D}_\varepsilon$ [7.19] or [7.23] for the involution σ_ε [7.20] interchanging the sheets of the double covering φ_ε [7.2].

PROOF.– Let $(a_1, a_2, a_3, b_1, b_2, b_3)$ be a canonical homology basis of $\mathcal{D}_\varepsilon$, such that $\sigma(a_1) = a_3$, $\sigma(b_1) = b_3$, $\sigma(a_2) = -a_2$, $\sigma(b_2) = -b_2$ for the involution σ_ϵ.

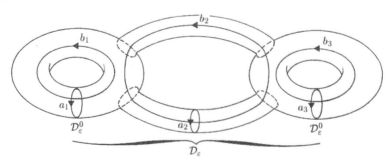

Figure 7.6. *Genus 3curve with a canonical basis of cycles*

As a basis of holomorphic differentials $\omega_1, \omega_2, \omega_3$ on the curve $\mathcal{D}_\epsilon$, we take,

$$\omega_1 = \frac{k_1(\alpha^2-1)\beta^2 d\beta}{\alpha u}, \qquad \omega_2 = \frac{k_2 d\beta}{\alpha u}, \qquad \omega_3 = \frac{d\beta}{u}, \qquad [7.26]$$

and obviously $\sigma^*(\omega_1) = -\omega_1$, $\sigma^*(\omega_2) = -\omega_2$, $\sigma^*(\omega_3) = \omega_3$. Recall that the Prym variety $Prym(\mathcal{D}_\varepsilon/\mathcal{D}_\varepsilon^0)$ is a subabelian variety of the Jacobi variety $Jac(\mathcal{D}_\varepsilon) = Pic^0(\mathcal{D}_\varepsilon) = H^1(\mathcal{O}_{\mathcal{D}_\varepsilon})/H^1(\mathcal{D}_\varepsilon, \mathbb{Z})$, constructed from the double cover φ_ε (7.22), the involution σ_ε on $\mathcal{D}_\varepsilon$ interchanging sheets, extends by linearity to a map $\sigma_\varepsilon : Jac(\mathcal{D}_\varepsilon) \to Jac(\mathcal{D}_\varepsilon)$ and up to some points of order two, $Jac(\mathcal{D}_\varepsilon)$ splits into an even part and an odd part: the even part is an elliptic curve (the quotient of $\mathcal{D}_\varepsilon$ by the involution σ_ε, i.e. $\mathcal{D}_\varepsilon^0$ [7.21]) and the odd part is a two-dimensional Abelian surface $Prym(\mathcal{D}_\varepsilon/\mathcal{D}_\varepsilon^0)$. We consider the period matrix Ω of $Jac(\mathcal{D}_\varepsilon)$

$$\Omega = \begin{pmatrix} \int_{a_1}\omega_1 & \int_{a_2}\omega_1 & \int_{a_3}\omega_1 & \int_{b_1}\omega_1 & \int_{b_2}\omega_1 & \int_{b_3}\omega_1 \\ \int_{a_1}\omega_2 & \int_{a_2}\omega_2 & \int_{a_3}\omega_2 & \int_{b_1}\omega_2 & \int_{b_2}\omega_2 & \int_{b_3}\omega_2 \\ \int_{a_1}\omega_3 & \int_{a_2}\omega_3 & \int_{a_3}\omega_3 & \int_{b_1}\omega_3 & \int_{b_2}\omega_3 & \int_{b_3}\omega_3 \end{pmatrix}.$$

Then,

$$\Omega = \begin{pmatrix} \int_{a_1}\omega_1 & \int_{a_2}\omega_1 & -\int_{a_1}\omega_1 & \int_{b_1}\omega_1 & \int_{b_2}\omega_1 & -\int_{b_1}\omega_1 \\ \int_{a_1}\omega_2 & \int_{a_2}\omega_2 & -\int_{a_1}\omega_2 & \int_{b_1}\omega_2 & \int_{b_2}\omega_2 & -\int_{b_1}\omega_2 \\ \int_{a_1}\omega_3 & 0 & \int_{a_1}\omega_3 & \int_{b_1}\omega_3 & 0 & \int_{b_1}\omega_3 \end{pmatrix},$$

and therefore the period matrices of $Jac(\mathcal{D}_\varepsilon^0)$ (i.e. $\mathcal{D}_\varepsilon^0$), $Prym(\mathcal{D}_\varepsilon/\mathcal{D}_\varepsilon^0)$ and $Prym^*(\mathcal{D}_\varepsilon/\mathcal{D}_\varepsilon^0)$ are, respectively, $\Delta = \begin{pmatrix} \int_{a_1}\omega_3 & \int_{b_1}\omega_3 \end{pmatrix}$,

$$\Gamma = \begin{pmatrix} 2\int_{a_1}\omega_1 & \int_{a_2}\omega_1 & 2\int_{b_1}\omega_1 & \int_{b_2}\omega_1 \\ 2\int_{a_1}\omega_2 & \int_{a_2}\omega_2 & 2\int_{b_1}\omega_2 & \int_{b_2}\omega_2 \end{pmatrix},$$

$$\Gamma^* = \begin{pmatrix} \int_{a_1}\omega_1 & \int_{a_2}\omega_1 & \int_{b_1}\omega_1 & \int_{b_2}\omega_1 \\ \int_{a_1}\omega_2 & \int_{a_2}\omega_2 & \int_{b_1}\omega_2 & \int_{b_2}\omega_2 \end{pmatrix}.$$

Let

$$L_\Omega = \left\{ \sum_{i=1}^3 m_i \int_{a_i} \begin{pmatrix} \omega_1 \\ \omega_2 \\ \omega_3 \end{pmatrix} + n_i \int_{b_i} \begin{pmatrix} \omega_1 \\ \omega_2 \\ \omega_3 \end{pmatrix} : m_i, n_i \in \mathbb{Z} \right\},$$

be the period lattice associated with Ω. Let us also denote by L_Δ, the period lattice associated Δ. We have the following diagram (N_{ρ_ε} is the norm mapping, surjective),

$$\begin{array}{ccccccccc}
& & & & 0 & & & & \\
& & & & \downarrow & & & & \\
& & & & \mathcal{D}_\varepsilon^0 & & & \mathcal{D}_\varepsilon & \\
& & & & \downarrow \rho_\varepsilon^* & \swarrow & & \downarrow \rho_\varepsilon & \\
0 & \longrightarrow & \ker N_{\rho_\varepsilon} & \longrightarrow & Prym(\mathcal{D}_\varepsilon/\mathcal{D}_\varepsilon^0) \oplus \mathcal{D}_\varepsilon^0 = Jac(\mathcal{D}_\varepsilon) & \xrightarrow{N_{\rho_\varepsilon}} & \mathcal{D}_\varepsilon^0 & \longrightarrow & 0 \\
& & & \searrow \tau_\varepsilon & \downarrow & & & & \\
& & & & \widetilde{M_c} \cup 2\mathcal{D}_\varepsilon \simeq \mathbb{C}^2/Lattice & & & & \\
& & & & \downarrow & & & & \\
& & & & 0 & & & &
\end{array}$$

The polarization map $\tau_\varepsilon : Prym(\mathcal{D}_\varepsilon/\mathcal{D}_\varepsilon^0) \longrightarrow \widetilde{M_c} = Prym^*(\mathcal{D}_\varepsilon/\mathcal{D}_\varepsilon^0)$ has kernel $(\rho^*\mathcal{D}_\varepsilon^0) \simeq \mathbb{Z}_2 \times \mathbb{Z}_2$ and the induced polarization on $Prym(\mathcal{D}_\varepsilon/\mathcal{D}_\varepsilon^0)$ is of type (1,2). Let $\widetilde{M_c} \longrightarrow \mathbb{C}^2/L_\Lambda : p \longmapsto \int_{p_0}^p \begin{pmatrix} dt_1 \\ dt_2 \end{pmatrix}$ be the uniformizing map where dt_1, dt_2 are two differentials on $\widetilde{M_c}$ corresponding to the flows generated, respectively, by H_1, H_2 such that: $dt_1|_{\mathcal{D}_\varepsilon} = \omega_1$ and $dt_2|_{\mathcal{D}_\varepsilon} = \omega_2$, $L_\Lambda = \left\{ \sum_{k=1}^4 n_k \begin{pmatrix} \int_{\nu_k} dt_1 \\ \int_{\nu_k} dt_2 \end{pmatrix} : n_k \in \mathbb{Z} \right\}$, is the lattice associated with the period matrix

$$\Lambda = \begin{pmatrix} \int_{\nu_1} dt_1 & \int_{\nu_2} dt_1 & \int_{\nu_3} dt_1 & \int_{\nu_4} dt_1 \\ \int_{\nu_1} dt_2 & \int_{\nu_2} dt_2 & \int_{\nu_3} dt_2 & \int_{\nu_4} dt_2 \end{pmatrix},$$

and $(\nu_1, \nu_2, \nu_3, \nu_4)$ is a basis of $H_1(\widetilde{M_c}, \mathbb{Z})$. By the Lefschetz theorem on hyperplane section, the map $H_1(\mathcal{D}_\varepsilon, \mathbb{Z}) \longrightarrow H_1(\widetilde{M_c}, \mathbb{Z})$ induced by the inclusion $\mathcal{D}_\varepsilon \hookrightarrow \widetilde{M_c}$ is surjective and we can find four cycles $\nu_1, \nu_2, \nu_3, \nu_4$ on the curve $\mathcal{D}_\varepsilon$ such that

$$\Lambda = \begin{pmatrix} \int_{\nu_1} \omega_1 & \int_{\nu_2} \omega_1 & \int_{\nu_3} \omega_1 & \int_{\nu_4} \omega_1 \\ \int_{\nu_1} \omega_2 & \int_{\nu_2} \omega_2 & \int_{\nu_3} \omega_2 & \int_{\nu_4} \omega_2 \end{pmatrix},$$

and $L_\Lambda = \left\{ \sum_{k=1}^4 n_k \begin{pmatrix} \int_{\nu_k} \omega_1 \\ \int_{\nu_k} \omega_2 \end{pmatrix} : n_k \in \mathbb{Z} \right\}$.

The cycles $\nu_1, \nu_2, \nu_3, \nu_4$ that we look for in $\mathcal{D}_\varepsilon$ are a_1, a_2, b_1, b_2 and they generate $H_1(\widetilde{M_c}, \mathbb{Z})$, such that

$$\Lambda = \begin{pmatrix} \int_{a_1} \omega_1 & \int_{a_2} \omega_1 & \int_{b_1} \omega_1 & \int_{b_2} \omega_1 \\ \int_{a_1} \omega_2 & \int_{a_2} \omega_2 & \int_{b_1} \omega_2 & \int_{b_2} \omega_2 \end{pmatrix},$$

is a Riemann matrix. So $\Lambda = \Gamma^*$, that is, the period matrix of $Prym^*(\mathcal{D}_\varepsilon/\mathcal{D}_\varepsilon^0)$ dual of $Prym(\mathcal{D}_\varepsilon/\mathcal{D}_\varepsilon^0)$. Consequently, $\widetilde{M_c}$ and $Prym^*(\mathcal{D}_\varepsilon/\mathcal{D}_\varepsilon^0)$ are two Abelian varieties analytically isomorphic to the same complex torus $\mathbb{C}^2/L_\Lambda$. By Chow's theorem (Griffiths and Harris 1978), $\widetilde{M_c}$ and $Prym^*(\mathcal{D}_\varepsilon/\mathcal{D}_\varepsilon^0)$ are then algebraically isomorphic. $\square$

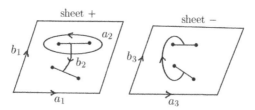

Figure 7.7. *Sheet $+$ and sheet $-$*

In summary, the use of the Laurent series (and the study of convergence of various series intervening in the problem via the majorant method and the concept of normally generated line bundle) show, that the orbits of the vector field [3.12] going through the curve $\mathcal{D}$ form a smooth surface Σ near $\mathcal{D}$, such that $\Sigma \backslash M_c \subseteq \widetilde{M_c}$ and the variety $\widetilde{M_c} = M_c \cup \Sigma$ is smooth, compact and connected. More precisely, for generic c, the affine surface M_c completes into an Abelian surface $\widetilde{M_c}$ of polarization $(1,2)$, by adding a singular divisor $\mathcal{D}$ consisting of two isomorphic genus 3 curves $\mathcal{D}_{\varepsilon=i}$ and $\mathcal{D}_{\varepsilon=-i}$ intersecting in four points. Each $\mathcal{D}_{\varepsilon=\pm i}$ is a double cover of an elliptic curve $\mathcal{D}_\varepsilon^0$ ramified at four points. It defines a line bundle and a polarization $(1,2)$ on $\widetilde{M_c}$. The divisors $2\mathcal{D}_{\varepsilon=i}$, $2\mathcal{D}_{\varepsilon=-i}$ or $\mathcal{D}_{\varepsilon=i} + \mathcal{D}_{\varepsilon=-i}$ (of geometric genus 9) are all very ample and define a polarization $(2,4)$; moreover, the eight-dimensional space of sections of the corresponding line bundle embeds the Abelian surface into $\mathbb{P}^7(\mathbb{C})$. In other words, the line bundle $[\mathcal{D}]$ defines a polarization of type $(2,4)$ on $\widetilde{M_c}$ and leads to an embedding of $\widetilde{M_c}$ in $\mathbb{P}^7(\mathbb{C})$; it is not normally generated, but the line bundle $[2\mathcal{D}_{\varepsilon=i} + \mathcal{D}_{\varepsilon=-i}]$ is. The Abelian surface $\widetilde{M_c}$ is equipped with two everywhere independent commuting vector fields. Otherwise, the Abelian surface $\widetilde{M_c}$ can also be identified as the dual of the Prym variety $(\mathcal{D}_\varepsilon/\mathcal{D}_\varepsilon^0)$ and the problem linearizes on this variety.

7.6. The Hénon–Heiles system

Consider the system

$$\dot{x}_1 = y_1, \qquad \dot{y}_1 = -Ax_1 - 2x_1x_2, \qquad [7.27]$$
$$\dot{x}_2 = y_2, \qquad \dot{y}_2 = -Bx_2 - x_1^2 - \varepsilon x_2^2,$$

corresponding to a generalized Hénon–Heiles Hamiltonian

$$H = \frac{1}{2}(y_1^2 + y_2^2) + \frac{1}{2}(Ax_1^2 + Bx_2^2) + x_1^2 x_2 + \frac{\varepsilon}{3}x_2^3, \qquad [7.28]$$

where A, B, ε are constant parameters and x_1, x_2, y_1, y_2 are canonical coordinates and momenta, respectively. First studied as a mathematical model to describe the chaotic motion of a test star in an axisymmetric galactic mean gravitational field (Hénon and Heiles 1964), this system is widely explored in other branches of physics. It is well known from applications in stellar dynamics, statistical mechanics and quantum mechanics. It provides a model for the oscillations of atoms in a three-atomic molecule (Berry 1978). Usually, the Hénon–Heiles system is not integrable and represents a classical example of chaotic behavior. Nevertheless, at some special values of the parameters, it is integrable; to be precise, there are three known integrable cases: i) $\varepsilon = 6$, A and B arbitrary. The second integral of motion is

$$H_2 = x_1^4 + 4x_1^2 x_2^2 - 4y_1^2 x_2 + 4y_1 y_2 y_1 + 4Ax_1^2 x_2 + (4A - B)y_1^2 + A(4A - B)x_1^2. \quad [7.29]$$

ii) $\varepsilon = 1$, $A = B$. The second integral of motion is

$$H_2 = y_1 y_2 + \frac{1}{3}x_1^3 + x_1 x_2^2 + Ax_1 x_2.$$

iii) $\varepsilon = 16$, $B = 16A$. The second integral of motion is

$$H_2 = 3y_1^4 + 6Ay_1^2 x_1^2 + 12y_1^2 x_1^2 x_2 - 4y_1 y_2 x_1^3 - 4Ax_1^4 x_2 - 4x_1^4 x_2^2$$
$$+ 3A^2 x_1^4 - \frac{2}{3}x_1^6.$$

In the two cases (i) and (ii), the system [7.27] has been integrated by making use of genus one and genus two theta functions. For case (i), it was shown (Ankiewicz and Pask 1983) that this case separates in translated parabolic coordinates. Solving the problem in case (ii) is not difficult (this case trivially separates in Cartesian coordinates). In case (iii), the system can also be integrated (Ravoson et al. 1993) by making use of elliptic functions. The general solutions of the equations of motion for Hamiltonian [7.28], for cases (i) and (ii), have the Painlevé propriety, that is, they only admit poles in the complex time variable. This section deals with case (i) (case (iii) will be studied in section 8.3). The system [7.27] can be written in the form

$$\dot{z} = f(z) = J\frac{\partial H}{\partial z}, \quad z = (x_1, x_2, y_1, y_2)^\mathsf{T}, \quad [7.30]$$

where $H = H_1[7.28]$, $\frac{\partial H}{\partial z} = \left(\frac{\partial H}{\partial x_1}, \frac{\partial H}{\partial x_2}, \frac{\partial H}{\partial y_1}, \frac{\partial H}{\partial y_2}\right)^\mathsf{T}$, $J = \begin{pmatrix} O & I \\ -I & O \end{pmatrix}$. The second flow commuting with the first is regulated by the equations: $\dot{z} = J\frac{\partial H_2}{\partial z}$, $z = (x_1, x_2, y_1, y_2)^\mathsf{T}$, with H_2 defined by [7.29]. The invariant (or level) variety

$$M_c = \bigcap_{i=1}^{2}\{z : H_i(z) = c_i\} \subset \mathbb{C}^4, \quad [7.31]$$

is a smooth affine surface for generic $c = (c_1, c_2) \in \mathbb{C}^2$. As before, the question is: how do we find the compactification of M_c into an Abelian surface? Now the system [7.30] is integrable in the sense of Liouville, but how does one effectively integrate the problem? We check the existence near each movable singularity of a Laurent series that represents the general solution. In fact, the Laurent decomposition of such asymptotic solutions have the following form $z(t) \equiv (x_1(t), x_2(t), y_1(t), y_2(t))$, where

$$x_1 = \sum_{k=0}^{\infty} x_1^{(k)} t^{k-1}, \quad x_2 = \sum_{k=0}^{\infty} x_2^{(k)} t^{k-2}, \quad y_1 = \dot{x}_1, \quad y_2 = \dot{x}_2,$$

which depend on three free parameters. Putting these expressions into [7.30], solving inductively for the $z^{(k)}$, one finds a nonlinear equation at the 0th step,

$$z^{(0)} + f(z^{(0)}) = 0, \qquad [7.32]$$

and a linear system of equations at the kth step :

$$(L - kI)z^{(k)} = \begin{cases} 0 \text{ for } k = 1, \\ \text{quadratic polynomial in } z^{(1)}, ..., z^{(k-1)} \text{ for } k > 1, \end{cases} \qquad [7.33]$$

where L denotes the Jacobian map of [7.32]. One parameter appears at the 0th step, that is, in the resolution of [7.32] and the two remaining ones at the kth step, $k = 3$, $k = 6$. The resolution of these systems is a linear algebra problem, which is a straight forward computation. Using the majorant method (see theorem 7.1), we can show that the formal Laurent series solutions are convergent. Consequently, we have the following theorem:

THEOREM 7.19.– The non-identically zero solutions of [7.32] define a line. The system [7.33] has 1 degree of freedom for $k = 3$ and $k = 6$. The flow X_{H_1} [7.27] possesses Laurent series solutions that depend on three free parameters α, β and γ. The first free parameter α appears in the resolution of [7.32] and the two remaining ones β, γ at the kth step, $k = 3$, $k = 6$ of [7.33]. These Laurent series solutions are explicitly given by

$$x_1 = \frac{x_1^{(0)}}{t} + x_1^{(2)} t + x_1^{(3)} t^2 + x_1^{(4)} t^3 + x_1^{(5)} t^4 + x_1^{(6)} t^5 + \cdots$$

$$x_2 = \frac{x_2^{(0)}}{t^2} + x_2^{(2)} + x_2^{(4)} t^2 + x_2^{(5)} t^3 + x_2^{(6)} t^4 + \cdots \qquad [7.34]$$

$$y_1 = \dot{x}_1, \quad y_2 = \dot{x}_2$$

with leading terms given explicitly by $x_1^{(0)} = \alpha =$ free parameter, $x_2^{(0)} = -1$, $x_1^{(1)} = x_2^{(1)} = 0$, $x_1^{(2)} = \frac{\alpha^3}{12} + \frac{\alpha A}{2} - \frac{\alpha B}{12}$, $x_2^{(2)} = \frac{\alpha^2}{12} - \frac{B}{12}$, $x_1^{(3)} = \beta =$ free parameter, $x_2^{(3)} = 0$,

$$x_1^{(4)} = \frac{\alpha AB}{24} - \frac{\alpha^5}{72} + \frac{11\alpha^3 B}{720} - \frac{11\alpha^3 A}{120} - \frac{\alpha B^2}{720} - \frac{\alpha A^2}{8}, x_2^{(4)} = \frac{\alpha^4}{48} + \frac{\alpha^2 A}{10} - \frac{\alpha^2 B}{60} - \frac{B^2}{240},$$
$$x_1^{(5)} = -\frac{B\alpha^2}{12} + \frac{\beta B}{60} - \frac{A\beta}{10}, x_2^{(5)} = \frac{\alpha\beta}{3}, x_1^{(6)} = -\frac{\alpha\gamma}{9} - \frac{\alpha^7}{15552} - \frac{\alpha^5 A}{2160} + \frac{\alpha^5 B}{12960} + \frac{\alpha^3 B^2}{25920} +$$
$$\frac{\alpha^3 A^2}{1440} - \frac{\alpha^3 AB}{4320} + \frac{\alpha AB^2}{1440} - \frac{\alpha B^3}{19440} - \frac{\alpha A^2 B}{288} + \frac{\alpha A^3}{144} \text{ and } x_2^{(6)} = \gamma = \text{ free parameter.}$$

Consider points at infinity that are limit points of trajectories of the flow. We search for the set of Laurent solutions that remain confined to the fixed affine invariant surface M_c[7.31] related to specific values of c_1 and c_2.

THEOREM 7.20.– The pole solutions [7.34] restricted to the surface M_c[7.31] is a smooth genus 3 hyperelliptic curve $\mathcal{D}$[7.35], which is a double ramified cover of an elliptic curve $\mathcal{E}$[7.37].

PROOF.– By substituting [7.34] in the constants of the motion $H_1 = c_1$ and $H_2 = c_2$, we eliminate the parameter γ linearly, leading to algebraic relation between the two remaining parameters, which is the equation of the divisor $\mathcal{D}$ along which the $z(t) \equiv (x_1(t), x_2(t), y_1(t), y_2(t))$ blow up. So, $\mathcal{D}$ is the closure of the continuous components of $\{$Laurent series solutions $z(t)$ such that $H_i(z(t)) = c_i, 1 \leq i \leq 2\}$, that is, $\mathcal{D} = t^0 -$ coefficient of $\bigcap_{k=1}^{2} \{z \in \mathbb{C}^4 : H_k(z(t)) = c_k\}$. Thus, we find an algebraic curve defined by

$$\mathcal{D} = \overline{\{(\beta, \alpha) : \beta^2 = P_8(\alpha)\}}, \quad [7.35]$$

where

$$P_8(\alpha) = -\frac{7}{15,552}\alpha^8 - \frac{1}{432}\left(5A - \frac{13}{18}B\right)\alpha^6$$
$$- \frac{1}{36}\left(\frac{671}{15,120}B^2 + \frac{17}{7}A^2 - \frac{943}{1,260}BA\right)\alpha^4$$
$$- \frac{1}{36}\left(4A^3 - \frac{1}{2,520}B^3 - \frac{13}{6}A^2 B + \frac{2}{9}AB^2 - \frac{10}{7}c_1\right)\alpha^2 + \frac{1}{36}c_2.$$

The curve $\mathcal{D}$ determined by an eight-order equation is smooth, hyperelliptic and its genus is 3. Moreover, the map

$$\sigma : \mathcal{D} \longrightarrow \mathcal{D}, (\beta, \alpha) \longmapsto (\beta, -\alpha), \quad [7.36]$$

is an involution on $\mathcal{D}$ and the quotient $\mathcal{E} = \mathcal{D}/\sigma$ is an elliptic curve defined by

$$\mathcal{E} = \overline{\{(\beta, \zeta) : \beta^2 = P_4(\zeta)\}}, \quad [7.37]$$

where $P_4(\zeta)$ is the degree 4 polynomial in $\zeta = \alpha^2$ obtained from the polynomial $P_8(\alpha)$ above. The hyperelliptic curve $\mathcal{D}$ is thus a two-sheeted ramified covering of the elliptic curve $\mathcal{E}$ [7.37],

$$\rho : \mathcal{D} \longrightarrow \mathcal{E}, (\beta, \alpha) \longmapsto (\beta, \zeta), \quad [7.38]$$

ramified at the four points covering $\zeta = 0$ and ∞. □

Let T be a smooth surface compactifying M_c[7.31]. Consider a basis $1, f_1, ..., f_N$ of the vector space $\mathcal{L}(\mathcal{D}) \equiv \{f : f \text{ meromorphic on } T, (f) \geq -\mathcal{D}\}$, of meromorphic functions on T with at worst a simple pole along $\mathcal{D}$ and the map $T \to \mathbb{P}^N(\mathbb{C})$, $p \curvearrowright [1, f_1(p), ..., f_N(p)]$, considered projectively, because if at p some $f_i(p) = \infty$, we divide by f_i having the highest order pole near p, which makes every element finite. The Kodaira embedding theorem tells us that if the line bundle associated with the divisor is positive, then for $k \in \mathbb{N}$, the functions of $\mathcal{L}(k\mathcal{D})$ embed smoothly T into $\mathbb{P}^N(\mathbb{C})$ and then by Chow's theorem, T can be realized as an algebraic variety, that is, $T = \bigcap_i \{Z \in \mathbb{P}^N(\mathbb{C}) : P_i(Z) = 0\}$, where $P_i(Z)$ are homogeneous polynomials. In fact in our case, $k = 2$ and $N = 7$ suffice, that is, the divisor $2\mathcal{D}$ provides a smooth embedding into $\mathbb{P}^7(\mathbb{C})$, via the meromorphic section of $\mathcal{L}(2\mathcal{D})$. As in previous section, it is easy to find a set of polynomial functions $\{1, f_1, ..., f_N\}$ in $\mathcal{L}(2\mathcal{D})$ such that the embedding of $2\mathcal{D}$ with those functions into $\mathbb{P}^N$ yields a curve of genus $N + 2$. Straightforward calculation, using asymptotic expansions, shows that the space $\mathcal{L}(2\mathcal{D})$ is spanned by the following functions:

$$\mathcal{L}(2\mathcal{D}) = \{f_0, f_1, ..., f_7\},$$
$$= \{1, x_1, x_1^2, x_2, y_1, y_1^2 + x_1^2 x_2, y_2 x_1 - 2y_1 x_2, y_1 y_2 + 2A x_1 x_2 + 2x_1 x_2^2\},$$
$$= \left\{1, \frac{\alpha}{t}, \frac{\alpha^2}{t^2}, -\frac{1}{t^2}, -\frac{\alpha}{t^2}, -\frac{\alpha^2(\alpha^2 - 8A - B)}{t^2}, \frac{\alpha(\alpha^2 + 4A - B)}{2t^2}, \frac{6\beta}{t^2}\right\}$$

+higher order terms in t.

The application

$$2\mathcal{D} \longrightarrow \mathbb{P}^7, \quad p \longmapsto \lim_{t \to 0} t^2 [f_0(p), f_1(p), ..., f_7(p)] = \left[0, 0, f_2^{(0)}(p), ..., f_7^{(0)}(p)\right],$$

maps the curve $2\mathcal{D}$ into $\widetilde{2\mathcal{D}} \subseteq \mathbb{P}^7$ and the genus of $2\mathcal{D}$ is 9.

THEOREM 7.21.– The orbits of the vector field [7.7] running through $2\mathcal{D}$ form a smooth surface Σ near $2\mathcal{D}$ such that $\Sigma \backslash 2\mathcal{D} \subseteq T$. Moreover, the variety $T = M_c \cup \Sigma$, is smooth, compact, connected and it comes equipped with two everywhere independent commuting vector fields, which extend holomorphically on T. The system [7.7] is algebraic complete integrable and the corresponding flow evolves on an Abelian surface $T \simeq \mathbb{C}^2/Lattice$.

PROOF.– Let $\psi(t, p) = \{z(t) = (x_1(t), x_2(t), y_1(t), y_2(t)) : t \in \mathbb{C}, 0 < |t| < \varepsilon\}$ be the orbit of the vector field [7.7] going through the point $p \in 2\mathcal{D}$ and consider the surface element $\Sigma_p \subset \mathbb{P}^7(\mathbb{C})$ formed by the divisor $2\mathcal{D}$ and the orbits going through p. Consider the curve $\mathcal{S} = H \cap \Sigma$ where $H \subset \mathbb{P}^7(\mathbb{C})$ is a hyperplane transversal to the direction of the flow and $\Sigma \equiv \bigcup_{p \in 2\mathcal{D}} \Sigma_p$. If $\mathcal{S}$ is smooth, then using the implicit function theorem, the surface Σ is smooth. But if $\mathcal{S}$ is singular at 0, then Σ would be singular along the trajectory (t–axis) which immediately goes into the affine part M_c. Hence, M_c would be singular, which is a contradiction (M_c is the fiber of a morphism from $\mathbb{C}^4$ to $\mathbb{C}^2$ and so smooth for almost all of the two constants of the motion c_i). Let $\overline{M}_c$ be the projective closure of M_c into $\mathbb{P}^4(\mathbb{C})$, $Z = [Z_0, Z_1 = z_1 Z_0, ..., Z_4 = z_4 Z_0] \in \mathbb{P}^4(\mathbb{C})$, and let $M_\infty = \overline{M}_c \cap \{Z_0 = 0\}$ be the locus at infinity. Consider the map $\overline{M}_c \subseteq \mathbb{P}^4 \longrightarrow \mathbb{P}^7(\mathbb{C})$, $Z \longmapsto f(Z)$, where $f = (f_0, f_1, ..., f_7) \in L(2\mathcal{D})$ and let $T = f(\overline{M}_c)$. In a neighborhood $V(p) \subseteq \mathbb{P}^7(\mathbb{C})$ of p, we have $\Sigma_p = T$ and $\Sigma_p \backslash 2\mathcal{D} \subseteq M_c$. Otherwise, there would exist an element of surface $\Sigma'_p \subseteq T$ such that $\Sigma_p \cap \Sigma'_p = t-$axis, orbit $\psi(t,p) = t-$axis$\backslash p \subseteq M_c$, and hence M_c would be singular along the t–axis, which is impossible. Since the variety $\overline{M}_c \cap \{Z_0 \neq 0\}$ is irreducible and the generic hyperplane section $H_{gen.}$ of $\overline{M}_c$ is also irreducible, all hyperplane sections are connected and hence M_∞ is also connected. Consider the graph $\Gamma_f \subseteq \mathbb{P}^4 \times \mathbb{P}^7(\mathbb{C})$ of the map f, which is irreducible together with $\overline{M}_c$. It follows from the irreducibility of M_∞ that a generic hyperplane section $\Gamma_f \cap \{H_{gen.} \times \mathbb{P}^7(\mathbb{C})\}$ is irreducible, hence the special hyperplane section $\Gamma_f \cap \{\{Z_0 = 0\} \times \mathbb{P}^7(\mathbb{C})\}$ is connected and therefore the projection map $proj_{\mathbb{P}^7}[\Gamma_f \cap \{\{Z_0 = 0\} \times \mathbb{P}^7(\mathbb{C})] = f(M_\infty) \equiv 2\mathcal{D}$ is connected. Hence, the variety $M_c \cup \Sigma = T$ is compact, connected and embeds smoothly into $\mathbb{P}^7(\mathbb{C})$ via f. Let g^{t_1} and g^{t_2} be the flows generated, respectively, by vector fields X_{H_1} and X_{H_2}. For $p \in 2\mathcal{D}$ and for small $\varepsilon > 0$, $g^{t_1}(p), \forall t_1, 0 < |t_1| < \varepsilon$, is well defined and $g^{t_1}(p) \in M_c$. We define g^{t_2} on M_c by $g^{t_2}(q) = g^{-t_1} g^{t_2} g^{t_1}(q)$, $q \in U(p) = g^{-t_1}(U(g^{t_1}(p)))$, where $U(p)$ is a neighborhood of p. By commutativity one can see that g^{t_2} is independent of t_1; $g^{-t_1 - \varepsilon_1} g^{t_2} g^{t_1 + \varepsilon_1}(q) = g^{-t_1} g^{-\varepsilon_1} g^{t_2} g^{t_1} g^{\varepsilon_1} = g^{-t_1} g^{t_2} g^{t_1}(q)$. We affirm that $g^{t_2}(q)$ is holomorphic away from $2\mathcal{D}$. This is because $g^{t_2} g^{t_1}(q)$ is holomorphic away from $2\mathcal{D}$ and that g^{t_1} is holomorphic in $U(p)$ and maps bi-holomorphically $U(p)$ onto $U(g^{t_1}(p))$. Since the flows g^{t_1} and g^{t_2} are holomorphic and independent on $2\mathcal{D}$, we can show along the same lines as in theorem 7.10 that T is a torus by considering the holomorphic map $\mathbb{C}^2 \longrightarrow T$, $(t_1, t_2) \longmapsto g^{t_1} g^{t_2}(p)$, for a fixed origin $p \in M_c$. The additive subgroup $Lattice = \{(t_1, t_2) \in \mathbb{C}^2 : g^{t_1} g^{t_2}(p) = p\}$ is a lattice of $\mathbb{C}^2$ and hence $\mathbb{C}^2/lattice \longrightarrow T$ is a biholomorphic diffeomorphism. Therefore, $T \subseteq \mathbb{P}^7$ is conformal to a complex torus $\mathbb{C}^2/lattice$ and an Abelian surface as a consequence of the Chow theorem. $\square$

REMARK 7.5.– Recall that a Kähler variety is a variety with a Kähler metric, that is, a Hermitian metric whose associated differential 2-form of type $(1,1)$ is closed. The complex torus $\mathbb{C}^2/lattice$ with the Euclidean metric $\sum dz_i \otimes d\bar{z}_i$ is a Kähler variety and any compact complex variety that can be embedded in projective space is also a Kähler variety. A compact complex Kähler variety having as many independent meromorphic functions as its dimension is a projective variety. We have shown that T is a complex torus $\mathbb{C}^2/lattice$, and so, in particular, T is a Kähler variety with Kähler metric given by $dt_1 \otimes d\bar{t}_1 + dt_2 \otimes d\bar{t}_2$. As mentioned above, a compact complex Kähler variety having the required number as (its dimension) of independent meromorphic functions is a projective variety. Thus, T is both a projective variety and a complex torus $\mathbb{C}^2/Lattice$ and hence an Abelian surface as a consequence of the Chow theorem.

THEOREM 7.22.– The flow [7.27] evolves on Abelian surface $T \subseteq \mathbb{P}^7(\mathbb{C})$ of period matrix $\begin{pmatrix} 2 & 0 & a & c \\ 0 & 4 & c & b \end{pmatrix}$, $\mathrm{Im}\begin{pmatrix} a & c \\ c & b \end{pmatrix} > 0$, $(a, b, c \in \mathbb{C})$, and is expressed in terms of Abelian integrals (involving the differentials [7.40]).

PROOF.– Note that the affine invariant surface M_c [7.31] has the following involution $\sigma : (x_1, x_2, y_1, y_2) \longmapsto (x_1, x_2, -y_1, -y_2)$, which maps X_{H_j} into $-X_{H_j}$, $j = 1, 2$, where X_{H_1} is the flow [7.27] and X_{H_2} the other flow commuting with the first, thus σ amounts to a reflection about some appropriately chosen origin on T. This map acts on the parameters of the Laurent solution [7.34] as follows: $(t, \beta, \alpha) \longmapsto (-t, \beta, -\alpha)$. Since $\mathcal{L}$ is symmetric ($\sigma^*\mathcal{L} \simeq \mathcal{L}$), σ can be lifted to $\mathcal{L}$ as an involution $\tilde{\sigma}$ in two ways, differing in sign and for each section (theta-function) $s \in H^0(\mathcal{L})$, we therefore have $\tilde{\sigma}s = \pm s$. Recall that a section $s \in H^0(\mathcal{L})$ is called even (respectively, odd) if $\tilde{\sigma}s = +s$ (respectively, $\tilde{\sigma}s = -s$). Under $\tilde{\sigma}$, the vector space $H^0(\mathcal{L})$ splits into an even and odd subspace $H^0(L) = H^0(\mathcal{L})^{even} \oplus H^0(\mathcal{L})^{odd}$, with $H^0(\mathcal{L})^{even}$ containing all of the even sections and $H^0(\mathcal{L})^{odd}$ all of the odd ones. Using the inverse formula (Mumford 1967a, p. 331), we see after a small computation that

$$h^0(\mathcal{L})^{even} \equiv \dim H^0(\mathcal{L})^{even} = \frac{\delta_1\delta_2}{2} + 2^{-1+\# \text{ even } \delta_k}, k = 1, 2 \qquad [7.39]$$

$$h^0(\mathcal{L})^{odd} \equiv \dim H^0(\mathcal{L})^{odd} = \frac{\delta_1\delta_2}{2} - 2^{-1+\# \text{ even } \delta_k}, k = 1, 2$$

By the classification theory of ample line bundles on Abelian varieties and theorem 7.21, $T \simeq \mathbb{C}^2/\mathcal{L}_\Omega$ with period lattice given by the columns of the following matrix $\begin{pmatrix} \delta_1 & 0 & a & c \\ 0 & \delta_2 & c & b \end{pmatrix}$, $\mathrm{Im}\begin{pmatrix} a & c \\ c & b \end{pmatrix} > 0$, $(a, b, c \in \mathbb{C})$, with $\delta_1\delta_2 = \dim H^0(\mathcal{L}^{\otimes 2}) = \dim \mathcal{L}(2\mathcal{D}) = g(2\mathcal{D}) - 1 = 8$. We have two possibilities: (i) $\delta_1 = 1$, $\delta_2 = 8$ and (ii) $\delta_1 = 2$, $\delta_2 = 4$. From [7.39], the corresponding line bundle $L^{\otimes 2}$ has five even sections and three odd sections in case (i) and six even sections and two odd sections in case (ii). The space $L^{\otimes 2}$ splits into two subspaces $\left(\mathcal{L}^{\otimes 2}\right)^{even}$ and $\left(\mathcal{L}^{\otimes 2}\right)^{odd}$ of even and odd functions for the σ-involution

$$\mathcal{L}^{\otimes 2} = \left(\mathcal{L}^{\otimes 2}\right)^{even} \oplus \left(\mathcal{L}^{\otimes 2}\right)^{odd} = \{f_0, f_1, f_2, f_3, f_5, f_7\} \oplus \{f_4, f_6\},$$
$$= \{1, x_1, x_1^2, x_2, y_1^2 + x_1^2x_2, y_1y_2 + 2Ax_1x_2 + 2x_1x_2^2\} \oplus \{y_1, y_2x_1 - 2y_1x_2\}.$$

Among the functions of $\mathcal{L}^{\otimes 2}$, there are six even and two odd functions for the involution σ, showing that case (ii) is the only alternative and the period matrix has the form $\begin{pmatrix} 2 & 0 & a & c \\ 0 & 2 & c & b \end{pmatrix}$, $\mathrm{Im}\begin{pmatrix} a & c \\ c & b \end{pmatrix} > 0$, $(a, b, c \in \mathbb{C})$. On T, let the holomorphic 1-forms

dt_1 and dt_2 defined by $dt_i(X_{H_j}) = \delta_{ij}$. The 1-forms $\omega_0 = \frac{\alpha d\alpha}{\beta}$, $\omega_1 = \frac{\alpha^2 d\alpha}{\beta}$, $\omega_2 = \frac{d\alpha}{\beta}$, yield a basis of $H^0(\mathcal{D}, \Omega_{\mathcal{D}}^1)$. Using the Laurent solutions, the differentials dt_1 and dt_2 corresponding to the flows generated, respectively, by H_1 and H_2, restricted to the curve $\mathcal{D}$[7.35], descend to two differentials on $\mathcal{D}$:

$$dt_1|_{\mathcal{D}} = \frac{\alpha^2 d\alpha}{\beta} = \omega_1, \qquad dt_2|_{\mathcal{D}} = \frac{d\alpha}{\beta} = \omega_2. \qquad [7.40]$$

This concludes the proof. □

REMARK 7.6.– The divisor $\mathcal{D}$ defines on the Abelian surface T a polarization $(1,2)$. The eight-dimensional space of sections of the line bundle $L^{\otimes 2}$ splits into two subspaces $(L^{\otimes 2})^{even}$ and $(L^{\otimes 2})^{odd}$ of even and odd sections (theta functions). In other words, we have $L^{\otimes 2} = (L^{\otimes 2})^{even} \oplus (L^{\otimes 2})^{odd} = \{f_0, f_1, f_2, f_3, f_5, f_7\} \oplus \{f_4, f_6\}$. The remarkable property is that $W\left((L^{\otimes 2})^{even}, (L^{\otimes 2})^{odd}\right) \subset \left((L^{\otimes 2})^{even}\right)^{\otimes 2}$, where $W(,)$ is the Wronskian $W(f_i, f_j) \equiv f_i X(f_j) - f_j X(f_i)$ between two theta functions, with respect to an arbitrary holomorphic vector field X on T. The Abelian surface T, as embedded in $\mathbb{P}^7(\mathbb{C})$ can be described by six quadratic relations between the theta functions, three of which only involve even sections and another three involving even and odd sections. From the divisor $\mathcal{D}$ and the line bundle $\mathcal{L}$, a lot of information can be obtained. For example, Adler and van Moerbeke (1988) have obtained a one-dimensional family of birationally maps between the Hénon-Heiles system (case i), Kowalewski's top and the geodesic flow on $SO(4)$ for the Manakov metric. Such birationally maps are given by identifying the three eight-dimensional spaces $\mathcal{L}$ of Hénon–Heiles, Kowalewski and Manakov.

THEOREM 7.23.– The Abelian surface T which completes the affine surface M_c is the dual Prym variety $Prym^*(\mathcal{D}/\mathcal{E})$ of the genus 3 hyperelliptic curve $\mathcal{D}$ [7.35] for the involution σ [7.36] interchanging the sheets of the double covering ρ [7.38] and the problem linearizes on this variety.

PROOF.– Simply follow the same reasoning that was provided in theorem 7.18, while taking account of the notations used here. □

We will now show that T can also be seen as a double unramified cover of the Jacobian variety $Jac(\mathcal{C})$ of the 2-genus hyperelliptic curve $\mathcal{C}$ [7.41]. The remarkable property of this curve is that the flow of solutions of the equations of motion is linearized on its Jacobian $Jac(\mathcal{C})$ and so, the solutions can be expressed in terms of theta-functions of two variables. We also discuss the Kummer surface associated with T; it will be constructed explicitly.

THEOREM 7.24.– The torus T can be regarded as a double unramified cover of the Jacobian variety $Jac(\mathcal{C})$ of the 2-genus hyperelliptic curve $\mathcal{C}$ [7.41] and the system

[7.27] can be integrated in terms of genus 2 hyperelliptic functions of time. The involution σ [7.36] on the Jacobian $Jac(\mathcal{C})$ leads to a singular surface $K_\mathcal{C}$, which after desingularization, defines a K_3 surface $\widetilde{K_\mathcal{C}}$.

PROOF.– Remember that the Laurent solutions [7.34] restricted to the surface M_c [7.31] are parameterized by an hyperelliptic curve $\mathcal{D}$ [7.35] of genus 3. The latter is a double ramified cover of an elliptic curve $\mathcal{E}$ [7.37] and can also be seen as a two-sheeted unramified cover $\pi : \mathcal{D} \longrightarrow \mathcal{C}$, $(\beta, \alpha) \longmapsto (\eta, \zeta)$ of the following hyperelliptic curve $\mathcal{C}$ of genus 2:

$$\mathcal{C} : \eta^2 = \zeta P_4(\zeta). \qquad [7.41]$$

We show, using the results obtained above and applying the method explained in Vanhaecke (2001), that the linearized flow can be realized on the Jacobian variety $Jac(\mathcal{C})$ of the 2-genus hyperelliptic curve $\mathcal{C}$. The torus T can be regarded as a double unramified cover of the Jacobian variety $Jac(\mathcal{C})$ and the system [7.27] can be integrated in terms of genus 2 hyperelliptic functions of time. The differentials dt_1 and dt_2, corresponding to the flows X_{H_1} and X_{H_2}, restricted to the curve $\mathcal{D}$, go down to $\mathcal{C}$. Indeed, using [7.40], $dt_1|_\mathcal{D} = \omega_1 = \frac{\alpha^2 d\alpha}{\beta} = \frac{\zeta d\zeta}{\eta}$, $dt_2|_\mathcal{D} = \omega_2 = \frac{d\alpha}{\beta} = \frac{d\zeta}{\eta}$, yielding the two hyperelliptic differentials on $\mathcal{C}$. The map ρ [7.38] extends to a map $\widetilde{\rho} : T \to Jac(\mathcal{C})$. We denote by $|\mathcal{C}|$ the linear system, that is, the set of all effective divisors linearly equivalent to $\mathcal{C}$. We have $|\mathcal{C}| = \mathbb{P}(L(\mathcal{D}))$; associating to each non-zero function $f \in \mathcal{L}(\mathcal{C})$. An Abelian variety $T \simeq \mathbb{C}^n/lattice$ has a natural involution σ, induced by the sign flip $(z_1, \ldots, z_n) \longmapsto (-z_1, \ldots, -z_n)$, in $\mathbb{C}^n$. Its fixed points are exactly the 2^{2n} half-periods of T. The quotient T/σ is called the Kummer surface. In our case, $n = 2$ and the involution σ [7.36] on the Jacobian $Jac(\mathcal{C})$ leads to a singular surface K_C, which after desingularization, at the 16 fixed points of the involution σ, defines a K_3 surface $\widetilde{K_\mathcal{C}}$. It has no holomorphic 1-forms and it has a trivial canonical divisor. The Kummer surface $K_\mathcal{C}$ of $Jac(\mathcal{C})$ can be considered as the image of $\psi_{\widetilde{\rho}^*|2\mathcal{C}|} : T \to \mathbb{P}^3(\mathbb{C})$, with $\widetilde{\rho}^*|2\mathcal{C}| \subset |2\mathcal{D}|$. Since $\mathcal{C}$ is hyperelliptic of genus 2, it has six distinct Weierstrass points (indeed, let Γ a smooth hyperelliptic curve of genus $g \geq 2$ with $\phi : \Gamma \to \mathbb{P}^1(\mathbb{C})$ the two-sheeted map, then all of the branch points of ϕ are the only Weierstrass points of Γ. By the Riemann–Hurwitz formula, the number of these points is equal to $2g + 2$). Choose a Weierstrass point P on the curve $\mathcal{C}$ and coordinates $[Z_0, Z_1, Z_2, Z_3]$ for $\mathbb{P}^3$ such that $\psi_{\widetilde{\rho}^*|2\mathcal{C}|}(P) = [0, 0, 0, 1]$. Then this point will be a singular point for the Kummer surface K of equation $a(Z_0, Z_1, Z_2) Z_3^2 + b(Z_0, Z_1, Z_2) Z_3 + c(Z_0, Z_1, Z_2) = 0$, where a, b and c are polynomials of degree, 2, 3 and 4, respectively. After a projective transformation that fixes $[0, 0, 0, 1]$ we may assume that $a(Z_0, Z_1, Z_2) = Z_1^2 - 4 Z_0 Z_2$. We can construct an algebraic map from M_c to the Jacobi variety $Jac(C)$: $M_c \longrightarrow Jac(\mathcal{C})$, $p \in M_c \longmapsto (\zeta_1 + \zeta_2) \in Jac(\mathcal{C})$, and the flows generated by the constants of the motion are straight lines on $Jac(\Gamma)$, that is, the linearizing equations are given by

$$\int_{\zeta_1(0)}^{\zeta_1(t)} \omega_1 + \int_{\zeta_2(0)}^{\zeta_2(t)} \omega_1 = c_1 t, \qquad \int_{\zeta_1(0)}^{\zeta_1(t)} \omega_2 + \int_{\zeta_2(0)}^{\zeta_2(t)} \omega_2 = c_2 t,$$

where ω_1, ω_2[7.40] span the 2-dimensional space of holomorphic differentials on the curve $\mathcal{C}$ and ζ_1, ζ_2, two appropriate variables given by

$$\zeta_1 = \frac{-Z_1 + \sqrt{a(Z_0, Z_1, Z_2)}}{2Z_0}, \qquad \zeta_2 = \frac{-Z_1 - \sqrt{a(Z_0, Z_1, Z_2)}}{2Z_0},$$

algebraically related to the ones originally given, for which the Hamilton–Jacobi equation could be solved by the separation of variables. □

7.7. The Manakov geodesic flow on the group $SO(4)$

Let $x \in \mathbb{C}^6$, $t \in \mathbb{C}$ and $\mathcal{U} \subset \mathbb{C}^6$ a non-empty Zariski open set. The map $\Psi : (H_1, ..., H_4) : \mathbb{C}^6 \longrightarrow \mathbb{C}^4$ (see section 4.5) is submersive on $\mathcal{U}$, that is, $dH_1(x), ..., dH_4(x)$ are linearly independent on $\mathcal{U}$. Let $I = \Psi(\mathbb{C}^6 \backslash \mathcal{U}) = \{c = (c_i) \in \mathbb{C}^4 : \exists x \in \Psi^{-1}(c), dH_1(x) \wedge ... \wedge dH_4(x) = 0\}$, be the set of critical values of Ψ and $\overline{I}$ the Zariski closure of I in $\mathbb{C}^4$. The non-empty Zariski open set $\mathcal{U}$ can be chosen as the set $\mathcal{U} = \{x \in \mathbb{C}^6 : \Psi(x) \in \mathbb{C}^4 \backslash \overline{I}\}$. The invariant variety defined by $M_c = \Psi^{-1}(c) = \bigcap_{i=1}^{4}\{x \in \mathbb{C}^6 : H_i(x) = c_i\}$, is the fiber of a morphism from $\mathbb{C}^6$ to $\mathbb{C}^4$, thus M_c is a smooth affine surface for generic $c = (c_1, ..., c_4) \in \mathbb{C}^4$ and the main problem will be to complete M_c into an Abelian surface. Now, how do we find the compactification of M_c into an Abelian surface? This compactification is not trivial and the simplest one obtained as a closure: $\overline{M}_c = \bigcap_{i=1}^{4}\{H_i(x) = c_i x_0^2\} \subset \mathbb{P}^6(\mathbb{C})$, that is,

$$x_1 x_4 + x_2 x_5 + x_3 x_6 = c_1 x_0^2, \qquad \lambda_1 x_1^2 + \lambda_2 x_2^2 + \cdots + \lambda_6 x_6^2 = c_3 x_0^2,$$

$$x_1^2 + x_2^2 + \cdots + x_6^2 = c_2 x_0^2, \qquad \mu_1 x_1^2 + \mu_2 x_2^2 + \cdots + \mu_6 x_6^2 = c_4 x_0^2,$$

where $[x_0 : x_1 : ... : x_6]$ are homogeneous coordinates on $\mathbb{P}^6(\mathbb{C})$, does not lead to this result (in the following we will not distinguish between x_1 as a homogeneous coordinates $[x_0 : x_1]$ and as an affine coordinate x_1/x_0). An Abelian surface is not simply connected and therefore cannot be a projective complete intersection. In other words, if M_c is to be the affine part of an Abelian surface, M_c must have a singularity somewhere along the locus at infinity $C = \overline{M}_c \cap \{x_0 = 0\}$. A direct calculation shows that C is an ordinary double curve of $\overline{M}_c$ except at 16 ordinary pinch points of $\overline{M}_c$; the variety $\overline{M}_c$ has a local analytic equation $x^2 = yz^2$. The reduced curve C_r is a smooth elliptic curve. Now, it is only after blowing up $\overline{M}_c$ along the curve C_r that one gets the desired Abelian surface.

THEOREM 7.25.– The divisor of poles of the functions $x_1, x_2, ..., x_6$ is a Riemann surface $\mathcal{D}$ of genus 9. For generic constants, the surface M_c is the affine part of an Abelian surface $\widetilde{M_c}$ obtained by gluing to M_c the divisor $\mathcal{D}$.

PROOF.– Consider points at infinity which are limit points of trajectories of the flow. There is a Laurent decomposition of such asymptotic solutions,

$$X(t) = t^{-1}\left(X^{(0)} + X^{(1)}t + X^{(2)}t^2 + \cdots\right), \qquad [7.42]$$

which depend on $\dim(\text{phase space}) - 1 = 5$ free parameters. Putting [7.42] into [4.8], solving inductively for the $X^{(k)}$, one finds a nonlinear equation at the 0th step,

$$X^{(0)} + \left[X^{(0)}, \Lambda.X^{(0)}\right] = 0, \qquad [7.43]$$

and a linear system of equations at the kth step :

$$(L - kI)\left(X^{(k)}\right) = \begin{cases} 0 \text{ for } k = 1 \\ \text{quadratic polynomial in } X^{(1)}, ..., X^{(k-1)} \text{ for } k \geq 2 \end{cases}$$
$$[7.44]$$

where L denotes the linear map

$$L(Y) = \left[Y, \Lambda.X^{(0)}\right] + \left[X^{(0)}, \Lambda.Y\right] + Y = \text{Jacobian map of [7.43]}.$$

One parameter appears at the 0th step, that is, in the resolution of [7.43] and the four remaining ones at the kth step, $k = 1, ..., 4$. By only taking into account solution trajectories lying on the surface M_c, we obtain one-parameter families that are parameterized by a Riemann surface. To be precise, we search for the set $\mathcal{D}$ of Laurent solutions [7.42] restricted to the affine invariant surface M_c, that is

$$\mathcal{D} = \text{closure of the continuous components of}$$
$$\{\text{Laurent solutions } X(t) \text{ such that } H_i\left(X(t)\right) = c_i, \, 1 \leq i \leq 4\},$$
$$= \bigcap_{i=1}^{4} \{t^0 - \text{coefficient of } H_i\left(X(t)\right) = c_i\},$$
$$= \text{a Riemann surface (algebraic curve) whose affine equation is}$$

$$w^2 + c_1\left(x_5^{(0)}x_6^{(0)}\right)^2 + c_2\left(x_4^{(0)}x_6^{(0)}\right)^2 + c_3\left(x_4^{(0)}x_5^{(0)}\right)^2 + c_4 x_4^{(0)} x_5^{(0)} x_6^{(0)}$$
$$\equiv w^2 + F\left(x_4^{(0)}, x_5^{(0)}, x_6^{(0)}\right) = 0, \qquad [7.45]$$

where w is an arbitrary parameter and $x_4^{(0)}, x_5^{(0)}, x_6^{(0)}$ parameterizes the elliptic curve

$$\mathcal{E}: \begin{cases} \left(x_4^{(0)}\right)^2 + \left(x_5^{(0)}\right)^2 + \left(x_6^{(0)}\right)^2 = 0, \\ \left(\beta x_5^{(0)} + \alpha x_6^{(0)}\right)\left(\beta x_5^{(0)} - \alpha x_6^{(0)}\right) = 1, \end{cases} \quad [7.46]$$

with (α, β) such that: $\alpha^2 + \beta^2 + 1 = 0$. The Riemann surface $\mathcal{D}$ is a two-sheeted ramified covering of the elliptic curve $\mathcal{E}$ and it easy to check that the elliptic curve $\mathcal{E}$ is exactly the reduced curve C_r. The branch points are defined by the 16 zeroes of $F\left(x_4^{(0)}, x_5^{(0)}, x_6^{(0)}\right)$ on $\mathcal{E}$. The Riemann surface $\mathcal{D}$ is unramified at infinity and by Riemann–Hurwitz's formula, $2g(\mathcal{D}) - 2 = N(2g(\mathcal{E}) - 2) + R$, where N is the number of sheets and R the ramification index, the genus $g(\mathcal{D})$ of $\mathcal{D}$ is 9. To show that M_c is the affine part of an Abelian surface $\widetilde{M_c}$ with $\widetilde{M_c} \setminus M_c = \mathcal{D}$, we can use the same method of Laurent's developments used previously (see Haine (1983)). Here, by following Mumford (see appendix of Adler and van Moerbeke (1982)), we give an abstract algebro-geometrical proof that the four quadrics in this problem intersect in the affine part of an Abelian surface using Enriques classification of algebraic surfaces. For this, we will compute the invariants of $\widetilde{M_c}$ and use Enriques classification of algebraic surfaces (Griffiths and Harris 1978, p. 590). Let $K_{\widetilde{M_c}}$ be the canonical bundle, $\chi(\mathcal{O}_{\widetilde{M_c}})$ the Euler characteristic and $q(\widetilde{M_c})$ the irregularity of $\widetilde{M_c}$. Now if $\phi : \widetilde{M_c} \longrightarrow \overline{M_c} \subset \mathbb{P}^6(\mathbb{C})$ is the normalization of $\overline{M_c}$, then the pullback map on sections $\phi^* : H^0\left(\overline{M_c}, \mathcal{O}_{\overline{M_c}}\right) \longrightarrow H^0\left(\widetilde{M_c}, \mathcal{O}_{\widetilde{M_c}}\right)$, is an isomorphism and $K_{\widetilde{M_c}} = \widetilde{K_{\overline{M_c}}} - \mathcal{D}$, where $\widetilde{K_{\overline{M_c}}} = \phi^*\left(K_{\overline{M_c}}\right)$ and so for H a hyperplane in $\mathbb{P}^6(\mathbb{C})$, $K_{\widetilde{M_c}} = \phi^*\left(\overline{M_c}.K_{\mathbb{P}^6(\mathbb{C})} + (\sum_{i=1}^4 \deg H_i).H\right) - \mathcal{D} = 0$. Also $\chi\left(\mathcal{O}_{\widetilde{M_c}}\right) = \chi\left(\phi_*\mathcal{O}_{\widetilde{M_c}}/\mathcal{O}_{\overline{M_c}}\right) + \chi\left(\mathcal{O}_{\overline{M_c}}\right) = \chi(\phi_*\mathcal{O}_\mathcal{D}/\mathcal{O}_\mathcal{E}) + \chi\left(\mathcal{O}_{\overline{M_c}}\right)$. The Riemann surface $\mathcal{D}$[7.45] of genus 9 is a double cover ramified over 16 points of the elliptic curve $\mathcal{E}$[7.46]. We use the Koszul complex to compute $\chi\left(\mathcal{O}_{\overline{M_c}}\right)$. In the local ring at each point of $\mathbb{P}^6(\mathbb{C})$ the localizations of the 4 homogeneous polynomials H_i give a regular sequence, and the Koszul complex gives a canonical resolution

$$0 \to \mathcal{O}_{\mathbb{P}^6(\mathbb{C})}(-8) \to \mathcal{O}_{\mathbb{P}^6(\mathbb{C})}(-6)^4 \to \mathcal{O}_{\mathbb{P}^6(\mathbb{C})}(-4)^6$$
$$\to \mathcal{O}_{\mathbb{P}^6(\mathbb{C})}(-2)^4 \to \mathcal{O}_{\mathbb{P}^6(\mathbb{C})} \to \mathcal{O}_{\overline{M_c}} \to 0$$

Thus $\chi\left(\mathcal{O}_{\overline{M_c}}\right) = 8$, hence $\chi\left(\mathcal{O}_{\widetilde{M_c}}\right) = 0$, $q\left(\widetilde{M_c}\right) = 2$. By Enriques–Kodaira's classification theorem (Griffiths and Harris 1978), it follows that $\widetilde{M_c}$ is an Abelian surface. $\square$

THEOREM 7.26.– *The flow* [4.8] *evolves on an Abelian surface* $\widetilde{M_c} \cong \mathbb{C}^2/lattice$ *of polarization* $\begin{pmatrix} 2 & 0 & a & c \\ 0 & 4 & c & b \end{pmatrix}$, $\mathrm{Im}\begin{pmatrix} a & c \\ c & b \end{pmatrix} > 0$.

PROOF.– Let $L \equiv \{f : f \text{ meromorphic on } \widetilde{M}_c, (f) + \mathcal{D} \geq 0\}$ be the vector space of meromorphic functions on $\widetilde{M}_c$ with at worst a simple pole along $\mathcal{D}$, and let $\chi(\mathcal{D}) = \dim H^0\left(\widetilde{M}_c, \mathcal{O}(\mathcal{D})\right) - \dim H^1\left(\widetilde{M}_c, \mathcal{O}(\mathcal{D})\right)$ be the Euler characteristic of $\mathcal{D}$. The adjunction formula and the Riemann–Roch theorem for divisors on Abelian surfaces imply that $g(\mathcal{D}) = \frac{K_{\widetilde{M}_c}.\mathcal{D} + \mathcal{D}.\mathcal{D}}{2} + 1$, and $\chi(\mathcal{D}) = p_a\left(\widetilde{M}_c\right) + 1 + \frac{1}{2}\left(\mathcal{D}.(\mathcal{D} - K_{\widetilde{M}_c})\right)$, where $g(\mathcal{D})$ is the geometric genus of $\mathcal{D}$ and $p_a\left(\widetilde{M}_c\right)$ is the arithmetic genus of $\widetilde{M}_c$. Since $\widetilde{M}_c$ is an Abelian surface $\left(K_{\widetilde{M}_c} = 0, p_a\left(\widetilde{M}_c\right) = -1\right)$, $g(\mathcal{D}) - 1 = \frac{\mathcal{D}.\mathcal{D}}{2} = \chi(\mathcal{D})$. Using Kodaira–Serre duality (Griffiths and Harris 1978, p. 153), the Kodaira–Nakano vanishing theorem (Griffiths and Harris 1978, p. 154) and a theorem on theta-functions (Griffiths and Harris 1978, p. 317), it easy to see that

$$g(\mathcal{D}) - 1 = \dim L(\mathcal{D}) \left(\equiv h^0(L)\right) = \delta_1 \delta_2, \qquad [7.47]$$

where $\delta_1, \delta_2 \in \mathbb{N}$, are the elementary divisors of the polarization $c_1(L)$ of $\widetilde{M}_c$. Note that $\sigma \equiv -id : (x_0, x_1, \ldots, x_6) \longmapsto (-x_0, x_1, \ldots, x_6)$, is the reflection about the origin of $\mathbb{C}^2$ and has 16 fixed points on $\widetilde{M}_c$, given by the 16 branch points on $\mathcal{D}$ covering the 16 roots of the polynomial $F(x_4^0, x_5^0, x_6^0)$ [7.45]. Since L is symmetric ($\sigma^* L \simeq L$), σ can be lifted to L as an involution $\widetilde{\sigma}$ in two ways differing in sign and for each section (theta-function) $s \in H^0(L)$, we therefore have $\widetilde{\sigma} s = \pm s$. A section $s \in H^0(L)$ is called even (respectively, odd) if $\widetilde{\sigma} s = +s$ (respectively, $\widetilde{\sigma} s = -s$). Under $\widetilde{\sigma}$ the vector space $H^0(L)$ splits into an even and odd subspace $H^0(L) = H^0(L)^{even} \oplus H^0(L)^{odd}$, with $H^0(L)^{even}$ containing all of the even sections and $H^0(L)^{odd}$ all of the odd ones. Note that $c_1(L) = \phi^*(H)$ and $(c_1(L)^2) = 16$ (since the degree of $\widetilde{M}_c$ is 16). By the classification theory of ample line bundles on Abelian varieties, $\widetilde{M}_c \simeq \mathbb{C}^2/L_\Omega$ with period lattice given by the columns of the matrix $\Omega = \begin{pmatrix} \delta_1 & 0 & a & c \\ 0 & \delta_2 & c & b \end{pmatrix}$, $\mathrm{Im}\begin{pmatrix} a & c \\ c & b \end{pmatrix} > 0$, according to [7.47], with $\delta_1 \delta_2 = h^0(L) = g(\mathcal{D}) - 1 = 8$, $\delta_1 \mid \delta_2$, $\delta_i \in \mathbb{N}^*$. Hence, we have two possibilities: (i) $\delta_1 = 1, \delta_2 = 8$ and (ii) $\delta_1 = 2, \delta_2 = 4$. From formula [7.39], the corresponding line bundle L has five even sections and three odd sections in case (i), and six even sections and two odd sections in case (ii). Now $x_1, \ldots, x_6$ are six even sections, showing that case (ii) is the only alternative and the period matrix has the form $\begin{pmatrix} 2 & 0 & a & c \\ 0 & 4 & c & b \end{pmatrix}$, $\mathrm{Im}\begin{pmatrix} a & c \\ c & b \end{pmatrix} > 0$. $\square$

THEOREM 7.27.– *The Abelian surface $\widetilde{M}_c$ that completes the affine surface M_c is the Prym variety $Prym_\alpha(\Gamma)$ of the genus 3 Riemann surface Γ:*

$$\Gamma : \begin{cases} w^2 = -c_1 \left(x_5^0 x_6^0\right)^2 - c_2 \left(x_6^0\right)^2 z - c_3 \left(x_5^0\right)^2 z + c_4 y, \\ y^2 = z \left(\alpha^2 z - 1\right)\left(\beta^2 z + 1\right), \end{cases} \qquad [7.48]$$

for the involution $\sigma : \Gamma \longrightarrow \Gamma$, $(w, y, z) \longmapsto (-w, y, z)$, interchanging the two sheets of the double covering $\Gamma \longmapsto \Gamma_0$, $(w, y, z) \longmapsto (y, z)$, Γ_0 is the elliptic curve,

$$\Gamma_0 : y^2 = z\left(\alpha^2 z - 1\right)\left(\beta^2 z + 1\right). \qquad [7.49]$$

PROOF.– After substitution $z \equiv \left(x_4^0\right)^2$, the curve $\mathcal{D}$ can also be seen as a four-sheeted unramified covering of another curve Γ, determined by the equation

$$\Gamma : G(w, z) \equiv \left(w^2 + c_1 \left(x_5^0 x_6^0\right)^2 + c_2 \left(x_6^0\right)^2 z + c_3 \left(x_5^0\right)^2 z\right)^2 - c_4^2 \left(x_5^0 x_6^0\right)^2 z = 0.$$

Equations [7.46] are equivalent to $\left(x_5^0\right)^2 = \beta^2 z + 1$ and $\left(x_6^0\right)^2 = \alpha^2 z - 1$. The curve Γ is invariant under an involution

$$\sigma : \Gamma \longrightarrow \Gamma, \quad (w, z) \longmapsto (-w, z). \qquad [7.50]$$

Consider a map $\rho : \Gamma \longrightarrow \Gamma_0 \equiv \Gamma/\sigma$, $(w, y, z) \longmapsto (y, z)$, of the curve Γ onto an elliptic curve $\Gamma_0 \equiv \Gamma/\sigma$, that is given by the equation [7.49]. The genus of Γ [7.48] is calculated easily by means of the map ρ. The latter is a two-sheeted ramified covering of Γ_0 and it has four branch points. Using the Riemann–Hurwitz formula, we obtain $g(\Gamma) = 3$. We will now proceed to show that the Abelian surface $\widetilde{M}_c$ can be identified as Prym variety $Prym_\sigma(\Gamma)$. Let $(a_1, a_2, a_3, b_1, b_2, b_3)$ be a basis of cycles in Γ with the intersection indices $a_i o a_j = b_i o b_j = 0$, $a_i o b_j = \delta_{ij}$, such that $\sigma(a_1) = a_3$, $\sigma(b_1) = b_3$, $\sigma(a_2) = -a_2$, $\sigma(b_2) = -b_2$ for the involution σ [7.50]. By the Poincaré residue formula, the three holomorphic 1-forms $\omega_0, \omega_1, \omega_2$ in Γ are the differentials $P(w,z) \frac{dz}{(\partial G/\partial w)(w,z)}\Big|_{G(w,z)=0} = P(w,z)\frac{dz}{4wy}$, for P a polynomial of degree $\leq \deg G - 3 = 1$. Therefore, $\omega_0 = \frac{dz}{y}$, $\omega_1 = \frac{z \, dz}{wy}$, $\omega_2 = \frac{dz}{wy}$ form a basis of holomorphic differentials on Γ and obviously $\sigma^*(\omega_0) = \omega_0$, $\sigma^*(\omega_k) = -\omega_k$, $(k = 1, 2)$, for the involution σ [7.50]. It is well known that the period matrix Ω of $Prym_\sigma(\Gamma)$ can be written as follows:

$$\Omega = \begin{pmatrix} 2\int_{a_1}\omega_1 & \int_{a_2}\omega_1 & 2\int_{b_1}\omega_1 & \int_{b_2}\omega_1 \\ 2\int_{a_1}\omega_2 & \int_{a_2}\omega_2 & 2\int_{b_1}\omega_2 & \int_{b_2}\omega_2 \end{pmatrix}.$$

Let (dt_1, dt_2) be a basis of holomorphic 1-forms on the Abelian surface $\widetilde{M}_c$, such that $dt_j|_\mathcal{D} = \omega_j$, $(j=1,2)$, $L_{\Omega'} = \left\{ \sum_{k=1}^{2} m_k \int_{a'_k} \binom{dt_1}{dt_2} + n_k \int_{b'_k} \binom{dt_1}{dt_2} : m_k, n_k \in \mathbb{Z} \right\}$, is the lattice associated with the period matrix

$$\Omega' = \begin{pmatrix} \int_{a'_1}dt_1 & \int_{a'_2}dt_1 & \int_{b'_1}dt_1 & \int_{b'_2}dt_1 \\ \int_{a'_1}dt_2 & \int_{a'_2}dt_2 & \int_{b'_1}dt_2 & \int_{b'_2}dt_2 \end{pmatrix},$$

where (a'_1, a'_2, b'_1, b'_2) is a basis of $H_1(\widetilde{M_c}, \mathbb{Z})$ and $\widetilde{\mathcal{A}} \longrightarrow \mathbb{C}^2/L_{\Omega'} : p \longmapsto \int_{p_0}^p \binom{dt_1}{dt_2}$, is the uniformizing map. By the Lefschetz theorem on hyperplane section (Griffiths and Harris 1978, p. 156), the map $H_1(\mathcal{D}, \mathbb{Z}) \longrightarrow H_1(\widetilde{M_c}, \mathbb{Z})$ induced by the inclusion $\mathcal{D} \hookrightarrow \widetilde{M_c}$ is surjective and consequently we can find four cycles a'_1, a'_2, b'_1, b'_2 on the Riemann surface $\mathcal{D}$ such that

$$\Omega' = \begin{pmatrix} \int_{a'_1} \omega_1 & \int_{a'_2} \omega_1 & \int_{b'_3} \omega_1 & \int_{b'_4} \omega_1 \\ \int_{a'_1} \omega_2 & \int_{a'_2} \omega_2 & \int_{b'_3} \omega_2 & \int_{b'_4} \omega_2 \end{pmatrix},$$

and $L_{\Omega'} = \left\{ \sum_{k=1}^2 m_k \int_{a'_k} \binom{\omega_1}{\omega_2} + n_k \int_{b'_k} \binom{\omega_1}{\omega_2} : m_k, n_k \in \mathbb{Z} \right\}$. Recall that $F(x_4^0, x_5^0, x_6^0)$[7.45] has four zeroes on Γ_0[7.49] and 16 zeroes on $\mathcal{E}$[7.46], and it follows that the four cycles a'_1, a'_2, b'_1, b'_2 on $\mathcal{D}$ that we are looking for are $2a_1, a_2, 2b_1, b_2$ and they form a basis of $H_1(\widetilde{M_c}, \mathbb{Z})$, such that

$$\Omega' = \begin{pmatrix} 2\int_{a_1} \omega_1 & \int_{a_2} \omega_1 & 2\int_{b_1} \omega_1 & \int_{b_2} \omega_1 \\ 2\int_{a_1} \omega_2 & \int_{a_2} \omega_2 & 2\int_{b_1} \omega_2 & \int_{b_2} \omega_2 \end{pmatrix} = \Omega$$

is a Riemann matrix. Thus, $\widetilde{M_c}$ and $Prym_\sigma(\Gamma)$ are two Abelian varieties analytically isomorphic to the same complex torus $\mathbb{C}^2/L_\Omega$. By Chow's theorem, $\widetilde{M_c}$ and $Prym_\sigma(\Gamma)$ are then algebraically isomorphic. $\square$

REMARK 7.7.– Strange as it may seem, the use of the Lax spectral curve technique may not give the tori correctly, but perhaps with period doubling, in contrast with the statement that the correct tori would be obtained by the Kowalewski–Painlevé analysis. This indicated a need for caution in the interpretation of the result for tori calculated from the Lax spectral curve technique. A striking example of this phenomenon appears in the Euler equations associated with a class of geodesic flow on $SO(4)$ for a left-invariant diagonal metric. We know from section 4.5 that the linearization of the Euler–Arnold equations [4.8] takes place on the Prym variety $Prym_\sigma(\mathcal{C})$ of the genus 3 Riemann surface $\mathcal{C}$ [4.16]; the latter is a double ramified cover of an elliptic curve $\mathcal{C}_0$. Also, from the asymptotic analysis (section 7.7) of equations [4.8], the affine variety M_c completes into an Abelian surface $\widetilde{M_c}$ upon adding a Riemann surface $\mathcal{D}$ [7.45] of genus 9, which is a fourfold unramified cover of a Riemann surface Γ [7.48] of genus 3; the latter is a double ramified cover of an elliptic curve Γ_0. The Abelian surface $\widetilde{M_c}$ can also be identified as the Prym variety $Prym_\sigma(\Gamma)$ and the problem linearizes on $Prym_\sigma(\Gamma)$. From the fundamental exponential sequence $0 \to \mathbb{Z} \to \mathcal{O}_{\widetilde{M_c}} \stackrel{\exp}{\to} \mathcal{O}^*_{\widetilde{M_c}} \to 0$, we get the map $\cdots \to H^1\left(\widetilde{M_c}, \mathcal{O}^*_{\widetilde{M_c}}\right) \to H^2\left(\widetilde{M_c}, \mathbb{Z}\right) \to \cdots$, that is, the first Chern class of a line bundle on $\widetilde{M_c}$. Any line bundle with Chern class zero can be realized by constant multipliers. Therefore, the group $Pic^o\left(\widetilde{M_c}\right)$ of holomorphic line bundles on $\widetilde{M_c}$

with Chern class zero is given by $Pic^o\left(\widetilde{M_c}\right) = H^1\left(\widetilde{M_c}, \mathcal{O}_{\widetilde{M_c}}\right)/H^1\left(\widetilde{M_c}, \mathbb{Z}\right)$ and is naturally isomorphic to the dual Abelian surface $\widetilde{M_c}^*$ of $\widetilde{M_c}$ (* means the dual Abelian surface). The relationship between $\widetilde{M_c}$ and $\widetilde{M_c}^*$ is symmetric like the relationship between two vectors spaces set up a bilinear pairing. It is interesting to observe that the Abelian surfaces $\widetilde{M_c} = Prym_\sigma(\Gamma)$ obtained from the asymptotic analysis of the differential equations and $Prym_\sigma(\mathcal{C})$ obtained from the orbits in the Kac–Moody Lie algebra are not identical but only isogenous, that is, one can be obtained from the other by doubling some periods and leaving other unchanged. The precise relation between these two Abelian surfaces is $\widetilde{M_c} = (Prym_\sigma(\mathcal{C}))^*$, that is, they are dual of each other. The functions $x_1, ..., x_6$ are themselves meromorphic on $\widetilde{M_c}$, while only their squares are on $Prym_\sigma(\mathcal{C})$. The relationship between the Riemann surfaces Γ and $\mathcal{C}$ is quite intricate. As usual we let Θ the theta divisor on $Jac(\Gamma)$, we have $Prym_\sigma(\mathcal{C})\backslash\Pi = \Theta \cap Prym_\sigma(\mathcal{C}) = \Gamma$, with Π a Zariski open set of $Prym_\sigma(\mathcal{C})$. Also $\Theta \cap \widetilde{M_c} = \mathcal{C}$, where Θ is a translate of the theta divisor of $Jac(\mathcal{C})$ invariant under the involution σ. Moser (1980) was aware of a similar situation in the context of the Jacobi's geodesic flow problem on ellipsoids.

7.8. Geodesic flow on $SO(4)$ with a quartic invariant

Often, when studying the geodesic flow on $SO(4)$, it is more convenient to use the coordinates $u = (x_1, x_2, x_3)$ and $v = (x_4, x_5, x_6)$, they correspond to the decomposition $u \oplus v \in so(4) \simeq so(3) \oplus so(3)$. In these coordinates, the geodesic flow on the group $SO(4)$ can be written as

$$X_H : \dot{u} = u \times \frac{\partial H}{\partial u}, \qquad \dot{v} = v \times \frac{\partial H}{\partial v},$$

for invariant metric defined by the quadratic form

$$H = \frac{1}{24}\sum_{i=1}^{3}\left(3(3c_i + d_i)x_i^2 + (c_i + 3d_i)x_{i+3}^2 + 6(d_i - c_i)x_i x_{i+3}\right), \quad [7.51]$$

with coefficients $c_i = \frac{b_i}{a_i}$, $d_i = \frac{b_j - b_k}{a_j - a_k}$, $\sum_{i=1}^{3} a_i = 0$, $\sum_{i=1}^{3} b_i = 0$, and ijk permutations of 123. This geodesic flow has three quadratic invariants, namely, the Casimir functions $\|u\|^2$ and $\|v\|^2$, and the metric [7.51], and one quartic invariant will be given later. The invariants $\|u\|^2$ and $\|v\|^2$ define the four-dimensional non-degenerate symplectic leaves of Hamiltonian structure, which are therefore parameterized by the values of $\|u\|^2$ and $\|v\|^2$. After the following linear change of coordinates (motivated by the equations of a curve of rank three quadrics, see Adler and van Moerbeke (1987)), which is meaningful insofar $a \neq 0, -1, 1, -1/3, 1/3$,

$$\begin{pmatrix} x_1 \\ x_4 \end{pmatrix} = \sqrt{-1} \begin{pmatrix} a-1 & -1 \\ 3a+1 & 1 \end{pmatrix} \begin{pmatrix} (a-1)z_1 \\ (3a-1)(a+1)z_4 \end{pmatrix},$$

$$\begin{pmatrix} x_2 \\ x_5 \end{pmatrix} = -\sqrt{-1} \begin{pmatrix} a+1 & -1 \\ 3a-1 & 1 \end{pmatrix} \begin{pmatrix} (a+1)z_2 \\ (3a+1)(a-1)z_5 \end{pmatrix},$$

$$\begin{pmatrix} x_3 \\ x_6 \end{pmatrix} = \sqrt{-1} \begin{pmatrix} a-1 & a+1 \\ 3a+1 & 3a-1 \end{pmatrix} \begin{pmatrix} (a-1)z_3 \\ (a+1)z_6 \end{pmatrix},$$

the geodesic flow takes (after rescaling time) on the simple form:

$$\frac{dz_1}{dt} = z_3 z_5, \qquad \frac{dz_4}{dt} = \frac{2a}{3a-1} z_5 z_6 + \frac{a-1}{3a-1} z_2 z_3,$$

$$\frac{dz_2}{dt} = z_4 z_6, \qquad \frac{dz_5}{dt} = \frac{2a}{3a+1} z_5 z_6 + \frac{a+1}{3a+1} z_2 z_3, \qquad [7.52]$$

$$\frac{dz_3}{dt} = \frac{1-a}{2} z_4 z_5 + z_1 z_5 + \frac{1+a}{2} z_1 z_2,$$

$$\frac{dz_6}{dt} = \frac{1+a}{2} z_4 z_5 + z_2 z_4 + \frac{1-a}{2} z_1 z_2,$$

with three quadratic invariants (in z):

$$Q_1 \equiv a(z_5^2 - z_1 z_4) + \frac{1-a}{3a+1}(z_1^2 - z_3^2 + z_1 z_4) = A_1,$$

$$Q_2 \equiv -a(z_4^2 - z_2 z_5) - \frac{a+1}{3a-1} = A_2,$$

$$Q_3 \equiv \frac{2(z_1 z_4 + z_2 z_5 - z_3 z_6)}{(3a-1)(3a+1)} - \frac{z_4^2 - z_2 z_5}{3a+1} + \frac{z_5^2 - z_1 z_4}{3a-1} = A_3,$$

and a quartics invariant (in z):

$$Q_4 \equiv -\frac{1-a}{3a+1}\left((z_4^2 - z_2 z_5)^2 + (\frac{2}{3a-1}(z_2 z_3 - z_5 z_6))^2\right)$$

$$+ \frac{1+a}{3a-1}\left((z_5^2 - z_1 z_4)^2 + (\frac{-2}{3a+1}(z_1 z_6 - z_3 z_4))^2\right) \qquad [7.53]$$

$$+ \frac{3(1-a^2)}{(3a-1)(3a+1)}\left(2(z_4^2 - z_2 z_5)(z_5^2 - z_1 z_4) - (z_1 z_2 - z_4 z_5)^2\right)$$

$$+ \frac{4(1+a)}{(3a-1)(3a+1)}(z_5^2 - z_1 z_4)(z_1 z_4 + z_2 z_5 - z_3 z_6 + z_2^2 - z_6^2 + z_2 z_5)$$

$$+ \frac{4(1-a)}{(3a-1)(3a+1)}(z_4^2 - z_2 z_5)(z_1 z_4 + z_2 z_5 - z_3 z_6 + z_1^2 - z_3^2 + z_1 z_4) = A_4.$$

The geodesic flow in question admits one family of Laurent solutions,

$$z = \frac{\zeta}{t}\left(\mathbf{1} + UY^1 t + \frac{1}{\gamma Z^2 + \delta}\left(U^2 Y_0^2 + \sum_{i=1}^{3} A_i Y_i^2\right) t^2 + o(t^3)\right),$$

where $\mathbf{1} = (1, 1, ..., 1)^\top$, Y^1, Y_0^2, Y_i^2 are appropriate vectors depending on Y, Z and a only, $\gamma \equiv 4a$, $\delta \equiv (a-1)(3a+1)$ and $\zeta = \text{diag}\left(\frac{Y^2}{Z}, \frac{Z^2}{Y}, -\frac{Y}{Z}, Z, Y, -\frac{Z}{Y}\right)$, with $Y, Z \in \mathbb{C}$ such that $Y^2 + Z^2 = 1$. The five-dimensional family of Laurent solutions depend on the parameters Z, U, A_1, A_2 and A_3. The vectors Y_i^2 can be chosen such that $Q_i(z(t)) = A_i$ for $i = 1, 2, 3$. Confining the five-dimensional family of Laurent solutions to the invariant manifold $\mathcal{A} = \{z : Q_i(z) = A_i\}$ yields a relation between the free parameters, defining a curve

$$\mathcal{D}: \begin{cases} (U, V, Y, Z) \text{ such hat } Z^2 = V, Y^2 = 1 - V \text{ and} \\ P(U, V) = \left(U^2(1-V)V(\alpha V + \beta)\right)^2 \\ \qquad -2U^2(1-V)VP(V) + Q(V) = 0, \end{cases}$$

where $\alpha = 16a^3$, $\beta = (a-1)^3(3a+1)$,

$$P(V) = (\alpha V + \beta)\left[((3a^2+1)A_3 - A_1 - A_2)(V-1) + A_1 V - A_2(V-1)\right]$$
$$-2V(V-1)\left[A_1(1-a)^3(1+3a) + A_2(1+a^3(1-3a)\right.$$
$$\left.-A_3(1-a^2)(1-9a^2)\right],$$

$$Q(V) = \left[((3a^2+1)A_3 - A_1 - A_2)V(V-1) - A_1 V + A_2(V-1)\right]^2$$
$$+V(V-1)\left[(4aV + (a-1)(3a+1))A_4 + 4A_1 A_2\right.$$
$$\left.-(a-1)(3a+1)(a+1)(3a-1)A_3^2\right].$$

Note that

$$P^2(V) - (\alpha V + \beta)^2 Q(V) = V(1-V)R(V), \qquad [7.54]$$

with $R(V)$ being a cubic polynomial. The curve $\mathcal{D}$ is an unramified $4-1$ cover of the curve $\mathcal{C}: P(U, V) = 0$. In view of [7.54], the curve $\mathcal{C}$ itself is a double cover of the hyperelliptic curve $\mathcal{H}: W^2 = V(1-V)R(V)$, of genus 2, ramified over four points where $Q(V) = 0$. Therefore, $\mathcal{C}$ has genus 5 and $\mathcal{D}$ has genus 17. The curve $\mathcal{D}$ must be thought of as being a very ample divisor on some Abelian surface $\widetilde{\mathcal{A}}$, to be constructed according to the method described in the preceding problems (for more details, see Adler and van Moerbeke (2004)). The curve $\mathcal{D}$, wrapped around $\widetilde{\mathcal{A}}$, intersects itself transversally in eight points, adding 8 to the genus 17. Therefore, the torus $\widetilde{\mathcal{A}} \simeq \mathbb{C}^2/L_\Omega$ on which the geodesic flow linearizes is defined by a period lattice

Ω given by the columns of the following matrix $\Omega = \begin{pmatrix} \delta_1 & 0 & a & c \\ 0 & \delta_2 & c & b \end{pmatrix}$, $\operatorname{Im} \begin{pmatrix} a & c \\ c & b \end{pmatrix} > 0$, according to [7.47], with $\delta_1 \delta_2 = g(\mathcal{D}) - 1 = 24$, $\delta_1 \mid \delta_2$, $\delta_i \in \mathbb{N}^*$. We have two possibilities: (i) $\delta_1 = 1$, $\delta_2 = 24$ and (ii) $\delta_1 = 2$, $\delta_2 = 12$. The line bundle $L(\mathcal{D}) = \{1, z_1, ..., z_6, F_1, ..., F_5, G_1, ..., G_8, H_1, ..., H_4\}$ is specified as follows $F_1 = z_4^2 - z_2 z_5$, $F_2 = z_5^2 - z_1 z_4$, $F_3 = z_1 z_2 - z_4 z_5$, $F_4 = \frac{2}{3a-1}(z_2 z_3 - z_5 z_6)$, $F_5 = \frac{-2}{3a+1}(z_1 z_6 - z_3 z_4)$, $F_6 = z_1 z_4 + z_2 z_5 - z_3 z_6$, $F_7 = z_1^2 - z_3^2 + z_1 z_4$, $F_8 = z_2^2 - z_6^2 + z_2 z_5$, $G_1 = -2az_2 F_2 - (1-a)z_4 F_3$, $G_2 = -2az_1 F_1 - (1+a)z_5 F_3$, $G_3 = (1-a)z_5 F_4 + (1+a)z_4 F_5$, $G_4 = (1+a)z_5 F_5 + (1-a)z_1 F_4$, $G_5 = (1-a)z_4 F_4 + (1+a)z_2 F_5$, $G_6 = -(1-a)z_3 F_1 - (1+a)z_6 F_2$, $G_7 = 2az_5 F_2 - (1-a)z_1 F_3$, $G_8 = -2az_4 F_1 - (1+a)z_2 F_3$, $H_1 = 4a^2 F_1 F_2 + (1-a^2)F_3^2$, $H_2 = 2a^2 F_1 F_5 - (1-a)F_3 F_4$, $H_3 = -2a^2 F_2 F_4 - (1+a)F_3 F_5$, $H_4 = 2a^2 F_4 F_5 + F_3((1+a)F_2 - (1-a)F_1)$. The reflection about the origin on the Abelian surface amounts to flipping the time for each linear flow on it, but since the flow $\frac{dz}{dt_1}$ given by [7.52] is quadratic and since the other flow $\frac{dz}{dt_2}$ (commuting with the first) is quartic (as it derives from the quartic Hamiltonian [7.53]), flipping the signs of t_1 and t_2 for each of the flows amounts to the flip $(z_1, ..., z_6) \longmapsto (-z_1, ..., -z_6)$. From formula [7.39], the above line bundle $L(\mathcal{D})$ has 11 even sections and 13 odd sections in case (i) and 10 even sections and 14 odd sections in case (ii), showing that case (ii) is the only alternative and the period matrix has the form $\begin{pmatrix} 2 & 0 & a & c \\ 0 & 12 & c & b \end{pmatrix}$, $\operatorname{Im} \begin{pmatrix} a & c \\ c & b \end{pmatrix} > 0$. Differentiating $\frac{1}{z_1}$ and $\frac{z_2}{z_1}$ with respect to t_1 (corresponding to the flow [7.52]) and t_2 (corresponding to the quartic flow generated by the invariant Q_4 [7.53]) yields two differentials ω_1 and ω_2 defined on the curve $\mathcal{C}$:

$$\omega_1 = \frac{\varphi(V) dV}{U\sqrt{V(1-V)R(V)}}, \qquad \omega_2 = \frac{dV}{U\sqrt{V(1-V)R(V)}},$$

where $\varphi(V)$ is a rational function in V having the form

$$\varphi(V) = \frac{4aV + (a-1)(3a+1)}{V(1-V)} \left[(\alpha V + \beta) U^2 V(1-V) \right.$$

$$\left. + (A_3(3a^2+1) - A_1 - A_2)(1-V)V - A_1 V - A_2(1-V)) \right].$$

The restriction of the differentials dt_1 and dt_2 to the curve $\mathcal{D}$ is

$$\omega_1 = dt_1|_{\mathcal{D}} = \varphi(Z^2) \omega_2, \qquad \omega_2 = dt_2|_{\mathcal{D}} = \frac{dZ}{UY\sqrt{R}}.$$

Recall that $\mathcal{C}$ is a double ramified cover of a hyperelliptic curve $\mathcal{H}$ of genus 2, whose sheets are interchanged by the involution $(V, U) \longmapsto (V, -U)$. Hence $Jac(\mathcal{C}) = Prym(\mathcal{C}/\mathcal{H}) \oplus Jac(\mathcal{H})$. Since ω_1 and ω_2 are both odd differentials for that involution, the flows evolve on the three-dimensional $Prym(\mathcal{C}/\mathcal{H})$ and therefore $\widetilde{\mathcal{A}} \subset Prym(\mathcal{C}/\mathcal{H})$. This shows that $Prym(\mathcal{C}/\mathcal{H})$ splits further, up to isogenies, into an elliptic curve $\mathcal{E}$ and the two-dimensional invariant torus $\widetilde{\mathcal{A}}$, $\widetilde{\mathcal{A}} \oplus \mathcal{E} = Prym(\mathcal{C}/\mathcal{H})$. In summary, the affine invariant surface $\mathcal{A}$ for the Adler-van Moerbeke geodesic flow completes into a generic Abelian surface $\widetilde{\mathcal{A}}$ of polarization $(1,6)$, that is, defined by a period matrix of the form $\begin{pmatrix} 1 & 0 & a & c \\ 0 & 6 & c & b \end{pmatrix}$, $\mathrm{Im} \begin{pmatrix} a & c \\ c & b \end{pmatrix} > 0$, by adjoining at infinity a curve of genus 25, with eight normal crossings and smooth version $\mathcal{D}$. There exists an elliptic curve $\mathcal{E}$ such that $\widetilde{\mathcal{A}}$ satisfies $\widetilde{\mathcal{A}} \oplus \mathcal{E} = Prym(\mathcal{C}/\mathcal{H})$. More precisely

$$\bigcap_{j=1}^{4} \{x \in \mathbb{C}^6 : Q_j(x) = c_j\} = \widetilde{\mathcal{A}} \setminus \left\{ \begin{array}{c} \text{a curve of genus 25} \\ \text{with 8 singular points} \end{array} \right\}.$$

Put in a more geometrical language, the tori $\widetilde{\mathcal{A}}$ contain a very ample and projectively normal curve of geometric genus 25, with eight normal crossings whose smooth version $\mathcal{D}$ is a $4-1$ unramified cover of a curve $\mathcal{C}$ of genus 5. The curve $\mathcal{C}$ itself is a double cover ramified over four points of a genus 2 hyperelliptic curve $\mathcal{H}$. The linearization takes place on a two-dimensional subtorus of the three-dimensional Prym variety $Prym(\mathcal{C}/\mathcal{H})$ with $Prym(\mathcal{C}/\mathcal{H}) = \widetilde{\mathcal{A}} \oplus \mathcal{E}$, where $\mathcal{E}$ is an elliptic curve (for more details, see Adler and van Moerbeke (2004)). This situation provides a full description of the moduli for the Abelian surfaces of polarization $(1,6)$.

7.9. The geodesic flow on $SO(n)$ for a left invariant metric

Consider the group $SO(n)$ and its Lie algebra $so(n)$ paired with itself, via the customary inner product $\langle X, Y \rangle = -\frac{1}{2} tr(X.Y)$, where $X, Y \in so(n)$. A left invariant metric on $SO(n)$ is defined by a non-singular symmetric linear map $\Lambda : so(n) \longrightarrow so(n), X \longmapsto \Lambda.X$, and by the following inner product; given two vectors gX and gY in the tangent space $SO(n)$ at the point $g \in SO(n)$, $\langle gX, gY \rangle = \langle X, \Lambda^{-1}.Y \rangle$. The question of classifying the metrics for which geodesic flow on $SO(n)$ is algebraically completely integrable is difficult.

Case $n = 3$: The Euler problem of a rigid body (section 7.4) is always algebraically completely integrable and can be regarded as geodesic flow on $SO(3)$. *Case* $n = 4$: In the classification Adler and van Moerbeke (1984) of algebraic integrable geodesic flow on $SO(4)$, three cases come up; two are linearly equivalent to cases of rigid body motion in a perfect fluid studied last century, respectively, by Clebsch and Lyapunov–Steklov, and there is a third new case, namely the Kostant–Kirillov Hamiltonian flow on the dual of $so(4)$. As mentioned in the previous section, it is more convenient to use the coordinates $u = (x_1, x_2, x_3)$ and $v = (x_4, x_5, x_6)$, they correspond to the decomposition $u \oplus v \in so(4) \simeq so(3) \oplus so(3)$. In these coordinates, the geodesic flow on the group $SO(4)$ can be written as $\dot{u} = u \times \frac{\partial H}{\partial u}$, $\dot{v} = v \times \frac{\partial H}{\partial v}$, for the metric defined by the quadratic form $H = \frac{1}{2} \sum_{j=1}^{6} \lambda_j x_j^2 + \sum_{j=1}^{3} \mu_j x_j x_{j+3}$, where $\lambda_1, ..., \lambda_6, \mu_1, \mu_2, \mu_3 \in \mathbb{C}$ and $\lambda_{12} \lambda_{23} \lambda_{31} \lambda_{45} \lambda_{56} \lambda_{64} \mu_1 \mu_2 \mu_3 \neq 0$ with $\lambda_{jk} \equiv \lambda_j - \lambda_k$. Explicitly, the equations above are written as

$$\frac{dx_1}{dt} = \lambda_{32} x_2 x_3 + \mu_3 x_2 x_6 - \mu_2 x_3 x_5, \quad \frac{dx_4}{dt} = \lambda_{65} x_5 x_6 + \mu_3 x_3 x_5 - \mu_2 x_2 x_6,$$

$$\frac{dx_2}{dt} = \lambda_{13} x_3 x_1 + \mu_1 x_3 x_4 - \mu_3 x_1 x_5, \quad \frac{dx_5}{dt} = \lambda_{46} x_6 x_4 + \mu_1 x_1 x_6 - \mu_3 x_3 x_4,$$

$$\frac{dx_3}{dt} = \lambda_{21} x_1 x_2 + \mu_2 x_1 x_5 - \mu_1 x_2 x_4, \quad \frac{dx_6}{dt} = \lambda_{54} x_4 x_5 + \mu_2 x_2 x_4 - \mu_1 x_1 x_5.$$

Besides the energy $Q_1 = H$, the equations have two trivial constants of the motion $Q_2 = x_1^2 + x_2^2 + x_3^2$, $Q_3 = x_4^2 + x_5^2 + x_6^2$. Adler and van Moerbeke (2004) have shown that the geodesic flow on $SO(4)$ for the metric defined by the above quadratic form is algebraically completely integrable if and only if: **(a)** The quadratic form H is diagonal with regard to the customary $so(4)$ coordinates (Manakov metric), that is, $2H = \sum_{\substack{j,k=1 \\ j<k}}^{4} \Lambda_{jk} X_{jk}^2$, $(X_{jk})_{1 \le j,k \le 4} \in so(4)$, with $\Lambda_{jk} = \frac{\beta_j - \beta_k}{\alpha_j - \alpha_k}$, $(\alpha_j, \beta_j \in \mathbb{C}, 1 \le j \le 4)$, all Λ_{jk} distinct. The extra invariant Q_4 is quadratic and the flow evolves on Abelian surfaces $\mathbb{C}^2/lattice \subseteq \mathbb{P}^7(\mathbb{C})$, having period matrix $\begin{pmatrix} 2 & 0 & a & c \\ 0 & 4 & c & b \end{pmatrix}$, Im $\begin{pmatrix} a & c \\ c & b \end{pmatrix} > 0$, $(a, b, c \in \mathbb{C})$, and also the linearization takes place on a Prym variety as discussed earlier in section 7.8. The periods of this Prym variety provide the exact periods of the motion in terms of Abelian integrals. The problem of the solid body in a fluid in the case of Clebsch is a particular case of this metric. See sections 3.3.1 and 4.5 where the problem has been studied via the spectral method and section 7.7 via the Kowalewski-Painlevé analysis. **(b)** The quadratic form H satisfies the conditions

$$(\mu_1^2, \mu_2^2, \mu_3^2) = \frac{\lambda_{12} \lambda_{23} \lambda_{31} \lambda_{45} \lambda_{56} \lambda_{64}}{(\lambda_{46} \lambda_{32} - \lambda_{65} \lambda_{13})^2} \left(\frac{(\lambda_{23} - \lambda_{56})^2}{\lambda_{23} \lambda_{56}}, \frac{(\lambda_{31} - \lambda_{64})^2}{\lambda_{31} \lambda_{64}}, \frac{(\lambda_{12} - \lambda_{45})^2}{\lambda_{12} \lambda_{45}} \right),$$

with the product $\mu_1 \mu_2 \mu_3$ being rational in $\lambda_1, ..., \lambda_6$ and with the following sign specification $\mu_1 \mu_2 \mu_3 = \frac{\lambda_{12} \lambda_{23} \lambda_{31} \lambda_{45} \lambda_{56} \lambda_{64}}{(\lambda_{46} \lambda_{32} - \lambda_{65} \lambda_{13})^3} (\lambda_{12} - \lambda_{45})(\lambda_{23} - \lambda_{56})(\lambda_{31} - \lambda_{64})$. The extra invariant Q_4 is quadratic and the flow linearizes on two-dimensional hyperelliptic Jacobians. More precisely

$$\bigcap_{j=1}^{4} \{x \in \mathbb{C}^6 : Q_j(x) = c_j\} = Jac \text{ (hyperelliptic curve } \mathcal{C} \text{ of genus 2)} \backslash \mathcal{D},$$

where $\mathcal{D}$ is a divisor of genus 17, which contains four translates of the Θ-divisor in $Jac(\mathcal{C})$, each of which is isomorphic to $\mathcal{C}$. The hyperelliptic curve $\mathcal{C}$ is a double cover of the curve $\mathcal{C}_0$ defined as $\{t_1 : t_2 : t_3 : t_4\} \in \mathbb{P}^3(\mathbb{C})$ such that $\sum t_j Q_j$ has rank 3$\}$, isomorphic to $\mathbb{P}^1(\mathbb{C})$. The periods of the motion are given by the periods of the hyperelliptic curve $\mathcal{C}$. When studying the differential systems in this case, as well as the invariants via Kowalewski's analysis, it is advantageous to rewrite them in a simpler form in order to reduce the notations and thus avoid too much calculation (see exercise 7.6 for explicit notation and Adler and van Moerbeke (2004) for more details). The problem of the solid body in a fluid in the case of Lyapunov–Steklov is a particular case of this metric (see section 3.3.2). **(c)** The form H satisfies

$$(\mu_1^4, \mu_2^4, \mu_3^4) = \lambda_{13}\lambda_{46}\lambda_{21}\lambda_{54}\lambda_{32}\lambda_{65}\left(\frac{1}{\lambda_{32}\lambda_{65}}, \frac{1}{\lambda_{13}\lambda_{46}}, \frac{1}{\lambda_{21}\lambda_{54}}\right).$$

The quantities ζ, ξ and η defined by $\zeta^2 \equiv \frac{\lambda_{46}}{\lambda_{13}}$, $\xi^2 \equiv \frac{\lambda_{54}}{\lambda_{21}}$, $\eta^2 \equiv \frac{\lambda_{65}}{\lambda_{32}}$, satisfy the quadratic relations $\zeta\xi + \xi\eta + \eta\zeta + 1 = 0$, $3\xi\eta + \eta - \xi + 1 = 0$. The geodesic flow has a quartic invariant, evolves on Abelian surfaces $A \subseteq \mathbb{P}^{23}(\mathbb{C})$ having period matrix $\begin{pmatrix} 2 & 0 & a & c \\ 0 & 12 & c & b \end{pmatrix}$, $\text{Im}\begin{pmatrix} a & c \\ c & b \end{pmatrix} > 0$, $(a, b, c \in \mathbb{C})$, and it will be expressed in terms of Abelian integrals. *Case $n \geq 5$*: We have seen that if a system is algebraically completely integrable, then it has a family of meromorphic Laurent series depending on "dim (phase space) $- 1$" free parameters. Now, trying to generalize the result to the geodesic flow on $SO(n)$ for $n \geq 5$ using the same method leads to insurmountable calculations. As shown by Haine (1984), for $n \geq 5$ Manakov's metrics are the only left invariant diagonal metrics on $SO(n)$, for which the geodesic flow is algebraically completely integrable. Note that it turns out that the geodesic flow on $SO(n)$ admits a lot of invariant manifolds on which they reduce to geodesic flow on $SO(3)$ and the solutions of the differential equation with initial conditions on these manifolds are elliptic functions and this without any condition on the metric. Haine (1984) has shown that looking at solutions near these special *a priori* known solutions and imposing these solutions to be single-valued functions of $t \in \mathbb{C}$, suffices to single out the left invariant diagonal metrics for which the geodesic flow is algebraically completely integrable. This criterion was first used, without proof by Lyapunov (1893) (the proof is due to Haine (1984)), who showed that the only integrable tops whose solutions have analytic properties belong to the classical known cases: Euler top, Lagrange top and Kowalewski top.

7.10. The periodic five-particle Kac–van Moerbeke lattice

The periodic five-particle Kac–van Moerbeke lattice (Kac and van Moerbeke 1975) is given by the quadratic vector field $\dot{x}_j = x_j(x_{j-1} - x_{j+1})$, $j = 1, ..., 5$, where $(x_1, ..., x_5) \in \mathbb{C}^5$ and $x_j = x_{j+5}$. This system forms a Hamiltonian vector field for the Poisson structure $\{x_j, x_k\} = x_j x_k (\delta_{j,k+1} - \delta_{j+1,k})$, $1 \leq j,k \leq 5$, and admits three independent first integrals

$$H_1 = x_1 x_3 + x_2 x_4 + x_3 x_5 + x_4 x_1 + x_5 x_2,$$
$$H_2 = x_1 + x_2 + x_3 + x_4 + x_5,$$
$$H_3 = x_1 x_2 x_3 x_4 x_5.$$

Let us show that this system is algebraically completely integrable. We easily check that H_1 and H_2 are in involution while H_3 is a Casimir. The system in question is therefore integrable in the Liouville sense. In addition, it is shown (Adler and van Moerbeke 2004) that the affine variety $\bigcap_{j=1}^{3} \{x \in \mathbb{C}^5 : H_j(x) = c_j\}$, $(c_1, c_2, c_3) \in \mathbb{C}^3$, $c_3 \neq 0$, defined by the intersection of the constants of motion is isomorphic to $\text{Jac}(\mathcal{C}) \backslash \mathcal{D}$ where $\mathcal{C} = \{(z,w) \in \mathbb{C}^2 : w^2 = (z^3 - c_1 z^2 + c_2 z)^2 - 4z\}$, is a smooth curve of genus 2 and $\mathcal{D}$ consists of five copies of $\mathcal{C}$ in the Jacobian variety $\text{Jac}(\mathcal{C})$. The flows generated by H_1 and H_2 are linearized on $\text{Jac}(\mathcal{C})$ and the system in question is algebraically completely integrable. The reader interested in the study of this system via various methods can find further information with more detail in Adler and van Moerbeke (2004), as well as in Teschil (2000).

7.11. Generalized periodic Toda systems

Let $e_0, ..., e_l$ be linearly dependent vectors in the following Euclidean vector space $(\mathbb{R}^{l+1}, \langle . | . \rangle)$, $l \geq 1$, such that they are l to l linearly independent (i.e. for all j, the vectors $e_0, ..., \widehat{e_j}, ..., e_l$ are linearly independent). Suppose that the non-zero reals $\xi_0, ..., \xi_l$ satisfying $\sum_{j=0}^{l} \xi_j e_j = 0$ are non-zero sum; that is, $\sum_{j=0}^{l} \xi_j \neq 0$. Let $A = (a_{ij})_{0 \leq i,j \leq l}$ be the matrix whose elements are defined by $a_{ij} = 2\frac{\langle e_i | e_j \rangle}{\langle e_j | e_j \rangle}$, $0 \leq i,j \leq l$. We consider the vector field X_A on $\mathbb{C}^{2(l+1)}$, $X_A : \begin{cases} \dot{x} = x.y \\ \dot{y} = Ax \end{cases}$, where $x, y \in \mathbb{C}^{l+1}$ and $x.y = (x_0 y_0, ..., x_l y_l)$. Using the method described in this chapter, we show (Adler and van Moerbeke 2004) that if X_A is an integrable vector field of an irreducibly algebraically completely integrable system, then A is the Cartan matrix of a possibly twisted affine Lie algebra. This system was studied by many authors (see Adler and van Moerbeke (2004) and the references therein). Specific detailed results can be found in the technical paper (Adler and van Moerbeke 1991) (and also in Adler and van Moerbeke (2004), concerning the link between the Toda lattice, Dynkin diagrams, singularities and Abelian varieties). The periodic $l+1$ particle Toda lattices are associated with extended Dynkin diagrams. They have $l+1$ polynomial invariants, as many as there are dots in the Dynkin diagram and are integrable Hamiltonian systems. The complex invariant manifold defined by putting these invariants equal to generic constants completes into an Abelian variety by gluing on a specific divisor $\mathcal{D}$. The latter is entirely described by the extended

Dynkin diagram: each point of the diagram corresponds to a component of the divisor and each subdiagram determines the intersection of the corresponding divisors. The global geometry of the complex invariant tori (Abelian varieties), such as polarization, divisor equivalences, dimension of certain linear systems, etc., is also entirely given by the extended Dynkin diagram and the linear equivalence between them is expressed in Lie-theoretic terms. More precisely, the divisor $\mathcal{D}$ consists of $l+1$ irreducible components $\mathcal{D}_j$ each associated with a root α_j of the Dynkin diagram Δ. The intersection of k components $\mathcal{D}_{j_1},...,\mathcal{D}_{j_k}$ satisfies the following relation: the intersection multiplicity of the intersection of k components of the divisor equals $\frac{\text{order}(W)}{\det(A)}$ where W and A are the Weyl group and the Cartan matrix going with the sub-Dynkin diagram $\alpha_{j_1},...,\alpha_{j_k}$ associated with the k components. The intersection of all the divisors is empty and the intersection of all divisors but one is a discrete set of points whose number is explicitly determined. we have the following expression for the set-theoretical number of points in terms of the Dynkin diagram

$$\sharp\left(\bigcap_{\beta\neq\alpha}\mathcal{D}_\beta\right) = \frac{p_\alpha}{p_0}\left(\frac{\text{order(Weyl group of the Dynkin diagram }\Delta\setminus\alpha_0)}{\text{order(Weyl group of the Dynkin diagram }\Delta\setminus\alpha)}\right),$$

where the integers p_α are given by the null vector of the Cartan matrix going with Δ. The singularities of the divisor are canonically associated with semi-simple Dynkin diagrams. The singularities of each component only occur at the intersections with other components, and their multiplicities at the intersection with other divisors are expressed in terms of how a corresponding root is located in the sub-Dynkin diagram determined by this root and those of the members of the above divisor intersection. The following inclusion holds for the singular locus $\text{sing}(\mathcal{D}_k)$ of $\mathcal{D}_k$: $\text{sing}(\mathcal{D}_k) \subseteq \mathcal{D}_k \cap \sum_{\substack{0\leq j\leq l \\ j\neq k}} \mathcal{D}_j$, $k=0,...,l$. The multiplicity of the singularity of a particular component $\mathcal{D}_k$, at its intersection with m other divisors, that is, $\text{sing}(\mathcal{D}_k) \cap (\mathcal{D}_{j_1},...,\mathcal{D}_{j_m})$, all $j_1,...,j_m \neq k$, is entirely specified by the way the corresponding root α_k sits in the sub-Dynkin diagram $\alpha_k, \alpha_{j_1},...,\alpha_{j_m}$ (for proof of these results, as well as other information, see Adler and van Moerbeke (1991)).

7.12. The Gross–Neveu system

The Gross–Neveu Hamiltonian system plays an important role in particle physics and written as $\dot{x}_j = \frac{\partial H}{\partial y_j}$, $\dot{y}_j = -\frac{\partial H}{\partial x_j}$, with $j=1,...,n$ and whose energy of the form $H = \frac{1}{2}\sum_{j=1}^n y_j^2 + \sum_{\alpha\in\mathbb{R}} e^{ic\langle\alpha,x\rangle}$, where c is a constant, $\alpha=(\alpha_1,...,\alpha_n)$ and $\langle\alpha,x\rangle = \sum_{j=1}^n \alpha_j x_j$. The sum on α above extends over the root system of a simple Lie algebra $\mathcal{L}$. We show (Adler and van Moerbeke 1982) that (a) the Hamiltonian system above for $j=1,2,3$, $\mathcal{L}=sl(3)$ and

$$H = \frac{1}{2}\sum_{j=1}^{3} y_j^2 + \sum_{j,k=1}^{3} e^{i(x_j - x_k)},$$

with Abelian functions y_j, e^{ix_j}, $1 \leq j \leq 3$, is not algebraically completely integrable. (b) The same conclusion is obtained for the Hamiltonian system in (a) where $j = 1, 2, 3, 4$ and $\mathcal{L} = sl(4) \simeq o(6)$.

7.13. The Kolossof potential

Let us study the integrability of the following Kolossof Hamiltonian system (Gavrilov *et al.* 1992): $\dot{q}_1 = \frac{\partial H}{\partial p_1}, \dot{p}_1 = -\frac{\partial H}{\partial q_1}, \dot{q}_2 = \frac{\partial H}{\partial p_2}, \dot{p}_2 = -\frac{\partial H}{\partial q_2}$, where $H = \frac{1}{2}(p_1^2 + p_2^2) + r + \frac{1}{r} - a\cos\theta$, $a \in \mathbb{R}$, is the Hamiltonian and $q_1 = r\cos\theta$, $q_2 = r\sin\theta$. This system describes the motion of a particle of mass unit in the plane (q_1, q_2) under the effect of the Kolossof potential $V(q_1, q_2) = r + \frac{1}{r} - a\cos\theta$, $a \in \mathbb{R}$. The Kolossof system admits a second first integral

$$F = -(a^2 + q_2^2)p_1^2 + (q_1 - a)(2q_2 p_1 - q_1 p_2 + ap_2)p_2$$
$$- 2a(q_1 - a)(aq_1 - 1)(q_1^2 + q_2^2)^{-1/2},$$

and it is Liouville integrable (but it is not algebraically completely integrable). Let $M_c = \{(q_1, q_2, p_1, p_2, z) \in \mathbb{C}^5 : H = c_1, F = c_2, q_1^2 + q_2^2 = z^2, z \neq 0\}$ (where $c = (c_1, c_2)$ is not a critical value) be the invariant affine variety by the Kolossof flow. Let us consider $P(u) = -2u^3 + 2c_1 u^2 - 2(1 - a^2)u + c_2$. We show (see Gavrilov *et al.* (1992)) that if the polynomial $(u^2 - a^2)P(u)$ does not have a double root, then the variety M_c is smooth and biholomorphic to the complex manifold $\widetilde{M_c}\backslash \mathcal{D}$ where $\widetilde{M_c}$ is an Abelian variety and $\mathcal{D}$ a divisor on $\widetilde{M_c}$. In addition, $\widetilde{M_c}$ is an unbranched double cover of the Jacobian variety Jac($\mathcal{C}$) where $\mathcal{C}$ is a 2-genus curve defined by $\mathcal{C}: w^2 = (u^2 - a^2)P(u)$. The trajectories of the Hamiltonian flow generated by H in M_c are straight lines, but the motion is nonlinear. The trajectories of the flow generated by $H + sF$, $s \neq 0$ in M_c are not linear. As a result, the flows generated by H and F do not linearize on $\widetilde{M_c}$.

7.14. Exercises

EXERCISE 7.1.– We consider the system described by the Hamiltonian (Roekaerts 1987):

$$H = \frac{1}{2}\left(y_1^2 + y_2^2\right) + y_1 x_1 x_2 + y_2 \left(\frac{1}{32}x_1^2 + 2x_2^2\right).$$

Is this system algebraically completely integrable? A second first integral is

$$F = 256y_1^4 y_2 + 32y_1^2\left(4y_1^2\left(x_1^2 + 8x_2^2\right) + y_2^2 x_1^2\right)$$

$$+x_1^2\left(64y_1^3 x_1 x_2 + 8y_1^2 y_2\left(x_1^2 + 16x_2^2\right) + y_2^3 x_1^2\right) + x_1^4\left(y_1 x_1 + 2y_2 x_2\right)^2.$$

EXERCISE 7.2.– Consider a system of differential equations (Haine 1984):

$$\dot{z} = f(z), \quad z \in \mathbb{C}^m, \qquad [7.55]$$

where f is holomorphic on $\mathbb{C}^m$. Assume that all solutions of [7.55] with initial conditions in a dense set $A \subset \mathbb{C}^m$ are (analytic) single-valued functions of $t \in \mathbb{C}$.

a) Show that all solutions of [7.55] are single-valued functions of $t \in \mathbb{C}$.

b) Let $\varphi(t)$ be a particular solution of [7.55] holomorphic along some closed path l in the complex t-plane. Show that the analytic continuation along l of any solution of the variational (linearized) equations: $\dot{\delta} = \frac{\partial f}{\partial x}(\varphi(t))\delta$, has to be single-valued.

c) We use the notation from definition 7.1 (section 7.2). Show that if the system [7.9] is algebraically completely integrable with Abelian functions z_i, then the conclusions of questions (a) and (b) hold.

EXERCISE 7.3.– Suppose that the Hamiltonian system [7.9] is algebraically completely integrable with Abelian functions z_i. Can we conclude that the analytic continuation of any solution of this system can at worst lead to pole singularities? (In other words, all of its solutions are meromorphic functions of $t \in \mathbb{C}$). Justify your answer.

EXERCISE 7.4.– Consider the following system of differential equations:

$$\dot{X} = [X, \lambda X], \qquad [7.56]$$

where $X = (x_{ij}) \equiv \sum_{i<j} x_{ij} e_{ij} \in so(n)$, $(\lambda X)_{ij} = \lambda_{ij} X_{ij}$, $\lambda_{ij} = \lambda_{ji}$. Show that under the non-degeneracy assumption on the diagonal metric λ that all λ_{ij} be distinct, the system [7.56] is algebraically completely integrable with Abelian functions x_{ij} if and only if the metric λ satisfies: $\lambda_{ij} = \frac{\beta_i - \beta_j}{\alpha_i - \alpha_j} \iff [X, \beta] + [\alpha, \lambda X] = 0, \forall X$, with $\alpha = \text{diag}(\alpha_1, ..., \alpha_n)$, $\beta = \text{diag}(\beta_1, ..., \beta_n)$, $\prod_{i<j}(\alpha_i - \beta_j) \neq 0$ (hint: use exercise 7.2 c)).

EXERCISE 7.5.– Consider the geodesic flow on $SO(4)$ for a left invariant metric (section 7.9, case of the second metric, i.e. case (b)) and the problem is to show, among other things, that this is a weight homogeneous algebraic complete integrable system.

a) Show that this geodesic flow X_1 and a commuting flow X_2 can be written, respectively, in the form

$$\dot{x}_1 = x_2 x_6, \qquad \dot{x}_2 = \frac{1}{2} x_3 (x_1 + x_4), \qquad \dot{x}_3 = \frac{1}{2} x_2 (x_1 + x_4),$$

$$\dot{x}_4 = x_3 x_5, \qquad \dot{x}_5 = x_3 x_4, \qquad \dot{x}_6 = x_1 x_2,$$

and

$$\dot{x}_1 = x_5 x_6, \qquad \dot{x}_2 = x_3 x_4, \qquad \dot{x}_3 = x_2 x_4,$$

$$\dot{x}_4 = x_5 (2x_3 - x_6), \qquad \dot{x}_5 = x_4 (2x_3 - x_6), \qquad \dot{x}_6 = x_1 x_5.$$

The Hamiltonian structure being determined by the following Poisson bracket:
$\{H, F\} = \langle \frac{\partial H}{\partial x}, J \frac{\partial F}{\partial x} \rangle = \sum_{i,j} J_{ij} \frac{\partial H}{\partial x_i} \frac{\partial F}{\partial x_j}$, where

$$J = \begin{pmatrix} 0 & x_3 x_2 & 0 & 0 & 2x_2 - x_5 \\ -x_3 & 0 & 0 & 0 & 0 & 0 \\ -x_2 & 0 & 0 & 0 & 0 & 0 \\ 0 & 0 & 0 & 0 & 0 & x_5 \\ 0 & 0 & 0 & 0 & 0 & x_4 \\ -2x_2 + x_5 & 0 & 0 & -x_5 & -x_4 & 0 \end{pmatrix}.$$

b) Show that this geodesic flow has four quadric invariants:

$$H_1 = -x_4^2 + x_5^2 = c_1, \quad H_2 = -x_1^2 + x_6^2 = c_2, \quad H_3 = -x_2^2 - x_3^2 = \frac{c_3}{4},$$

$$H_4 = -(x_1 - x_4)^2 + 2(x_2 - x_5)^2 + 2(x_3 - x_6)^2 = 4c_4,$$

(with generic $(c_1, c_2, c_3, c_4) \in \mathbb{C}^4$), and evolves on some hyperelliptic Jacobians. The hyperelliptic curve is a double cover of the curve of rank four quadrics (isomorphic to $\mathbb{P}^1(\mathbb{C})$):

$$\left\{ t \in \mathbb{P}^3 \text{ such that } t_1(H_1 - c_1 x_0^2) + t_2(H_2 - c_2 x_0^2) + t_3(H_3 - \frac{c_3}{4} x_0^2) \right.$$

$$\left. + t_4(H_4 - 4c_4 x_0^2) \text{ has rank } 4 \right\},$$

ramified at the six points where the rank drops to 3.

EXERCISE 7.6.– We again consider the problem of geodesic flow on $SO(4)$ mentioned in the previous exercise.

a) Show that the system in question possesses Laurent solutions depending on five free parameters to be determined explicitly.

b) Prove that the affine surface defined by the constants of motion can be completed into a torus T by adjoining a singular divisor $\mathcal{D}$. The latter consists of four copies $\mathcal{H}_1,...,\mathcal{H}_4$ of the genus two hyperelliptic curves. Analyze the points of intersection of these curves.

c) Show that all these curves are translates of the Θ-divisor by $\frac{1}{2}$-periods and that three of these curves form a very ample and projectively normal divisor, which results in the embedding of the Jacobian in $\mathbb{P}^8(\mathbb{C})$ and the functions having poles there form a closed system of quadratic equations under differentiations, as well as their ratios.

d) Show that the line bundle $[\mathcal{D}]$ defines a polarization of type $(4,4)$ on T and leads to an embedding in $\mathbb{P}^{15}(\mathbb{C})$.

e) Show that the three flows X_1, X_2 and $2X_1 - X_2$ are doubly tangent to each of the four curves $\mathcal{H}_1,...,\mathcal{H}_4$ at four points $P_1,...,P_4$ and that the 16 half-periods on the torus are given by the total set of branch points of these hyperelliptic curves.

f) Deduce that the values of the constants of motion c_1/c_4, c_2/c_4, c_3/c_4 provide the three moduli for the full family of two-dimensional hyperelliptic Jacobians (hint: see Adler and van Moerbeke (1985)).

EXERCISE 7.7.– Prove the results mentioned in section 7.12 concerning the Gross–Neveu system.

EXERCISE 7.8.– Show that the system of differential equations [4.20] admit Laurent series solutions depending on three free parameters and study its complete algebraic integrability in detail.

EXERCISE 7.9.– Consider the following system of five differential equations:

$$\dot{z}_1 = 2z_4, \quad \dot{z}_3 = -4az_2 - 6z_1z_2 - 16z_2^3,$$
$$\dot{z}_2 = z_3, \quad \dot{z}_4 = -az_1 - z_1^2 - 8z_1z_2^2 + z_5, \quad [7.57]$$
$$\dot{z}_5 = -8z_2^2 z_4 - 2az_4 - 2z_1 z_4 + 4z_1 z_2 z_3,$$

where a is a constant.

a) Show that the following three quartics:

$$F_1 = \frac{1}{2}z_5 + 2z_1z_2^2 + \frac{1}{2}z_3^2 + \frac{1}{2}az_1 + 2az_2^2 + \frac{1}{4}z_1^2 + 4z_2^4,$$
$$F_2 = az_1z_2 + z_1^2z_2 + 4z_1z_2^3 - z_2z_5 + z_3z_4,$$ [7.58]
$$F_3 = z_1z_5 - 2z_1^2z_2^2 - z_4^2.$$

are constants of motion for this system and it is completely integrable in the sense of Liouville.

b) Show that the complex affine variety $M_c = \bigcap_{k=1}^{2}\{z : F_k(z) = c_k\} \subset \mathbb{C}^5$, defined by putting these invariants equal to generic constants, is a double cover of a Kummer surface to be determined explicitly. Deduce that the system [7.57] can be integrated in genus 2 hyperelliptic functions.

c) Show that the system [7.57] possesses Laurent series solutions (which depend on four free parameters) to be determined explicitly and deduce that these meromorphic solutions restricted to the surface M_c are parameterized by two isomorphic smooth hyperelliptic curves $\mathcal{H}_{\varepsilon=\pm i}$ of genus 2.

d) Show that the variety M_c is generically the affine part of an Abelian surface $\widetilde{M}_c$, more precisely the Jacobian of a genus 2 curve. The reduced divisor at infinity $\widetilde{M}_c \backslash M_c = \mathcal{H}_i + \mathcal{H}_{-i}$ consists of two smooth isomorphic genus 2 curves $\mathcal{H}_\varepsilon$.

e) Deduce that the system of differential equations [7.57] is algebraic complete integrable and the corresponding flows evolve on $\widetilde{M}_c$. The system [7.57] and the invariants [7.58] will play a crucial role in exercise 8.8 (hint: see Lesfari (2020)).

8

Generalized Algebraic Completely Integrable Systems

Some interesting cases of integrable systems, to be discussed in this chapter, appear as coverings of algebraic completely integrable systems. The manifolds in variant by the complex flows are coverings of Abelian varieties and these systems are called generalized algebraic completely integrable. The latter are completely integrable in the sense of Arnold–Liouville and so generically, the compact connected manifolds invariant by the real flows are tori, the real parts of complex affine coverings of Abelian varieties. Also, we will see how some algebraic completely integrable systems can be constructed from known algebraic completely integrable in the generalized sense. We will see that a large class of algebraic completely integrable systems in the generalized sense are part of new algebraic completely integrable systems. We consider (as examples of applications) the Hénon–Heiles problem, the Ramani–Dorizzi–Grammaticos (RDG) potential, the Yang–Mills system, Goryachev–Chaplygin and Lagrange tops, as well as other interesting systems related to these examples.

8.1. Generalities

There are many examples of differential equations $\dot{z} = f(z)$, $z \in \mathbb{C}^m$, which have the weak Painlevé property that all movable singularities of the general solution have only a finite number of branches, and some integrable systems appear as coverings of algebraic completely integrable systems. The manifold invariants by the complex flows are coverings of Abelian varieties and these systems are called *generalized algebraic completely integrable*. They are Liouville integrable and the compact connected manifolds invariant by the real flows are tori, the real parts of complex affine coverings of Abelian varieties. Most of these systems possess solutions that are Laurent series of $t^{1/n}$ ($t \in \mathbb{C}$), and whose coefficients depend rationally on certain algebraic parameters. These parameters come from an $(n-1)$-dimensional family of affine varieties (where n is the dimension of the invariant manifolds) and a number of independent constants of the motion.

DEFINITION 8.1.– *Let $(M, \{.,.\}, \varphi)$ be a complex integrable system where M is a non-singular affine variety and $\varphi = (H_1, ..., H_s)$ where each H_i in φ is a regular function. We say that this system is a generalized algebraic completely integrable system if each generic fiber of H is a Zariski open subset of a commutative algebraic group, on which the Hamiltonian vector fields generated by H_i are translation invariant.*

The integrable systems we are going to deal with here are complex integrable systems where $M = \mathbb{C}^m$. Therefore, for the generalized algebraic completely integrable systems, it suffices to replace the condition (ii) in remark 7.4 (a), by the following: (iii) *the invariant manifolds* $\bigcap_{i=1}^{n+k} \{z \in \mathbb{C}^m : H_i(z) = c_i\}$ *are related to an l-fold cover* $\widetilde{T}^n$ *of the complex algebraic torus T^n ramified along a divisor $\mathcal{D}$ in T^n as follows:* $\bigcap_{i=1}^{n+k} \{z \in \mathbb{C}^m : H_i(z) = c_i\} = \widetilde{T}^n \backslash \mathcal{D}$.

EXAMPLE 8.1.– The following differential equations

$$\dot{x} = y^3, \qquad \dot{y} = -x^3, \qquad [8.1]$$

are written in the form of a Hamiltonian vector field $\dot{z} = J\frac{\partial H}{\partial z}$, where $z = (x,y)^\mathsf{T}$, $H = \frac{1}{4}(x^4 + y^4) = a$ is the Hamiltonian and $J = \begin{pmatrix} 0 & -1 \\ 1 & 0 \end{pmatrix}$. This system is completely integrable and can be solved in terms of Abelian integrals. We deduce from the equations $\dot{x} = y^3$, $\frac{1}{4}(x^4 + y^4) = a$, the integral form $t = \int \frac{dx}{(a-x^4)^{3/4}} + t_0$. The system [8.1] admits four one-dimensional families of Laurent solutions in $\sqrt{t}$, depending on one free parameter, $x = \frac{1}{\sqrt{t}}(x_0 + x_1 t + x_2 t^2 + \cdots)$, $y = \frac{1}{\sqrt{t}}(y_0 + y_1 t + y_2 t^2 + \cdots)$, where $x_0 + 2y_0^3 = 0$, $-y_0 + 2x_0^3 = 0$, $x_1 = y_1 = 0$, $-x_2 + 2y_0^2 y_2 = 0$, $y_2 + 2x_0^2 x_2 = 0$. Hence, $(2x_0 y_0)^2 = -1$, $\left(\frac{y_0}{x_0}\right)^4 = -1$, $x_1 = y_1 = 0$ and x_2, y_2 depend on one free parameter. We have seen that it is possible for the variables x, y to contain square root terms of the type $\sqrt{t}$, which are strictly not allowed by the Painlevé test. However, these terms are trivially removed by introducing some new variables z_1, z_2, z_3, which restores the Painlevé property to the system. A simple inspection of the Laurent series mentioned above suggests choosing $z_1 = x^2$, $z_2 = y^2$, $z_3 = xy$. By using the first integrals $H = a$, and the differential equations [8.1], we obtain a new system of differential equations in three unknowns z_1, z_2, z_3, having two quadrics invariants F_1, F_2: $\dot{z}_1 = 2z_2 z_3$, $\dot{z}_2 = -2z_1 z_3$, $\dot{z}_3 = z_2^2 - z_1^2$, and $F_1 = z_1^2 + z_2^2 = 4a$, $F_2 = z_1^2 - z_2^2 + z_3^2 = b$. The intersection $\mathcal{A} = \{z \equiv (z_1, z_2, z_3) \in \mathbb{C}^3 : F_1(z) = 4a, F_2(z) = b\}$ is an elliptic curve $\mathcal{E} : \{z_2^2 = -z_1^2 + 4a, \ z_3^2 = -2z_1^2 + 4a + b\}$. Note that the equation $x^4 + y^4 = 4a$ defines a Riemann surface of genus 3, but is not a torus. An equivalent description of $x^4 + y^4 = 4a$ is given by $\{z_2^2 = -z_1^2 + 4a, z_3^2 = -2z_1^2 + 4a + b\}$ and $\{x^2 = z_1, y^2 = z_2, xy = z_3\}$; as a double cover of $\mathcal{E}$ ramified at the four points where $z_i = \infty$. Consequently, the invariant surface completes into a double cover of an elliptic curve ramified at the points where the variables blow up. This example corresponds to definition (i), (iii) and we shall see more complicated examples, but very interesting problems, later. Consider the change of variable: $z_1 = \frac{1}{2}(m_2 - m_1)$, $z_2 = \frac{1}{2}(m_1 + m_2)$, $z_3 = m_3$. Taking the derivative and using the differential equations above for z_1, z_2, z_3 leads to the following system of differential equations: $\dot{m}_1 = -2m_2 m_3, \dot{m}_2 = 2m_1 m_3, \dot{m}_3 = m_1 m_2$. We see the resemblance with the equations of the Euler rigid body motion.

It was shown in series of publications (Abenda and Fedorov 2000; Vanhaecke 2001) that the θ-divisor can serve as a carrier of integrability. Let $\mathcal{H}$ be a hyperelliptic curve of genus g and $\mathrm{Jac}(\mathcal{H}) = \mathbb{C}^g/\Lambda$ its Jacobian variety, where Λ is a lattice of maximal rank in $\mathbb{C}^g$. Let

$$\mathcal{A}_k : \mathrm{Sym}^k(\mathcal{H}) \longrightarrow \mathrm{Jac}(\mathcal{H}), (P_1, ..., P_k) \longmapsto \sum_{j=1}^k \int_\infty^{P_j} (\omega_1, ..., \omega_g) \ \mathrm{mod}.\Lambda,$$

$(0 \leq k \leq g)$, be the Abel map where $(\omega_1, ..., \omega_g)$ is a canonical basis of the space of differentials of the first kind on $\mathcal{H}$. The theta divisor Θ is a subvariety of $\mathrm{Jac}(\mathcal{H})$ defined as $\Theta \equiv \mathcal{A}\left[\mathrm{Sym}^{g-1}(\mathcal{H})\right]/\Lambda$. By Θ_k, we will denote the subvariety (called strata) of $\mathrm{Jac}(\mathcal{H})$ defined by $\Theta_k \equiv \mathcal{A}_k \left[\mathrm{Sym}^k(\mathcal{H})\right]/\Lambda$ and we have the following stratification: $\{O\} \subset \Theta_0 \subset \Theta_1 \subset \Theta_2 \subset ... \subset \Theta_{g-1} \subset \Theta_g = \mathrm{Jac}(\mathcal{H})$, where O is the origin of $\mathrm{Jac}(\mathcal{H})$. It was shown in Vanhaecke (2001) that these stratifications of the Jacobian are connected with stratifications of the Sato Grassmannian via an extension of Krichever's map and some remarks on the relation between Laurent solutions for the Master systems and stratifications of the Jacobian of a hyperelliptic curve. In Vanhaecke (2001) there is a study about Lie–Poisson structure in the Jacobian, which indicates that invariant manifolds associated with Poisson brackets can be identified with these strata. Some problems were considered in Vanhaecke (2001) and Abenda and Fedorov (2000), where a connection was established with the flows on these strata. Such varieties or their open subsets often appear as coverings of complex invariants manifolds of finite dimensional integrable systems (Hénon–Heiles and Neumann systems). Let us consider the RDG series of integrable potentials (Ramani 1982; Hietarinta 1987):

$$V(x,y) = \sum_{k=0}^{[m/2]} 2^{m-2i} \binom{m-i}{i} x^{2i} y^{m-2i}, \quad m = 1, 2, ...$$

It can be straightforwardly proven that a Hamiltonian H:

$$H = \frac{1}{2}(p_x^2 + p_y^2) + \alpha_m V_m, \quad m = 1, 2, \ldots$$

containing V is Liouville integrable, with an additional first integral:

$$F = p_x(xp_y - yp_x) + \alpha_m x^2 V_{m-1}, \quad m = 1, 2, \ldots$$

The study of cases $m = 1$ and $m = 2$ is easy. The study of other cases is not obvious. For the case $m = 3$, we obtain the Hénon–Heiles system that we will see in section 8.3. The case $m = 4$ corresponds to the system that will be studied in section 8.2. However, the case $m = 5$ corresponds to a system with an Hamiltonian of the form $H = \frac{1}{2}(p_x^2 + p_y^2) + y^5 + x^2 y^3 + \frac{3}{16} x^4 y$. The corresponding Hamiltonian system admits a second first integral: $F = -p_x^2 y + p_x p_y x - \frac{1}{2} x^2 y^4 + \frac{3}{8} x^4 y^2 + \frac{1}{32} x^6$, and admits three-dimensional family solutions x, y, which are Laurent series of $t^{1/3}$: $x = at^{-\frac{1}{3}}$, $x = bt^{-\frac{2}{3}}$, $b^3 = -\frac{2}{9}$, but for which there are no polynomials P, such that $P(x(t), y(t), \dot{x}(t), \dot{y}(t))$ is Laurent series in t.

We introduce a method for generating new integrable systems from known ones. For the algebraic integrable systems in the generalized sense, the Laurent series solutions contain square root terms of the type $t^{-1/n}$ that are strictly not allowed by the Painlevé test (i.e. the general solutions should have no movable singularities other than poles in the complex plane). However, for some problems these terms are trivially removed by introducing new variables, which restores the Painlevé property to the system. By inspection of the Laurent solutions of the algebraic integrable systems in the generalized sense, we look for polynomials in the variables defining these systems, without fractional exponents. For many problems, obtaining these new variables is not a problem, just use (by simple inspection) the first terms of the Laurent solutions. These new variables belong to the space $\mathcal{L}(\mathcal{D})$, where $\mathcal{D}$ is a divisor on an Abelian variety T^n, which completes the affine defined by the intersection of the invariants of the new algebraically completely integrable system. In all of the problems we have studied, we find that the known algebraically integrable systems in the generalized sense are part of new algebraically integrable systems. Let $\dot{x} = J \frac{\partial H}{\partial x}$, $x \in \mathbb{C}^m$, be an algebraically integrable system in the generalized sense. The Laurent series solutions of this system contain fractional exponents and the manifolds invariant by the complex flows are coverings of Abelian varieties. We might conjecture (with some additional conditions to be determined) from the problems discussed in this chapter that this system is part of a new algebraically integrable system in $m + 1$ variables. In other words, there is a new algebraically integrable system $\dot{z} = J \frac{\partial \mathbf{H}}{\partial z}$, $z \in \mathbb{C}^{m+1}$, that is, whose solutions expressible in terms of theta functions are associated with an Abelian variety with a divisor on it, and the Hamiltonian flows are linear on this Abelian variety.

8.2. The RDG potential and a five-dimensional system

Consider the RDG system (Ramani 1982),

$$\ddot{q}_1 - q_1\left(q_1^2 + 3q_2^2\right) = 0, \qquad \ddot{q}_2 - q_2\left(3q_1^2 + 8q_2^2\right) = 0, \qquad [8.2]$$

corresponding to the Hamiltonian

$$H_1 = \frac{1}{2}(p_1^2 + p_2^2) - \frac{3}{2}q_1^2 q_2^2 - \frac{1}{4}q_1^4 - 2q_2^4,$$

where p_1 and p_2 are the momenta conjugate to q_1 and q_2, respectively. This system is integrable, with the second first integral (of degree 8) being

$$H_2 = p_1^4 - 6q_1^2 q_2^2 p_1^2 + q_1^4 q_2^4 - q_1^4 p_1^2 + q_1^6 q_2^2 + 4q_1^3 q_2 p_1 p_2 - q_1^4 p_2^2 + \frac{1}{4}q_1^8.$$

The first integrals H_1 and H_2 are in involution, that is, $\{H_1, H_2\} = 0$. Recall that a system $\dot{z} = f(z)$ is weight-homogeneous with a weight ν_k going with each variable z_k if $f_k(\lambda^{\nu_i} z_1, \ldots, \lambda^{\nu_n} z_n) = \lambda^{\nu_k + 1} f_k(z_1, \ldots, z_n)$, for all $\lambda \in \mathbb{C}$. The system [8.2] is weight-homogeneous, with q_1, q_2 having weight 1 and p_1, p_2 having weight 2, so that H_1 and H_2 have weight 4 and 8, respectively. When we examine all of the possible singularities, we find that it is possible for the variable q_1 to contain square root terms of the type $\sqrt{t}$, which are strictly not allowed by the Painlevé test. However, we will see later that these terms are trivially removed by introducing new variables $z_1, \ldots, z_5$, which restores the Painlevé property to the system. Let $\mathcal{B}$ be the affine variety defined by

$$\mathcal{B} = \bigcap_{k=1}^{2} \{z \in \mathbb{C}^4 : H_k(z) = b_k\}, \qquad [8.3]$$

for generic $(b_1, b_2) \in \mathbb{C}^2$.

THEOREM 8.1.– a) The system [8.2] admits Laurent solutions depending on three free parameters: u, v and w. These solutions restricted to the surface $\mathcal{B}$ [8.3] are parameterized by two copies Γ_1 and Γ_{-1} of the Riemann surface Γ [8.4] of genus 16.

b) The system [8.2] is algebraic complete integrable in the generalized sense and extends to a new system [8.5] of five differential equations algebraically completely integrable with three quartics invariants. Generically, the invariant manifold $\mathcal{A}$ [8.6] defined by the intersection of these quartics form the affine part of an Abelian surface $\widetilde{\mathcal{A}}$. The reduced divisor at infinity $\left(\widetilde{\mathcal{A}} \setminus \mathcal{A}\right) = \mathcal{C}_1 + \mathcal{C}_{-1}$ is very ample and consists of two components $\mathcal{C}_1$ and $\mathcal{C}_{-1}$ of a genus 7 curve $\mathcal{C}$ [8.7]. In addition, the invariant surface $\mathcal{B}$ can be completed as a cyclic double cover $\overline{\mathcal{B}}$ of the Abelian surface $\widetilde{\mathcal{A}}$, ramified along the divisor $\mathcal{C}_1 + \mathcal{C}_{-1}$. Moreover, $\overline{\mathcal{B}}$ is smooth except at the point lying over the singularity (of type A_3) of $\mathcal{C}_1 + \mathcal{C}_{-1}$ and the resolution $\widetilde{\mathcal{B}}$ of $\overline{\mathcal{B}}$ is a surface of general type with invariants: $\mathcal{X}(\widetilde{\mathcal{B}}) = 1$ and $p_g(\widetilde{\mathcal{B}}) = 2$.

PROOF.– a) The system [8.2] possesses a three-dimensional family of Laurent solutions (principal balances) depending on three free parameters u, v and w. There are precisely two such families, labeled by $\varepsilon = \pm 1$, and they are given as follows:

$$(q_1, q_2, p_1, p_2) = (t^{-1/2}, t^{-1}, t^{-3/2}, t^{-2}) \times \text{a Taylor series in } t,$$

and it is not difficult to determine them explicitly. These formal series solutions are convergent as a consequence of the majorant method. By substituting these series in the constants of the motion $H_1 = b_1$ and $H_2 = b_2$, we eliminate the parameter w linearly, leading to an equation connecting the two remaining parameters u and v:

$$\Gamma : \frac{65}{4} uv^3 + \frac{93}{64} u^6 v^2 + \frac{3}{8192} \left(-9829 u^8 + 26112 H_1\right) u^3 v \quad [8.4]$$
$$- \frac{10299}{65536} u^{16} - \frac{123}{256} H_1 u^8 + H_2 + \frac{1536298731}{52} = 0.$$

According to Hurwitz's formula, this defines a Riemann surface Γ of genus 16. The Laurent solutions restricted to the affine surface $\mathcal{B}$ [8.3] are thus parameterized by two copies Γ_{-1} and Γ_1 of the same Riemann surface Γ.

b) Let $\varphi : \mathcal{B} \longrightarrow \mathbb{C}^5$, $(q_1, q_2, p_1, p_2) \longmapsto (z_1, z_2, z_3, z_4, z_5)$, be the morphism defined on the affine variety $\mathcal{B}$ (8.3) by $z_1 = q_1^2$, $z_2 = q_2$, $z_3 = p_2$, $z_4 = q_1 p_1$, $z_5 = p_1^2 - q_1^2 q_2^2$. These variables are easily obtained by simple inspection of the above Laurent series. By using the variables $z_1, ..., z_5$ and differential equations [8.2], we obtain

$$\begin{aligned} \dot{z}_1 &= 2z_4, & \dot{z}_3 &= z_2(3z_1 + 8z_2^2), \\ \dot{z}_2 &= z_3, & \dot{z}_4 &= z_1^2 + 4z_1 z_2^2 + z_5, \\ \dot{z}_5 &= 2z_1 z_4 + 4z_2^2 z_4 - 2z_1 z_2 z_3. & & \end{aligned} \quad [8.5]$$

This new system on $\mathbb{C}^5$ admits the following three first integrals:

$$F_1 = \frac{1}{2} z_5 - z_1 z_2^2 + \frac{1}{2} z_3^2 - \frac{1}{4} z_1^2 - 2 z_2^4,$$
$$F_2 = z_5^2 - z_1^2 z_5 + 4 z_1 z_2 z_3 z_4 - z_1^2 z_3^2 + \frac{1}{4} z_1^4 - 4 z_2^2 z_4^2,$$
$$F_3 = z_1 z_5 + z_1^2 z_2^2 - z_4^2.$$

The first integrals F_1 and F_2 are in involution, while F_3 is trivial (Casimir function). The invariant variety $\mathcal{A}$ defined by

$$\mathcal{A} = \bigcap_{k=1}^{3} \{z : F_k(z) = c_k\} \subset \mathbb{C}^5, \quad [8.6]$$

is a smooth affine surface for generic values of $(c_1, c_2, c_3) \in \mathbb{C}^3$. The system [8.5] is completely integrable and possesses Laurent series solutions,

$$(z_1, z_2, z_3, z_4, z_5) = (t^{-1}, t^{-1}, t^{-2}, t^{-2}, t^{-2}) \times \text{a Taylor series in } t,$$

labeled by $\varepsilon = \pm 1$, depending on four free parameters $\alpha, \beta, \gamma, \theta$. The convergence of these series is guaranteed by the majorant method. By substituting these developments in equations $F_k(z) = c_k$, $k = 1, 2, 3$, we obtain three polynomial relations between $\alpha, \beta, \gamma, \theta$. Eliminating γ and θ from these equations leads to an equation connecting the two remaining parameters α and β:

$$\mathcal{C}: \quad 64\beta^3 - 16\alpha^3\beta^2 - 4\left(\alpha^6 - 32\alpha^2 c_1 - 16 c_3\right)\beta \quad [8.7]$$
$$+\alpha \left(32 c_2 - 32\alpha^4 c_1 + \alpha^8 - 16\alpha^2 c_3\right) = 0.$$

The Laurent solutions restricted to the surface $\mathcal{A}$ [8.6] are parameterized by two copies $\mathcal{C}_{-1}$ and $\mathcal{C}_1$ of the same Riemann surface $\mathcal{C}$ [8.7]. According to the Riemann–Hurwitz formula, the genus of $\mathcal{C}$ is 7. Applying the method used in Chapter 7, we embed these curves in a hyperplane of $\mathbb{P}^{15}(\mathbb{C})$ using the 16 functions:

1, z_1, z_2, $5 - z_1^2$, $z_3 + 2\varepsilon z_2^2$, $z_4 + \varepsilon z_1 z_2$, $W(f_1, f_2)$, $f_1(f_1 + 2\varepsilon f_4)$,

$f_2(f_1 + 2\varepsilon f_4)$, $z_4(f_3 + 2\varepsilon f_6)$, $z_5(f_3 + 2\varepsilon f_6)$, $f_5(f_1 + 2\varepsilon f_4)$, $f_1 f_2(f_3 + 2\varepsilon f_6)$,

$f_4 f_5 + W(f_1, f_4)$, $W(f_1, f_3) + 2\varepsilon W(f_1, f_6)$, $f_3 - 2z_5 + 4f_4^2$,

where $W(s_j, s_k) \equiv \dot{s}_j s_k - s_j \dot{s}_k$ (Wronskian) and we show that these curves have two points in common in which $\mathcal{C}_1$ is tangent to $\mathcal{C}_{-1}$. The system [8.2] is algebraic complete integrable in the generalized sense. The invariant surface $\mathcal{B}$ [8.3] can be completed as a cyclic double cover $\overline{\mathcal{B}}$ of the Abelian surface $\widetilde{\mathcal{A}}$, ramified along the divisor $\mathcal{C}_1 + \mathcal{C}_{-1}$. Moreover, $\overline{\mathcal{B}}$ is smooth except at the point lying over the singularity (of type A_3) of $\mathcal{C}_1 + \mathcal{C}_{-1}$ (double points of intersection of the curves $\mathcal{C}_1$ and $\mathcal{C}_{-1}$) and the resolution $\widetilde{\mathcal{B}}$ of $\overline{\mathcal{B}}$ is a surface of general type. We shall resume the proof of these results (already used previously in other similar problems) with more detail. Observe that the morphism φ is an unramified cover. The Riemann surface Γ [8.4] plays an important role in the construction of a compactification $\overline{\mathcal{B}}$ of $\mathcal{B}$. Let us denote by G a cyclic group of two elements $\{-1, 1\}$ on $V_\varepsilon^j = U_\varepsilon^j \times \{\tau \in \mathbb{C} : 0 < |\tau| < \delta\}$, where $\tau = t^{1/2}$ and U_ε^j is an affine chart of Γ_ε for which the Laurent solutions are defined. The action of G is defined by $(-1) \circ (u, v, \tau) = (-u, -v, -\tau)$ and is without fixed points in V_ε^j. So we can identify the quotient V_ε^j / G with the image of the smooth map $h_\varepsilon^j : V_\varepsilon^j \longrightarrow \mathcal{B}$ defined by the Laurent expansions. We have $(-1, 1).(u, v, \tau) = (-u, -v, \tau)$, $(1, -1).(u, v, \tau) = (u, v, -\tau)$, that is, $G \times G$ acts separately on each coordinate. Thus, identifying $V\varepsilon^j / G^2$ with the image of $\varphi \circ h_\varepsilon^j$ in

$\mathcal{A}$. Note that $\mathcal{B}^j_\varepsilon = V^j_\varepsilon / G$ is smooth (except for a finite number of points) and the coherence of the $\mathcal{B}^j_\varepsilon$ follows from the coherence of V^j_ε and the action of G. By taking $\mathcal{B}$ and by gluing on various varieties $\mathcal{B}^j_\varepsilon \setminus \{\text{some points}\}$, we obtain a smooth complex manifold $\widehat{\mathcal{B}}$, which is a double cover of the Abelian variety $\widetilde{\mathcal{A}}$ ramified along $\mathcal{C}_1 + \mathcal{C}_{-1}$, and can therefore be completed to an algebraic cyclic cover of $\widetilde{\mathcal{A}}$. To see what happens to the missing points, we investigate the image of $\Gamma \times \{0\}$ in $\cup \mathcal{B}^j_\varepsilon$. The quotient $\Gamma \times \{0\}/G$ is birationally equivalent to the Riemann surface Υ of genus 7:

$$\Upsilon: \frac{65}{4}y^3 + \frac{93}{64}x^3 y^2 + \frac{3}{8192}\left(-9,829 x^4 + 26,112 b_1\right) x^2 y$$

$$+x\left(-\frac{10,299}{65,536}x^8 - \frac{123}{256}b_1 x^4 + b_2 + \frac{15,362\,98,731}{52}\right) = 0, y \equiv uv, x \equiv u^2.$$

The Riemann surface Υ is birationally equivalent to $\mathcal{C}$. The only points of Υ fixed under $(u, v) \longmapsto (-u, -v)$ are the points at ∞, which correspond to the ramification points of the map $\Gamma \times \{0\} \xrightarrow{2-1} \Upsilon : (u, v) \longmapsto (x, y)$, and coincide with the points at ∞ of the Riemann surface $\mathcal{C}$. Then, the variety $\widehat{\mathcal{B}}$ constructed above is birationally equivalent to the compactification $\overline{\mathcal{B}}$ of the generic invariant surface $\mathcal{B}$. So $\overline{\mathcal{B}}$ is a cyclic double cover of the Abelian surface $\widetilde{\mathcal{A}}$ ramified along the divisor $\mathcal{C}_1 + \mathcal{C}_{-1}$, where $\mathcal{C}_1$ and $\mathcal{C}_{-1}$ have two points in common at which they are tangent to each other. It follows that the system [8.2] is algebraic complete integrable in the generalized sense. Moreover, $\overline{\mathcal{B}}$ is smooth except at the point lying over the singularity (of type A_3) of $\mathcal{C}_1 + \mathcal{C}_{-1}$. In term of an appropriate local holomorphic coordinate system (X, Y, Z), the local analytic equation about this singularity is $X^4 + Y^2 + Z^2 = 0$. Now, let $\widetilde{\mathcal{B}}$ be the resolution of singularities of $\overline{\mathcal{B}}$, $\mathcal{X}(\widetilde{\mathcal{B}})$ be the Euler characteristic of $\widetilde{\mathcal{B}}$ and $p_g(\widetilde{\mathcal{B}})$ the geometric genus of $\widetilde{\mathcal{B}}$. Then $\widetilde{\mathcal{B}}$ is a surface of general type with invariants: $\mathcal{X}(\widetilde{\mathcal{B}}) = 1$ and $p_g(\widetilde{\mathcal{B}}) = 2$. This completes the proof. $\square$

On the Abelian variety $\widetilde{\mathcal{A}}$, consider the holomorphic 1-forms dt_1 and dt_2 defined by $dt_i(X_{F_j}) = \delta_{ij}$, where X_{F_1} and X_{F_2} are the vector fields generated, respectively, by F_1 and F_2. Taking the differentials of $\zeta = 1/z_2$ and $\xi = \frac{z_1}{z_2}$ viewed as functions of t_1 and t_2, using the vector fields and the Laurent series and solving linearly for dt_1 and dt_2, we obtain the holomorphic differentials (where $\Delta \equiv \frac{\partial \zeta}{\partial t_1}\frac{\partial \xi}{\partial t_2} - \frac{\partial \zeta}{\partial t_2}\frac{\partial \xi}{\partial t_1}$):

$$\omega_1 = dt_1|_{\mathcal{C}_\varepsilon} = \frac{1}{\Delta}\left(\frac{\partial \xi}{\partial t_2}d\zeta - \frac{\partial \zeta}{\partial t_2}d\xi\right)\Big|_{\mathcal{C}_\varepsilon} = \frac{8}{\alpha\left(-4\beta + \alpha^3\right)}d\alpha,$$

$$\omega_2 = dt_2|_{\mathcal{C}_\varepsilon} = \frac{1}{\Delta}\left(\frac{-\partial \xi}{\partial t_1}d\zeta - \frac{\partial \zeta}{\partial t_1}d\xi\right)\Big|_{\mathcal{C}_\varepsilon} = \frac{2}{\left(-4\beta + \alpha^3\right)^2}d\alpha.$$

The zeroes of ω_2 provide the points of tangency of the vector field X_{F_1} to $\mathcal{C}_\varepsilon$. We have $\frac{\omega_1}{\omega_2} = \frac{4}{\alpha}\left(-4\beta + \alpha^3\right)$ and X_{F_1} is tangent to $\mathcal{H}_\varepsilon$ at the point covering $\alpha = \infty$.

Note that the reflection σ on the affine variety $\mathcal{A}$ amounts to the flip $\sigma : (z_1, z_2, z_3, z_4, z_5) \longmapsto (z_1, -z_2, z_3, -z_4, z_5)$, changing the direction of the commuting vector fields. It can be extended to the (-Id)-involution about the origin of $\mathbb{C}^2$ to the time flip $(t_1, t_2) \longmapsto (-t_1, -t_2)$ on the Abelian variety $\widetilde{\mathcal{A}}$, where t_1 and t_2 are the time coordinates of each of the flows X_{F_1} and X_{F_2}. The involution σ acts on the parameters of the Laurent solution as follows: $\sigma : (t, \alpha, \beta, \gamma, \theta) \longmapsto (-t, -\alpha, -\beta, -\gamma, \theta)$, interchanges the Riemann surfaces $\mathcal{C}_\varepsilon$ and the linear space $\mathcal{L}$ can be split into a direct sum of even and odd functions. Geometrically, this involution interchanges $\mathcal{C}_1$ and $\mathcal{C}_{-1}$, that is, $\mathcal{C}_{-1} = \sigma \mathcal{C}_1$.

8.3. The Hénon–Heiles problem and a five-dimensional system

As discussed in section 7.6 of the Hénon–Heiles study, the present section deals with the case (iii) (see section 7.6). We will briefly study this problem knowing that the method is the same as the one described previously. When we examine all of the possible singularities, we find that it is possible for the variable y_1 to contain square root terms of the type $\sqrt{t}$, which are strictly not allowed by the Painlevé test. However, these terms are trivially removed by introducing some new variables $z_1, \ldots, z_5$, which restores the Painlevé property to the system. And as mentioned above, we obtain a new algebraically completely integrable system. The system [7.27] for case (iii), that is,

$$\dot{x}_1 = y_1, \quad \dot{y}_1 = -Ax_1 - 2x_1 x_2,$$
$$\dot{x}_2 = y_2, \quad \dot{y}_2 = -16Ax_2 - x_1^2 - 16x_2^2, \qquad [8.8]$$

can be written in the form $\dot{u} = J \frac{\partial H}{\partial u}$, $u = (x_1, x_2, y_1, y_2)^\top$, $J = \begin{pmatrix} O & I \\ -I & O \end{pmatrix}$, where

$$H \equiv H_1 = \frac{1}{2}(y_1^2 + y_2^2) + \frac{A}{2}(x_1^2 + 16x_2^2) + x_1^2 x_2 + \frac{16}{3}x_2^3.$$

The functions H_1 and

$$H_2 = 3y_1^4 + 6Ay_1^2 x_1^2 + 12y_1^2 x_1^2 x_2 - 4y_1 y_2 x_1^3 - 4Ax_1^4 x_2 - 4x_1^4 x_2^2 + 3A^2 x_1^4 - \frac{2}{3}x_1^6,$$

commute, that is, $\{H_1, H_2\} = 0$. The second flow commuting with the first is regulated by the equations $\dot{u} = J \frac{\partial H_2}{\partial u}$. The system [8.8] admits Laurent solutions in $\sqrt{t}$,

$$(x_1, x_2, y_1, y_2) = (t^{-1/2}, t^{-2}, t^{-3/2}, t^{-3}) \times \text{a Taylor series in } t,$$

depending on three free parameters : α, β, γ. These formal series solutions are convergent as a consequence of the majorant method. By substituting these series in the constants of the motion $H_1 = b_1$ and $H_2 = b_2$, one eliminates the parameter γ linearly, leading to an equation connecting the two remaining parameters α and β:

$$144\alpha\beta^3 - \frac{294A^2}{5}\alpha^3\beta + \frac{143}{504}\alpha^8 - \frac{4}{21}\alpha^6 + \frac{44}{21}\left(4A^3 - 3b_1\right)\alpha^4 + b_2 = 0$$

which is the equation of an algebraic curve $\mathcal{D}$, along which the $u(t)$ blow up. To be more precise, $\mathcal{D}$ is the closure of the continuous components of $\{$Laurent series solutions $u(t)$ such that $H_k(u(t)) = b_k$, $1 \leq k \leq 2\}$, that is to say, $\mathcal{D} = t^0$ − coefficient of $\{u \in \mathbb{C}^4 : H_1(u(t)) = b_1\} \cap \{u \in \mathbb{C}^4 : H_2(u(t)) = b_2\}$. The invariant variety $\mathcal{A} = \bigcap_{k=1}^{2}\{z \in \mathbb{C}^4 : H_k(z) = b_k\}$ is a smooth affine surface for generic $(b_1, b_2) \in \mathbb{C}^2$. The Laurent solutions restricted to the surface $\mathcal{A}$ are parameterized by the curve $\mathcal{D}$. We show that the system [8.8] is part of a new system of differential equations in five unknowns having one quartic and two cubic invariants (constants of motion). By inspection of the above Laurent expansions, we look for polynomials in (x_1, x_2, y_1, y_2) without fractional exponents. Let $\varphi : \mathcal{A} \longrightarrow \mathbb{C}^5$, $(x_1, x_2, y_1, y_2) \longmapsto (z_1, z_2, z_3, z_4, z_5)$, be a morphism on the affine variety $\mathcal{A}$, where $z_1, \ldots, z_5$ are defined as $z_1 = x_1^2$, $z_2 = x_2$, $z_3 = y_2$, $z_4 = x_1 y_1$, $z_5 = 3y_1^2 + 2x_1^2 x_2$. Using the two first integrals H_1, H_2 and differential equations [8.8], we obtain a system of differential equations in five unknowns,

$$\dot{z}_1 = 2z_4, \qquad \dot{z}_3 = -z_1 - 16A_1 z_2 - 16z_2^2,$$
$$\dot{z}_2 = z_3, \qquad \dot{z}_4 = -A_1 z_1 + \frac{1}{3}z_5 - \frac{8}{3}z_1 z_2, \qquad [8.9]$$
$$\dot{z}_5 = -6A_1 z_4 + 2z_1 z_3 - 8z_2 z_4,$$

having one quartic and two cubic invariants (constants of motion),

$$F_1 = \frac{1}{2}A_1 z_1 + \frac{1}{6}z_5 + 8A_1 z_2^2 + \frac{1}{2}z_3^2 + \frac{2}{3}z_1 z_2 + \frac{16}{3}z_2^3,$$
$$F_2 = 9A_1^2 z_1^2 + z_5^2 + 6A_1 z_1 z_5 - 2z_1^3 - 24A_1 z_1^2 z_2 - 12z_1 z_3 z_4 + 24z_2 z_4^2 - 16z_1^2 z_2^2,$$
$$F_3 = z_1 z_5 - 3z_4^2 - 2z_1^2 z_2.$$

This new system is completely integrable and the Hamiltonian structure is defined by the Poisson bracket $\{F, H\} = \left\langle \frac{\partial F}{\partial z}, J\frac{\partial H}{\partial z} \right\rangle = \sum_{k,l=1}^{5} J_{kl}\frac{\partial F}{\partial z_k}\frac{\partial H}{\partial z_l}$, and

$$J = \begin{pmatrix} 0 & 0 & 0 & 2z_1 & 12z_4 \\ 0 & 0 & 1 & 0 & 0 \\ 0 & -1 & 0 & 0 & -2z_1 \\ -2z_1 & 0 & 0 & 0 & -8z_1 z_2 + 2z_5 \\ -12z_4 & 0 & 2z_1 & 8z_1 z_2 - 2z_5 & 0 \end{pmatrix},$$

is a skew-symmetric matrix for which the corresponding Poisson bracket satisfies the Jacobi identities. The system [8.9] can be written as $\dot{z} = J\frac{\partial H}{\partial z}$, $z = (z_1, z_2, z_3, z_4, z_5)^\top$, where $H = F_1$. The second flow commuting with the first is regulated by the equations $\dot{z} = J\frac{\partial F_2}{\partial z}$, $z = (z_1, z_2, z_3, z_4, z_5)^\top$. These vector fields are in involution, that is, $\{F_1, F_2\} = \left\langle \frac{\partial F_1}{\partial z}, J\frac{\partial F_2}{\partial z} \right\rangle = 0$ and the remaining one is Casimir, that is, $J\frac{\partial F_3}{\partial z} = 0$. The system [8.9] is integrable in the sense of Liouville. The invariant variety $\mathcal{B} = \bigcap_{k=1}^{3} \{z \in \mathbb{C}^5 : F_k(z) = c_k\}$ is a smooth affine surface for generic values of c_1, c_2, c_3. The system [8.9] possesses Laurent series solutions that depend on four free parameters. These meromorphic solutions restricted to the surface $\mathcal{B}$ can be read off from the above expansions and the change of variable φ. Following the methods previously used, we find the compactification of $\mathcal{B}$ into an Abelian surface $\widetilde{\mathcal{B}}$, the system [8.9] is algebraic complete integrable and the corresponding flows evolve on $\widetilde{\mathcal{B}}$. Also, we show (as in the proof of theorem 8.1b)) that the invariant surface $\mathcal{A}$ can be completed as a cyclic double cover $\overline{\mathcal{A}}$ of an Abelian surface $\widetilde{\mathcal{B}}$. The system [8.8] is algebraic complete integrable in the generalized sense. Moreover, $\overline{\mathcal{A}}$ is smooth except at the point lying over the singularity of type A_3 and the resolution $\widetilde{\mathcal{A}}$ of $\overline{\mathcal{A}}$ is a surface of general type. We have shown that the morphism φ maps the vector field [8.8] into an algebraic completely integrable system [8.9] in five unknowns, and the affine variety $\mathcal{A}$ onto the affine part $\mathcal{B}$ of an Abelian variety $\widetilde{\mathcal{B}}$. This explains (among other) why the asymptotic solutions to the differential equations [8.8] contain fractional powers. All of this is summarized as follows:

THEOREM 8.2.– The system [8.8] admits Laurent solutions containing square root terms of the type $\sqrt{t}$, depending on three free parameters, and is algebraic complete integrable in the generalized sense. The morphism φ maps this system into a new algebraic completely integrable system [8.9] in five unknowns.

8.4. The Goryachev–Chaplygin top and a seven-dimensional system

The Goryachev–Chaplygin top mentioned in section 3.2.4 is a rigid body rotating about a fixed point, for which the principal moments of inertia I_1, I_2, I_3 satisfy the relation: $I_1 = I_2 = 4I_3$, the center of mass lies in the equatorial plane through the fixed point and the principal angular momentum is perpendicular to the direction of gravity. The equations of the motion can be written in the form

$$\begin{aligned}
\dot{m}_1 &= 3m_2 m_3, & \dot{\gamma}_1 &= 4m_3\gamma_2 - m_2\gamma_3, \\
\dot{m}_2 &= -3m_1 m_3 - 4\gamma_3, & \dot{\gamma}_2 &= m_1\gamma_3 - 4m_3\gamma_1, \\
\dot{m}_3 &= 4\gamma_2, & \dot{\gamma}_3 &= m_2\gamma_1 - m_1\gamma_2,
\end{aligned} \quad [8.10]$$

where $m_1, m_2, m_3, \gamma_1, \gamma_2, \gamma_3$ are the coordinates of the phase space. The following four quadrics are constants of motion for this system:

$$H_1 = m_1^2 + m_2^2 + 4m_3^2 - 8\gamma_1 = 6b_1, \quad H_2 = (m_1^2 + m_2^2)m_3 + 4m_1\gamma_3 = 2b_2,$$

$$H_3 = \gamma_1^2 + \gamma_2^2 + \gamma_3^2 = b_3, \quad H_4 = m_1\gamma_1 + m_2\gamma_2 + m_3\gamma_3 = 0,$$

for generic $b_1, b_2, b_3 \in \mathbb{C}$. This system is completely integrable, and H_1 (energy) and H_4 are in involution, while H_2, H_3 are Casimir invariants. The Goryachev–Chaplygin system has asymptotic solutions that are meromorphic in $\sqrt{t}$,

$$(m_1, m_2, m_3, \gamma_1, \gamma_2, \gamma_3) = (t^{-3/2}, t^{-3/2}, t^{-1}, t^{-2}, t^2, t^{-1/2}) \times \text{a Taylor series in } t,$$

depending on four free parameters b_1, b_2, b_3 and u, v. Let $\mathcal{A}$ be the affine variety defined by

$$\mathcal{A} = \{x : H_1(x) = 6b_1, H_2(x) = 2b_2, H_3(x) = b_3, H_4(x) = 0\}, \qquad [8.11]$$

where $x = (m_1, m_2, m_3, \gamma_1, \gamma_2, \gamma_3)$. These solutions restricted to $\mathcal{A}$ are parameterized by two copies $\mathcal{C}_{\varepsilon=+i}$ and $\mathcal{C}_{\varepsilon=-i}$ of the curve $\mathcal{C}$ of genus 4:

$$\mathcal{C} : 16b_3 u^4 + \varepsilon u^2 (b_2 + 6b_1 v - 16v^3) - v^2 = 0.$$

The asymptotic solutions of the system [8.10] contain fractional powers, that is, square root terms of the type $\sqrt{t}$, which are strictly not allowed by the Painlevé test, but the new variables $(z_1, z_2, z_3, z_4, z_5, z_6, z_7)$ defined by $z_1 = m_1^2 + m_2^2$, $z_2 = m_3$, $z_3 = \gamma_3^2$, $z_4 = \gamma_1$, $z_5 = \gamma_2$, $z_6 = m_1\gamma_3$, $z_7 = m_2\gamma_3$ restore the Painlevé property to the system. *These variables are easily obtained by a simple inspection of the Laurent series above; we use the first terms of the Laurent series and it is generally sufficient to choose combinations (often obvious) of the initial variables so as not to have terms of the type $\sqrt{t}$.* Now let $\varphi : \mathcal{A} \longrightarrow \mathbb{C}^7$, $(m_1, m_2, m_3, \gamma_1, \gamma_2, \gamma_3) \longmapsto (z_1, z_2, z_3, z_4, z_5, z_6, z_7)$, be a morphism on the affine variety $\mathcal{A}$. These affine variables were originally used in Bechlivanidis and van Moerbeke (1987) without any discussion of their origin and algebraic properties. The morphism φ maps the vector field [8.10] into the system in seven unknowns $(z_1, z_2, z_3, z_4, z_5, z_6, z_7) \in \mathbb{C}^7$ (Bechlivanidis and van Moerbeke 1987),

$$\dot{z}_1 = -8z_7, \quad \dot{z}_2 = 4z_5, \quad \dot{z}_3 = 2(z_4 z_7 - z_5 z_6),$$
$$\dot{z}_4 = 4z_2 z_5 - z_7, \quad \dot{z}_5 = z_6 - 4z_2 z_4, \qquad [8.12]$$
$$\dot{z}_6 = -z_1 z_5 + 2z_2 z_7, \quad \dot{z}_7 = z_1 z_4 - 2z_2 z_6 - 4z_3,$$

having five quadrics invariants

$$F_1 = z_1 - 8z_4 + 4z_2^2 = 6c_1, \quad F_2 = z_1 z_2 + 4z_6 = 2c_2, \quad F_3 = z_3 + z_4^2 + z_5^2 = c_3,$$

$$F_4 = z_2 z_3 + z_4 z_6 + z_5 z_7 = c_4, \qquad F_5 = z_6^2 + z_7^2 - z_1 z_3 = c_5,$$

where c_1, c_2, c_3, c_4, c_5 are generic constants. To obtain these invariants, we used the first integrals H_1, H_2, H_3, H_4 and differential equations [8.10]. This system is completely integrable and the symplectic structure is defined by the Poisson bracket $\{F, H\} = \langle \frac{\partial F}{\partial z}, J \frac{\partial H}{\partial z} \rangle$, where

$$J = \begin{pmatrix} 0 & 0 & -A & z_7 & -z_6 & B & -C \\ 0 & 0 & 0 & -\frac{1}{2} z_5 & \frac{1}{4} z_4 & -\frac{1}{2} z_7 & \frac{1}{2} z_6 \\ A & 0 & 0 & 0 & 0 & -z_3 z_5 & z_3 z_4 \\ -z_7 & \frac{1}{2} z_5 & 0 & 0 & 0 & 0 & -\frac{1}{2} z_3 \\ z_6 & -\frac{1}{2} z_4 & 0 & 0 & 0 & \frac{1}{2} z_3 & 0 \\ -B & \frac{1}{2} z_7 & z_3 z_5 & 0 & -\frac{1}{2} z_3 & 0 & -z_2 z_3 \\ C & -\frac{1}{2} z_6 & -z_3 z_4 & \frac{1}{2} z_3 & 0 & z_2 z_3 & 0 \end{pmatrix},$$

$$A \equiv 2 z_4 z_7 - 2 z_5 z_6, \qquad B \equiv z_1 z_5 + 2 z_2 z_7, \qquad C \equiv z_1 z_4 + 2 z_2 z_6,$$

is a skew-symmetric matrix whose elements polynomial satisfy the Jacobi identity. The system [8.12] is written in the form $\dot{z} = J \frac{\partial H}{\partial z}$, where $H = F_1$ and $x = (x_1, x_2, x_3, x_4, x_5, x_6, x_7)^\top$. The two first integrals F_1 and F_2 are in involution, that is, $\{F_1, F_2\} = 0$, while F_3, F_4 and F_5 are Casimir invariants, that is, $J \frac{\partial F_k}{\partial x} = 0$, where $k = 3, 4, 5$. The first fact to observe is that if the system [8.12] is to have Laurent solutions depending on six free parameters, the Laurent decomposition of such asymptotic solutions must have the following form: $z_i = \sum_{k=0}^{\infty} z_i^{(k)} t^{k-1}$ for $i = 1, 2, 3$ and $z_i = \sum_{k=0}^{\infty} z_i^{(k)} t^{k-2}$ for $i = 4, 5, 6, 7$. By putting these expansions into the five quadrics invariants above, solving inductively for the $z_i^{(k)}$, we explicitly find the following Laurent solutions depending on six free parameters:

$$z_1 = -\frac{2\varepsilon\alpha}{t} + 2\alpha^2 - 2\varepsilon(\alpha(\alpha^2 - 2c_1) + \zeta)t - (2\xi + \zeta)\alpha t^2 + o(t^3),$$

$$z_2 = -\frac{\varepsilon}{2t} - \frac{\alpha}{2} - \frac{\varepsilon}{2}(\alpha^2 - 2c_1)t - \frac{1}{4}(2\xi + \zeta)\alpha t^2 + o(t^3),$$

$$z_3 = \frac{\varepsilon}{8t}(\xi + \zeta) + \frac{3\alpha}{8}(\xi + \zeta) - \frac{\varepsilon}{8}((5\alpha^2 - c_1)(\xi + \zeta) - 8(2c_3\alpha + c_4))t + o(t^2),$$

$$z_4 = -\frac{1}{8t^2} + \frac{1}{8}(\alpha^2 - 2c_1) + \frac{\varepsilon}{8}(2\xi + \zeta)t + o(t^2), \qquad [8.13]$$

$$z_5 = \frac{\varepsilon}{8t^2} - \frac{\varepsilon}{8}(\alpha^2 - 2c_1) - \frac{1}{8}(2\xi + 3\zeta)t + o(t^2),$$

$$z_6 = \frac{\alpha}{4t^2} + \frac{1}{4}(2\xi - (\alpha^2 - 2c_1)\alpha + \zeta) - \frac{\varepsilon\alpha}{4}(2\xi + 3\zeta)t + o(t^2),$$

$$z_7 = -\frac{\varepsilon\alpha}{4t^2} + \frac{\varepsilon}{4}(\alpha(\alpha^2 - 2c_1) + \zeta) + \frac{\alpha}{4}(2\xi + \zeta)t + o(t^2),$$

where $\varepsilon = \pm i$, $\xi(\alpha) = 2\alpha^3 - 3c_1\alpha + c_2$ and the parameters α, ζ belong to a genus 2 hyperelliptic curve,

$$\mathcal{H}: \zeta^2 = (2\alpha^3 - 3c_1\alpha + c_2)^2 - 4(4c_3\alpha^2 + 4c_4\alpha + c_5). \qquad [8.14]$$

Using the majorant method, we can show that the formal Laurent series solutions are convergent. In fact, the Laurent solutions are parameterized by two copies $\mathcal{H}_{+i}$ and $\mathcal{H}_{-i}$ of the genus 2 hyperelliptic curve $\mathcal{H}$ for $\varepsilon = \pm i$. In order to embed $\mathcal{H}$ into some projective space, we search for functions $z_0, z_1, ..., z_N$ of increasing degree in the original variables, having the property that the embedding $\mathcal{D}$ of $\mathcal{H}_i + \mathcal{H}_{-i}$ into $\mathbb{P}^N(\mathbb{C})$ via those functions satisfies the relation geometric genus$(2\mathcal{D}) = N + 2$. As in the previous chapter, we show that this occurs for the first time for $N = 15$, that is, the embedding $\mathcal{D}$ of $\mathcal{H}_i + \mathcal{H}_{-i}$ into $\mathbb{P}^{15}(\mathbb{C})$ is done via the original functions $z_0, z_1, ..., z_7$ enlarged by adjoining the following eight other functions: $z_8 \equiv z_2 z_3$, $z_9 \equiv z_1 z_3$, $z_{10} \equiv z_1 z_4 + 2 z_2 z_6$, $z_{11} \equiv z_1 z_5 + 2 z_2 z_7$, $z_{12} \equiv -z_5 z_6 + z_4 z_7$, $z_{13} \equiv z_2 z_{12} - 2 z_3 z_5$, $z_{14} \equiv z_3^2$, $z_{15} \equiv z_1 z_{12} + 4 z_3 z_7$. Using these functions, we embed the curves $\mathcal{H}_i$ and $\mathcal{H}_{-i}$ into a hyperplane of $\mathbb{P}^{15}(\mathbb{C})$. Thus embedded, these curves have one point in common at which they are tangent to each other.

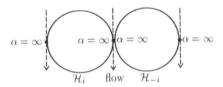

Figure 8.1. Curves $\mathcal{H}_{\pm i}$

In the neighborhood of $\alpha = \infty$, the curve $\mathcal{H}$ has two points at which $\xi + \zeta$ behaves as follows: $\xi + \zeta = 4\alpha^3 + o(\alpha)$, picking the $+$ sign for ζ and $\xi + \zeta = \frac{4c_3\alpha^2 + 4c_4\alpha + c_5}{\alpha^3} +$ lower order terms, and picking the $-$ sign for ζ. So when choosing the $+$ sign for ζ and dividing the vector $(1, z_1, ..., z_{15})$ by $z_{14} = z_3^2$, the corresponding point is sent to the point $[0 : \cdots : 1 : 0] \in \mathbb{P}^{15}(\mathbb{C})$, which is independent of ε. The choice of the sign $-$ for ζ conducts to two different points, taking into account the sign of ε. Therefore, the divisor $\mathcal{D}$ obtained in this way has genus 5 and $2\mathcal{D}$ has genus 17, satisfying the relation: geometric genus of $2\mathcal{D} = N + 2$, that is, $2\mathcal{D} \subset \mathbb{P}^{15}(\mathbb{C}) = \mathbb{P}^{g-2}(\mathbb{C})$. Following the method explained and used in detail in the previous chapter, we show that the affine surface

$$\mathcal{B} = \bigcap_{k=1}^{5} \{z = (z_1, z_2, z_3, z_4, z_5) : F_k(z) = c_k\} \subset \mathbb{C}^7, \qquad [8.15]$$

can be completed into an Abelian surface $\widetilde{\mathcal{B}}$ by adjoining the divisor $\mathcal{D} = \mathcal{H}_i + \mathcal{H}_{-i}$ at infinity. The variety $\widetilde{\mathcal{B}}$ is equipped with two commuting, linearly independent vector fields. Let dt_1, dt_2 be two holomorphic 1-forms on $\widetilde{\mathcal{B}}$ corresponding, respectively, to the vector fields X_{F_1}, X_{F_2}. By letting $y_1 = \frac{x_1}{x_2}$, $y_2 = \frac{1}{x_2}$, we obtain the forms

$$\omega_1 = dt_1|_{\mathcal{H}_\varepsilon} = \frac{1}{\Delta}\left(\frac{\partial y_1}{\partial t_2}dy_2 - \frac{\partial y_2}{\partial t_2}dy_1\right) = \frac{a\alpha}{\zeta}d\alpha,$$

$$\omega_2 = dt_2|_{\mathcal{H}_\varepsilon} = -\frac{1}{\Delta}\left(\frac{\partial y_1}{\partial t_1}dy_2 - \frac{\partial y_2}{\partial t_1}dy_1\right) = \frac{b}{\zeta}d\alpha,$$

where a and b are constants and $\Delta = \frac{\partial y_2}{\partial t_1}\frac{\partial y_1}{\partial t_2} - \frac{\partial y_2}{\partial t_2}\frac{\partial y_1}{\partial t_1}$. The points where the vector field X_{F_1} is tangent to the curves $\mathcal{H}_i$ and $\mathcal{H}_{-i}$ on $\widetilde{\mathcal{B}}$ are provided by the zeros of the form ω_2. Note that X_{F_1} is tangent to $\mathcal{H}_i$ and $\mathcal{H}_{-i}$ at the point where both curves touch; this point corresponds to $\alpha = \infty$. The involution $\sigma : (z_1, z_2, z_3, z_4, z_5, z_6, z_7) \longmapsto (z_1, z_2, z_3, z_4, -z_5, z_6, -z_7)$ on the variety $\mathcal{B}$ acts on the free parameters (keeping the same notation) as follows: $\sigma : (t, \alpha, \zeta, \varepsilon, c_1, c_2, c_3, c_4, c_5) \longmapsto (-t, \alpha, \zeta, -\varepsilon, c_1, c_2, c_3, c_4, c_5)$. Therefore, we have $\mathcal{H}_i = \sigma\mathcal{H}_{-i}$ and geometrically, this means that $\mathcal{H}_i$ and $\mathcal{H}_{-i}$ are deduced from one another by a translation in the Abelian variety $\widetilde{\mathcal{B}}$. Following the same reasoning used in the proof of theorem 8.1b), we show that the invariant variety $\mathcal{A}$ [8.11] can be completed as a cyclic double cover $\overline{\mathcal{A}}$ of the Jacobian of a genus two curve, and $\overline{\mathcal{A}}$ is smooth except at the point (tacnode) lying over the singularity of type A_3. Therefore, we have the following result (Bechlivanidis and van Moerbeke 1987; Piovan 1992):

THEOREM 8.3.– a) The system [8.12] is algebraically completely integrable. The Laurent solution [8.13] depends on six free parameters. The affine surface $\mathcal{B}$ [8.15] completes into an Abelian variety $\widetilde{\mathcal{B}}$ by adjoining a divisor $\mathcal{H}_i + \mathcal{H}_{-i}$, where $\mathcal{H}_{+i}$ and $\mathcal{H}_{-i}$ are two copies of the same genus 2 hyperelliptic curve $\mathcal{H}$ [8.14] for $\varepsilon = \pm 1$ that intersect each other in a tacnode belonging to $\mathcal{H}_i + \mathcal{H}_{-i}$.

b) The invariant variety $\mathcal{A}$[8.11] of the Goryachev–Chaplygin top can be compactified as a cyclic double cover $\overline{\mathcal{A}}$ of the Jacobian of a genus 2 curve, ramified along the divisor $\mathcal{H}_i + \mathcal{H}_{-i}$. Moreover, $\overline{\mathcal{A}}$ is smooth except at the point (tacnode) lying over the singularity (of type A_3) of $\mathcal{H}_i + \mathcal{H}_{-i}$, and the resolution $\widetilde{\mathcal{A}}$ of $\overline{\mathcal{A}}$ is a surface of general type with invariants: Euler characteristic of $\widetilde{\mathcal{A}} = \mathcal{X}(\widetilde{\mathcal{A}}) = 1$ and geometric genus of $\widetilde{\mathcal{A}} = p_g(\widetilde{\mathcal{A}}) = 2$. The system [8.10] is algebraic completely integrable in the generalized sense.

The extended system [8.12] includes some other known integrable systems. It is shown (Bechlivanidis and van Moerbeke 1987) that the system [8.10] is rationally related to the three-body Toda system.

8.5. The Lagrange top

We show that the equations governing the motion of the Lagrange top (see sections 3.2.2 and 4.7) form an algebraic completely integrable system in the generalized sense. These equations are explicitly written in the form

$$\lambda_1 \dot{m}_1 = \lambda_1(\lambda_3 - \lambda_1)m_2 m_3 - \gamma_2, \qquad \dot{\gamma}_1 = \lambda_3 m_3 \gamma_2 - \lambda_1 m_2 \gamma_3,$$
$$\lambda_1 \dot{m}_2 = \lambda_1(\lambda_1 - \lambda_3)m_1 m_3 + \gamma_1, \qquad \dot{\gamma}_2 = \lambda_1 m_1 \gamma_3 - \lambda_3 m_3 \gamma_1,$$
$$\dot{m}_3 = 0, \qquad \dot{\gamma}_3 = \lambda_1(m_2 \gamma_1 - m_1 \gamma_2).$$

This system admits the following four first integrals:

$$H_1 = \frac{\lambda_1^2}{2}(m_1^2 + m_2^2) + \frac{\lambda_1 \lambda_3}{2}m_3^2 - \gamma_3, \qquad H_2 = \gamma_1^2 + \gamma_2^2 + \gamma_3^2,$$
$$H_3 = \lambda_1(m_1 \gamma_1 + m_2 \gamma_2 + m_3 \gamma_3), \qquad H_4 = \lambda_3 m_3$$

and is integrable in the sense of Liouville. The Poisson structure is given by $\{m_i, m_j\} = -\epsilon_{ijk} m_k$, $\{m_i, \gamma_j\} = -\epsilon_{ijk} \gamma_k$, $\{\gamma_i, \gamma_j\} = 0$, where $1 \leq i, j, k \leq 3$ and ϵ_{ijk} is the total antisymmetric tensor for which $\epsilon_{ijk} = 1$. Let

$$\mathcal{M}_c = \left\{(m_1, m_2, m_3, \gamma_1, \gamma_2, \gamma_3) \in \mathbb{C}^6 : H_1 = c_1, H_2 = 1, H_3 = c_3, H_4 = c_4\right\},$$

be the affine variety defined by the intersection of the constants of the motion and let $\mathbb{C}^* \sim \mathbb{C}/2\pi i \mathbb{Z}$ be the group of rotations defined by the flow of the vector field generated by H_4, that is, $\dot{m}_1 = m_2$, $\dot{m}_2 = -m_1$, $\dot{m}_3 = 0$, $\dot{\gamma}_1 = \gamma_2$, $\dot{\gamma}_2 = -\gamma_1$, $\dot{\gamma}_3 = 0$. The quotient $\mathcal{M}_c/\mathbb{C}^*$ is an elliptic curve. We show that the algebraic variety $\mathcal{M}_c$ is not isomorphic to the direct product of the curve $\mathcal{M}_c/\mathbb{C}^*$ and $\mathbb{C}^*$. For generic constants c_j, the complex invariant manifold $\mathcal{M}_c$ is biholomorphic to an affine subset of $\mathbb{C}^2/\Lambda$ where $\Lambda \subset \mathbb{C}^2$ is a lattice of rank 3,

$$\Lambda = \mathbb{Z}\begin{pmatrix} 2\pi i \\ 0 \end{pmatrix} \oplus \mathbb{Z}\begin{pmatrix} 0 \\ 2\pi i \end{pmatrix} \oplus \mathbb{Z}\begin{pmatrix} \tau_1 \\ \tau_2 \end{pmatrix}, \quad \text{Re}(\tau_1) < 0.$$

Hence, $\mathbb{C}^2/\Lambda$ is an non-compact algebraic group and can be considered as a non-trivial extension of the elliptic curve $\mathbb{C}/\{2\pi i \mathbb{Z} \oplus \tau_1 \mathbb{Z}\}$ by $\mathbb{C}^* \sim \mathbb{C}/2\pi i \mathbb{Z}$,

$$0 \longrightarrow \mathbb{C}/2\pi i \mathbb{Z} \longrightarrow \mathbb{C}^2/\Lambda \overset{\varphi}{\longrightarrow} \mathbb{C}/\{2\pi i \mathbb{Z} \oplus \tau_1 \mathbb{Z}\} \longrightarrow 0, \quad \varphi(z_1, z_2) = z_1.$$

The algebraic group $\mathbb{C}/2\pi i \mathbb{Z}$ is the generalized Jacobian of an elliptic curve with two points identified at infinity. We have the following result (Gavrilov and Angel Zhivkov 1998):

THEOREM 8.4.– *The differential system governing the Lagrange top form an algebraic completely integrable system in the generalized sense.*

8.6. Exercises

EXERCISE 8.1.– Let $\mathcal{A}$ be the complexified invariant surface of a generalized algebraic completely integrable system, with $\mathcal{A}$ being an unramified m-covering of the affine part of an Abelian variety $\mathcal{B}$. Assume that the system possesses Laurent solutions $z(\tau, p)$, depending on the parameter $\tau = t^{\frac{1}{m}}$, ($p \in \mathcal{D}$, where $\mathcal{D}$ is a generalized "Painlevé" divisor going with such solutions), and a $(r-1)$-dimensional family of divisors. Moreover, assume that there is a cyclic automorphism of $\mathcal{A}$ exchanging sheets and that there is a properly discontinuous action of the cyclic group μ_m of m elements on the space $U_\varepsilon = \{|\tau| < \varepsilon\} \times \mathcal{D}$, which is free on $U_\varepsilon \backslash \mathcal{D}$ and leaves the Laurent expansions invariant.

a) Show (Piovan 1992) that the degree m holomorphic map

$$\psi_\varepsilon = (\tau, p) \in \{\tau \in \mathbb{C}, 0 < |\tau| < \varepsilon\} \times \mathcal{D} = U_\varepsilon \backslash \mathcal{D} \longrightarrow z(\tau, p) \in \mathcal{A},$$

is full rank and induces an isomorphism $\psi_\varepsilon : (U_\varepsilon \backslash \mathcal{D})|_{\mu_\varepsilon} \longrightarrow \psi_\varepsilon(U_\varepsilon \backslash \mathcal{D}) \subset \mathcal{A}$.

b) Show (Piovan 1992) that $\mathcal{A}$ is birationally equivalent to an m-cyclic covering of $\mathcal{B}$ and the coordinate functions extend meromorphically.

EXERCISE 8.2.– Let $\mathcal{A}$ be a m-cyclic cover of an Abelian surface $\mathcal{B}$ ramified along the smooth reduced divisor $\mathcal{D}$, such that $\mathcal{D}$ is linearly equivalent to $m\mathcal{S}$, for some ample divisor $\mathcal{S}$. Show (Piovan 1992) that $\mathcal{A}$ is a surface of general type.

EXERCISE 8.3.– Let $\mathcal{A}$ be a m-cyclic cover of an Abelian surface $\mathcal{B}$ ramified along the reduced divisor D. Assume D is linearly equivalent to $m\mathcal{S}$, for some ample divisor $\mathcal{S}$. Show (Piovan 1992) that $\mathcal{A}$ has the following invariants:

a) The genus $p_g(\mathcal{A})$ of $\mathcal{A}$ is $p_g(\mathcal{A}) = 1 + \frac{\theta^2}{12}(m-1)m(2m-1)$.

b) $\mathcal{X}(\mathcal{A}) = \frac{\theta^2}{12}(m-1)m(2m-1)$, where $\mathcal{X}(\mathcal{A})$ is the holomorphic Euler characteristic $\mathcal{X}(\mathcal{A})$ of $\mathcal{A}$.

EXERCISE 8.4.– Let X be a double cover of an Abelian surface Y ramified along the divisor B. We assume that B only has simple singularities, that is, double or triple points of type $A - D - E$, and B is linearly equivalent to 2θ, for some ample divisor θ. Then, a resolution of singularities of X is a surface of general type with invariants: $\mathcal{X}(X) = \frac{\theta^2}{12}$, $p_g(X) = 1 + \frac{\theta^2}{12}$.

EXERCISE 8.5.– We consider the differential system described by the Hamiltonian (Roekaerts 1987):

$$H = \frac{1}{2}\left(y_1^2 + y_2^2\right) + y_1 x_1 x_2 + y_2 \left(\frac{1}{8}x_1^2 + 2x_2^2\right).$$

Is this system algebraically completely integrable in the generalized sense? A second first integral is provided by

$$F = y_1^4 + y_1^2 x_1 \left(4y_1 x_2 - \frac{3}{2} y_2 x_1\right) + x_1^3 \left(-\frac{1}{2} y_1^2 x_1 - y_1 y_2 x_2 + \frac{1}{16} y_2^2 x_1\right).$$

EXERCISE 8.6.– Same question for the following Hamiltonian (Hietarinta 1985):

$$H = \frac{1}{2}\left(y_1^2 + y_2^2\right) + y_1 x_1 x_2 + y_2 \left(-\frac{1}{4}x_1^2 + \frac{1}{2}x_2^2\right).$$

A second first integral is provided by $F = \left(2y_1^2 + y_2^2\right)\left(2y_2 + x_1^2 + 2x_2^2\right)^2$.

EXERCISE 8.7.– Same question for the following Hamiltonian (Roekaerts 1987):

$$H = \frac{1}{2}\left(y_1^2 + y_2^2\right) + y_1 x_1 x_2 - y_2 x_2^2.$$

A second first integral is provided by $F = y_1^2 \left(2y_2 + x_1^2\right)$.

EXERCISE 8.8.– Consider the following Hamiltonian:

$$H_1 = \frac{1}{2}(p_1^2 + p_2^2) + \frac{a}{2}(q_1^2 + 4q_2^2) + \frac{1}{4}q_1^4 + 4q_2^4 + 3q_1^2 q_2^2,$$

where a is a constant. The corresponding system, that is,

$$\ddot{q}_1 = -(a + q_1^2 + 6q_2^2)q_1, \qquad \ddot{q}_2 = -2(2a + 3q_1^2 + 8q_2^2)q_2, \qquad [8.16]$$

is integrable and the second integral is

$$H_2 = aq_1^2 q_2 + q_1^4 q_2 + 2q_1^2 q_2^3 - q_2 p_1^2 + q_1 p_1 p_2.$$

As mentioned in section 4.8, this case can be deduced from the paper (Dorizzi et al. 1983). Let $\mathcal{A} = \bigcap_{k=1}^{2}\{z = (q_1, q_2, p_1, p_2) \in \mathbb{C}^4 : H_k(z) = b_k\}$ be the invariant surface defined (for generic $(b_1, b_2) \in \mathbb{C}^2$) by the two constants of motion.

a) Show that the system [8.16] possesses a three-dimensional family of Laurent solutions (depending on three free parameters)

$$(q_1, q_2, p_1, p_2) = \left(t^{-1/2}, t^{-1}, t^{-3/2}, t^{-2}\right) \times \text{a Taylor series},$$

and moreover, these solutions restricted to the surface $\mathcal{A}$ are parameterized by two smooth curves of genus 4.

b) Show that the system of differential equations [8.16] can be written as follows:

$$\frac{ds_1}{\sqrt{P_6(s_1)}} - \frac{ds_2}{\sqrt{P_6(s_2)}} = 0, \qquad \frac{s_1 ds_1}{\sqrt{P_6(s_1)}} - \frac{s_2 ds_2}{\sqrt{P_6(s_2)}} = dt,$$

where $P_6(s) = s\left(-8s^5 - 4as^3 + 2b_1 s + b_2\right)$, and the flow can be linearized in terms of genus 2 hyperelliptic functions.

c) Let $\varphi : A \longrightarrow \mathbb{C}^5$, $(q_1, q_2, p_1, p_2) \longmapsto (z_1, z_2, z_3, z_4, z_5)$, be a morphism on the affine variety A, where $z_1, \ldots, z_5$ are defined as $z_1 = q_1^2$, $z_2 = q_2$, $z_3 = p_2$, $z_4 = q_1 p_1$, $z_5 = 2q_1^2 q_2^2 + p_1^2$. Show that this morphism maps the vector field [8.16] into the system [7.57] in five unknowns z_1, z_2, z_3, z_4, z_5 (we have seen in exercise 7.9, that the system [7.57] possesses three quartic invariants F_1, F_2, F_3 [7.58] and the affine variety $\mathcal{B} = \bigcap_{k=1}^{3}\{z : F_k(z) = c_k\} \subset \mathbb{C}^5$, defined by putting these invariants equal to generic constants, is the affine part of an Abelian surface $\widetilde{\mathcal{B}}$, more precisely the Jacobian of a genus 2 curve and the system of differential equations [7.57] is algebraic complete integrable and the corresponding flows evolve on $\widetilde{\mathcal{B}}$).

d) Show that the invariant surface $\mathcal{A}$ can be completed as a cyclic double cover $\overline{\mathcal{A}}$ of the Abelian surface $\widetilde{\mathcal{B}}$ and the system [8.16] is algebraic complete integrable in the generalized sense. Also analyze the resolution $\widetilde{\mathcal{A}}$ of $\overline{\mathcal{A}}$ as a surface of general type with invariants to be specified (hint: see Lesfari (2020)).

9

The Korteweg–de Vries Equation

The Korteweg–de Vries (KdV) equation, which is a nonlinear partial differential equation of the third order, is a universal mathematical model for the description of weakly nonlinear long wave propagation in dispersive media. It is a most remarkable nonlinear partial differential equation in $1 + 1$ dimensions whose solutions can be exactly specified; it has a soliton-like solution or a solitary wave of sech^2 form. Various physical systems of dispersive waves admit solutions in the form of generalized solitary waves. The study of this equation is the archetype of an integrable system and is one of the most fundamental equations of soliton phenomena and a topic of active mathematical research. Our purpose here is to give a motivated and brief overview of this interesting subject. One of the objectives of this chapter is to study the KdV equation and the inverse scattering method (based on Schrödinger and Gelfand–Levitan equations) used to solve it.

9.1. Historical aspects and introduction

Korteweg and de Vries established a nonlinear partial differential equation describing the gravitational wave propagating in a shallow channel (Korteweg and de Vries 1895) and possessing remarkable mathematical properties:

$$\frac{\partial u}{\partial t} - 6u\frac{\partial u}{\partial x} + \frac{\partial^3 u}{\partial x^3} = 0, \qquad [9.1]$$

where $u(x, t)$ is the amplitude of the wave at the point x and the time t. The equation thus bearing their name (abbreviated KdV) presents a solution: the soliton or solitary wave. This model was obtained from Euler's equations (assuming irrotational flow) by Boussinesq around 1877 (see Boussinesq 1877, p. 360) and rediscovered by Korteweg and de Vries in 1890. The solution to this equation was obtained and interpreted rigorously only in the early 1970s, even though a solitary wave had already been observed in 1834 by the engineer Scott-Russell riding along the

Edinburgh Glasgow Canal in Scotland; he described his observation of a hydrodynamic phenomenon as follows: "I was observing the motion of a boat which was rapidly drawn along a narrow channel by a pair of horses, when the boat suddenly stopped – not so the mass of water in the channel which it had put in motion; it accumulated round the prow of the vessel in a state of violent agitation, then suddenly leaving it behind, rolled forward with great velocity, assuming the form of a large solitary elevation, a rounded, smooth and well-defined heap of water, which continued its course along the channel apparently without change of form or diminution of speed. I followed it on horseback, and overtook it still rolling on at a rate of some eight or nine miles an hour, preserving its original figure some 30-feet long and a foot to a foot and a half in height. Its height gradually diminished, and after a chase of one or two miles I lost it in the windings of the channel. Such, in the month of August 1834, was my first chance interview with that singular and beautiful phenomenon which I have called the Wave of Translation". Fascinated by this phenomenon, Scott-Russell built a wave pool in his garden and worked to generate and study these waves more carefully. This led to a paper (Scott-Russell 1844) named "The report on waves" published in 1844 by the British Association for the Advancement of Science. A little later, Boussinesq, then Korteweg and de Vries, proposed equation [9.1] to explain this phenomenon. The KdV equation preserves mass, momentum, energy and many other quantities. Many experiments have uncovered the astonishing properties of the solutions of this equation satisfying zero boundary conditions: when $|t| \longrightarrow \infty$, these solutions are decomposed into solitons, that is, in to waves of defined forms progressing at different speeds. These waves propagate over long distances without deformation and one of the remarkable characteristics of solitons is that they are exceptionally stable with respect to disturbances; the term $u\frac{\partial u}{\partial x}$ leads to shock waves while the term $\frac{\partial^3 u}{\partial x^3}$ produces a scattering effect. Everyone can contemplate solitons where the tide comes to die on the beaches. In the field of hydrodynamics, for example, tsunamis (tidal waves) are manifestations of solitons. Generally, we group together under the term soliton, solutions of nonlinear wave equations presenting the following characteristic properties: they are localized in space, last indefinitely and retain their amplitude and velocity even at the end of several collisions with other solitons. Solitons have become indispensable for the study of several phenomena, in particular, the study of wave propagation in hydrodynamics and the study of localized waves in astrophysical plasmas. They are involved in the study of signals in optical fibers, charge transport phenomena in conductive polymers, localized modes in magnetic crystals, etc. Industrialized societies have developed, after soliton studies, what may be called solitary lasers. The latter play an important role in the field of telecommunications. Ultra-short light signals sent in certain optical fibers made from a specific material can travel long distances without lengthening or fading. The construction of memories with ultra-fast communication time and low energy consumption is based on the movement of magnetic vortices in the dielectric junction between two superconductors. At the molecular level, the theory of solitons can elucidate the contraction mechanism of striated muscles, the dynamics of biological macromolecules such as DNA and proteins. In the peptide and hydrogen chain of proteins, solitons arise from the marriage of dispersion due to intrapeptide vibrations and the nonlinearity due to the interaction of these vibrations with the displacements of peptide groups around their position being balanced. But the theory of solitons has also had an impact on pure mathematics; for example, it provides the answer to the famous Schottky problem, posited a century ago, on the relationships between the periods coming from a Riemann surface. Roughly, it is a question of finding criteria so that a matrix of the periods belonging to the Siegel half-space is the matrix of the periods of a Riemann surface. Geometrically, Schottky's problem consists of characterizing the Jacobians among all the Abelian mainly polarized varieties. In addition to the KdV equation, examples that may be mentioned among the nonlinear equations having soliton-type solutions are as follows: the nonlinear Kadomtsev–Petviashvili equation: $\frac{\partial^2 u}{\partial y^2} - \frac{\partial}{\partial x}\left(4\frac{\partial u}{\partial t} - 12u\frac{\partial u}{\partial x} - \frac{\partial^3 u}{\partial x^3}\right) = 0$, the nonlinear Schrödinger equation: $i\frac{\partial \psi}{\partial t} + \frac{\partial^2 \psi}{\partial x^2} + |\psi|^2\psi = 0$, the Sine Gordon equation: $\frac{\partial^2 u}{\partial t^2} - \frac{\partial^2 u}{\partial x^2} + \sin u = 0$, the Boussinesq equation: $\frac{\partial^2 u}{\partial t^2} - \frac{\partial^2 u}{\partial x^2} + \frac{\partial^4 u}{\partial x^4} + \frac{\partial^2 u^2}{\partial x^2} = 0$, the Camassa–Holm equation: $\frac{\partial u}{\partial t} - \frac{\partial^3 u}{\partial t \partial x^2} + 3u\frac{\partial u}{\partial x} = 2\frac{\partial u}{\partial x}\frac{\partial^2 u}{\partial x^2} + u\frac{\partial^3 u}{\partial x^3} - 2\alpha\frac{\partial u}{\partial x}$, $\alpha \in \mathbb{R}$, the Toda lattice (previously studied) described by a system: $\frac{dx_j}{dt} = y_j$, $\frac{dy_j}{dt} = -e^{x_j - x_{j+1}} + e^{x_{j-1} - x_j}$, consisting of vibrating masses arranged on a circle and interconnected by springs whose return force is exponential. Solitons have appeared in many other fields; in particular the nonlinear Klein Gordon equation, the Zabusky–Kruskal equation for the Fermi–Pasta–Ulam model of phonons in anharmonic lattice, and so on.

9.2. Stationary Schrödinger and integral Gelfand–Levitan equations

Since the method (used later) of solving the KdV equation is based on the idea of studying it in the form of an equation of a certain operator and using the analogy with quantum mechanics, we will expose certain mathematical notions of this mechanic. The terminology of physicists will be used to describe the properties of the solutions of the stationary Schrödinger equation,

$$\frac{\hbar}{2m}\psi'' + (\lambda - u(x))\psi = 0, \quad ' \equiv \frac{d}{dx},$$

without stopping on the physical motivations of the introduced notions. We will see that the method of the inverse diffusion is reduced to the solution of a linear integral equation (Gelfand–Levitan equation). In the following, we will simplify the notation by using a system of units in which the Planck constant is $\hbar = 1$ and the mass of the particle is $m = \frac{1}{2}$. So consider the equation

$$\psi'' + (\lambda - u(x))\psi = 0, \quad -\infty < x < \infty \qquad [9.2]$$

where ψ (unknown) is the wave function of the particle, the spectral parameter λ is the energy of the particle, the function $u(x)$ is the potential or potential energy of the particle. This potential is assumed to have a compact support, that is, it is different from zero only in some domain. When the particle is free (i.e. $u = 0$) and has a positive energy (i.e. $\lambda = k^2$), then equation [9.2] is reduced to

$$\psi''(x) + k^2\psi = 0, \qquad [9.3]$$

and admits two linearly independent solutions e^{ikx} (describing the particle moving to the right) and e^{-ikx} (describing the particle moving to the left). Let us denote by $E^{rs}_{(2)}$ (respectively, $E^{cs}_{(2)}$) the space of the real (respectively, complex) solutions of equation [9.2] and by $E^{rs}_{(3)}$ (respectively, $E^{cs}_{(3)}$) the space of the real (respectively, complex) solutions of equation [9.3]. The space $E^{cs}_{(2)}$ (of states of the particle) is the complexification of $E^{rs}_{(2)}$ and all four states of the particle arriving and departing to the right and left belong to the space $E^{cs}_{(2)}$. The space $E^{rs}_{(3)}$ (respectively, $E^{cs}_{(3)}$) has the following natural basis $(\cos kx, \sin kx)$ (respectively, (e^{ikx}, e^{-ikx})).

Let $[\alpha, \beta]$ be the bounded support of u. The monodromy operator, denoted by $\mathcal{M}$, of equation [9.2] with a potential of compact support is a linear operator mapping the state space of a free particle with energy $\lambda = k^2$ into itself. It is defined in the following way: to a solution of equation [9.3] of a free particle, we assign a solution of the Schrödinger equation coinciding with it to the left of the support, and to this solution, in turn, we assign its value to the right of the support.

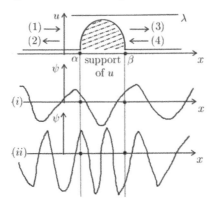

Figure 9.1. *Particle moving*

(We denote by (1) particle moving from left, (2) particle moving to left, (3) particle moving to right and (4) particle moving from right.) *This figure does not reflect reality but it is a good approximation.* In Figure 9.1, case (i) corresponds to equation [9.3] whose solution is $a\cos kx + b\sin kx$ where a, b are constants. Case (ii) corresponds to equation [9.2] whose solution is $a\cos kx + b\sin kx$ if $x < \alpha$, solution of equation

[9.2] if $\alpha \leq x \leq \beta$ and $c\cos kx + d\sin kx$ if $x > \beta$ where $(c,d) = \mathcal{M}_u(a,b)$. So we have,

$$\mathcal{M} : E_{(3)}^{rs} \longrightarrow E_{(3)}^{rs}, \quad a\cos kx + b\sin kx \longmapsto \begin{cases} a\cos kx + b\sin kx & \text{if } x < \alpha \\ c\cos kx + d\sin kx & \text{if } x > \beta \end{cases}$$

where a, b are constants and $(c,d) = \mathcal{M}_u(a,b)$. This means that for each solution of equation [9.3] is associated: (•) the solution of [9.2] which is to the left of α; in this region the solution of [9.3] coincides with that of [9.2], and (••) the solution of [9.2] which is to the right of β. Similarly, the complex monodromy operator of equation [9.2] is defined by

$$\mathcal{M} : E_{(3)}^{cs} \longrightarrow E_{(3)}^{cs}, \quad ae^{ikx} + be^{-ikx} \longmapsto \begin{cases} ae^{ikx} + be^{-ikx} & \text{if } x < \alpha \\ ce^{ikx} + de^{-ikx} & \text{if } x > \beta \end{cases}$$

THEOREM 9.1.– Let W be the phase plane formed by the representative points (i.e. pairs of real numbers) (ψ, ψ'). Let $\mathcal{B}_{(2)}^{x_1} : E_{(2)}^{rs} \longrightarrow W$, $\psi \longmapsto \mathcal{B}_{(2)}^{x_1}\psi = (\psi(x_1), \psi'(x_1))$, be an operator with ψ a solution of equation [9.2] whose initial conditions for $x = x_1 \in \mathbb{R}$ are $(\psi(x_1), \psi'(x_1))$. Then, the space $E_{(2)}^{rs}$ is isomorphic to W and

$$g_{x_1}^{x_2} \equiv \mathcal{B}_{(2)}^{x_2}\left(\mathcal{B}_{(2)}^{x_1}\right)^{-1} : W \longrightarrow W, \quad (\psi(x_1), \psi'(x_1)) \longmapsto (\psi(x_2), \psi'(x_2)),$$

is a linear isomorphism.

PROOF.– It is clear that $\mathcal{B}_{(2)}^{x_1}$ is linear. In addition, for any representative point $(\psi, \psi') \in W$, there exists from the existence theorem a solution ψ satisfying the initial condition $(\psi(x_1), \psi'(x_1))$. Then $\operatorname{Im} \mathcal{B}_{(2)}^{x_1} \equiv \left\{\mathcal{B}_{(2)}^{x_1}\psi : \psi \in E_{(2)}^{rs}\right\} = W$. Finally, $\operatorname{Ker} \mathcal{B}_{(2)}^{x_1} \equiv \left\{\psi : \psi \in E_{(2)}^{rs}, \mathcal{B}_{(2)}^{x_1}\psi = 0\right\} = 0$ follows from the uniqueness theorem because the solution satisfying the initial condition at the point x_1 is equal to zero. The fact that $g_{x_1}^{x_2}$ is a linear isomorphism follows from the fact that the inverse of an isomorphism is one. More specifically, if ψ_1 and ψ_2 are two solutions of equation [9.2], then

$$(\psi(x_1), \psi'(x_1)) = \mathcal{B}_{(2)}^{x_1}\psi = \mathcal{B}_{(2)}^{x_1}\psi_1 + \mathcal{B}_{(2)}^{x_1}\psi_2 = (\psi_1(x_1), \psi_1'(x_1))$$
$$+ (\psi_2(x_1), \psi_2'(x_1)),$$

and this is equivalent to

$$\left(\mathcal{B}_{(2)}^{x_1}\right)^{-1}\left((\psi_1(x_1), \psi_1'(x_1)) + (\psi_2(x_1), \psi_2'(x_1))\right)$$

$$= \left(\mathcal{B}_{(2)}^{x_1}\right)^{-1}(\psi(x_1), \psi'(x_1)) = \psi = \psi_1 + \psi_2,$$

$$= \left(\mathcal{B}_{(2)}^{x_1}\right)^{-1}(\psi_1(x_1), \psi_1'(x_1)) + \left(\mathcal{B}_{(2)}^{x_1}\right)^{-1}(\psi_2(x_1), \psi_2'(x_1)).$$

This completes the proof. □

The isomorphism $g_{x_1}^{x_2}$ is called the phase transformation from x_1 to x_2.

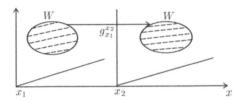

Figure 9.2. *Phase transformation*

In the same way, we can define an operator $\mathcal{B}_{(3)}^{x_1}$ of $E_{(3)}^{rs}$ in W that associates with each solution of equation [9.3], its initial condition at the point x_1. In this case, instead of "phase transformation", there will be "phase point". A particle propagating from $x = -\infty$ crosses a potential barrier with a transmission coefficient T and a reflection coefficient R if the equation [9.2] where $\lambda = k^2$ admits a solution ψ such that:

$$\psi = \begin{cases} Te^{ikx}, & \text{to the right of the barrier} \\ e^{ikx} + Re^{-ikx}, & \text{to the left of the barrier} \end{cases}$$

THEOREM 9.2.– If equation [9.2] where $\lambda = k^2$ has a confounded solution with ae^{ikx} for $x \ll 0$ and with be^{-ikx} for $x \gg 0$, then this solution is null. In addition, for all $k > 0$ the ψ, T and R defined above exist and are unique.

PROOF.– Consider in $E_{(2)}^{cs}$ the Hermitian forms $\langle ae^{ikx}, ae^{ikx} \rangle$, $\langle be^{-ikx}, be^{-ikx} \rangle$ and $\langle ae^{ikx}, ae^{-ikx} \rangle$. Let us designate by $[.,.]$ the left scalar product[1], then

$$\langle ae^{ikx}, ae^{ikx} \rangle = \frac{i}{2}[ae^{ikx}, \bar{a}e^{-ikx}] = \frac{i}{2}\begin{vmatrix} a & ia \\ \bar{a} & -i\bar{a} \end{vmatrix} = |a|^2.$$

Similarly, we have $\langle be^{-ikx}, be^{-ikx} \rangle = -|b|^2$ and $\langle ae^{ikx}, ae^{-ikx} \rangle = 0$. Therefore, by setting $z = z_1 e^{ikx} + z_2 e^{-ikx}$ where z_1 and z_2 are the coordinates of the vector

---

[1] $[\xi, \eta]$ is the oriented area of the parallelogram constructed on the vectors: $\xi = \xi_1 e_1 + \xi_2 e_2$, $\eta = \eta_1 e_1 + \eta_2 e_2$ of the real plane where (e_1, e_2) is a fixed basis in which $[e_1, e_2] = 1$. We show that $[\xi, \eta] = \begin{vmatrix} \xi_1 & \xi_2 \\ \eta_1 & \eta_2 \end{vmatrix}$.

z in the basis (e^{ikx}, e^{-ikx}), we obtain $\langle z, z \rangle = |z_1|^2 - |z_2|^2$, that is, $\langle ., . \rangle$ is of type $(1,1)$. Since the monodromy operator retains this Hermitian form, we deduce that $|a|^2 = -|b|^2$ and so $a = b = 0$. Consider now a particle going to $+\infty$ and let e^{ikx} be a solution to the right of the barrier. To the left of the barrier this solution becomes

$$e^{ikx} \curvearrowright ae^{ikx} + be^{-ikx}. \qquad [9.4]$$

From what precedes, the coefficient a is non-zero. To get the solution in question, simply divide the two members of [9.4] by a, $\frac{1}{a}e^{ikx} \curvearrowright e^{ikx} + \frac{b}{a}e^{-ikx}$. Taking $T = \frac{1}{a}$, $R = \frac{b}{a}$, this shows that T and R are uniquely defined. $\square$

We will now demonstrate the Liouville theorem, which will be useful later.

THEOREM 9.3.– Let $\frac{dx}{dt} = f(x)$, $x = (x_1, ..., x_n)$, be a system of differential equations whose solutions extend to the whole time axis. Let $\{g^t\}$ be the corresponding group of transformations: $g^t x = x + f(x)t + o(t^2)$, for t small. We denote by D a domain in phase space, $D(t) \equiv g^t D(0)$ and by $v(t)$ the volume of $D(t)$. If $\operatorname{div} f = \sum_{j=1}^n \frac{\partial f_j}{\partial x_j} = 0$, then $v(t) = v(0)$, that is, g^t preserves the volume of any domain.

PROOF.– We have $v(t) = \int_{D(t)} dx = \int_{D(0)} \frac{\partial g^t x}{\partial x} dx$, where $\frac{\partial g^t x}{\partial x}$ is the Jacobian matrix, $\frac{\partial g^t x}{\partial x} = I + \frac{\partial f}{\partial x}t + o(t^2)$. The determinant of the operator $I + \frac{\partial f}{\partial x}t$ is equal to the product of the eigenvalues. These (taking into account their multiplicities) are equal to $1 + t\frac{\partial f_j}{\partial x_j}$ where $\frac{\partial f_j}{\partial x_j}$ are the eigenvalues of $\frac{\partial f}{\partial x}$. Then

$$\det \frac{\partial g^t x}{\partial x} = 1 + t \sum_{j=1}^n \frac{\partial f_j}{\partial x_j} + o(t^2) = 1 + t \operatorname{div} f + o(t^2).$$

Therefore,

$$v(t) = \int_{D(0)} (1 + t \operatorname{div} f + o(t^2)) dx \implies \left. \frac{dv(t)}{dt} \right|_{t=0} = \int_{D(0)} \operatorname{div} f dx.$$

Since $t = t_0$ is not worse than $t = 0$, we also have $\left. \frac{dv(t)}{dt} \right|_{t=t_0} = \int_{D(t_0)} \operatorname{div} f dx$. $\square$

REMARK 9.1.– Liouville's theorem is easily generalized to the case of non-autonomous systems ($f = f(x, t)$). Indeed, the terms of first degree in the expression of $\frac{\partial g^t x}{\partial x}$ remain the same. But the terms of degree greater than one do not intervene in the demonstration. In other words, Liouville's theorem is a first-order theorem.

Let $SL(2,\mathbb{R})$ be the real unimodular group, that is, the set of all real 2×2 matrices with determinant one. In other words, $SL(2,\mathbb{R})$ is the group of all linear transformations of $\mathbb{R}^2$ that preserve oriented area $[.,.]$ (see the notation used in the proof of theorem 9.2). Consider the group $SU(1,1)$ of $(1,1)$-unitary unimodular matrices. This is the set of all complex 2×2 matrices with determinant one preserving the Hermitian form $|z_1|^2 - |z_2|^2$ (see the notation used in the proof of theorem 9.2). In other words, they are matrices of the form $\begin{pmatrix} a & b \\ c & d \end{pmatrix}$ for which $|a|^2 - |b|^2 = |c|^2 - |d|^2 = 1, a\bar{c} - b\bar{d} = 0, ad - bc = 1$.

THEOREM 9.4.– In the basis $(\cos kx, \sin kx)$ (respectively, (e^{ikx}, e^{-ikx})), the matrix of the monodromy operator $\mathcal{M}$ belongs to the group $SL(2,\mathbb{R})$ (respectively, $SU(1,1)$).

PROOF.– We show that the determinant of the monodromy operator of the Schrödinger equation is equal to one. Note that $(\cos kx, \sin kx)$ is a basis on the space $E^{sr}_{(3)}$. As $\mathcal{B}^x_{(3)} \cos kx = (\cos kx, -k\sin kx)$, $\mathcal{B}^x_{(3)} \sin kx = (\sin kx, k\cos kx)$, so W is provided with a basis in which the matrix of the operator (we use here the same notation for the operator and the matrix) is written as $\mathcal{B}^x_{(3)} = \begin{pmatrix} \cos kx & \sin kx \\ -k\sin kx & k\cos kx \end{pmatrix}$, hence $\det \mathcal{B}^x_{(3)} = k$, independent of x. Let us denote by x^+ the point x to the left of the support of the potential and by x^- the one on the right. We have the following situation:

$$\mathcal{M} : E^{sr}_{(3)} \longrightarrow E^{sr}_{(3)},$$

$$a\cos kx + b\sin kx \longmapsto c\cos kx + d\sin kx, \quad (c,d) = \mathcal{M}_u(a,b),$$

$$\mathcal{B}^{x^-}_{(3)} : E^{sr}_{(3)} \longrightarrow W,$$

$$a\cos kx + b\sin kx \longmapsto (a\cos kx^- + b\sin kx^-, -ak\sin kx^- + bk\cos kx^-),$$

$$\mathcal{B}^{x^+}_{(3)} : E^{sr}_{(3)} \longrightarrow W,$$

$$c\cos kx + d\sin kx \longmapsto (a\cos kx^+ + b\sin kx^+, -ak\sin kx^+ + bk\cos kx^+),$$

$$g^{x^+}_{x^-} : W \longrightarrow W,$$

$$(a\cos kx^- + b\sin kx^-, -ak\sin kx^- + bk\cos kx^-)$$
$$\longmapsto (a\cos kx^+ + b\sin kx^+, -ak\sin kx^+ + bk\cos kx^+).$$

We verify directly that: $g^{x^+}_{x^-} o \mathcal{B}^{x^-}_{(3)} = \mathcal{B}^{x^+}_{(3)} o \mathcal{M}$, and since $\det \mathcal{B}^{x^+}_{(3)} = \mathcal{B}^{x^-}_{(3)}$, so we have $\det \mathcal{M} = \det g^{x^+}_{x^-}$. Now g^x preserves the areas according to Liouville's theorem 9.3. Indeed, by putting $\psi_1 = \psi$, $\psi_2 = \psi'$, we rewrite equation [9.2] in the form $\psi'_1 = \psi_2 \equiv f_1$, $\psi'_2 = (u(x) - \lambda)\psi_1 \equiv f_2$. Here, we have $f = (f_1, f_2)$, $t = x$

and div $f = \frac{\partial \psi_2}{\partial \psi_1} + \frac{\partial (u(x)-\lambda)\psi_1}{\partial \psi_2} = 0$. Therefore, $\det g_{x^-}^{x^+} = 1$ and consequently $\det \mathcal{M} = 1$. For the case of $SU(1,1)$, we will show that the matrix (also denoted $\mathcal{M}$) of an operator is real and unimodular in the basis $(\cos kx, \sin kx)$ if and only if it is special $(1,1)$-unitary in complex conjugate basis (e^{ikx}, e^{-ikx}). By setting as in the proof of theorem 9.2, $z = z_1 e^{ikx} + z_2 e^{-ikx}$ where z_1 and z_2 are the coordinates of the vector z in the basis (e^{ikx}, e^{-ikx}), we obtain $\langle z, z \rangle = |z_1|^2 - |z_2|^2$, that is, $\langle .,. \rangle$ is of type $(1,1)$. The monodromy operator conserves this Hermitian form. Say that $\mathcal{M}$ is real and unimodular in the basis $(\cos kx, \sin kx)$ which is equivalent to $\mathcal{M} \in GL(2,\mathbb{R}) \cap SL(2,\mathbb{C})$ or what amounts to the same $\mathcal{M} \in SU(1,1)$ or what is equivalent $\mathcal{M}$ is $(1,1)$-unitary and unimodular in the basis (e^{ikx}, e^{-ikx}). $\square$

REMARK 9.2.– It is well known that the sum of the transmission and reflection coefficients is equal to one. This property of the Schrödinger equation can be obtained as a corollary of the preceding theorem and thus without the use of the theory of probabilities. Indeed, we use the definition of the monodromy operator, $e^{ikx} + R e^{-ikx} \longmapsto T e^{ikx}$, $e^{-ikx} + \overline{R} e^{ikx} \longmapsto \overline{T} e^{-ikx}$. Divide the first expression by $T \neq 0$ and the second by $\overline{T} \neq 0$, we get

$$\begin{pmatrix} e^{ikx} \\ e^{-ikx} \end{pmatrix} \begin{pmatrix} \frac{1}{T} & \frac{\overline{R}}{\overline{T}} \\ \frac{R}{T} & \frac{1}{\overline{T}} \end{pmatrix} \longmapsto \begin{pmatrix} e^{ikx} \\ e^{-ikx} \end{pmatrix}.$$

So in the basis (e^{ikx}, e^{-ikx}), the matrix of the inverse of the monodromy operator is $\mathcal{M}^{-1} = \begin{pmatrix} \frac{1}{T} & \frac{\overline{R}}{\overline{T}} \\ \frac{R}{T} & \frac{1}{\overline{T}} \end{pmatrix}$, and therefore, the matrix of the monodromy operator in the basis (e^{ikx}, e^{-ikx}) is $\mathcal{M} = \begin{pmatrix} \frac{1}{\overline{T}} & -\frac{\overline{R}}{\overline{T}} \\ -\frac{R}{T} & \frac{1}{T} \end{pmatrix}$. According to theorem 9.4, we have $\mathcal{M} \in SU(1,1)$ ($\det \mathcal{M} = 1$) and consequently $|T|^2 + |R|^2 = 1$.

Define the solutions $\psi_1(x,\lambda)$ and $\psi_2(x,\lambda)$ of equation [9.2] by the initial conditions: $\psi_1(0,\lambda) = 1$, $\psi_1'(0,\lambda) = 0$, and $\psi_2(0,\lambda) = 0$, $\psi_2'(0,\lambda) = 1$. For the simple case $u(x) = 0$, we have

$$\begin{cases} \psi_1(x,\lambda) = \cos \sqrt{\lambda} x = 1 + \left(-\frac{1}{2}\lambda\right) x^2 + \left(\frac{1}{24}\lambda^2\right) x^4 + O\left(x^6\right), \\ \psi_2(x,\lambda) = \frac{1}{\sqrt{\lambda}} \sin \sqrt{\lambda} x = x + \left(-\frac{1}{6}\lambda\right) x^3 + \left(\frac{1}{120}\lambda^2\right) x^5 + O\left(x^7\right). \end{cases} \quad [9.5]$$

For $\sqrt{\lambda}$, we choose, for example, the determination $\sqrt{\lambda} = \sqrt{r} e^{i\frac{\theta}{2}}$ where $\lambda = r e^{i\theta}$, $r > 0$, $-\pi < \theta < \pi$. Let α be an arbitrary real number. The function $\psi(x,\lambda) = \psi_1(x,\lambda) + \alpha \psi_2(x,\lambda)$ is also solution of equation [9.2] and satisfies the boundary condition $\psi'(0,\lambda) - \alpha \psi(0,\lambda) = 0$. For $\alpha = 0$, we have $\psi(x,\lambda) = \psi_1(x,\lambda)$ and for $\alpha = \infty$, we put $\psi(x,\lambda) = \psi_2(x,\lambda)$. We assume that for $\lambda \in \mathbb{C}$ and $x \geq 0$, we have

$$\psi(x,\lambda) = \cos \sqrt{\lambda} x + \int_0^x K(x,t) \cos \sqrt{\lambda} t \, dt, \qquad [9.6]$$

where K is to be determined, subject to the condition of having partial derivatives of order one and order two continuous in the set of real pairs (x,t) such that: $0 \leq t \leq x$. In other words, we look for $\psi(x,.)$ as a perturbation of the function $x \mapsto \psi(x,\lambda) = \cos\sqrt{\lambda}x$ and, precisely, as a transform $(I+K)\psi_1(x,.)$ where K is a Volterra operator in $[0,+\infty[$. We look for the conditions that $K(x,t)$ must satisfy for the function [9.6] to be a solution of the differential equation [9.2]. From equation [9.6], we get

$$\frac{\partial^2 \psi}{\partial x^2}(x,\lambda) = -\lambda \cos\sqrt{\lambda}x + \frac{dK(x,x)}{dx}\cos\sqrt{\lambda}x - \sqrt{\lambda}K(x,x)\sin\sqrt{\lambda}x$$

$$+ \left.\frac{\partial K(x,t)}{\partial x}\right|_{t=x}\cos\sqrt{\lambda}x + \int_0^x \frac{\partial^2 K(x,t)}{\partial x^2}\cos\sqrt{\lambda}t\,dt. \quad [9.7]$$

Let us calculate the expression $\lambda \int_0^x K(x,t)\cos\sqrt{\lambda}t\,dt$, by doing two integrations in parts, we get

$$\lambda \int_0^x K(x,t)\cos\sqrt{\lambda}t\,dt = \sqrt{\lambda}K(x,x)\sin\sqrt{\lambda}x$$

$$+ \left.\frac{\partial K(x,t)}{\partial t}\right|_{t=x}\cos\sqrt{\lambda}x - \left.\frac{\partial K(x,t)}{\partial t}\right|_{t=0}$$

$$- \int_0^x \frac{\partial^2 K(x,t)}{\partial t^2}\cos\sqrt{\lambda}t\,dt. \quad [9.8]$$

To calculate expression [9.2], substitute [9.7] and [9.8],

$$0 = \psi'' + (\lambda - u(x))\psi$$

$$= \frac{dK(x,x)}{dx}\cos\sqrt{\lambda}x + \left(\frac{\partial K(x,t)}{\partial t} + \frac{\partial K(x,t)}{\partial x}\right)_{x=t}\cos\sqrt{\lambda}x$$

$$- \left.\frac{\partial K(x,t)}{\partial t}\right|_{t=0} - u(x)\cos\sqrt{\lambda}x$$

$$+ \int_0^x \left(\frac{\partial^2 K(x,t)}{\partial x^2} - \frac{\partial^2 K(x,t)}{\partial t^2} - u(x)K(x,t)\right)\cos\sqrt{\lambda}t\,dt.$$

We have

$$\frac{\partial^2 K(x,t)}{\partial x^2} - u(x)K(x,t) = \frac{\partial^2 K(x,t)}{\partial t^2}, \quad [9.9]$$

with the boundary conditions

$$\left.\frac{\partial K(x,t)}{\partial t}\right|_{t=0} = 0, \quad [9.10]$$

$$\frac{dK(x,x)}{dx} = \frac{1}{2}u(x). \quad [9.11]$$

For the initial conditions, we have $\psi(0,\lambda) = 1$ and $\psi'(0,\lambda) = K(0,0)$. As $\psi'(0,\lambda) - \alpha\psi(0,\lambda) = 0$, then $K(0,0) = \alpha$. Therefore,

$$K(x,x) = \alpha + \frac{1}{2}\int_0^x u(t)dt. \qquad [9.12]$$

If $u(x)$ has a continuous derivative, then there exists a unique solution of [9.9], satisfying conditions [9.10] and [9.12]. Hence, there exists a satisfying function $K(x,t)$ [9.6]. Let us solve equation [9.6] as an equation of Volterra, we get

$$\cos\sqrt{\lambda}x = \psi(x,\lambda) - \int_0^x K_1(x,t)\psi(t,\lambda)dt, \qquad [9.13]$$

and in the same way as before, we show that $K_1(x,t)$ is solution of the equation

$$\frac{\partial^2 K_1(x,t)}{\partial x^2} = \frac{\partial^2 K_1(x,t)}{\partial t^2} - u(t)K_1(x,t),$$

with the conditions $\left(\frac{\partial K_1}{\partial t} - \alpha K_1\right)_{t=0} = 0$, $K_1(x,x) = \alpha + \frac{1}{2}\int_0^x u(t)dt$.

For the case $\alpha = \infty$, we look for $\psi(x,\lambda)$ as a perturbation of the function $x \mapsto \psi(x,\lambda) = \frac{1}{\sqrt{\lambda}}\sin\sqrt{\lambda}x$ (see expression [9.5]) or what is equivalent as a transform $(I+K)\psi_1(x,.)$ where K is a Volterra operator in $[0,+\infty[$. In other words, we set $\lambda \in \mathbb{C}$ and $x \geq 0$,

$$\psi(x,\lambda) = \frac{\sin\sqrt{\lambda}x}{\sqrt{\lambda}} + \int_0^x L(x,t)\frac{\sin\sqrt{\lambda}x}{\sqrt{\lambda}t}dt, \qquad [9.14]$$

where L is a function to be determined, subject to the condition of having partial derivatives of order one and order two continuous in the set of real pairs (x,t) such that: $0 \leq t \leq x$. By reasoning as before, we obtain the relation

$$\frac{\partial^2 L(x,t)}{\partial x^2} - u(x)L(x,t) = \frac{\partial^2 L(x,t)}{\partial t^2},$$

with the conditions: $L(x,x) = 0$, $L(x,x) = \frac{1}{2}\int_0^x u(t)dt$. By solving equation [9.14], we obtain

$$\frac{\sin\sqrt{\lambda}x}{\sqrt{\lambda}} = \psi(x,\lambda) + \int_0^x L_1(x,t)\psi(t,\lambda)dt. \qquad [9.15]$$

The functions $L(x,t)$ and $L_1(x,t)$ have the same properties as the functions $K(x,t)$ and $K_1(x,t)$ previously obtained. Recall that for every function $f(x) \in L^2(\mathbb{R})$, we have the Parseval identity, $\int_0^\infty f^2(x)dx = \int_{-\infty}^\infty F^2(\lambda).d\rho(\lambda)$, where $F(\lambda) = \int_0^\infty f(x)\psi(x,\lambda)dx$, is the Fourier transform of $f(x)$ and $\rho(\lambda)$ a monotone function, bounded on any finite interval. The sequence of functions $F_n(\lambda) = \int_0^n f(x)\psi(x,\lambda)dx$ converges in quadratic mean (with respect to the spectral measure $\rho(\lambda)$) to $F(\lambda)$, that is, $\lim_{n\to\infty} \int_{-\infty}^\infty (F(\lambda) - F_n(\lambda))^2 d\rho(\lambda) = 0$. We choose $\rho(\lambda)$ in the form $\rho(\lambda) = \frac{2}{\pi}\sqrt{\lambda} + \sigma(\lambda)$ if $\lambda > 0$, and $\rho(\lambda) = \sigma(\lambda)$ if $\lambda < 0$, where $\sigma(\lambda)$ is a measure with compact support satisfying the condition: $\int_{-\infty}^\infty |\lambda|.|d\sigma(\lambda)| < +\infty$. For $0 < b < y < a < x$, the functions $\int_a^x \psi(t,\lambda)dt$ and $\int_b^y \cos\sqrt{\lambda}t\, dt$ are orthogonal with respect to $\rho(\lambda)$. In other words, we have the orthogonality relation: $I \equiv \int_{-\infty}^\infty \left(\int_a^x \psi(t,\lambda)dt\right)\left(\int_b^y \cos\sqrt{\lambda}t\, dt\right) d\rho(\lambda) = 0$. Indeed, by integrating equation [9.13] from b to y, we obtain

$$\int_b^y \cos\sqrt{\lambda}t\, dt = \int_b^y \psi(t,\lambda)dt - \int_b^y dt \int_0^t K_1(t,s)\psi(s,\lambda)ds,$$

$$= \int_b^y \psi(t,\lambda)dt - \int_0^b \psi(s,\lambda)ds \int_b^y K_1(t,s)dt$$

$$- \int_b^y \psi(s,\lambda)dt \int_s^y K_1(t,s)dt.$$

By definition, this function is expressed using the transform (in $\psi(t,\lambda)$) of a null function outside the interval $]b,y[$. Since $]b,y[\cap]a,x[= \emptyset$, we deduce from Parseval's equality that $I = 0$. To obtain the Gelfand–Levitan integral equation, we proceed as follows: according to equation [9.6], we have

$$\int_a^x \psi(t,\lambda)dt = \int_a^x \cos\sqrt{\lambda}t\, dt + \int_a^x dt \int_0^t K(t,s)\cos\sqrt{\lambda}s\, ds,$$

$$= \int_a^x \cos\sqrt{\lambda}t\, dt + \int_0^a \cos\sqrt{\lambda}s\, ds \int_a^x K(t,s)dt$$

$$+ \int_a^x \cos\sqrt{\lambda}s\, ds \int_s^x K(t,s)dt,$$

by virtue of Lebesgue–Fubini's theorem. Therefore,

$$I = \int_{-\infty}^\infty \left(\int_a^x \cos\sqrt{\lambda}t\, dt\right)\left(\int_b^y \cos\sqrt{\lambda}t\, dt\right) d\rho(\lambda)$$

$$+ \int_{-\infty}^\infty \left(\int_0^a \cos\sqrt{\lambda}s\, ds \int_a^x K(t,s)dt + \int_a^x \cos\sqrt{\lambda}s\, ds \int_s^x K(t,s)dt\right)$$

$$\times \left(\int_b^y \cos\sqrt{\lambda}t\, dt\right) d\rho(\lambda),$$

$$= 0.$$

This expression can be written using the definition of $\rho(\lambda)$ in the form

$$I = \int_{-\infty}^{\infty} \left(\int_{a}^{x} \cos\sqrt{\lambda}t\, dt \right) \left(\int_{b}^{y} \cos\sqrt{\lambda}t\, dt \right) d\sigma(\lambda)$$

$$+ \int_{-\infty}^{\infty} \left(\int_{0}^{a} \cos\sqrt{\lambda}s\, ds \int_{a}^{x} K(t,s)dt + \int_{a}^{x} \cos\sqrt{\lambda}s\, ds \int_{s}^{x} K(t,s)dt \right)$$

$$\times \left(\int_{b}^{y} \cos\sqrt{\lambda}t\, dt \right) d\sigma(\lambda)$$

$$+ \frac{2}{\pi} \int_{-\infty}^{\infty} \left(\int_{a}^{x} \cos\sqrt{\lambda}t\, dt \right) \left(\int_{b}^{y} \cos\sqrt{\lambda}t\, dt \right) d\sigma(\lambda)$$

$$+ \frac{2}{\pi} \int_{-\infty}^{\infty} \left(\int_{0}^{a} \cos\sqrt{\lambda}s\, ds \int_{a}^{x} K(t,s)dt + \int_{a}^{x} \cos\sqrt{\lambda}s\, ds \int_{s}^{x} K(t,s)dt \right)$$

$$\times \left(\int_{b}^{y} \cos\sqrt{\lambda}t\, dt \right) d\sigma(\lambda),$$

$$= 0.$$

Since $b < y < a < x$, then given the Parseval identity, the third term is equal to zero while the fourth is equal to

$$\int_{b}^{y} \left(\int_{0}^{a} \cos\sqrt{\lambda}s\, ds \int_{a}^{x} K(t,s)dt + \int_{a}^{x} \cos\sqrt{\lambda}s\, ds \int_{s}^{x} K(t,s)dt \right) ds$$

$$= \int_{b}^{y} ds \int_{a}^{x} K(t,s)dt.$$

Therefore,

$$I = \int_{-\infty}^{\infty} \frac{(\sin\sqrt{\lambda}x - \sin\sqrt{\lambda}a)(\sin\sqrt{\lambda}y - \sin\sqrt{\lambda}b)}{\lambda} d\sigma(s)$$

$$+ \int_{-\infty}^{\infty} \left(\int_{0}^{a} \cos\sqrt{\lambda}s\, ds \int_{a}^{x} K(t,s)dt + \int_{a}^{x} \cos\sqrt{\lambda}s\, ds \int_{s}^{x} K(t,s)dt \right)$$

$$\times \left(\int_{b}^{y} \cos\sqrt{\lambda}s\, ds \right) d\sigma(\lambda)$$

$$+ \int_{b}^{y} ds \int_{a}^{x} K(t,s)dt,$$

$$= 0.$$

By setting

$$F(x,y) \equiv \int_{-\infty}^{\infty} \frac{\sin\sqrt{\lambda}x \sin\sqrt{\lambda}y}{\lambda} d\sigma(\lambda), \quad G(x,s) \equiv \begin{cases} \int_a^x K(t,s)dt, & 0 \leq s \leq a \\ \int_s^x K(t,s)dt, & a \leq s \leq x \\ 0, & s > x \end{cases}$$

the equation above becomes

$$F(x,y) - F(x,b) - F(a,y) + F(a,b)$$
$$+ \int_{-\infty}^{\infty} \left(\int_0^x G(x,s) \cos\sqrt{\lambda}s \, ds \right) \left(\int_b^y \cos\sqrt{\lambda}s \, ds \right) d\sigma(\lambda)$$
$$+ \int_b^y ds \int_a^x K(t,s)dt = 0.$$

This last equation can still be written, doing an integration by parts and noticing that $G(x,x) = 0$,

$$F(x,y) - F(x,b) - F(a,y) + F(a,b)$$
$$+ \int_{-\infty}^{\infty} \left(\int_0^x \frac{\partial G(x,s)}{\partial s} \frac{\sin\sqrt{\lambda}s}{\sqrt{\lambda}} ds \right) \left(\frac{\sin\sqrt{\lambda}y - \sin\sqrt{\lambda}b}{\sqrt{\lambda}} \right) d\sigma(\lambda)$$
$$+ \int_b^y ds \int_a^x K(t,s)dt = 0. \qquad [9.16]$$

But

$$\int_{-\infty}^{\infty} \left(\int_0^x \frac{\partial G(x,s)}{\partial s} \frac{\sin\sqrt{\lambda}s}{\sqrt{\lambda}} ds \right) \left(\frac{\sin\sqrt{\lambda}y - \sin\sqrt{\lambda}b}{\sqrt{\lambda}} \right) d\sigma(\lambda),$$
$$= \int_0^x \frac{\partial G(x,s)}{\partial s} \left(\int_{-\infty}^{\infty} \left(\frac{\sin\sqrt{\lambda}s \sin\sqrt{\lambda}y - \sin\sqrt{\lambda}s \sin\sqrt{\lambda}b}{\lambda} \right) d\sigma(\lambda) \right) ds,$$
$$= \int_0^x \frac{\partial G(x,s)}{\partial s} (F(s,y) - F(s,b)) \, ds,$$
$$= -\int_0^x G(x,s) \left(\frac{\partial F(s,y)}{\partial s} - \frac{\partial F(s,b)}{\partial s} \right) ds,$$
$$= -\int_0^a \left(\frac{\partial F(s,y)}{\partial s} - \frac{\partial F(s,b)}{\partial s} \right) ds \left(\int_a^x K(t,s)dt \right)$$

$$-\int_a^x \left(\frac{\partial F(s,y)}{\partial s} - \frac{\partial F(s,b)}{\partial s}\right) ds \left(\int_s^x K(t,s)dt\right),$$

$$= \int_a^x dt \int_0^t \left(\frac{\partial F(s,y)}{\partial s} - \frac{\partial F(s,b)}{\partial s}\right) ds,$$

so equation [9.16] becomes

$$F(x,y) - F(x,b) - F(a,y) + F(a,b) + \int_a^x dt \int_0^t \left(\frac{\partial F(s,y)}{\partial s} - \frac{\partial F(s,b)}{\partial s}\right) ds$$

$$+ \int_b^y ds \int_a^x K(t,s)dt = 0.$$

Deriving this expression with respect to y and then with respect to x (the support of the measure σ is compact), we obtain

$$\frac{\partial^2 F}{\partial x \partial y} + \int_0^x K(x,s) \frac{\partial^2 F(s,y)}{\partial s \partial y} + K(x,y) = 0.$$

By setting $f(x,y) \equiv \frac{\partial^2 F}{\partial x \partial y}$, we finally obtain the Gelfand–Levitan integral equation for the function $x \longmapsto K(x,y)$ valid for $0 < y < x$,

$$f(x,y) + K(x,y) + \int_0^x K(x,s)f(s,y)ds = 0, \qquad y \leq x. \qquad [9.17]$$

For the case $\alpha = \infty$, that is, $\psi(x,\lambda) = \psi_2(x,\lambda)$, just integrate the two members of equation [9.15] from 0 to x and use a similar reasoning. Under the continuity assumption of K, equation [9.17] must be checked for $x = 0$ and $x = y$. Note also that if we set x in the previous equation, then we will obtain the so-called Fredholm linear integral equation. We can prove that, conversely, equation [9.17] gives a single continuous solution in the set of pairs of real numbers such that: $0 \leq t \leq x$. We will not look for the solution at this level, it will be done later (in the next section) when we treat the KdV equation.

9.3. The inverse scattering method

Let us first examine some particular solutions of the KdV equation [9.1], of the kind of progressive waves $u(x,t) = s(x - ct)$, where c is the phase velocity. By replacing this expression in [9.1], we obtain $-c\frac{\partial s}{\partial x} - 6s\frac{\partial s}{\partial x} + \frac{\partial^3 s}{\partial x^3} = 0$. By integrating this equation with respect to x and imposing the boundary condition that s and its derivatives decrease for $|x| \longrightarrow \infty$, we get $-cs - 3s^2 + \frac{\partial^2 s}{\partial x^2} = 0$, hence $-cs - 2s^3 + \left(\frac{\partial s}{\partial x}\right)^2 = 0$, and the exact expression of the solution s requires the use of elliptic functions. Suppose that $\frac{\partial s}{\partial x}(0) = 0$, in which case the solution of this last equation is $s(x - ct) = -\frac{c}{2}\operatorname{sech}^2\frac{\sqrt{c}}{2}(x - ct)$, where sech denotes the hyperbolic secant, that is, $\frac{1}{\cosh}$. Therefore, $u(x, 0) = u_0 \operatorname{sech}^2\frac{x}{l}$, $u_0 \equiv -\frac{c}{2}$, $l^2 \equiv \frac{4}{c}$. This expression shows that u is removed for an infinitely long time in the position $u \simeq 0$, then it reaches the value u_0, is reflected on this point and returns again in the position of $u \simeq 0$. This solution is called a soliton. To obtain this solution, we can use the so-called Bäcklund transformations for the KdV equation. When solitons collide, dimensions and speeds of solutions do not change after collision. This remarkable phenomenon has suggested the idea of conservation laws. And indeed, Kruskal, Zabusky, Lax, Gardner, Green and Miura (Gardner et al. 1967; Lax 1968) have been able to find a whole series of first integrals for the KdV equation. These integrals are of the form $\int P_n\left(u, ..., u^{(n)}\right) dx$, where P_n is a polynomial. Indeed, the conservation equations that can be deduced from the KdV equation take the following general form: $\frac{\partial P_n}{\partial t} + \frac{\partial Q_n}{\partial x} = 0$, where P_n and Q_n form a series of functions of which we discuss the first three as follows: (i) The KdV equation can be written in the form $\frac{\partial u}{\partial t} + \frac{\partial}{\partial x}\left(-3u^2 + \frac{\partial^2 u}{\partial x^2}\right) = 0$. Hence, $P_1 = u$, $Q_1 = -3u^2 + \frac{\partial^2 u}{\partial x^2}$. (ii) Multiply the KdV equation by u, this gives

$$u\frac{\partial u}{\partial t} - 6u^2\frac{\partial u}{\partial x} + u\frac{\partial^3 u}{\partial x^3} = 0,$$

$$\frac{\partial}{\partial t}\left(\frac{u^2}{2}\right) + \frac{\partial}{\partial x}\left(-2u^3 + u\frac{\partial^2 u}{\partial x^2} - \frac{1}{2}\left(\frac{\partial u}{\partial x}\right)^2\right) = 0.$$

Hence, $P_2 = \frac{u^2}{2}$, $Q_2 = -2u^3 + u\frac{\partial^2 u}{\partial x^2} - \frac{1}{2}\left(\frac{\partial u}{\partial x}\right)^2$. (iii) We have

$$\left(3u^2 - \frac{\partial^2 u}{\partial x^2}\right)\left(\frac{\partial u}{\partial t} - 6u\frac{\partial u}{\partial x} + \frac{\partial^3 u}{\partial x^3}\right) = 0,$$

$$\left(3u^2\frac{\partial u}{\partial t} + \frac{\partial u}{\partial x}\frac{\partial^2 u}{\partial x \partial t}\right) +$$

$$\left(-18u^3\frac{\partial u}{\partial x} + 3u^2\frac{\partial^3 u}{\partial t^3} + 6u\frac{\partial u}{\partial x}\frac{\partial^2 u}{\partial x^2} - \frac{\partial^2 u}{\partial x^2}\frac{\partial^3 u}{\partial x^3} - \frac{\partial^2 u}{\partial x^2}\frac{\partial u}{\partial t} - \frac{\partial u}{\partial x}\frac{\partial^2 u}{\partial x \partial t}\right) = 0.$$

Therefore,

$$\frac{\partial}{\partial t}\left(u^3 + \frac{1}{2}\left(\frac{\partial u}{\partial x}\right)^2\right) + \frac{\partial}{\partial t}\left(-\frac{9}{2}u^4 + 3u^2\frac{\partial^2 u}{\partial x^2} - \frac{1}{2}\left(\frac{\partial^2 u}{\partial x^2}\right)^2 - \frac{\partial u}{\partial x}\frac{\partial u}{\partial t}\right) = 0.$$

Consequently, $P_3 = u^3 + \frac{1}{2}\left(\frac{\partial u}{\partial x}\right)^2$, $Q_3 = -\frac{9}{2}u^4 + 3u^2\frac{\partial^2 u}{\partial x^2} - \frac{1}{2}\left(\frac{\partial^2 u}{\partial x^2}\right)^2 - \frac{\partial u}{\partial x}\frac{\partial u}{\partial t}$. If u vanishes for $x \to \infty$, we get $\frac{\partial}{\partial t}\int P_n dx = 0$, then $\int P_n dx$ are first integrals of the KdV equation. Let $u(x,t) = \frac{\partial y}{\partial x}(x,t)$ and suppose that $\frac{\partial y}{\partial t}, \frac{\partial y}{\partial x}, \frac{\partial^3 y}{\partial t^3}$ decay when $|x| \to \infty$. The KdV equation is written as $\frac{\partial y}{\partial t} - 3\left(\frac{\partial y}{\partial x}\right)^2 + \frac{\partial^3 y}{\partial t^3} = 0$. Hence,

$$\frac{\partial}{\partial t}\int_{-\infty}^{\infty} y(x,t)dx = 3\int_{-\infty}^{\infty}\left(\frac{\partial y}{\partial x}\right)^2(x,t)dx = 3\int_{-\infty}^{\infty} u^2(x,t)dx = \text{constant}.$$

Since $u = \frac{\partial y}{\partial x}$, we also have

$$\frac{\partial}{\partial t}\int_{-\infty}^{\infty} y(x,t)dx = \frac{\partial}{\partial t}\int_{-\infty}^{\infty}\int_{-\infty}^{x} u(z,t)dzdx,$$

$$= x\frac{\partial}{\partial t}\int_{-\infty}^{x} u(z,t)dz\bigg|_{-\infty}^{\infty} - \frac{\partial}{\partial t}\int_{-\infty}^{\infty} u(z,t)dx,$$

$$= -\frac{\partial}{\partial t}\int_{-\infty}^{\infty} xu(x,t)dx,$$

because by hypothesis u^2 and $\frac{\partial^2 u}{\partial x^2}$ tend to 0 when $|x| \to \infty$. Comparing the two expressions obtained, we obtain a new first integral $\frac{\partial}{\partial t}\int_{-\infty}^{\infty} xu(x,t)dx = \text{constant}$.

Lax (1968) showed that the equation of KdV is equivalent to the following equation: $\frac{dA}{dt} = [B,A] = BA - AB$, where $A = -\frac{\partial^2}{\partial x^2} + u(x,t)$, (Sturm–Liouville operator), $B = -4\frac{\partial^3}{\partial x^3} + 6u\frac{\partial}{\partial x} + 3\frac{\partial}{\partial x}$. We deduce that the spectrum of A is conserved: if A is a symmetric operator ($A^\top = A$) and T an orthogonal transformation ($T^\top = T^{-1}$), then the spectrum of $T^{-1}AT$ coincides with that of A. The appearance of an infinite series of first integrals is easily explained by the Lax equation. The Sturm–Liouville equation $A\psi = \lambda\psi$, where λ is a real parameter, can be written in the form

$$\frac{\partial^2 \psi}{\partial x^2} + (\lambda - u(x,t))\psi = 0. \qquad [9.18]$$

This equation reminds us of the unidimensional and stationary Schrödinger equation. In the following, we will see that the complete solution of the KdV equation is closely related to the solution of this equation. We will look at solutions for which u decreases fast enough for $x \longrightarrow \pm\infty$. It should be noted that there are other interesting conditions to know: the case where $u(x,t)$ tends to different constants for $|x| \longrightarrow \infty$ and the one where $u(x,t)$ is periodic in x. So consider equation [9.18] where $u(x,t)$ is the solution of the KdV equation [9.1]. It is assumed that after a certain time, equation [9.18] has N bound states with energy $\lambda_n = -k_n^2$, $n = 1, 2, ..., N$ and continuous states with for energy $\lambda = k^2$. We draw u from equation [9.18] and replace it in equation [9.1]. After a long calculation, after multiplying by ψ^2, we get the expression

$$\frac{\partial \lambda}{\partial t}.\psi^2 + \frac{\partial}{\partial x}\left(\psi\frac{\partial \Upsilon}{\partial x} - \frac{\partial \psi}{\partial x}\Upsilon\right) = 0, \qquad [9.19]$$

where $\Upsilon \equiv \frac{\partial \psi}{\partial t} + \frac{\partial^3 \psi}{\partial x^3} - 3(u+\lambda)\frac{\partial \psi}{\partial x}$. For the study of the discrete part of the spectrum $\lambda_n(t) = -k_n^2(t)$, we show the following result:

THEOREM 9.5.– If ψ_n (measurable and square integrable function) and $\frac{\partial \psi_n}{\partial x}$ tend to zeros when $|x|$ goes to infinity, then $\lambda_n(t) = $ constant and the solution of equation [9.18] is given by $\psi_n(t) = c_n(0)e^{k_n(x-4k_n^2 t)}$, where $c_n(0)$ is determined by the initial condition $u(x,0) = u_0(x)$ of the KdV equation.

PROOF.– Just integrate equation [9.19], this gives $\frac{\partial \lambda_n}{\partial t}.\int_{-\infty}^{\infty}\psi_n^2 dx + \psi_n\frac{\partial \Upsilon}{\partial x} - \frac{\partial \psi_n}{\partial x}\Upsilon = 0$. By hypothesis, $\psi_n \in L^2$ and ψ_n, $\frac{\partial \psi_n}{\partial x}$ tend to zero when $|x|$ goes to infinity, so $\psi_n\frac{\partial \Upsilon}{\partial x} - \frac{\partial \psi_n}{\partial x}\Upsilon$ tends to 0 for $|x| \to \infty$ and we deduce that $\lambda_n(t) = $ constant. Now, since $\frac{\partial \lambda}{\partial t} = 0$, then equation [9.19] becomes $\frac{\partial}{\partial x}\left(\psi\frac{\partial \Upsilon}{\partial x} - \frac{\partial \psi}{\partial x}\Upsilon\right) = 0$. Let us integrate this expression twice, $\frac{(\psi\frac{\partial \Upsilon}{\partial x} - \frac{\partial \psi}{\partial x}\Upsilon)}{\psi^2} = \frac{A}{\psi^2}$, that is, $\left(\frac{\Upsilon}{\psi}\right)' = \frac{A}{\psi^2}$, hence, $\Upsilon = \psi\int\frac{A(t)}{\psi^2}dx + B(t)\psi$, where $A(t)$ and $B(t)$ are integration constants. So we have

$$\frac{\partial \psi_n}{\partial t} + \frac{\partial^3 \psi_n}{\partial x^3} - 3(u+\lambda_n)\frac{\partial \psi_n}{\partial x} = \psi_n\int\frac{A_n}{\psi_n^2}dx + B_n\psi_n. \qquad [9.20]$$

Note that $A_n(t) = 0$ because ψ_n satisfies [9.20] and decreases to zero for $t \to -\infty$. Let us consider $u \cong 0$ for $x \to -\infty$ because otherwise ψ_n would not have the decay assumption. Multiply [9.20] by ψ_n and integrate

$$\int_{-\infty}^{\infty}\psi_n\frac{\partial \psi_n}{\partial t}dx + \int_{-\infty}^{\infty}\left(\psi_n\frac{\partial^3 \psi_n}{\partial x^3} - 3\lambda_n\psi_n\frac{\partial \psi_n}{\partial x}\right)dx = B_n\int_{-\infty}^{\infty}\psi_n^2 dx.$$

This expression can be written by adding and subtracting $\frac{\partial \psi_n}{\partial x}\frac{\partial^2 \psi_n}{\partial x^2}$,

$$\int_{-\infty}^{\infty}\frac{1}{2}\frac{\partial \psi_n^2}{\partial t}dx + \int_{-\infty}^{\infty}\frac{\partial}{\partial x}\left(\psi_n\frac{\partial^2 \psi_n}{\partial x^2} - \frac{3}{2}\lambda_n\psi_n^2 - \frac{1}{2}\left(\frac{\partial \psi_n}{\partial x}\right)^2\right)dx$$

$$= B_n\int_{-\infty}^{\infty}\psi_n dx.$$

We have $B_n(t) = 0$ because $\psi_n \in L^2$ and decreases to zero when $x \to -\infty$. Since $u \cong 0$ for $x \to -\infty$, then from equation [9.18], it comes $\psi_n(x,t) = c_n(t)e^{k_n x}$,

$x \to -\infty$. By replacing the latter in equation [9.20], we obtain $\left(\frac{\partial c_n}{\partial t} + 4c_n k_n^3\right) e^{k_n x} = 0$, hence $c_n(t) = c_n(0) e^{-4k_n^3 t}$. Consequently, $\psi_n(x,t) = c_n(0) e^{k_n(x - 4k_n^2 t)}$. □

For the study of the continuous part of the spectrum $\lambda(t) = k^2(t)$, we proceed as follows: we assume that a stationary plane wave propagates from $x = -\infty$ and meets a potential $u(x,t)$ with a transmission coefficient T and a reflection coefficient R. In this case, equation [9.18] admits a solution ψ such that:

$$\psi = \begin{cases} T(k,t) e^{ikx}, & x \to +\infty \text{ (i.e. to the right of the potential barrier)} \\ e^{ikx} + R(k,t) e^{-ikx}, & x \to +-\infty \text{ (i.e. to the left of the potential barrier)} \end{cases}$$

where $|R|^2 + |T|^2 = 1$.

THEOREM 9.6.– If $u \simeq 0$ for $|x| \to \infty$, then we have $T(k,t) = T(k,0)$ and $R(k,t) = R(k,0) e^{-8ik^3 t}$, where $R(k,0)$ and $T(k,0)$ are determined by the initial condition $u(x,0) = u_0(x)$ of the KdV equation.

PROOF.– Choose λ = constant since the spectrum for $\lambda > 0$ is continuous. So equation [9.20] remains valid,

$$\frac{\partial \psi}{\partial t} + \frac{\partial^3 \psi}{\partial x^3} - 3(u + \lambda) \frac{\partial \psi}{\partial x} = \psi \int \frac{A}{\psi^2} dx + B\psi. \qquad [9.21]$$

For $u \cong 0$, when $x \to +\infty$, we replace $\psi = T(k,t) e^{ikx}$, $\lambda = k^2$ in equation [9.21] and we get $\frac{\partial T}{\partial t} - 4ik^3 T = \frac{A}{T} \int e^{-2ikx} dx + BT$. For this equation to preserve meaning when $x \to +\infty$, we must have $A = 0$, hence

$$\frac{\partial T}{\partial t} - (4ik^3 + B) T = 0. \qquad [9.22]$$

Similarly, for $u \cong 0$, when $x \to -\infty$, we replace $\psi = e^{ikx} + R(k,t) e^{-ikx}$, $\lambda = k^2$ in equation [9.21] and we get

$$\left(\frac{\partial R}{\partial t} + 4ik^3 R - BR\right) e^{-ikx} - (4ik^3 + B) e^{ikx}$$
$$= A(e^{ikx} + R e^{-ikx}) \int \frac{dx}{e^{2ikx} + R^2 e^{-2ikx} + 2R}.$$

For $x \to +\infty$, the equation above preserves a sense if $A = 0$ and is written as

$$\left(\frac{\partial R}{\partial t} + 4ik^3 R - BR\right) e^{-ikx} - (4ik^3 + B) e^{ikx} = 0.$$

For $4ik^3 + B = 0$, that is, $B = -4ik^3$, equation [9.22] implies that $T(k,t) = T(k,0)$ while the condition $\frac{\partial R}{\partial t} + 4ik^3 R - BR = 0$ gives us $R(k,t) = R(k,0)e^{-8ik^3 t}$. □

The knowledge of $c_n(t)$, $k_n(t)$, $n = 1, 2, ..., N$ and $R(k,t)$ allows us to express $u(x,t)$ for any time; it is the problem of the inverse diffusion. The latter is reduced to the solution $K(x,y;t)$ (to simplify the notations, the reader can obviously use $K(x,y)$ instead of $K(x,y;t)$) of the Gelfand–Levitan linear integral equation:

$$K(x,y;t) + I(x+y,t) + \int_{-\infty}^{x} I(y+z,t)K(x,z;t)dz = 0, \quad y \leq x \quad [9.23]$$

where $I(x+y,t) = \frac{1}{2\pi}\int_{-\infty}^{\infty} R(k,t)e^{-ik(x+y)}dk + \sum_{n=1}^{N} c_n^2(t)e^{k_n(t)(x+y)}$. The solution $u(x,t)$ of the KdV equation is then given (see [9.11]) by

$$u(x,t) = 2\frac{d}{dx}K(x,x;t). \quad [9.24]$$

The nonlinear KdV equation is transformed into the linear Gelfand–Levitan equation. The initial problem is thus completely solved. This method presents two major simplifications. First, in the analytical approach of the solution of the KdV equation, it suffices at each stage to solve only linear equations. Then t only appears parametrically and more than for all t the Gelfand–Levitan equation seems superficially to be an integral equation of two variables, actually x intervenes as a parameter and so we have to do to a family of integral equations for the functions $K(x,y)$ of a single variable y. Before dealing with the general case, that is, the case of distinct N solitons, let us return first to the case of a soliton and therefore consider the solution $u(x,t) = -\frac{c}{2}\text{sech}^2\frac{\sqrt{c}}{2}(x - ct)$, of the KdV equation obtained previously with the following initial condition: $u(x,0) = -2\text{sech}^2 x$, where by convention we put $c = 4$. The Schrödinger equation [9.18] is written as

$$\frac{\partial^2 \psi}{\partial x^2} + (2\text{sech}^2 x + \lambda)\psi = 0. \quad [9.25]$$

To study equation [9.25], one poses

$$\psi = A\,\text{sech}^\alpha x.w(x), \quad [9.26]$$

where A is an arbitrary amplitude, $\alpha^2 = -\lambda$ and w satisfies the following equation: $\frac{\partial^2 w}{\partial x^2} - 2\alpha\tanh x\frac{\partial w}{\partial x} + (2+\alpha-\alpha^2)\text{sech}^2 x.w = 0$. By doing the substitution $u = \frac{1}{2}(1-\tanh x)$, the last equation comes down to a hypergeometric differential equation or Gaussian equation: $u(1-u)\frac{\partial^2 w}{\partial u^2} + (c-(a+b+1)u)\frac{\partial w}{\partial u} - abw = 0$, where a, b, c denote constants and are equal to $a = 2+\alpha$, $b = -1+\alpha$, $c = 1+\alpha$. This equation presents three regular singular points: $u = 0$, $u = 1$, $u = \infty$. The solution of this equation for $u = 0$ is

$$w \equiv F(a,b,c,u) = 1 + \frac{ab}{c} \cdot \frac{u}{1!} + \frac{a(a+1)b(b+1)}{c(c+1)} \cdot \frac{u^2}{2!} \qquad [9.27]$$

$$+ \frac{a(a+1)...(a+n-1)b(b+1)...(b+n-1)}{c(c+1)...(c+n-1)} \cdot \frac{u^n}{n!} + \cdots$$

For $x \to \infty$ (i.e. when $u \to 0$), we have $w \to 1$. According to [9.26], we have $\psi = A2^\alpha(e^x + e^{-x})^{-\alpha}.w(x)$, and this one tends to $Ae^{2\alpha}e^{-\alpha x}$, $x \to \infty$. To represent a plane wave Ae^{ikx} going to $+\infty$, we will put $\alpha = -ik$. The asymptotic form of the wave function for $x \to -\infty$ ($u \to 1$) is obtained by transforming the hypergeometric function using the well-known functional relation:

$$F(a,b,c,u) = \frac{\Gamma(c)\Gamma(c-a-b)}{\Gamma(c-a)\Gamma(c-b)} F(a,b,a+b-c+1,1-u)$$

$$+ (1-u)^{c-a-b} \frac{\Gamma(c)\Gamma(a+b-c)}{\Gamma(a)\Gamma(b)} F(c-a,c-b,c-a-b+1,1-u),$$

where $\Gamma(z) = \int_0^\infty e^{-t} e^{z-1} dt$, Re $z > 0$, is the Euler Gamma function. Taking into account [9.27] and the expression above, relation [9.26] becomes

$$\psi = A \operatorname{sech}^\alpha x \left[\frac{\Gamma(c)\Gamma(c-a-b)}{\Gamma(c-a)\Gamma(c-b)} \left(1 + \frac{ab}{a+b-c+1}(1-u) + \cdots \right) \right.$$

$$\left. + (1-u)^{c-a-b} \frac{\Gamma(c)\Gamma(a+b-c)}{\Gamma(a)\Gamma(b)} \left(1 + \frac{(c-a)(c-b)}{c-a-b+1}(1-u) + \cdots \right) \right].$$

When $u \to 1$ ($x \to -\infty$), we have $(1-u)^{c-a-b} \to e^{-2\alpha x}$ and since $\alpha = -ik$, then

$$\psi \longrightarrow Ae^\alpha \frac{\Gamma(c)\Gamma(a+b-c)}{\Gamma(a)\Gamma(b)} \left(e^{ikx} + \frac{\Gamma(c-a-b)\Gamma(a)\Gamma(b)}{\Gamma(c-a)\Gamma(c-b)\Gamma(a+b-c)} \right).$$

This last expression combined with the fact (already seen) that ψ tends to $Ae^{2\alpha} e^{-\alpha}$ when $x \to \infty$ gives us the transmission coefficient $T = \frac{\Gamma(a)\Gamma(b)}{\Gamma(c)\Gamma(a+b-c)}$ and the reflection coefficient $R = \frac{\Gamma(c-a-b)\Gamma(a)\Gamma(b)}{\Gamma(c-a)\Gamma(c-b)\Gamma(a+b-c)}$. Here, we have $k_1 = 1$, $c(0) = \sqrt{2}$, $R(k,0) = 0$. For an individual soliton, equation [9.1] has a precise solution. It turns out that the soliton of amplitude u_0 has only one discrete level with eigenvalue $\lambda = \frac{u_0}{2}$, while the next level corresponds to the point $\lambda = 0$ (with the respective eigenfunction $\psi = \tanh x$) and already belongs to the continuous spectrum. The Gelfand–Levitan equation [9.23] where $I(\mu, t) = c_1^2(t)e^{k_1\mu} = c_1^2(0)e^{-8k_1 t}e^{k_1 t} = 2e^{-8t+\mu}$ is written as

$$K(x, y; t) + 2e^{-8t+x+y} + 2e^{-8t+y}\int_{-\infty}^{x} e^z K(x, z; t)dz = 0.$$

By putting $K(x, y, t) = f(x)e^y$, into this equation, we obtain $f(x) + 2e^{-8t+x} + e^{-8t+2x}f(x) = 0$, hence $f(x) = -2\frac{e^{-x}}{1+e^{8t-2x}}$. Therefore, solution [9.24] of the KdV equation in the case of a solitary wave is given as

$$u(x, t) = 2\frac{d}{dx}K(x, x, t) = -\frac{2}{\cosh^2(x - 4t)} = -2\operatorname{sech}^2(x - 4t).$$

This illustrates the method and the correspondence between eigenvalue and solution.

We will now look at the case of N-solitons through the procedure suggested by Gardner et al. (1967) and use the results of Wadati and Toda (1972). In order to solve the Gelfand–Levitan equation [9.23], where $R(k, t) = 0$, one poses

$$K(x, y) = \sum_{n=1}^{N} w_n(x, t)e^{k_n y}, \qquad [9.28]$$

where w_n are functions to be determined. By replacing this expression in the Gelfand–Levitan equation, we obtain the following linear system:

$$\begin{cases} w_1(x, t) + c_1^2(t)e^{k_1 x} + \sum_{m=1}^{N} c_1^2(t)\frac{e^{(k_1+k_m)x}}{k_1+k_m}w_m(x, t) = 0, \\ \vdots \\ w_N(x, t) + c_N^2(t)e^{k_N x} + \sum_{m=1}^{N} c_N^2(t)\frac{e^{(k_N+k_m)x}}{k_N+k_m}w_m(x, t) = 0. \end{cases}$$

Define the following notations:

$$A = \left(c_n^2(t)e^{(k_n+k_m)x}\right), \quad W = \begin{pmatrix} w_1 \\ \vdots \\ w_N \end{pmatrix}, \quad G = \begin{pmatrix} c_1^2(t)e^{k_1 x} \\ \vdots \\ c_N^2(t)e^{k_N x} \end{pmatrix},$$

$$P \equiv (P_{nm}) = \left(\delta_{nm} + c_n^2(t)\frac{e^{(k_n+k_m)x}}{k_n+k_m}\right) = I + A, \qquad [9.29]$$

where I is the unit matrix. The system above is written, $PW = -G$, and it is easy to show that it has a unique solution. From equation [9.28], we draw

$$K(x,x) = h^\top w = -h^\top P^{-1} G, \qquad h \equiv \begin{pmatrix} e^{k_1 x} \\ \vdots \\ e^{k_N x} \end{pmatrix}, \qquad P^{-1} = \frac{\alpha_{nm}}{\det P},$$

where α_{nm} is the cofactor of P. Or

$$\frac{d}{dx} P_{nm} = c_m^2 e^{k_n x} . e^{k_m x}, \qquad \det P = \sum_{n=1}^{N} \left(\delta_{nm} + c_n^2(t) \frac{e^{(k_n+k_m)x}}{k_n + k_m} \right) \alpha_{nm},$$

so

$$K(x,x) = -\sum_{n,m} \frac{\alpha_{nm}}{\det P} \frac{d}{dx} P_{nm} = -\frac{1}{\det P} \frac{d}{dx} (\det P) = -\frac{d}{dx} \ln \det P,$$

and according to [9.24], $u = 2\frac{d}{dx} K(x,x) = -2\frac{d^2}{dx^2} \ln \det P$. Therefore, we have the following theorem:

THEOREM 9.7.– The solution of the KdV equation is $u = -2\frac{d^2}{dx^2} \ln \det P$, where P is defined by [9.29] and whose $c_n(t) = c_n(0) e^{-4k_n^3 t}$, with $k_n > 0$ distinct.

The function obtained in theorem 9.7 is negative for all x, continuous and behaves like the exponential when $|x| \to \infty$. To get an idea of the behavior of solitons and in particular their asymptotic behavior, suppose that $k_1 < k_2 < \ldots < k_{N-1} < k_N$. But before this, we need the following lemma (Muir 1960) and the remark below:

LEMMA 9.1.–

$$\Delta \equiv \begin{vmatrix} \frac{1}{a_1-b_1} & \frac{1}{a_1-b_2} & \cdots & \frac{1}{a_1-b_n} \\ \frac{1}{a_2-b_1} & \frac{1}{a_2-b_2} & \cdots & \frac{1}{a_2-b_n} \\ \vdots & & \ddots & \\ \frac{1}{a_n-b_1} & \frac{1}{a_n-b_2} & \cdots & \frac{1}{a_n-b_n} \end{vmatrix} = (-1)^{\frac{n(n-1)}{2}} \frac{\prod_{j<k}(a_j - a_k) \prod_{j<k}(b_j - b_k)}{\prod_{j,k}(a_j - b_k)}.$$

PROOF.– Let $f(a_j) \equiv (a_j - b_1)(a_j - b_2)\ldots(a_j - b_n)$. We have

$$(f(a_1) f(a_2) \ldots f(a_n)) . \Delta = \left(\prod_{j,k} (a_j - b_k) \right) . \Delta = \left| \frac{f(a_j)}{a_j - b_k} \right|.$$

Let us do $a_1 = b_1$, $a_2 = b_2$,...,$a_n = b_n$, all the elements of $\left|\frac{f(a_j)}{a_j - b_k}\right|_{\substack{a_j = b_j \\ 1 \leq j \leq n}}$ are equal to zero except those on the diagonal. Therefore,

$$\left|\frac{f(a_j)}{a_j - b_k}\right|_{\substack{a_j = b_j \\ 1 \leq j \leq n}} = (b_j - b_1)(b_j - b_2)...(b_j - b_{j-1})(b_j - b_{j+1})...(b_j - b_n),$$

$$= (-1)^{\frac{n(n-1)}{2}}(b_1 - b_j)...(b_{j-1} - b_j)(b_j - b_{j+1})...(b_j - b_n).$$

So we have

$$\frac{f(a_1)}{a_1 - b_1} = (-1)^0 (b_1 - b_2)...(b_1 - b_n),$$

$$\frac{f(a_2)}{a_2 - b_2} = (-1)^1 (b_1 - b_2)(b_2 - b_3)...(b_2 - b_n),$$

$$\vdots$$

$$\frac{f(a_n)}{a_n - b_n} = (-1)^{n-1}(b_1 - b_n)...(b_n - b_{n-1}),$$

and therefore

$$\prod_{j=1}^{n} \frac{f(a_j)}{a_j - b_j} = (-1)^{\frac{n(n-1)}{2}} \left(\prod_{j<k}(b_j - b_k)\right)^2.$$

We use a similar reasoning for the a_j and we finally get the expression

$$\left(\prod_{j,k}(a_j - b_k)\right) . \Delta = (-1)^{\frac{n(n-1)}{2}} \prod_{j<k}(a_j - a_k)\prod_{j,k}(b_j - b_k).$$

This completes the demonstration. □

REMARK 9.3.– Consider the following determinant:

$$\Delta = \begin{vmatrix} 1 + c_1^2 b_{11} & c_1^2 b_{12} & c_1^2 b_{13} \\ c_2^2 b_{21} & 1 + c_2^2 b_{22} & c_2^2 b_{23} \\ c_3^2 b_{31} & c_3^2 b_{32} & 1 + c_3^2 b_{33} \end{vmatrix},$$

where $c_1 c_2 c_3 \neq 0$. If we divide the first line by c_1, the second line by c_2 and the third line by c_3, we will have

$$\Delta = c_1 c_2 c_3 \begin{vmatrix} \frac{1+c_1^2 b_{11}}{c_1} & c_1 b_{12} & c_1 b_{13} \\ c_2 b_{21} & \frac{1+c_2^2 b_{22}}{c_2} & c_2 b_{23} \\ c_3 b_{31} & c_3 b_{32} & \frac{1+c_3^2 b_{33}}{c_3} \end{vmatrix}.$$

Multiply the first column by c_1, the second column by c_2 and the third column by c_3, we get

$$\Delta = \begin{vmatrix} 1+c_1^2 b_{11} & c_1 c_2 b_{12} & c_1 c_3 b_{13} \\ c_2 c_1 b_{21} & 1+c_2^2 b_{22} & c_2 c_3 b_{23} \\ c_3 c_1 b_{31} & c_3 c_2 b_{32} & 1+c_3^2 b_{33} \end{vmatrix}.$$

So for the determinant of order N, we use the same procedure, that is, by dividing the jth row by c_j and multiply the jth column by c_j.

THEOREM 9.8.– The explicit solution of N-solitons of the KdV equation is given by

$$u(x,t) = \begin{cases} -2 \sum_{n=1}^{N} k_n^2 \operatorname{sech}^2(k_n \xi_n + \delta_n^+), & t \to +\infty \\ -2 \sum_{n=1}^{N} k_n^2 \operatorname{sech}^2(k_n \xi_n + \delta_n^-), & t \to -\infty \end{cases}$$

where $\delta_n^+ \equiv \frac{1}{2} \ln \frac{c_n^2}{2k_n} \left(\prod_{j=1}^{n-1} \frac{k_j - k_n}{k_j + k_n} \right)^2$, $\delta_n^- \equiv \frac{1}{2} \ln \frac{c_n^2}{2k_n} \left(\prod_{j=n+1}^{N} \frac{k_j - k_n}{k_j + k_n} \right)^2$, are the phase changes.

PROOF.– The determinant of the matrix P is written explicitly in the form

$\det P =$

$$\begin{vmatrix} 1 + \frac{c_1^2(t)}{2k_1} e^{2k_1 x} & \frac{c_1^2(t)}{k_1+k_2} e^{(k_1+k_2)x} & \cdots & \frac{c_1^2(t)}{k_1+k_j} e^{(k_1+k_j)x} & \cdots & \frac{c_1^2(t)}{k_1+k_N} e^{(k_1+k_N)x} \\ \frac{c_2^2(t)}{k_2+k_1} e^{(k_2+k_1)x} & 1 + \frac{c_2^2(t)}{2k_2} e^{2k_2 x} & & \frac{c_2^2(t)}{k_2+k_j} e^{(k_2+k_j)x} & \cdots & \frac{c_2^2(t)}{k_2+k_N} e^{(k_2+k_N)x} \\ \vdots & & \ddots & & & \\ \frac{c_N^2(t)}{k_N+k_1} e^{(k_N+k_1)x} & \frac{c_N^2(t)}{k_N+k_2} e^{(k_N+k_2)x} & \cdots & \frac{c_N^2(t)}{k_N+k_j} e^{(k_N+k_j)x} & & 1 + \frac{c_N^2(t)}{2k_N} e^{2k_N x} \end{vmatrix}$$

Applying the previous remark to the determinant $\det P$ above, we obtain

$$\det P = \begin{vmatrix} 1 + \frac{c_1^2(t)}{2k_1} e^{2k_1 x} & \frac{c_1(t)c_2(t)}{k_1+k_2} e^{(k_1+k_2)x} & \cdots & \frac{c_1(t)c_N(t)}{k_1+k_N} e^{(k_1+k_N)x} \\ \frac{c_2(t)c_1(t)}{k_2+k_1} e^{(k_2+k_1)x} & 1 + \frac{c_2^2(t)}{2k_2} e^{2k_2 x} & \cdots & \frac{c_2(t)c_N(t)}{k_2+k_N} e^{(k_2+k_N)x} \\ \vdots & & \ddots & \\ \frac{c_N(t)c_1(t)}{k_N+k_1} e^{(k_N+k_1)x} & \frac{c_N(t)c_2(t)}{k_N+k_2} e^{(k_N+k_2)x} & \cdots & 1 + \frac{c_N^2(t)}{2k_N} e^{2k_N x} \end{vmatrix}$$

Since $c_j(t) = c_j(0)e^{-4k_j^3 t}$, then

$$\det P = \begin{vmatrix} 1 + \frac{c_1^2(0)}{2k_1}e^{2k_1\xi_1} & \frac{c_1(0)c_2(0)}{k_1+k_2}e^{k_1\xi_1+k_2\xi_2} & \cdots & \frac{c_1(0)c_N(0)}{k_1+k_N}e^{k_1\xi_1+k_N\xi_N} \\ \frac{c_2(0)c_1(0)}{k_2+k_1}e^{k_2\xi_2+k_1\xi_1} & 1 + \frac{c_2^2(0)}{2k_2}e^{2k_2\xi_2} & \cdots & \frac{c_2(0)c_N(0)}{k_2+k_N}e^{k_2\xi_2+k_N\xi_N} \\ \vdots & & \ddots & \\ \frac{c_N(0)c_1(0)}{k_N+k_1}e^{k_N\xi_N+k_1\xi_1} & \frac{c_N(0)c_2(0)}{k_N+k_2}e^{k_N\xi_N+k_2\xi_2} & \cdots & 1 + \frac{c_N^2(0)}{2k_N}e^{2k_N\xi_N} \end{vmatrix},$$

where $\xi_n \equiv x - 4k_j^2 t$, $1 \leq j \leq N$. To get an idea of the behavior of solitons and their asymptotic behavior, suppose that $k_1 < k_2 < \ldots < k_{N-1} < k_N$. For $t \gg 0$, let us write ξ_j in the form $\xi_j \equiv \xi_n - \varepsilon_{jn}t$, $1 \leq j \leq N$ with $\varepsilon_{jn} \equiv 4k_j^2 - 4k_n^2$ and $c_j(0) \equiv c_j$. Note that $\varepsilon_{jn} < 0$ if $1 \leq j < n$, $\varepsilon_{nn} = 0$, $\varepsilon_{jn} > 0$ if $n < j \leq N$, and $\varepsilon_{jn} = -\varepsilon_{nj}$. We have $\varepsilon_{nm} > \varepsilon_{(n-1)m} > \ldots > \varepsilon_{(m+1)n} > 0$ if $n > m$, and $\varepsilon_{nm} < \varepsilon_{n(m-1)} < \ldots < \varepsilon_{n(m+1)} < 0$ if $n < m$. Replace these expressions in the determinant above and approximate the elements of the diagonal (for $j < n$) as follows: $1 + \frac{c_j^2}{2k_j}e^{2k_j(\xi_n - \varepsilon_{jn}t)} \simeq \frac{c_j^2}{2k_j}e^{2k_j(\xi_n - \varepsilon_{jn}t)}$, $j < n$, $t \to \infty$ (since for $j < n$, we have $\varepsilon_{jn} < 0$ and $1 + e^x \simeq e^x$ for $x \to \infty$). We put in factor the common expressions: $e^{2k_1(\xi_n - \varepsilon_{1n}t)}$, $e^{2k_2(\xi_n - \varepsilon_{2n}t)}, \ldots, e^{2k_{n-1}(\xi_n - \varepsilon_{(n-1)n}t)}$. By turning t to infinity, we have (since $\varepsilon_{jn} > 0$ for $n \leq j \leq N$):

$$\det P = C \begin{vmatrix} \frac{c_1^2}{2k_1} & \frac{c_1 c_2}{k_1+k_2} & \cdots & \frac{c_1 c_{n-1}}{k_1+k_{n-1}} & \frac{c_1 c_n}{k_1+k_n}e^{k_n\xi_n} & 0 & \cdots & 0 \\ \frac{c_2 c_1}{k_2+k_1} & \frac{c_2^2}{2k_2} & \cdots & \frac{c_2 c_{n-1}}{k_2+k_{n-1}} & \frac{c_2 c_n}{k_2+k_n}e^{k_n\xi_n} & 0 & \cdots & 0 \\ \vdots & & \ddots & \vdots & \vdots & \vdots & \ldots & \vdots \\ \frac{c_{n-1}c_1}{k_{n-1}+k_1} & \frac{c_{n-1}c_2}{k_{n-1}+k_2} & \cdots & \frac{c_{n-1}^2}{2k_{n-1}} & \frac{c_{n-1}c_n}{k_{n-1}+k_n}e^{k_n\xi_n} & 0 & \cdots & 0 \\ \frac{c_n c_1}{k_n+k_1}e^{k_n\xi_n} & \frac{c_n c_2}{k_n+k_2}e^{k_n\xi_n} & \cdots & \frac{c_n c_{n-1}}{k_n+k_{n-1}}e^{k_n\xi_n} & 1 + \frac{c_n^2}{2k_n}e^{2k_n\xi_n} & 0 & \cdots & 0 \\ 0 & 0 & \cdots & 0 & 0 & 1 & \cdots & 0 \\ \vdots & \vdots & \vdots & \vdots & \vdots & & \ddots & \vdots \\ 0 & 0 & \cdots & 0 & 0 & 0 & \cdots & 1 \end{vmatrix},$$

where $C \equiv \prod_{j=1}^{n-1} e^{2k_j(\xi_n - \varepsilon_{jn})}$. Obviously, we have

$$\det P = C \begin{vmatrix} \frac{c_1^2}{2k_1} & \frac{c_1 c_2}{k_1+k_2} & \cdots & \frac{c_1 c_{n-1}}{k_1+k_{n-1}} & \frac{c_1 c_n}{k_1+k_n}e^{k_n\xi_n} \\ \frac{c_2 c_1}{k_2+k_1} & \frac{c_2^2}{2k_2} & \cdots & \frac{c_2 c_{n-1}}{k_2+k_{n-1}} & \frac{c_2 c_n}{k_2+k_n}e^{k_n\xi_n} \\ \vdots & & \ddots & & \vdots \\ \frac{c_{n-1}c_1}{k_{n-1}+k_1} & \frac{c_{n-1}c_2}{k_{n-1}+k_2} & \cdots & \frac{c_{n-1}^2}{2k_{n-1}} & \frac{c_{n-1}c_n}{k_{n-1}+k_n}e^{k_n\xi_n} \\ \frac{c_n c_1}{k_n+k_1}e^{k_n\xi_n} & \frac{c_n c_2}{k_n+k_2}e^{k_n\xi_n} & \cdots & \frac{c_n c_{n-1}}{k_n+k_{n-1}}e^{k_n\xi_n} & 1 + \frac{c_n^2}{2k_n}e^{2k_n\xi_n} \end{vmatrix}.$$

This determinant is still written in the form

$$\det P = C \prod_{l=1}^{n-1} c_l^2 \begin{vmatrix} \frac{1}{2k_1} & \frac{1}{k_1+k_2} & \cdots & \frac{1}{k_1+k_{n-1}} \\ \frac{1}{k_2+k_1} & \frac{1}{2k_2} & \cdots & \frac{1}{k_2+k_{n-1}} \\ \vdots & & \ddots & \vdots \\ \frac{1}{k_{n-1}+k_1} & \frac{1}{k_{n-1}+k_2} & \cdots & \frac{1}{2k_{n-1}} \end{vmatrix}$$

$$+ C \prod_{l=1}^{n} c_l^2 \begin{vmatrix} \frac{1}{2k_1} & \frac{1}{k_1+k_2} & \cdots & \frac{1}{k_1+k_{n-1}} & \frac{1}{k_1+k_n} \\ \frac{1}{k_2+k_1} & \frac{1}{2k_2} & \cdots & \frac{1}{k_2+k_{n-1}} & \frac{1}{k_2+k_n} \\ \vdots & & \ddots & & \vdots \\ \frac{1}{k_{n-1}+k_1} & \frac{1}{k_{n-1}+k_2} & \cdots & \frac{1}{2k_{n-1}} & \frac{1}{k_{n-1}+k_n} \\ \frac{1}{k_n+k_1} & \frac{1}{k_n+k_2} & \cdots & \frac{1}{k_n+k_{n-1}} & \frac{1}{2k_n} \end{vmatrix},$$

for $n \geq 2$. For $n = 1$, it equals $1 + \frac{c_1^2}{2k_1}e^{2k_1\xi_1}$. Using the previous lemma, we get,

$\det P =$

$$\prod_{i=1}^{n-1} e^{2k_i(\xi_n - \varepsilon_{in}t)} \left(\prod_{j=1}^{n-1} c_j^2 \frac{(\prod_{i<j}(k_i - k_j))^2}{\prod_{i,j}(k_i + k_j)} + \prod_{j=1}^{n-1} c_j^2 \frac{(\prod_{i<j}(k_i - k_j))^2}{\prod_{i,j}(k_i + k_j)} e^{2k_n\xi_n} \right),$$

for $t \gg 0$. By replacing this expression in the solution obtained in theorem 9.7, we obtain the explicit solution of N-soliton:

$$u(x,t) = -2 \sum_{n=1}^{N} k_n^2 \operatorname{sech}^2 \left(k_n\xi_n + \frac{1}{2} \ln \frac{c_n^2}{2k_n} \left(\prod_{j=1}^{n-1} \frac{k_j - k_n}{k_j + k_n} \right)^2 \right), \quad t \to +\infty.$$

Similarly, it is shown that for $t \ll 0$

$$u(x,t) = -2 \sum_{n=1}^{N} k_n^2 \operatorname{sech}^2 \left(k_n\xi_n + \frac{1}{2} \ln \frac{c_n^2}{2k_n} \left(\prod_{j=n+1}^{N} \frac{k_j - k_n}{k_j + k_n} \right)^2 \right), \quad t \to -\infty.$$

This completes the demonstration. □

REMARK 9.4.– When $t \to \infty$, this result can be interpreted as follows:

$$\lim_{t \to \infty} u(x,t) = \lim_{t \to \infty} u(x - ct) = \begin{cases} -2k_n^2 \operatorname{sech}^2 \left(k_n(x - 4k_n^2 t) + \delta_n^+ \right) & \text{if } c = 4k_n^2 \\ 0 & \text{if } c \neq 4k_n^2 \end{cases}$$

This is the form of a solitary wave of amplitude $2k_n^2$, propagating on the right with a constant velocity equal to $4k_n^2$. The solution of the KdV equation actually splits into N-solitons at the limit for $|t| \to \infty$. This indicates that each soliton preserves its shape after collisions. These are analyzed by the phase changes δ_n^+ and δ_n^-. The relative phase change is determined by

$$\delta_n^+ - \delta_n^- = \frac{1}{2}\ln\frac{c_n^2}{2k_n}\left(\prod_{j=1}^{n-1}\frac{k_j-k_n}{k_j+k_n}\right)^2 - \frac{1}{2}\ln\frac{c_n^2}{2k_n}\left(\prod_{j=n+1}^{N}\frac{k_j-k_n}{k_j+k_n}\right)^2,$$

$$= \sum_{j=1}^{n-1}\ln\frac{k_j-k_n}{k_j+k_n} - \sum_{j=n+1}^{N}\ln\frac{k_j-k_n}{k_j+k_n},$$

and it is expressed in terms of k_j ($1 \le j \le N$). Since k_j are invariant with respect to time, then the $\delta_n^+ - \delta_n^-$ are also invariant. Recall that we assumed $k_1 < k_2 < ... < k_N$, then $\delta_1^+ - \delta_1^- = -\sum_{j=2}^{N}\ln\frac{k_j-k_1}{k_j+k_1} > 0$, $\delta_N^+ - \delta_N^- = \sum_{j=1}^{N-1}\ln\frac{k_N-k_1}{k_N+k_1} < 0$. In addition, it is easy to show that $\sum_{n=1}^{N}\delta_n^+ = \sum_{n=1}^{N}\delta_n^-$.

The KdV equation [9.1] is written in the form $\frac{\partial u}{\partial t} = \frac{\partial}{\partial x}\left(3u^2 - \frac{\partial^2 u}{\partial x^2}\right) = \frac{\partial}{\partial x}\frac{\delta H}{\delta u}$, where $H = \int_{-\infty}^{\infty} P_3 dx = \int_{-\infty}^{\infty}\left(u^3 + \frac{1}{2}\left(\frac{\partial u}{\partial x}\right)^2\right)dx$ is the first integral (Hamiltonian) obtained previously and the symbol $\frac{\delta H}{\delta u(x)}$ denotes the variational derivative (Fréchet derivative) of H, $\frac{\delta H}{\delta u(x)} = 3u^2 + 2\frac{\partial^2 u}{\partial x^2}$. The KdV equation forms an infinite dimensional Hamiltonian system, completely integrable and the Hamiltonian structure is defined by the Poisson bracket: $\{F, H\} = \int \frac{\delta F}{\delta u(x)}\frac{\partial}{\partial x}\frac{\delta H}{\delta u(x)}dx$. We check that the latter satisfies the Jacobi identity. We will discuss further (in the following chapter) the problem of studying the KdV equation via symplectic structures on operator algebra, the relation with the KP hierarchy, the Sato theory τ functions and the work of Jimbo–Miwa–Kashiwara. Moreover, the study of the periodic problem for the KdV equation has allowed some authors to discover an interesting class of completely integrable systems. The obtained solutions are endowed with remarkable properties: they define functions $u(x)$ for which equation [9.2] with periodic coefficients has a finite number of zones of parametric resonance on the axis λ. The spectrum of the Schrödinger operator is invariant by the Hamiltonian flow defined by the KdV equation. And as we have already pointed out, this spectrum provides an infinity of first integrals or invariants. The isospectral sets related to invariant manifolds defined by putting these invariants equal to generic constants are compact, connected and infinite-dimensional tori. Each of these isospectral sets is isomorphic to the real part of a Jacobi variety associated with a hyperelliptic curve of finite or infinite genus. The periods of this torus can be expressed using hyperelliptic integrals; in short, the explicit linearization of the flow of the KdV equation is made on this Jacobian variety using the Abel application, the Jacobi inversion problem and the theta functions.

9.4. Exercises

EXERCISE 9.1.– Consider a one dimensional nonlinear parabolic partial differential equation (the Burgers equation): $\frac{\partial w}{\partial t} + w\frac{\partial w}{\partial x} - \alpha\frac{\partial^2 w}{\partial x^2} = 0$, where α is a constant. Show that with the help of the Cole–Hopf transformation, $w = -2\frac{\alpha}{u}\cdot\frac{\partial u}{\partial x}$, there is a connection between the Burgers equation and the well-known heat equation: $\frac{\partial u}{\partial t} = \alpha\frac{\partial^2 u}{\partial x^2}$, where $u(x,t)$ is the solution of the heat equation and $w(x,t)$ is the solution of the Burgers equation. Determine a particular solution of the heat equation and deduce the solution of the Burgers equation.

EXERCISE 9.2.– Let $w(x)$ be a solution of the Schrödinger equation: $\frac{\partial^2 w}{\partial x^2} - \mathcal{U}(x)w = 0$, where the function $\mathcal{U}(x)$ is the potential energy of the particle (we also simplified the notation by choosing the Planck constant $\hbar = 1$ and the mass of the particle $m = \frac{1}{2}$). Show that the function $v(x)$ defined by $v = \frac{1}{w}\cdot\frac{\partial w}{\partial x}$ is a solution of the Riccati equation: $\frac{\partial v}{\partial x} + v^2 = \mathcal{U}$.

EXERCISE 9.3.– Let $w(x,t)$ be a solution of the modified KdV (mKdV) equation: $\frac{\partial w}{\partial t} - 6w^2\frac{\partial w}{\partial x} + \frac{\partial^3 w}{\partial x^3} = 0$.

a) Show that the function $u(x,t)$ defined by the Miura transformation (example of Bäcklund transformation, as mentioned in exercise 9.5): $u = w^2 \pm \frac{\partial w}{\partial x}$, $(x,t) \in \mathbb{R}^2$, satisfies the KdV equation: $\frac{\partial u}{\partial t} - 6u\frac{\partial u}{\partial x} + \frac{\partial^3 u}{\partial x^3} = 0$.

b) Show that all of the solutions of the mKdV equation, decaying sufficiently rapidly as $|x| \to \infty$, map into a sparse solution set of the KdV equation. How can we revert the process, that is, how can we transfer solutions of the KdV equation to solutions of the mKdV equation?

EXERCISE 9.4.– Consider the Gardner equation: $\frac{\partial w}{\partial t} + w\frac{\partial w}{\partial x} - \frac{\lambda^2}{6}w^2\frac{\partial w}{\partial x} + \frac{\partial^3 w}{\partial x^3} = 0$, where λ is a real parameter. Show that if w satisfies the Gardner equation, then the Gardner transformation: $u = w + \lambda\frac{\partial w}{\partial x} - \frac{\lambda^2}{6}w^2$, satisfies the KdV equation. Like the KdV equation, show that the Gardner equation is integrable by the inverse scattering method.

EXERCISE 9.5.– A Bäcklund transformation transforms a nonlinear partial differential equation into another partial differential equation, and a solution to the second partial differential equation must be compatible with the first partial differential equation. The simplest example of a Bäcklund transformation is given by the Cauchy–Riemann equations in complex analysis: $\frac{\partial u}{\partial x} = \frac{\partial v}{\partial y}$, $\frac{\partial u}{\partial y} = -\frac{\partial v}{\partial x}$. Both u and v solve Laplace's equations: $\frac{\partial^2 u}{\partial x^2} + \frac{\partial^2 u}{\partial y^2} = 0$, $\frac{\partial^2 v}{\partial x^2} + \frac{\partial^2 v}{\partial y^2} = 0$. Examples of well-known Bäcklund transformations are the Miura transformation (see exercise 9.3), which transforms the KdV equation into the modified KdV equation, and the Gardner transformation (see exercise 9.4), which transforms the KdV equation into the Gardner equation. The Bäcklund transformation is called the auto-Bäcklund transformation if the two nonlinear partial differential equations are the same. In other words, an auto-Bäcklund transformation is a transform that leaves a partial differential equation invariant. Let us reconsider the Burgers equation (see exercise 9.1): $\frac{\partial w}{\partial t} + w\frac{\partial w}{\partial x} - \alpha\frac{\partial^2 w}{\partial x^2} = 0$, where α is a constant. Show that the Bäcklund transformation that transforms the Burgers equation into the heat equation: $\frac{\partial u}{\partial t} = \frac{\partial^2 u}{\partial x^2}$ is given by $\frac{\partial u}{\partial x} = -\frac{uw}{2\alpha}$, $\frac{\partial u}{\partial t} = \frac{uw^2}{4\alpha} - \frac{w}{2}\frac{\partial u}{\partial x}$. Given a solution of the heat equation, show how one can derive a solution to the Burgers equation by using the Bäcklund transformation above.

EXERCISE 9.6.– Consider the sine-Gordon equation: $\frac{\partial^2 u}{\partial t^2} - \frac{\partial^2 u}{\partial x^2} + \sin u = 0$, $u = u(x,t)$, see exercise 4.9. Show that when looking for soliton solutions of this equation of the form $u(\zeta) \equiv u(x - vt)$ (where v is an arbitrary velocity and which $u \to 0$, $\frac{\partial u}{\partial \zeta} \to 0$, when $\zeta \to \pm\infty$), we obtain $u(x,t) = 4\arctan\left(\exp\frac{x-vt}{\sqrt{1-v^2}}\right)$ (hint: the following trigonometric formulas may be used, $\sin 2\alpha = \frac{2\tan\alpha}{1+\tan^2\alpha}$, $\tan 2\alpha = \frac{2\tan\alpha}{1-\tan^2\alpha}$). Show that the sine-Gordon equation admits more solutions of the form $u(x,t) = 4\arctan\left(\frac{f(x)}{g(t)}\right)$, where f and g are arbitrary functions.

EXERCISE 9.7.– The Camassa–Holm equation (mentioned in section 9.1) is a nonlinear partial differential equation defined by,

$$\frac{\partial u}{\partial t} - \frac{\partial^3 u}{\partial t \partial x^2} + 3u\frac{\partial u}{\partial x} = 2\frac{\partial u}{\partial x}\frac{\partial^2 u}{\partial x^2} + u\frac{\partial^3 u}{\partial x^3} - 2\alpha\frac{\partial u}{\partial x},$$

where the function $u(x,t)$ of the real variables x and t represents the fluid velocity in the x-direction at time by an observer moving at speed α (real constant). This equation was derived by Camassa and Holm by means of physical principles, showing that it models the propagation of unidirectional waves on shallow water. Let us put $v \equiv u - \frac{\partial^2 u}{\partial x^2} + \alpha$. Show that the above equation can be written in the form, $\frac{\partial v}{\partial t} + u\frac{\partial v}{\partial x} + 2v\frac{\partial u}{\partial x} = 0$. Determine the solutions of this last equation and deduce those of the Camassa–Holm equation. Determine a pair of Lax and analyze the results obtained in this problem.

EXERCISE 9.8.– The Davey–Stewartson equations (a "generalization" of the nonlinear Schrödinger equation) may be written as

$$i\frac{\partial \psi}{\partial t} + \frac{\partial^2 \psi}{\partial x^2} - a^2\frac{\partial^2 \psi}{\partial y^2} + 2\left(\varphi \pm |\psi|^2\right)\psi = 0,$$

$$\frac{\partial^2 \varphi}{\partial x^2} + a^2\frac{\partial^2 \varphi}{\partial y^2} \pm 2\left|\frac{\partial^2 \psi}{\partial x^2}\right| = 0,$$

where $a = i$ or $a = 1$. Here, $\psi(x,y,t)$ and $\varphi(x,y,t)$ are functions of the real variables x, y ant t, the latter being real-valued and the former being complex-valued. This model was first derived in the context of a water wave with purely physical considerations where ψ is the amplitude of a surface wave packet, while φ is the velocity potential of the mean flow interacting with the surface wave. Study the complete integrability of these equations. Determine and analyze the algebro-geometric solutions of these equations.

EXERCISE 9.9.– Let us reconsider the nonlinear Schrödinger equation (NLS) (see exercise 4.8): $i\frac{\partial \psi}{\partial t} + \frac{\partial^2 \psi}{\partial x^2} + 2|\psi|^2\psi = 0$, $\psi \in \mathbb{C}$. Show that one can solve it using the inverse scattering method and show that this equation can be asymptotically reduced to the Korteweg-de Vries equation in the unidirectional weakly nonlinear long-wave limit.

EXERCISE 9.10.– The sine-Gordon equation (exercise 9.6): $\frac{\partial^2 u}{\partial t^2} - \frac{\partial^2 u}{\partial x^2} + \sin u = 0$, $u = u(x,t)$, can be written in the form (where $\xi = \frac{1}{2}(x-t)$ and $\eta = \frac{1}{2}(x+t)$): $\frac{\partial^2 u}{\partial \xi \partial \eta} = \sin u$. Show that an auto-Bäcklund transformation (see exercise 9.5) for the latter differential equation can be defined by the following pair of coupled partial differential equations (where $\varepsilon \neq 0$, $u = u(\xi,\eta)$ and $v = v(\xi,\eta)$): $\frac{\partial v}{\partial \xi} = \frac{\partial u}{\partial \xi} + 2\varepsilon \sin\left(\frac{v+u}{2}\right)$, $\frac{\partial v}{\partial \eta} = -\frac{\partial u}{\partial t} + \frac{2}{\varepsilon}\varepsilon \sin\left(\frac{v-u}{2}\right)$. Determine a solution of the latter differential equation using the Bäcklund transformation method.

EXERCISE 9.11.– Using the change of coordinates: $\xi = \frac{1}{2}(x - iay)$, $\eta = \frac{1}{2}(x + iay)$, where $a = i$ or $a = 1$, show that the solutions of the Davey–Stewartson equations (exercise 9.8) can be obtained from solutions of the system

$$i\frac{\partial \psi}{\partial t} + \frac{1}{2}\left(\frac{\partial^2 \psi}{\partial \xi^2} + \frac{\partial^2 \psi}{\partial \eta^2}\right) + 2\phi\psi = 0,$$

$$-i\frac{\partial \psi^*}{\partial t} + \frac{1}{2}\left(\frac{\partial^2 \psi^*}{\partial \xi^2} + \frac{\partial^2 \psi^*}{\partial \eta^2}\right) + 2\phi\psi^* = 0,$$

$$\frac{\partial^2 \phi}{\partial \xi \partial \eta} + \frac{1}{2}\left(\frac{\partial^2 (\psi\psi^*)}{\partial \xi^2} + \frac{\partial^2 (\psi\psi^*)}{\partial \eta^2}\right) = 0,$$

where $\phi \equiv \varphi + \psi\psi^*$. Note that this system reduces to the Davey–Stewartson equations under the condition: $\psi^* = \pm\overline{\psi}$. (For more information, see Kalla 2011).

EXERCISE 9.12.– Consider the Dym equation: $\frac{\partial u}{\partial t} = \frac{\partial^3}{\partial x^3}\left(\frac{1}{\sqrt{u}}\right)$, where $u(x,t)$ is a real valued function, or under the following change of variables $v = \frac{1}{\sqrt{u}}$, this equation can be written in the equivalent form, $\frac{\partial v}{\partial t} = -\frac{1}{2}v^3\frac{\partial^3 v}{\partial x^3}$. This nonlinear partial differential equation possesses many properties typical for integrable systems and it is one of the most exotic soliton equations. Establish connections between this equation and KdV equation. Determine solutions of this equation from known solutions of the KdV equation.

EXERCISE 9.13.– Determine and analyze the algebro-geometric solutions of the Dym-type equation: $\frac{\partial^3 u}{\partial x^2 \partial t} + 2\frac{\partial u}{\partial x} \cdot \frac{\partial^2 u}{\partial x^2} + u\frac{\partial^3 u}{\partial x^3} - 2\alpha\frac{\partial u}{\partial x} = 0, \alpha \in \mathbb{C}$.

EXERCISE 9.14.– Let $i\frac{\partial \psi_k}{\partial t} + \frac{\partial^2 \psi_k}{\partial x^2} + 2\left(\sum_{j=1}^n s_j |\psi_j|^2\right)\psi_k = 0, 1 \leq k \leq n$ be the multi-component nonlinear Schrödinger equation, where $s = (s_1, ..., s_n)$, $s_j = \pm 1$, and $\psi_k(x,t)$ are complex valued functions of the real variables x, t. Study the integrability of this equation.

EXERCISE 9.15.– Consider the system

$$i\frac{\partial \psi_k}{\partial t} + \frac{\partial^2 \psi_k}{\partial x^2} + 2\left(\sum_{j=1}^n \psi_j \psi_j^*\right)\psi_k = 0,$$

$$i\frac{\partial \psi_k^*}{\partial t} + \frac{\partial^2 \psi_k^*}{\partial x^2} + 2\left(\sum_{j=1}^n \psi_j \psi_j^*\right)\psi_k^* = 0,$$

where $1 \leq k \leq n$ and $\psi_k(x,t)$, $\psi_k^*(x,t)$ are complex valued functions of the real variables x and t. Show that this system reduces to the multi-component nonlinear Schrödinger equation (see exercise 9.14) under the conditions: $\psi^* = s_k \pm \overline{\psi}$. Determine the solutions of this system and express them if possible in terms of theta functions.

EXERCISE 9.16.– We consider the integrable case proposed in section 4.8, that is,

$$\ddot{q}_1 + \left(a_1 + q_1^2 + q_2^2\right)q_1 = 0, \qquad \ddot{q}_2 + \left(a_2 + q_1^2 + q_2^2\right)q_2 = 0. \qquad [9.30]$$

(we use here the notations: q_1, q_2 instead of x_1, x_2 and p_1, p_2 instead of y_1, y_2, with $p_1 = \dot{q}_1, p_2 = \dot{q}_2$) and the problem is to show that at some special values of the parameters a_1 and a_2, the solutions may be expressed in terms of elliptic functions. In addition, we propose to solve the equations of motion with the help of finite-gap integration theory. The reader is referred to the book by Belokolos and Enol'skii (1994) for this theory and its applications to nonlinear integrable equations.

a) The elliptic solutions which we are looking for are given by the spectral problem (the Schrödinger equation):

$$\left(\frac{\partial^2}{\partial x^2} - \mathcal{U}(x)\right)\Psi = -z\Psi, \qquad [9.31]$$

depending in the elliptic potential

$$\mathcal{U}(x) = 2\sum_{i=1}^{N} \wp(x - x_i),\qquad [9.32]$$

where z is a spectral parameter. Here, $\Psi = \Psi(x, z)$ is an eigenfunction (elliptic Baker–Akhiezer function) of operator $\mathcal{L}$, $\wp$ is the elliptic Weierstrass function, $N > 2$ is a positive integer (the number of particles) and $x_1, ..., x_N$ are fixed on the closure of the locus, that is, the geometrical position of the points given by the equations, $\Delta = \{(x_1, ..., x_N) \in \mathbb{C}^N : \sum_{i \neq j} \wp'(x_i - x_j) = 0, \ x_i \neq x_j, j = 1, ..., N\}$. Show that Δ is non-empty (for triangular positive integers N, i.e. for numbers of the form $N = \frac{g(g+1)}{2}$ where g is the number of gaps in the spectrum or the genus of the corresponding algebraic curve) and study its geometry (see Airault et al. 1977).

b) Show that if $x_i = x_i(t)$, $j = 1, 2, 3$ evolve according to the law $\dot{x}_i = -12\sum_{i \neq j} \wp(x_i - x_j)$, then the function [9.32] is an elliptic solution of the KdV equation: $\frac{\partial u}{\partial t} - 6u\frac{\partial u}{\partial x} + \frac{\partial^3 u}{\partial x^3} = 0$, and is connected with the integrable Calogero–Moser system described by the Hamiltonian: $H = \frac{1}{2}\sum_{i=1}^{N} y_i^2 - 2\sum_{i \neq j} \wp(x_i - x_j)$, $N = \frac{g(g+1)}{2}$, with y_i, x_i, $i = 1, ..., N$, being canonical variables.

c) Show that the Jacobi inversion problem associated with two-gap solutions of the KdV equation is determined by $\int_{\infty}^{s_1} \frac{dz}{w} + \int_{\infty}^{s_2} \frac{dz}{w} = -8it + c_1$, $\int_{\infty}^{s_1} \frac{zdz}{w} + \int_{\infty}^{s_2} \frac{zdz}{w} = 2ix + c_2$, where c_1, c_2 are constants, $w^2 = \prod_{j=1}^{5}(z - z_j)$, $s_j = s_j(x, t), j = 1, 2$ and $z_j, j = 1, ..., 5$, are expressed in terms of integrals of motion.

d) Show that after an appropriate change of variables and corresponding choices of z_j, the Jacobi inversion problem above coincides with that associated with the Hénon–Heiles equations for the case (i), section 7.6 (hint: see Eilbeck and Enolskii 1994).

e) Discuss that equation [9.31] allows the coalescence of three particles x_i and show that the potential takes the form: $\mathcal{U}(x) = 6\sum_{i=1}^{n} \wp(x - x_i) + \sum_{i=1}^{m} \wp(x - x_i)$, $3n + m = N$.

f) Consider the two-gap potential for the above potential normalized by its expansion near $x = 0$ as $\mathcal{U}(x) = \frac{6}{x^2} + \alpha_1 x^2 + \alpha_2 x^4 + \alpha_3 x^6 + \alpha_4 x^8 + O(x^{10})$, where $\alpha_1, \alpha_2, \alpha_3, \alpha_4$ are functions of the moduli g_2, g_3 of the underlying elliptic curve. Show that this potential satisfies the Novikov equation Novikov (1974): $a_{-1}\frac{\delta S_{-1}}{\delta u} + a_0 \frac{\delta S_0}{\delta u} + a_1 \frac{\delta S_1}{\delta u} + a_2 \frac{\delta S_2}{\delta u} = 0$, where δ is the variational operator of the calculus of variations, a_1, a_2, are constants and $S_{-1} = \int u dx$, $S_0 = \int u^2 dx$, $S_1 = \int \left(\frac{1}{2}\left(\frac{\partial u}{\partial x}\right)^2 + u^3\right) dx$, $S_2 = \int \left(\frac{1}{2}\left(\frac{\partial^2 u}{\partial x^2}\right)^2 - \frac{5}{2}u^2\frac{\partial^2 u}{\partial x^2} + \frac{5}{2}u^4\right) dx$, are the first integrals of the KdV equation. Show that the algebraic curve associated with this potential has the form (Belokolos and Enol'skii 1989):

$$w^2 = z^5 - \frac{35}{2}\alpha_1 z^3 - \frac{63}{2}\alpha_2 z^2 + \frac{1}{4}\left(\frac{567}{2}\alpha_1^2 + 297\alpha_3\right)z + \frac{1377}{4}\alpha_1\alpha_2 - \frac{1287}{2}\alpha_4.$$

[9.33]

g) Consider the trace formulas (Novikov 1974) written for the elliptic potential in the form: $s_1 + s_2 = -\sum_{j=1}^{3} \wp(x-x_j) + \frac{1}{2}\sum_{j=1}^{5} z_j - \frac{C}{2}$ and $s_1 s_2 = 3\sum_{\substack{i,j=1 \\ i<j}}^{3} \wp(x-x_i)\wp(x-x_j) - \frac{3}{8}g_2 + \frac{1}{2}\sum_{\substack{i,j=1 \\ i<j}}^{5} z_i z_j - \frac{3}{8}\left(\sum_{j=1}^{5} z_j\right)^2 + \frac{3C}{2}\sum_{j=1}^{3}\wp(x-x_j) + \frac{3C^2}{8}$, where $x_1, x_2, x_3 \in \Delta$ and $z_1,...,z_5$ are the branching points of the genus 2 hyperelliptic curve [9.33]. We want the curve Γ [4.23] (section 4.8) to be associated with the genus 2 curve [9.33]. Let z_1, z_2 be two distinct branch points of the curve [9.33] which are the shifted values of the branch points a_1 a_2 of the curve Γ [4.23]. In the proof of theorem 4.8 (section 4.8), we introduced coordinates s_1 and s_2 such that: $q_1^2 = 2\frac{(a_1+s_1)(a_1+s_2)}{a_1-a_2}$, $q_2^2 = 2\frac{(a_2+s_1)(a_2+s_2)}{a_2-a_1}$, $a_1 \neq a_2$, hence, $q_1^2 + q_2^2 = 2(s_1 + s_2 + z_1 + z_2)$. Using the first trace formula above, show that [9.30] can be written in the following form: $\ddot{q}_1 - \mathcal{U}(x)q_1 = (-a_1 + 2(z_1+z_2))q_1$, $\ddot{q}_2 - \mathcal{U}(x)q_2 = (-a_2 + 2(z_1+z_2))q_2$, where we set without loss of generality $\sum_{i=1}^{5} z_i = 0$. Suppose that $a_1 = 3z_1 + 2z_2$ and $a_2 = 2z_1 + 3z_2$. Show that the formula for elliptic solutions of [9.30] is given by $q_1^2 = \frac{2z_1^2 + 2z_1 \Pi + \Phi}{z_2 - z_1}$, $q_2^2 = \frac{\Phi + 2z_2^2 + + 2z_2\Pi + +\Phi}{z_1 - z_2}$, where, $\Pi \equiv \sum_{i=1}^{N}\wp(x-x_i)$, $\Phi \equiv 6\sum_{1\leq i<j\leq N}\wp(x-x_i)\wp(x-x_j) - \frac{Ng_2}{4} + \sum_{1\leq i<j\leq 5} z_i z_j$, and $z_1,...,z_5$ are the branching points of the curve [9.33].

10

KP–KdV Hierarchy and Pseudo-differential Operators

In this chapter, we will study some generalities on the algebra of infinite-order differential operators. The algebras of Virasoro, Heisenberg and nonlinear evolution equations such as the Korteweg–de Vries (KdV), Boussinesq and Kadomtsev–Petviashvili (KP) equations play a crucial role in this study. We will make a careful study of a connection between pseudo-differential operators, symplectic structures, KP hierarchy and tau functions based on the Sato–Date–Jimbo–Miwa–Kashiwara theory. A few other connections and ideas concerning the KdV and Boussinesq equations, the Gelfand–Dickey flows, the Heisenberg and Virasoro algebras are also given.

10.1. Pseudo-differential operators and symplectic structures

Let L be a pseudo-differential operator with holomorphic coefficients. The set of these operators form a Lie algebra that we note $\mathcal{A}$. The algebra $\mathcal{A}$ decomposes in two subalgebras $\mathcal{A}_+$ and $\mathcal{A}_-$: $\mathcal{A} = \mathcal{A}_+ \oplus \mathcal{A}_-$, where $\mathcal{A}_+$ is the algebra of differential operators of the form $\zeta = \sum_{k \geq 0} u_k(x)\partial^k$, finite sum, $\partial = \frac{\partial}{\partial x}$, and $\mathcal{A}_-$ is the algebra of strictly pseudo-differential operators of the form

$$\eta = \sum_{k>0} u_{-k}(x)\partial^{-k} = \partial^{-1}v_0 + \partial^{-2}v_1 + \cdots, \quad \partial = \frac{\partial}{\partial x}.$$

The algebra $\mathcal{A}$ can be seen as an associative algebra for the product of two pseudo-differential operators L and L', $L.L' = \sum_{k=0}^{\infty} \frac{1}{k!} :\partial_\partial^k(L).\partial_x^k(L'):$, where $\partial = \frac{\partial}{\partial x}$ and the symbol :: denotes the normal order, that is, it means that the derivatives always appear on the right, independently of the commutation relations.

EXAMPLE 10.1.– For $m, n \in \mathbb{N}^*$, we have, for all functions u, v,

$$u\partial^m . v\partial^n = \sum_{k=0}^{m} \frac{m!}{k!(m-k)!} uv^{(k)} \partial^{m+n-k} \sum_{k=0}^{m} \frac{1}{k!} :\partial_\partial^k(u\partial^m).\partial_x^k(v\partial^n): \qquad [10.1]$$

$$\partial^{-1} u = u\partial^{-1} - u'\partial^{-2} + \cdots + (-1)^k u^{(k)} \partial^{-k-1} + \cdots = \sum_{k=0}^{\infty} :\partial_\partial^k(\partial^{-1}).\partial_x^k(u): \quad [10.2]$$

where ∂^{-1} is a formal inverse of ∂, that is, $\partial^{-1}.\partial = \partial.\partial^{-1} = 1$.

We define a coupling between $\mathcal{A}_+$ and $\mathcal{A}_-$ as follows: let $\operatorname{Res}(\zeta\eta)$ be the coefficient of ∂^{-1} in $\zeta\eta$. We have

$$\langle \zeta, \eta \rangle = \left\langle \sum_{k\geq 0} u_k \partial^k, \sum_{k>0} u_{-k} \partial^{-k} \right\rangle,$$

$$= \langle u_0 \partial^0 + u_1 \partial^1 + \cdots, \partial^{-1} v_0 + \partial^{-2} v_1 + \cdots \rangle,$$

$$= \int_{-\infty}^{\infty} \operatorname{Res}(\zeta\eta) dx = \int_{-\infty}^{\infty} \sum_{k\geq 0} u_k v_k dx.$$

Therefore, the Volterra group $(I + \mathcal{A}_-)$ acts on $\mathcal{A}_-$ by the adjoint action and on $\mathcal{A}_+$ by the coadjoint action. Let $\zeta \in \mathcal{A}_+$ and $\eta_k \in \mathcal{A}_-$. Taking into account the notation used in the definitions and theorems of Chapter 1, we obtain (Adler 1979),

$$\langle \operatorname{ad}^*_{\eta_1}(\zeta), \eta_2 \rangle = \langle \zeta, \operatorname{ad}_{\eta_1}(\eta_2) \rangle = \langle \zeta, [\eta_1, \eta_2] \rangle,$$

$$= \int \left(\partial^{-1} - \text{term of } (\zeta\eta_1\eta_2 - \zeta\eta_2\eta_1) \right) dx,$$

$$= \int \left(\partial^{-1} - \text{term of } (\zeta\eta_1 - \eta_1\zeta)_+ \eta_2 \right) dx,$$

$$= \langle [\zeta, \eta_1]_+, \eta_2 \rangle.$$

So the set $\mathcal{O}^*_{\mathcal{A}_+}(L)$ of the differential operators of the form

$$L = \partial^N + \sum_{k=0}^{N-2} u_k(x) \partial^k, \quad N \text{ fixed}, \qquad [10.3]$$

is a coadjoint orbit in $\mathcal{A}_+$. Let f be a function of class $\mathcal{C}^\infty$ in x and dependent on a finite number of derivatives $u_k^{(l)}$ of the coefficients u_k of L. Let

$$\nabla H(L) = \sum_{k=0}^{N-1} \partial^{-k-1} \sum_l (-1)^l \left(\frac{d}{dx} \right)^l \frac{\partial f}{\partial p_k^{(l)}} = \sum_{k=0}^{N-1} \partial^{-k-1} \frac{\delta H}{\delta u_k},$$

KP–KdV Hierarchy and Pseudo-differential Operators 277

be the gradient of the functional (defined on $\mathcal{A}_+$), $H(L) = \int_{-\infty}^{\infty} f(x, ..., u_k^{(l)}, ...) dx$, and such that: $dH = \int_{-\infty}^{\infty} \frac{\delta H}{\delta u_k} du_k = \left\langle \sum_{k=0}^{N} du_k . \partial^k, \nabla H \right\rangle = \langle dL, \nabla H \rangle$, where $dL = \sum_{k=0}^{N} du_k . \partial^k$. The scalar product between two pseudo-differential operators L and L' is defined by $\langle L, L' \rangle = -\int_{-\infty}^{\infty}(LL')_- dx = \int_{-\infty}^{\infty}(L'L)_- dx$. According to the Adler–Kostant–Symes theorem 4.2, the Hamiltonian vector fields on the coadjoint orbit $\mathcal{O}_{\mathcal{A}_+}^*$ define commutative flows and are given by

$$\frac{dL}{dt} = ad^*_{\nabla H(L)}(L) = [L, \nabla H(L)]_+ , \qquad [10.4]$$

where $H(L)$ is the Hamiltonian on $\mathcal{A}_+$. The operator L does not contain the coefficient u_{N-1}. Since the vector field [10.4] applied to the operator L [10.3] imposes the condition $\mathrm{Res}\,[L, \nabla H(L)] = 0$, we can replace the gradient $\frac{\delta H}{\delta p_{N-1}}$ by any expression satisfying this condition. Therefore a first bracket (Poisson bracket) is given by

$$\{H, F\}_1 = \langle L, [\nabla F, \nabla H] \rangle, \qquad [10.5]$$

and $\{H, F\}_1 = \int \mathrm{Res}\,(\nabla H[L, \nabla F])\,dx = \int \mathrm{Res}\,([\nabla H, L]\nabla F)\,dx$. Consider the Hamiltonians $H_{k+N} = \frac{N}{k+N} \int \left(\mathrm{Res}\, L^{\frac{k+N}{N}} \right) dx$, $k \in \mathbb{N}^*$. We have $\nabla H_{k+N}^{(L)} = \left(L^{\frac{k}{N}} \right)_-$, and the vector fields [10.4] applied to these Hamiltonians provide the integrable equations; N-reduction of Gelfand Dickey equations of KP hierarchy (see section 10.3) is given as:

$$\frac{dL}{dt} = [L, \nabla H_{k+N}(L)]_+ = -\left[(L^{\frac{k}{N}})_-, L\right]_+ = \left[(L^{\frac{k}{N}})_+, L\right]. \qquad [10.6]$$

Note that since $\left[(L^{\frac{k}{N}})_+, L\right]_+ = \left[L^{\frac{k}{N}} - (L^{\frac{k}{N}})_-, L\right]_+ = -\left[(L^{\frac{k}{N}})_-, L\right] \in \mathcal{A}^-$, equations [10.6] determine an infinite number of commutative vector fields (see theorem 10.5) on $\mathcal{A}^+ + \mathcal{A}^-$.

We will now study the existence of a second symplectic structure (Adler 1979; Gelfand and Dickey 1968; Dickey 1991, 1993, 1997). Let $\widetilde{L} = L + z$ where L is a differential operator of order n. We have

$$\frac{dL}{dt} = \left(\widetilde{L} \nabla H \right)_+ \widetilde{L} - \widetilde{L} \left(\nabla H \widetilde{L} \right)_+ . \qquad [10.7]$$

Note that [10.7] is a Hamiltonian vector field (generalizing [10.4]). Indeed, let $J : \mathcal{A}_-/\mathcal{A}_{-\infty, N-1} \longrightarrow \mathcal{D}_{0, N-1}$ be the function defined by

$$J(\zeta) = \left(\widetilde{L} \zeta \right)_+ \widetilde{L} - \widetilde{L} \left(\zeta \widetilde{L} \right)_+ = -\left(\widetilde{L} \zeta \right)_- \widetilde{L} + \widetilde{L} \left(\zeta \widetilde{L} \right)_-, \quad \zeta \in \mathcal{A}_-/\mathcal{A}_{-\infty, N-1}.$$

Hence, $\frac{dL}{dt} = \partial_{J(\zeta)}(L) \equiv \left(\tilde{L}\zeta\right)_+ \tilde{L} - \tilde{L}\left(\zeta\tilde{L}\right)_+$, which shows that it is indeed a vector field on the differential operators L of order n. Similarly, we have $\frac{dL}{dt} = -\left(\tilde{L}\nabla H\right)_- \tilde{L} + \tilde{L}\left(\nabla H \tilde{L}\right)_-$, and the same conclusion remains valid. In addition, we also have the relation $\frac{dL}{dt} = (L\nabla H)_+ L - L(\nabla HL)_+ + z[\nabla H, L]_+$, which shows that this vector field is an interpolation between [10.4] for $z = \infty$ and a new vector field for $z = 0$. Consider the two-differential form

$$\omega\left(\partial_{J(\zeta)}, \partial_{J(\eta)}\right) = \langle J(\zeta), \eta \rangle = \int \text{Res}\ (J(\zeta)\eta)dx.$$

We show that this form is closed, that it is antisymmetric $\langle J(\zeta), \eta \rangle = -\langle \zeta, J(\eta) \rangle$, and furthermore $\left[\partial_{J(\zeta)}, \partial_{J(\eta)}\right] = \partial_{J(\xi)}$, where

$$\xi = \left(-\zeta\left(\tilde{L}\eta\right)_+ + \left(\zeta\tilde{L}\right)_- \eta\right) - \left(-\eta\left(\tilde{L}\zeta\right)_+ + \left(\eta\tilde{L}\right)_- \zeta\right) + \partial_{J(\zeta)}\eta - \partial_{J(\eta)}\zeta.$$

DEFINITION 10.1.– *The functional algebra on the operator space of the form [10.3] for this symplectic form is called W algebra.*

THEOREM 10.1.– *The Hamiltonians $H_k, H_{k+N}, H_{k+2N}, ...$, defined in [10.6] are all in involution for the bracket [10.5].*

PROOF.– Indeed, let $J = J_1$ if $z = \infty$ and $= J_2$ if $z = 0$, where the Poisson brackets $\{.,.\}_1, \{.,.\}_2$ are given by $\{H_j, H_k\}_1 = \int \text{Res}\ (\nabla H_j J_1(\nabla H_k))$, and

$$\{H_j, H_k\}_2 = \langle \nabla H, J_2(\nabla F)\rangle = \int \text{Res}\ (\nabla H((L\nabla F)_+ L - L(\nabla FL)_+))dx,$$

$$= \int \text{Res}\ (L\nabla H(L\nabla F)_+ - \nabla HL(\nabla FL)_+)dx.$$

We deduce from the relation $\left(L\left(L^{\frac{r}{n}-1}\right)_-\right)_+ L - L\left(\left(L^{\frac{r}{n}-1}\right)_- L\right)_+ + \left[(L^{\frac{r}{n}})_-, L\right]_+ = 0$, the expression $\{H_j, H_k\}_1 = \int \text{Res}\ (\nabla H_j J_2 \nabla H_{k-N})$. Since the form ω is antisymmetric, we have

$$\{H_j, H_k\}_1 = -\int \text{Res}\ (\nabla H_{k-N} J_2(\nabla H_j)) = -\int \text{Res}\ (\nabla H_{k-N} J_1(\nabla H_{j+N})),$$

$$= -\int \text{Res}\ (\nabla H_{j+N} J_1(\nabla H_{k-N})) = \{H_{j+N}, H_{k-N}\}_1.$$

Then, $\{H_j, H_k\}_1 = \{H_{j+\alpha N}, H_{k-\alpha N}\}_1$. For α large enough, $J_1(\nabla H_{k-\alpha N}) = 0$, that is, $H_{k-\alpha N}$ is trivial and $\{H_j, H_k\}_1 = 0$, so H_j, H_k are in involution. □

10.2. KdV equation, Heisenberg and Virasoro algebras

THEOREM 10.2.– The operator $L = \partial^2 + q$, corresponding to the case $N = 2$ with $q \equiv u_0$, is related to the KdV equation and the Poisson bracket is provided in this case by $\{q(x), q(y)\}_1 = \frac{d}{dx}\delta(x - y)$.

PROOF.– Indeed, since $\nabla H(L) = \partial^{-1}\frac{\delta H}{\delta q} + \partial^{-2}\frac{1}{2}\left(\frac{\delta H}{\delta q}\right)'$, then the vector fields applied to the Hamiltonian $H = \int \left(q^3 - \frac{1}{2}q'^2\right) dx$ provide the KdV equation:

$$\frac{dq}{dt} = \frac{dL}{dt} = \frac{1}{2}[L, \nabla H]_+ = \frac{d}{dx}\frac{\delta H}{\delta q} = \frac{dq}{dt} = \frac{\partial^3 q}{\partial x^3} + 6q\frac{\partial q}{\partial x}. \qquad [10.8]$$

The Poisson bracket is $\{H, F\}_1 = \int \frac{\delta H}{\delta q} \frac{d}{dx} \frac{\delta F}{\delta q}$ and the result follows. □

REMARK 10.1.– Similarly, taking $N = 3$, $u \equiv u_2$, $v \equiv u_3$, $L = \partial^3 + u\partial + v$, $L^{\frac{2}{3}} = \partial^3 + \frac{2}{3}u$, then the flow [10.8] takes the form $\frac{\partial u}{\partial t_2} = -\frac{\partial^2 u}{\partial x^2} + 2\frac{\partial v}{\partial x}$ and $\frac{\partial v}{\partial t_2} = \frac{\partial^2 v}{\partial x^2} - \frac{2}{3}\frac{\partial^3 u}{\partial x^3} - \frac{2}{3}u\frac{\partial u}{\partial x}$. Eliminating v from this system yields the Boussinesq equation

$$3\left(\frac{\partial u}{\partial t_2}\right)^2 + \frac{\partial^2}{\partial x^2}\left(\frac{\partial^2 u}{\partial x^2} + 2u^2\right) = 0.$$

THEOREM 10.3.– In the previous theorem, by replacing $q(x)$ by the Fourier series

$$q(x) = \alpha \sum_{n=-\infty}^{\infty} e^{-inx} \varphi_n + \beta, \qquad -i\alpha^{-2} = 1, \qquad [10.9]$$

where $(\varphi_k)_{k \in \mathbb{Z}}$ are new coordinates (Fourier coefficients), one obtains the Heisenberg algebra and the Poisson bracket is provided by $\{\varphi_n, \varphi_m\}_1 = n\delta_{m+n,0}$.

PROOF.– Let $\mathcal{M}$ be the set of matrices (a_{kl}), $(k, l \in \mathbb{Z})$, with complex coefficients and $\mathcal{N} = \{(a_{kl}) \in \mathcal{M} : \exists r \text{ such that } a_{kl} = 0 \text{ for } |k - l| > r\}$, the $\mathbb{C}$-algebra, that is, the set of infinite matrices with support in a band around the diagonal (see Jacobi matrices, section 5.1). The product of two matrices belonging, respectively, to $\mathcal{N}$ and $\mathcal{M}$, is defined in the usual way. Note that $\mathcal{N}$ is a Lie algebra and $\mathcal{M}$ is a $\mathcal{N}$-module. Their extensions $\widetilde{\mathcal{N}}$ and $\widetilde{\mathcal{M}}$ are defined by

$$0 \longrightarrow \mathbb{C}c \longrightarrow \widetilde{\mathcal{N}} \longrightarrow \mathcal{N} \longrightarrow 0, \qquad 0 \longrightarrow \mathbb{C}c \longrightarrow \widetilde{\mathcal{M}} \longrightarrow \mathcal{M} \longrightarrow 0,$$

with $\widetilde{\mathcal{N}} = \mathcal{N} \oplus \mathbb{C}c$, $\widetilde{\mathcal{M}} = \mathcal{M} \oplus \mathbb{C}c$, where c is a central element, that is, $[c, A] = [c, B] = 0$, $\forall A \in \widetilde{\mathcal{N}}$, $\forall B \in \widetilde{\mathcal{M}}$. We note $e_{i,j} = (\delta_{ki}.\delta_{lj})_{kl}$ the elementary matrices, that is, the matrices whose coefficients are all zero except the one of the line i and the column j, which is equal to 1. Since a Jacobi matrix has no trace, we consider the matrix $A[J, B]$ where $A \in \mathcal{N}$, $B \in \mathcal{M}$ and J is the matrix defined by $J = \sum_{i \in \mathbb{Z}} \varepsilon(i) e_{i,i}$, where $\varepsilon(i) = +1$ if $i < 0$ and -1 if $i \geq 0$. The elements of the matrix $A[J, B]$ are null except for a finite number, so it is indeed a finite matrix and we define the cocycle of $A \in \mathcal{N}$ and $B \in \mathcal{M}$ using the formula $\rho(A, B) = \frac{1}{2} Tr(A[J, B]) = \frac{1}{2} \sum_{i,j} (\varepsilon(i) - \varepsilon(j)) a_{ij} b_{ji}$. Therefore, the bracket $\widetilde{[,]}$ of $A \in \mathcal{N}$ and $B \in \mathcal{M}$ is defined by $\widetilde{[A, B]} = [A + \alpha c, B + \beta c] = [A, B] + \rho(A, B)c$ and $\widetilde{\mathcal{N}}$ is a non-trivial central extension of N while $\widetilde{\mathcal{M}_f} = \mathcal{M}_f \oplus \mathbb{C} c$ is a trivial central extension of $\mathcal{M}_f = \{(a_{ij}) \in \mathcal{M} : (i, j) \longmapsto (a_{ij})$ with finished support$\}$. Let us put $E_i = \sum_{n \in \mathbb{Z}} e_{n, n+i}$, where $e_{i,i} = (\delta_{ki}.\delta_{ij})_{kl}$ are the elementary matrices defined above. The subspace $E = \bigoplus_{i \in \mathbb{Z}} \mathbb{C} E_i$ is a commutative subalgebra of $\mathcal{N}$. The subalgebra of $\mathcal{N}$ defined by setting $\widetilde{E} = E \oplus \mathbb{C} c$ is called Heisenberg subalgebra. We have

$$\widetilde{[E_i, E_j]} = i \delta_{i,-j} c. \qquad [10.10]$$

Consider the previous example and replace $q(x)$ with the Fourier series [10.9]. Let H be a functional of q. Its Fréchet derivative in terms of the coordinates φ_k is written as

$$\frac{\delta H}{\delta q} = \sum_{k=-\infty}^{\infty} \frac{\delta H}{\delta \varphi_k} \cdot \frac{\partial \varphi_k}{\partial q} = \alpha^{-1} \sum_{k=-\infty}^{\infty} \frac{\delta H}{\delta \varphi_k} e^{ikx}. \qquad [10.11]$$

We substitute [10.10] and [10.11] in [10.8] and we specify the Fourier coefficients; we get the relation $\alpha \frac{\partial \varphi_n}{\partial t} = -i\alpha^{-1} n \frac{\partial H}{\partial \varphi_n}$. Since the symplectic structure is given by the matrix of the Poisson brackets, we have $\frac{\partial \varphi_n}{\partial t} = \sum_{m=-\infty}^{\infty} \{\varphi_n, \varphi_m\}_1 \frac{\partial H}{\partial \varphi_m}$. So, $\{\varphi_n, \varphi_m\}_1 = -i\alpha^{-2} n \delta_{m+n,0}$. By putting $-i\alpha^{-2} = 1$, we obtain the Heisenberg algebra (where $\{,\}$ plays the role here of the bracket $\widetilde{[,]}$ [10.10]). $\square$

THEOREM 10.4.– In the case $N = 2$ (theorem 10.2), one obtains the Virasoro algebra, its structure is given by $\{\varphi_m, \varphi_n\}_2 = (m - n)\varphi_{m+n} + \frac{c}{12}(m^3 - m)\delta_{m+n,0}$.

PROOF.– Let $\text{Diff}(S^1)$ be the group of diffeomorphisms of the unit circle: $S^1 = \{z \in \mathbb{C} : |z| = 1\}$ and $F = \{f(z)\frac{d}{dz} : f(z) \in \mathbb{C}\left[z, \frac{1}{z}\right]\}$, the set of vector fields (Laurent's polynomials). F can be seen as the tangent space $\text{Diff}(S^1)$ at its unit point, so F is a Lie algebra with respect to the bracket $[,]$. By setting $\varphi_m = -z^{m+1}\frac{d}{dz}$, we obtain

$$[\varphi_m, \varphi_n] = \left((n+1)z^{m+n+1} - (m+1)^{m+n+1}\right)\frac{d}{dz} = -(m+n)z^{m+n+1}\frac{d}{dz},$$

$$= (m - n)\varphi_{m+n}.$$

We show that $H^2(F, \mathbb{C}) \cong \mathbb{C}$ and $\rho(\varphi_m, \varphi_n) = \frac{1}{12}(m^3 - m)\delta_{m,-n}$. The vector space $F \oplus \mathbb{C} c$ is called the Virasoro algebra, it is a central extension of the algebra of complex vector fields on the circle. The bracket is given by the formula

$$[\varphi_m, \varphi_n] = (m - n)\varphi_{m+n} + \frac{c}{12}(m^3 - m)\delta_{m,-n}. \qquad [10.12]$$

Let us now consider the example of the KdV equation. We have $N = 2$ and

$$\frac{dq}{dt} = \frac{dL}{dt} = (L\nabla H)_+ - L(\nabla H L)_+ = \left(\partial^3 + 2(\partial q + q\partial)\right)\frac{\delta H}{\delta q}.$$

In this case, the bracket is written $\{H, F\}_2 = \int \frac{\delta H}{\delta q}\left(\partial^3 + 2(\partial q + q\partial)\right)\frac{\delta F}{\delta q}$, and we have $\{q(x), q(y)\}_2 = \left(\partial^3 + 2(\partial q + q\partial)\right)\delta(x-y)$. By reasoning as previously, we obtain $\alpha \frac{\partial \varphi_m}{\partial t} = i \sum_n (n-m)\varphi_{m+n}\frac{\delta H}{\delta \varphi_n} + \frac{i}{2\alpha}(m^3 - 4\beta m)\frac{\delta H}{\delta \varphi_{-m}}$, where $(\varphi_k)_{k \in \mathbb{Z}}$ are the Fourier coefficients of q. By setting $4\beta = 1$, $\alpha = \frac{6i}{c}$ and taking into account the Fourier series [10.9], we obtain

$$\frac{\partial}{\partial t}\begin{pmatrix} \vdots \\ \varphi_m \\ \vdots \end{pmatrix} = \begin{pmatrix} & -\text{nth column} & \text{nth column} & \\ & \downarrow & \downarrow & \\ \text{mth line} \longrightarrow & \frac{c}{12}(m^3 - m) & \dots & (m-n)\varphi_{m+n} \end{pmatrix}\begin{pmatrix} \vdots \\ \frac{\delta H}{\delta \varphi_m} \\ \vdots \end{pmatrix}.$$

Consequently $\{\varphi_m, \varphi_n\}_2 = (m-n)\varphi_{m+n} + \frac{c}{12}(m^3 - m)\delta_{m+n,0}$, that is, the Virasoro structure (Gervais 1985) (where $\{,\}_2$ plays the role here of the bracket $[,]$ [10.12]). $\square$

10.3. KP hierarchy and vertex operators

Consider the pseudo-differential operator of infinite order

$$L = \partial + u_1 \partial^{-1} + u_2 \partial^{-2} + \cdots, \quad \partial \equiv \frac{\partial}{\partial x} \quad [10.13]$$

where $u_1, u_2, \dots$ are functions of class $\mathcal{C}^\infty$ depending on an infinity of independent variables $x \equiv t_1, t_2, \dots$ The compound operator L^n is calculated according to the rules [10.1] and [10.2]. We obtain

$$L^n = \partial^n + p_{n,2}\partial^{n-2} + \cdots + p_{n,n} + p_{n,n+1}\partial^{-1} + \cdots$$

$$= \partial^n + \sum_{j=2}^{n} p_{n,j}\partial^{n-j} + \sum_{j=1}^{\infty} p_{n,n+j}\partial^{-j},$$

where $p_{n,j}$ are polynomials in u_j and their derivatives in relation to x. The differential part L_+^n of L^n being equal to $L_+^n = \partial^n + \sum_{j=2}^{n} p_{n,j}\partial^{n-j}$, we deduce that

$$L_+^1 = \partial, \quad L_+^2 = \partial^2 + 2u_2, \quad L_+^3 = \partial^3 + 3u_2\partial + 3(u_3 + \partial u_2), \dots \quad [10.14]$$

The dependency between the functions $u_1, u_2, \dots$ and the variables $x = t_1, t_2, \dots$ is provided by the following system of partial differential equations:

$$\frac{\partial L}{\partial t_n} = [L_+^n, L], \quad n \in \mathbb{N}^* \quad [10.15]$$

The set of these equations is called the KP hierarchy. It is a hierarchy of isospectral deformations of the pseudo-differential operator [10.13]. We prove (see Date *et al.* 1983) the following result:

THEOREM 10.5.– There is an equivalence between [10.15] and the equations

$$\frac{\partial}{\partial t_n} L_+^m - \frac{\partial}{\partial t_m} L_+^n = [L_+^n, L_+^m], \qquad [10.16]$$

as well as their dual forms

$$\frac{\partial}{\partial t_n} L_-^m - \frac{\partial}{\partial t_m} L_-^n = -[L_-^n, L_-^m], \qquad [10.17]$$

where $L_-^n = L^n - L_+^n$. Equations [10.15] determine an infinite number of commutative vector fields on algebra $\mathcal{A} = \mathcal{A}_+ \oplus \mathcal{A}_-$.

PROOF.– Note that since $L^n = L_+^n + L_-^n$, then $\frac{\partial L}{\partial t_n} = [L_+^n, L] = -[L_-^n, L] \in \mathcal{A}_-$. Equations [10.15] define an infinite number of vector fields on $\mathcal{A}$. Since $\frac{\partial}{\partial t_n}$ and $[L_+^n, .]$ are derivations, then

$$\frac{\partial L^m}{\partial t_n} = [L_+^n, L_+^m] + [L_+^n, L_-^m] = -[L_-^n, L_+^m] - [L_-^n, L_-^m],$$

$$= \frac{1}{2}\left([L_+^n, L_+^m] - [L_-^n, L_+^m]\right) + \frac{1}{2}\left(-[L_-^n, L_-^m] + [L_+^n, L_-^m]\right),$$

$$= \frac{1}{2}\left([L_+^n, L_+^m] - [L_-^n, L_-^m]\right) + \frac{1}{2}\left([L_+^m, L_-^n] - [L_-^m, L_+^n]\right).$$

Similarly, we have (just swap n and m)

$$\frac{\partial L^n}{\partial t_m} = \frac{1}{2}\left([L_+^m, L_+^n] - [L_-^m, L_-^n]\right) + \frac{1}{2}\left([L_+^n, L_-^m] - [L_-^n, L_+^m]\right).$$

Hence, $\frac{\partial L^m}{\partial t_n} - \frac{\partial L^n}{\partial t_m} = [L_+^n, L_+^m] - [L_-^n, L_-^m]$. Or

$$\frac{\partial L^m}{\partial t_n} - \frac{\partial L^n}{\partial t_m} = \frac{\partial}{\partial t_n} L_+^m + \frac{\partial}{\partial t_n} L_-^m - \frac{\partial}{\partial t_m} L_+^n - \frac{\partial}{\partial t_m} L_-^n,$$

$$= \frac{\partial}{\partial t_n} L_+^m - \frac{\partial}{\partial t_m} L_+^n + \frac{\partial}{\partial t_n} L_-^m - \frac{\partial}{\partial t_m} L_-^n,$$

then $\frac{\partial}{\partial t_n} L_+^m - \frac{\partial}{\partial t_m} L_+^n - [L_+^n, L_+^m] = -\frac{\partial}{\partial t_n} L_-^m + \frac{\partial}{\partial t_m} L_-^n - [L_-^n, L_-^m]$. Since the expression on the left belongs to $\mathcal{A}_+$ and the one on the right belongs to $\mathcal{A}_-$, then the result comes from the decomposition $\mathcal{A} = \mathcal{A}_+ \oplus \mathcal{A}_-$ since obviously $\mathcal{A}_+ \cap \mathcal{A}_- = \emptyset$.

To show that the vector fields defined by these equations commute, we put $X(L) = [L_+^m, L]$ and $Y(L) = [L_+^n, L]$. Hence,

$$[X, Y](L) = (XY - YX)(L) = X\left([L_+^n, L]\right) - Y\left([L_+^m, L]\right),$$
$$= \left[X(L_+^n) - Y(L_+^m), L\right] + \left[L_+^n, X(L)\right] - \left[L_+^m, Y(L)\right],$$
$$= \left[X(L_+^n) - Y(L_+^m), L\right] + \left[L_+^n, [L_+^m, L]\right] - \left[L_+^m, [L_+^n, L]\right],$$
$$= \left[X(L_+^n) - Y(L_+^m) - [L_+^m, L_+^n], L\right],$$

according to Jacobi's identity and taking into account [10.16], we deduce that the vector fields in question commute. □

By specifying the quantifiers of ∂^k in [10.16], one obtains an infinity of nonlinear partial differential equations (Cherednick 1978) forming the KP hierarchy. These equations connect infinitely many functions u_j to infinitely many variables t_j. For example, for $m = 2$, $n = 3$, relations [10.16] and [10.14] determine two expressions based on u_2 and u_3. After eliminating u_3, we immediately obtain the KP equation:

$$3\frac{\partial^2 u_2}{\partial t_2^2} - \frac{\partial}{\partial t_1}\left(4\frac{\partial u_2}{\partial t_3} - 6u_2 \frac{\partial u_2}{\partial t_1} - \frac{\partial^3 u_2}{\partial t_1^3}\right) = 0. \qquad [10.18]$$

By solving the system: $\frac{\partial u_2}{\partial t_2} = 0$, $4\frac{\partial u_2}{\partial t_3} - 6u_2\frac{\partial u_2}{\partial t_1} - \frac{\partial^3 u_2}{\partial t_1^3} = 0$, we obtain particular solutions of [10.18]. Note that the last equation is precisely the KdV equation. The KP equation is therefore a generalization of the KdV equation, to which it is reduced when $\frac{\partial u_2}{\partial t_2} = 0$. Equations [10.15] and [10.16] imply the existence of the following pseudo-differential operator of degree 0 (wave operator) $W \in \mathcal{I} + \mathcal{A}_-$:

$$W = 1 + w_1(t)\partial^{-1} + w_2(t)\partial^{-2} + \cdots \qquad [10.19]$$

with $t = (t_1, t_2, \ldots) \in \mathbb{C}^\infty$. The inverse W^{-1} of W is also a pseudo-differential operator of the form $W^{-1} = 1 + v_1(t)\partial^{-1} + v_2(t)\partial^{-2} + \cdots$ and can be calculated term by term. Indeed, by definition, we have $WW^{-1} = 1$. Using the fact that $\partial^m u = \sum_{k=0}^{\infty} \frac{m!}{k!(m-k)!}(\partial^k u \partial^u)\partial^{m-k}u$, $\partial^m \partial^n = \partial^{m+n}$, as well as the formulas described in example 10.1, we specify the quantifiers of $\partial^{-1}, \partial^{-2}, \ldots$ in the equation $WW^{-1} = 1$ and we determine relations between w_m and v_m. We finally obtain for W^{-1} the following expression:

$$W^{-1} = 1 - w_1\partial^{-1} + (-w_2 + w_1^2)\partial^{-2} + (w_3 + 2w_1 w_2 - w_1 \partial w_1 - w_1^3)\partial^{-3} + \cdots$$

In terms of W, the operator L [10.13] can be written in the form

$$L = W.\partial.W^{-1}. \qquad [10.20]$$

According to [10.13] and [10.19], we deduce the relations: $u_2 = \partial w_2$, $u_3 = -\partial w_2 - w_1 \partial w_1$, $u_4 = -\partial w_3 + w_1 \partial w_2 + (\partial w_1)w_2 - w_1^2 \partial w_1 - (\partial w_1)^2$. We have the following theorem:

THEOREM 10.6.– Equations [10.15] or what amounts to the same (according to theorem 10.5, equations [10.16]) are equivalent to the existence of the wave operator W [10.19] such that the system of differential equations

$$LW = W\partial, \qquad [10.21]$$

$$\frac{\partial W}{\partial t_n} = -L_-^n W, \qquad [10.22]$$

has a solution (which can be inductively obtained).

THEOREM 10.7.– Let $\xi(t, z) = \sum_{j=1}^{\infty} t_j z^j$, $z \in \mathbb{C}$, be the phase function with $\partial^m \xi(t, z) = z^m$ and $\partial^m e^{\xi(t,z)} = z^m e^{\xi(t,z)}$. There is an equivalence between [10.18] and [10.22] and the following problem: there is a wave function ψ (Baker–Akhiezer function)

$$\Psi(t, z) = \left(1 + w_1(t)z^{-1} + w_2(t)z^{-2} + \cdots\right) e^{\xi(t,z)} = W \cdot e^{\xi(t,z)}, \qquad [10.23]$$

where $z \in \mathbb{C}$, W is identified as [10.19] and such that:

$$L\Psi = z\Psi, \qquad \frac{\partial \Psi}{\partial t_n} = L_+^n \Psi. \qquad [10.24]$$

PROOF.– Indeed, we have, from [10.23],

$$\begin{aligned}
\frac{\partial \Psi}{\partial t_n} &= \frac{\partial W}{\partial t_n} e^{\xi(t,z)} + W z^n e^{\xi(t,z)}, \\
&= -L_-^n W e^{\xi(t,z)} + z^n W e^{\xi(t,z)}, \quad \text{according to [10.20]} \\
&= -L_-^n \Psi + z^n \Psi, \quad \text{according to [10.21]} \\
&= -L_-^n \Psi + L^n \Psi, \quad \text{according to [10.22]} \\
&= L_+^n \Psi.
\end{aligned}$$

In other words, Ψ satisfies [10.23] and [10.24] is equivalent to the fact that W satisfies [10.19] and [10.22]. $\square$

Introduce the conjugation $\partial^* = -\partial$ and let $L^* = 1 + (-\partial)^{-1} u_1 + (-\partial)^{-2} u_2 + \cdots$, and $W^* = 1 + (-\partial)^{-1} w_1 + (-\partial)^{-2} w_2 + \cdots$, be the adjoints of L and W such that: $L^* = -(W^*)^{-1} . \partial . W^*$.

THEOREM 10.8.– The adjoint wave function $\Psi^*(t, z) = (W^*(t, \partial))^{-1} e^{-\xi(t,z)}$ satisfies the following relations: $L^* \Psi^* = z \Psi^*$, $\frac{\partial \Psi^*}{\partial t_n} = -(L_+^n)^* \Psi^*$.

PROOF.– The reason is as in the proof of the previous theorem. □

Therefore, the knowledge of Ψ implies the knowledge of W and also of W^* and L. Define the following residues: $\underset{z}{\mathrm{Res}} \sum a_k z^k = a_{-1}$, $\underset{\partial}{\mathrm{Res}} \sum a_k \partial^k = a_{-1}$, and consider the following lemma (Dickey 1997):

LEMMA 10.1.– We have $\underset{z}{\mathrm{Res}}((Pe^{xz}).(Qe^{-xz})) = \underset{\partial}{\mathrm{Res}}\, PQ^*$, where P and Q are two pseudo-differential operators and Q^* is the adjoint of Q.

PROOF.– Indeed, we have $\underset{z}{\mathrm{Res}}((Pe^{xz}).(Qe^{-xz})) = \underset{z}{\mathrm{Res}}\left(\sum p_k z^k \sum q_l(-z)^l\right) = \sum_{k+l=-1}(-1)^l p_k q_l$, and $\underset{\partial}{\mathrm{Res}}\, PQ^* = \underset{\partial}{\mathrm{Res}} \sum_{kl} p_k \partial^k (-\partial)^l q_l = \sum_{k+l=-1}(-1)^l p_k q_l$. □

Moreover, we have (Date et al. 1983; Dickey 1997):

$$\underset{z}{\mathrm{Res}}\,(\partial^k \Psi).\Psi^* = \underset{z}{\mathrm{Res}}\left(\partial^k W e^{\xi(t,z)}\right)(W^*)^{-1} e^{-\xi(t,z)},$$

$$= \underset{z}{\mathrm{Res}}\left(\partial^k W e^{xz}\right)(W^*)^{-1} e^{-xz}, \quad x \equiv t-1,$$

$$= \underset{\partial}{\mathrm{Res}}\,\partial^k W.W^{-1} = \underset{\partial}{\mathrm{Res}}\,\partial^k = 0.$$

This bilinear identity can be written in the following symbolic form: $\underset{z=\infty}{\mathrm{Res}}\,(\Psi(t,z).\Psi^*(t',z)) = 0 \quad \forall t, t'$. Therefore, we have

THEOREM 10.9.– $\Psi(t,z)$ is a wave function for the KP hierarchy if and only if the residue identity is satisfied:

$$\underset{z=\infty}{\mathrm{Res}}\,(\Psi(t,z).\Psi^*(t',z)) = 0 \quad \forall t, t' \qquad [10.25]$$

or what amounts to the same if and only if

$$\frac{1}{2\pi\sqrt{-1}} \int_\gamma \Psi(t,z).\Psi^*(t',z) dz = 0, \qquad [10.26]$$

with γ a closed path around $z = \infty$ (such that: $\int_\gamma \frac{dz}{2\pi\sqrt{-1}} = 1$).

DEFINITION 10.2.– A $\tau(t)$ function is defined by the Fay differential identity:

$$\{\tau(t-[y_1]), \tau(t-[y_2])\} + (y_1^{-1} - y_2^{-1})(\tau(t-[y_1])\tau(t-[y_2])$$
$$-\tau(t)\tau(t-[y_1]-[y_2])) = 0,$$

where $y_1, y_2 \in \mathbb{C}^*$ and $\{u,v\}$ is the Wronskian $u'v - uv'$.

THEOREM 10.10.– Let us put $[s] = \left(s, \frac{s^2}{2}, \frac{s^3}{3}, \ldots\right)$. The τ function satisfies the following identities:

i) Fay identity:

$$\mathcal{F}(t, y_0, y_1, y_2, y_3) \equiv (y_0 - y_1)(y_2 - y_3)\tau(t + [y_0] + [y_1])\tau(t + [y_2] + [y_3])$$
$$+ (y_0 - y_2)(y_3 - y_1)\tau(t + [y_0] + [y_2])\tau(t + [y_2] + [y_1])$$
$$+ (y_0 - y_3)(y_1 - y_2)\tau(t + [y_0] + [y_3])\tau(t + [y_1] + [y_2]),$$
$$= 0.$$

ii) Fay differential identity:

$$\{\tau(t - [y_1]), \tau(t - [y_2])\} + (y_1^{-1} - y_2^{-1})(\tau(t - [y_1])\tau(t - [y_2])$$
$$- \tau(t)\tau(t - [y_1] - [y_2])) = 0,$$

where $y_1, y_2 \in \mathbb{C}^*$ and $\{u, v\}$ is the Wronskian $u'v - uv'$. This identity can still be written in the form as $\partial^{-1}\psi(t, \lambda)\psi^*(t, \mu) = \frac{1}{\mu - \lambda} \frac{\tau(t - [\lambda^{-1}] + [\mu^{-1}])}{\tau(t)} e^{\sum_{j=1}^{\infty} t_j(\mu^j - \lambda^j)}$. The equation $\dot{\tau} = X(t, \lambda, \mu)\tau$ determines a vector field on the infinite dimension manifold of the τ functions where $X(t, \lambda, \mu)$ is the vertex operator (of Date–Jimbo–Kashiwara–Miwa) for the KP equation.

PROOF.– According to Sato theory (Sato 1989; Sato and Sato 1982), the functions Ψ and Ψ^* can be expressed in terms of a tau function as follows: $\Psi(t, z) = We^{\xi(t,z)} = \frac{\tau(t - [z^{-1}])}{\tau(t)} e^{\xi(t,z)}$, $\Psi^*(t, z) = (W^*)^{-1} e^{-\xi(t,z)} = \frac{\tau(t + [z^{-1}])}{\tau(t)} e^{-\xi(t,z)}$. By replacing these expressions in the residue formula [10.25] or [10.26], we obtain a bilinear relation for the τ functions. Indeed, equation [10.26] is written as: $\int_\gamma e^{\xi(t - t', z)} \tau(t - [z^{-1}])\tau(t' - [z^{-1}]) dz = 0$. Using the following change: $t \leftarrow t + s$ and $t' \leftarrow t + s$, we obtain $\int_\gamma e^{\xi(-2s, z)} \tau(t - s - [z^{-1}]) \tau(t + s + [z^{-1}]) dz = 0$. Again using the transformation, $s \leftarrow t + \frac{1}{2}([y_0] + [y_1] + [y_2] + [y_3])$, $t \leftarrow \frac{1}{2}([y_0] - [y_1] - [y_2] - [y_3])$, and taking into account that $e^{\sum_1^\infty (ab^{-1})^j \cdot j^{-1}} = 1 - ab^{-1}$, we obtain, via the residue theorem

$$0 = \int_\gamma \frac{1 - zy_0}{\prod_{j=1}^3 (1 - zy_j)} \tau(t - s - [z^{-1}])\tau(t + s + [z^{-1}]) dz,$$

$$= 2\pi\sqrt{-1} \sum \operatorname*{Res}_{y_1^{-1}, y_2^{-1}, y_3^{-1}} \left(\frac{1 - zy_0}{\prod_{j=1}^3 (1 - zy_j)} \tau(t - s - [z^{-1}])\tau(t + s + [z^{-1}]) \right),$$

$$= \frac{2\pi\sqrt{-1}}{(y_1 - y_2)(y_2 - y_3)(y_3 - y_1)} \mathcal{F}(t, y_0, y_1, y_2, y_3),$$

where

$$\mathcal{F}(t, y_0, y_1, y_2, y_3) \equiv (y_0 - y_1)(y_2 - y_3)\tau(t + [y_0] + [y_1])\tau(t + [y_2] + [y_3])$$
$$+ (y_0 - y_2)(y_3 - y_1)\tau(t + [y_0] + [y_2])\tau(t + [y_2] + [y_1])$$
$$+ (y_0 - y_3)(y_1 - y_2)\tau(t + [y_0] + [y_3])\tau(t + [y_1] + [y_2]).$$

The relation $\mathcal{F}(t, y_0, y_1, y_2, y_3) = 0$ is the Fay identity. In addition, by making the transformation in the expression $(y_1 y_2)^{-1} \frac{\partial \mathcal{F}}{\partial y_0}|_{y_0 = y_3 = 0}$ and replacing t by $t - [y_1] - [y_2]$, we obtain the Fay differential identity which allows to define the τ functions:

$$\{\tau(t - [y_1]), \tau(t - [y_2])\} + (y_1^{-1} - y_2^{-1})\left(\tau(t - [y_1])\tau(t - [y_2])\right.$$
$$\left. - \tau(t)\tau(t - [y_1] - [y_2])\right) = 0,$$

where $y_1, y_2 \in \mathbb{C}^*$ and $\{u, v\}$ is the Wronskian $u'v - uv'$. Consider the Fay differential identity above and replace t with $t + [y_1]$. We obtain

$$\{\tau(t), \tau(t + [y_1] - [y_2])\} + (y_1^{-1} - y_2^{-1})\left(\tau(t)\tau(t + [y_1] - [y_2])\right.$$
$$\left. - \tau(t)\tau(t - [y_2])\right) = 0.$$

By putting $\lambda = y_1^{-1}$, $\mu = y_2^{-1}$, we obtain, after having multiplied the expression obtained by $\frac{1}{\tau(t)} e^{\sum_1^\infty t_j (\mu^j - \lambda^j)}$, the following formula:

$$\frac{\tau(t + [\lambda^{-1}])}{\tau(t)} e^{-\sum t_j \lambda^j} \frac{\tau(t - [\mu^{-1}])}{\tau(t)} e^{\sum t_j \mu^j}$$
$$= \frac{1}{\mu - \lambda} \frac{\partial}{\partial x} \left(e^{\sum t_j (\mu^j - \lambda^j)} \frac{\tau(t + [\lambda^{-1}] - [\mu^{-1}])}{\tau(t)} \right).$$

Let $X(t, \lambda, \mu) = \frac{1}{\mu - \lambda} e^{\sum_1^\infty t_j (\mu^j - \lambda^j)} e^{\sum_1^\infty j^{-1}(\lambda^{-j} - \mu^{-j}) \frac{\partial}{\partial t_j}}$, $\lambda \neq \mu$, be the vertex operator (of Date–Jimbo–Kashiwara–Miwa) for the KP equation, then $X(t, \lambda, \mu)\tau$ and $\tau + X(t, \lambda, \mu)\tau$ are also τ functions. Therefore, $\dot{\tau} = X(t, \lambda, \mu)\tau$ determines a vector field on the infinite dimension manifold of functions τ. We deduce, according to Dickey (1997), that $\partial^{-1}\left(\Psi^*(t, \lambda)\Psi(t, \mu)\right) = \frac{1}{\tau(t)} X(t, \lambda, \mu)\tau(t)$. $\square$

REMARK 10.2.– Let $\Delta(s_1, ..., s_n) = \prod_{1 \leq j < i \leq n}(s_i^{-1} - s_j^{-1})$ be the Vandermonde determinant. Fay identities (theorem 10.10) are generalized as follows. The τ function satisfies identities:

$$\tau\left(t-\sum_{j=1}^{n}[y_j]\right)\Delta(y_1,...,y_n)\left(\left(t-\sum_{j=1}^{n}[y_j]\right)\Delta(x_1,...,x_n)\right)^{n-1}$$

$$=\det\left(\left(t-\sum_{j=1}^{n}[x_k]+[x_j]-[y_l]\right)\Delta(x_1,...,x_{j+1},y_j,x_{j+1},...,x_n)\right)_{1\leq j,l\leq n},$$

and

$$\{\psi(t,y_1^{-1}),...,\psi(t,y_n^{-1})\}$$
$$=e^{\sum_{j=1}^{\infty}t_j(y_1^{-j}+\cdots+y_n^{-j})}\frac{\tau(t-[y_l]-\cdots-[y_n])}{\tau(t)}\Delta(y_1,...,y_n),$$

where $\{u_1,...,u_n\}$ is the Wronskian $\det\left(\left(\frac{\partial}{\partial x}\right)^{j-1}u_j\right)_{1\leq i,j\leq n}$.

We will see that τ functions characterize the KP hierarchy. Let $s_j(t)$ denote the elementary Schur polynomials, that is, polynomials such that:

$$e^{\xi(t,z)}=e^{\sum_{j=1}^{\infty}t_jz_j}=\sum_{j=1}^{\infty}s_j(t)z_j,$$

$$=1+t_1z+\left(\frac{1}{2}t_1^2+t_2\right)z^2+\left(\frac{1}{6}t_1^3+t_1t_2+t_3\right)z^3+\cdots$$

with $s_j(t)=\frac{t_1^j}{j!}+\cdots+t_n$. By setting $\widetilde{\partial}=\left(\frac{\partial}{\partial t_1},\frac{1}{2}\frac{\partial}{\partial t_2},\frac{1}{3}\frac{\partial}{\partial t_3},...\right)$, we obtain

$$\Psi(t,z)=\frac{\tau\left(t_1-z^{-1},t_2-\frac{z^{-2}}{2},t_3-\frac{z^{-3}}{3},...\right)}{\tau(t)}e^{\xi(t,z)},$$

$$=\sum_{j=0}^{\infty}\frac{s_j(-\widetilde{\partial})\tau(t)}{\tau(t)}\partial^{-j}e^{\xi(t,z)}=W(t)e^{\xi(t,z)},$$

where

$$W(t)=\sum_{j=0}^{\infty}\frac{s_j(-\widetilde{\partial})\tau(t)}{\tau(t)}\partial^{-j},\qquad [10.27]$$

is the wave operator [10.19]. Similarly, we have $W^{-1}=\sum_{j=0}^{\infty}\partial^{-j}\frac{s_j(\widetilde{\partial})\tau(t)}{\tau(t)}$. In addition, it follows from [10.20] that $L^n=W.\partial^n.W^{-1}$, and therefore L^n is expressed in terms of the τ function, $L^n = \sum_{i,j=0}^{\infty} \frac{s_i(-\tilde{\partial})\tau}{\tau} \partial^{n-i-j} \frac{s_j(\tilde{\partial})}{\tau} \tau$. By developing this expression, we get

$$L^n = \partial^n + n(\log \tau)'' \partial^{n-2} + \cdots + \sum_{i+j=n+1} \frac{s_i(\tilde{\partial})\tau s_j(-\tilde{\partial})\tau}{\tau^2} + \cdots$$

The formula [10.22] is written taking into account this last expression of L^n and the relation [10.27] as follows: $\frac{\partial}{\partial t_n}\left(1 - \frac{\tau'}{\tau}\partial^{-1} + \cdots\right) = \left(-\sum_{i+j=n+1} \frac{s_i(-\tilde{\partial})\tau s_j(-\tilde{\partial})\tau}{\tau^2} \partial^{-1} + \cdots\right)\left(1 - \frac{\tau'}{\tau}\partial^{-1} + \cdots\right)$. Using the Hirota symbol[1], we have $\sum_{\substack{i+j=n+1 \\ i,j \geq 0}} \left(s_i(\tilde{\partial})\tau\right)\left(s_j(-\tilde{\partial})\tau\right) = s_{n+1}(\tilde{\partial})\tau.\tau$, and we obtain $\tau^2 \frac{\partial^2}{\partial t_n \partial t_1} \log \tau - \sum_{\substack{i+j=n+1 \\ i,j \geq 0}} s_i(\tilde{\partial})\tau s_j(-\tilde{\partial})\tau = 0$, $n \in \mathbb{N}^*$. These relations are called Hirota bilinear equations. They show that all the functions, $j \geq 2$, can be expressed in terms of the τ function. For example,

$$u_2 = \frac{\partial^2}{\partial t_1^2} \log \tau, \quad u_3 = \frac{1}{2}\left(\frac{\partial^3}{\partial t_1^3} + \frac{\partial}{\partial t_1}\frac{\partial}{\partial t_3}\right) \log \tau,$$

$$u_4 = \frac{1}{6}\left(\frac{\partial^4}{\partial t_1^4} - 3\frac{\partial^2}{\partial t_1^2}\frac{\partial}{\partial t_2} + 2\frac{\partial}{\partial t_1}\frac{\partial}{\partial t_2}\right)\log \tau - \left(\frac{\partial^2}{\partial t_1^2}\log \tau\right), \ldots$$

In particular, these equations provide the KP equation in the following bilinear form:

$$\frac{1}{12}\tau\left(\frac{\partial^4 \tau}{\partial t_1^4} - 4\frac{\partial^2 \tau}{\partial t_1 \partial t_3} + 3\frac{\partial^2 \tau}{\partial t_2^2}\right) - \frac{1}{3}\frac{\partial \tau}{\partial t_1}\left(\frac{\partial^3 \tau}{\partial t_1^3} - \frac{\partial \tau}{\partial t_3}\right)$$
$$+ \frac{1}{4}\left(\frac{\partial^2 \tau}{\partial t_1^2} + \frac{\partial \tau}{\partial t_2}\right)\left(\frac{\partial^2 \tau}{\partial t_1^2} - \frac{\partial \tau}{\partial t_2}\right) = 0.$$

Therefore, we have the following theorem.

THEOREM 10.11.– *The τ functions characterize the KP hierarchy.*

Equations of soliton theory play an important role in the characterization of Jacobian varieties. Let $\mathcal{H}_g = \{Z \in M_g(\mathbb{C}) : \Omega = \Omega^\top, \mathrm{I}\Omega > 0\}$ be the Siegel half-space, $\Lambda = \mathbb{Z}^g \oplus Z\mathbb{Z}^g$ a lattice in $\mathbb{C}^g$ and $T = \mathbb{C}^g/\Lambda$ a principally polarized Abelian variety. The following three conditions are equivalent (see Shiota 1986):

---

[1] $p(\partial_t)f(t).g(t) \equiv p\left(\frac{\partial}{\partial s_1}, \frac{\partial}{\partial s_2}, \ldots\right) f(t+s)g(t-s)\Big|_{s=0}$ where p is any polynomial, $f(t)$ and $g(t)$ are two differentiable functions.

(i) There are vector fields v_1, v_2, v_3 on $\mathbb{C}^g$ and a quadratic form $q(t) = \sum_{k,l=1}^{3} q_{kl}(t) t_k t_l$ such that for all $z \in \mathbb{C}^g$, the function $\tau(t) = e^{q(t)} \theta \left(\sum_{k=1}^{3} t_k v_k + z \right)$ satisfies the KP equation. The theta divisor does not contain an Abelian subvariety of T for which the vector v_1 is tangent. (ii) T is isomorphic to the Jacobian variety of a reduced non-singular complete curve of genus g. (iii) There is a matrix $\mathcal{V} = (v_1, v_2, ...)$ of order $g \times \infty$, $v_k \in \mathbb{C}^g$, of rank g and a quadratic form $Q(t) = \sum_{k,l=1}^{\infty} q_{kl}(t) t_k t_l$ such that for all $z \in \mathbb{C}^g$, $\widetilde{\tau}(t) = e^{Q(t)} \theta \left(\mathcal{V} t + z \right)$, is a τ function for KP hierarchy.

10.4. Exercises

EXERCISE 10.1.– Use [10.1] to compare $u(x).\partial$ and $\partial.u(x)$ (for some function $u(x)$) and show that these expressions are not equal to each other. What can be concluded about the multiplication of differential operators?

EXERCISE 10.2.– Let $P(x, \partial_x) = \sum_j a_j(x) \partial_x^j$, $Q(x, \partial_x) = \sum_j b_j(x)(-\partial)_x^j$, be two pseudo-differential operators with coefficients depending on x. Show that

$$\left(P(x, \partial_x) Q^\top(x, \partial_x) \right)_{-} \delta(x - y) = \int P(x, \partial_x) e^{xz} Q(y, \partial_y) e^{-yz} \frac{dz}{2\pi i} \mathcal{H}(x - y),$$

where the integral is taken over a small circle around $z = \infty$, δ is the customary δ-function and $\mathcal{H}$ the Heaviside function $\mathcal{H}(x) = \partial_x^{-1} \delta(x)$.

EXERCISE 10.3.– Given the pseudo-differential operator L, we can change the wave operator W (this operator is not unique) as follows, $W(t) \longmapsto W(t) W_0$, where W_0 is a constant coefficients pseudo-differential operator $W_0 = 1 + \sum_{j=1}^{\infty} a_j \partial^{-j}$, $a_j \in \mathbb{C}$. Show that this modification has the following effect on the wave function Ψ:

$$\Psi(t, z) = W(t) e^{\sum_{j=1}^{\infty} t_j z^j} \longmapsto \widetilde{\Psi} = \Psi(t, z) e^{-\sum_{j=1}^{\infty} \frac{b_j}{j} z^{-j}},$$

for some appropriate b_j, and the τ-function $\tau(t) \longmapsto \widetilde{\tau} = \tau(t) e^{\sum_{j=1}^{\infty} b_j t_j}$.

EXERCISE 10.4.– Show that $\partial^{-1}.(\partial.u(x)) = u(x)$, $(\partial.\partial^{-1})u(x) = \partial.(\partial^{-1}u(x))$, $\partial^2.(\partial^{-2}.(\partial^{-1} + x^2 \partial^{-3})) = \partial^{-1} + x^2 \partial^{-3}$.

EXERCISE 10.5.– Given an ordinary operator L, show (Dickey 1991) that there are two pseudo-differential operators L^{-1} and $L^{\frac{1}{n}}$ such that: $L.L^{-1} = L^{-1}.L = 1$, $\left(L^{\frac{1}{n}} \right)^n = L$.

EXERCISE 10.6.– Let $\mathcal{A}$ be the Lie algebra of a pseudo-differential operator and Res : $\mathcal{A} \longrightarrow \mathcal{C}^\infty(S^1)$, the residue given by $\left(\sum a_k \partial^{-k}\right) \longmapsto a_{-1}(x)$ (i.e. coefficient of ∂^{-1}). Show that for all $L_1, L_2 \in \mathcal{A}$, we have $Res[L_1, L_2] = 0$.

EXERCISE 10.7.– We use the notation from section 10.3. Consider the set of equations [10.15] (KP hierarchy). Show that $\int Res\, L^k dx$ are first integrals of equation [10.15] and vector fields defined by this equation commute.

References

Abenda, S. and Fedorov, Y. (2000). On the weak Kowalewski-Painlevé property for hyperelliptically separable systems. *Acta Appl. Math.*, 60, 137–178.

Abraham, R. and Marsden, J. (1978). *Foundations of Mechanics*, 2nd edition. Benjamin Cummings Publishing Company, Reading, MA.

Adler, M. (1979). On a trace functional for formal pseudo differential operators and the symplectic structure of the KdV type equations. *Invent. Math.*, 50(3), 219–248.

Adler, M. and van Moerbeke, P. (1980a). Completely integrable systems, Euclidean Lie algebras, and curves. *Adv. in Math.*, 38, 267–379.

Adler, M. and van Moerbeke, P. (1980b). Linearization of Hamiltonian systems, Jacobi varieties and representation theory. *Adv. in Math.*, 38, 318–379.

Adler, M. and van Moerbeke, P. (1982a). The algebraic complete integrability of geodesic flow on $SO(4)$. With an appendix by D. Mumford. *Invent. Math.*, 67, 297–331.

Adler, M. and van Moerbeke, P. (1982b). Kowalewski's asymptotic method, Kac-Moody lie algebras and regularization. *Comm. Math. Phys.*, 83(1), 83–106.

Adler, M. and van Moerbeke, P. (1984). Geodesic flow on $SO(4)$ and the intersection of quadrics. *Proc. Nat. Acad. Sci. U.S.A.*, 81, 4613–4616.

Adler, M. and van Moerbeke, P. (1985). Algebraic completely integrable systems: A systematic approach, I, II, III. *Séminaire de Mathématique*, Report no. 110, SC/MAPA, UCL, London.

Adler, M., and van Moerbeke, P. (1987). The intersection of four quadrics in $\mathbb{P}^6$. Abelian surfaces and their moduli. *Math. Ann.*, 279, 25–85.

Adler, M. and van Moerbeke, P. (1988). The Kowalewski and Hénon-Heiles motions as Manakov geodesic flows on SO(4) – A two-dimensional family of Lax pairs. *Comm. Math. Phys.*, 113(4), 659–700.

Adler, M. and van Moerbeke, P. (1989). The complex geometry of the Kowalewski-Painlevé analysis. *Invent. Math.*, 97, 3–51.

Adler, M. and van Moerbeke, P. (1991). The Toda lattice, Dynkin diagrams, singularities and abelian varieties. *Invent. Math.*, 103(2), 223–278.

Adler, M., van Moerbeke, P., Vanhaecke, P. (2004). *Algebraic Integrability, Painlevé Geometry and Lie Algebras*. Springer-Verlag, Berlin, Heidelberg.

Airault. H., McKean, H.P., Moser, J. (1977). Rational and elliptic solutions of the KdV equation and a related many-body problem. *Comm. Pure Appl. Math.*, 30, 94–148.

Ankiewicz, A. and Pask, C. (1983). The complete Whittaker theorem for two-dimensional integrable systems and its application. *J. Phys. A: Math. Gen.*, 16, 4203–4208.

Appel'rot, G. (1894). The problem of motion of a rigid body about a fixed point. *Uchenye Zap. Mosk. Univ. Otdel. Fiz. Mat. Nauk*, 11(3), 1–112.

Arbarello, E., Cornalba, M., Griffiths, P., Harris, J. (1985). *Geometry of Algebraic Curves*, Volume 1. Springer-Verlag, Berlin.

Arnold, V.I. (1988). *Geometrical Methods in the Theory of Ordinary Differential Equations*. Springer-Verlag, New York.

Arnold, V.I. (1989). *Mathematical Methods in Classical Mechanics*. Springer-Verlag, New York.

Arnold, V.I. and Givental, A.B. (1990). Symplectic geometry. In *Dynamical Systems IV*, Arnold, V.I. and Novikov, S.P. (eds). Springer, Berlin, Heidelberg.

Arutyunov, G. (2019). *Elements of Classical and Quantum Integrable Systems*. Springer, Cham.

Babelon, O., Bernard, D., Talon, M. (2003). *Introduction to Classical Integrable Systems*. Cambridge University Press, Cambridge.

Baker, S., Enolskii, V.Z., Fordy, A.P. (1995). Integrable quartic potentials and coupled KdV equations. *Phys. Lett.*, 201A, 167–174.

Bechlivanidis, C. and van Moerbeke, P. (1987). The Goryachev-Chaplygin top and the Toda lattice. *Comm. Math. Phys.*, 110(2), 317–324.

Belokolos, A.I. and Enol'skii, V.Z. (1989). Isospectral deformations of elliptic potentials. *Russ. Math. Surveys*, 44, 155–156.

Belokolos, A.I., Bobenko, V.Z., Enol'skii, V.Z., Its, A.R., Matveev, V.B. (1994). *Algebro-Geometric Approach to Nonlinear Integrable Equations*. Springer-Verlag, Berlin, Heidelberg.

Berry, M.L. (1978). Regular and irregular motions. In *Topics in Nonlinear Dynamics*, Jorna, S. (ed.). American Institute of Physics, New York.

Bialy, M. (2010). Integrable geodesic flows on surfaces. *Geom. Funct. Anal.*, 20, 357–367.

Bobylev, D. (1896). On a certain particular solution of the differential equations of rotation of a heavy rigid body about a fixed point. *Trudy Otdel. Fiz. Nauk Obsc. Estestvozn.*, 8, 21–25.

Bolsinov, A.V., Morales-Ruiz, J., Zung, N.T. (2016). *Geometry and Dynamics of Integrable Systems*. Springer, Basel.

Bolsinov, A.V., Izosimov, A.M., Tsonev, D.M. (2017). Finite-dimensional integrable systems: A collection of research problems. *J. Geom. Phys.*, 115, 2–15.

Boussinesq, J. (1877). Essai sur la théorie des eaux courantes. Mémoires présentés par divers savants. *Acad. des Sci. Inst. Nat. France*, XXIII, 1–680.

Chaplygin, S.A. (ed.) (1948). A new case of rotation of a rigid body, supported at one point. *Collected Works I*. Gostekhizdat, USSR. [in Russian].

Cherednick, I.V. (1978). Differential equations for the Baker-Akhiezer functions of algebraic curves. *Funct. Anal. Appl.*, 12, 45–54.

Christiansen, P.L., Eilbeck, J.C., Enolskii, V.Z., Kostov, N.A. (1995). Quasi-periodic of the coupled nonlinear Schrödinger equations. *Proc. R. Soc. Lond. A*, 451, 685–700.

Clebsch, A. (1871). Der Bewegung eines starren Körpers in einen Flüssigkeit. *Math. Ann.*, 3, 238–268.

Conte, R., Musette, M., Verhoeven, C. (2005). Completeness of the cubic and quartic Hénon-Heiles Hamiltonians. *Theor. Math. Phys.*, 144, 888–898.

Date, E., Jimbo, M., Kashiwara, M., Miwa, T. (1983). Transformation groups for soliton equations. *Proc. RIMS Symp. Nonlinear Integrable Systems, Classical and Quantum Theory (Kyoto, 1981)*, Jimbo, M., Miwa, T. (eds). World Scientific, Singapore.

Deift, P., Lund, F., Trubowitz, E. (1980). Nonlinear wave equations and constrained harmonic motion. *Comm. Math. Phys.*, 74, 141–188.

Dickey, L. (1991). *Soliton Equations and Integrable Systems*. World Scientific, Singapore.

Dickey, L. (1993). Additional symmetries of KP, Grassmannian and the string equation. *Mod. Phys. Lett. A*, 8, 1259–1272.

Dickey, L. (1997). Lectures on classical W-algebras (Cortona Lectures). *Acta Appl. Math.*, 47, 243–321.

Dorizzi, B., Grammaticos, B., Ramani, A. (1983). A new class of integrable systems. *J. Math. Phys.*, 24, 2282.

Dubrovin, B.A. (1981). Theta functions and non-linear equations. *Russian Math. Surv.*, 36(2), 11–92.

Dubrovin, B.A. and Novikov, S.P. (1974). Periodic and conditionally periodic analogues of multi-soliton solutions of the Korteweg-de Vries equation. *Soviet Physics JETP*, 40, 1058–1063.

Dubrovin, B.A., Novikov, S.P., Fomenko, A.T. (1984). *Modern Geometry–Methods and Applications. Part I*. Springer-Verlag, New York.

Dubrovin, B.A., Novikov, S.P., Fomenko, A.T. (1985). *Modern Geometry–Methods and Applications. Part II*. Springer-Verlag, New York.

Dubrovin, B.A., Novikov, S.P., Fomenko, A.T. (1990). *Modern Geometry–Methods and Applications. Part III*. Springer-Verlag, New York.

Eilbeck, J.C. and Enolskii, V.Z. (1994). Elliptic solutions and blow-up in an integrable Hénon-Heiles system. *Proc. Roy. Soc. Edinburgh, Sect. A*, 124, 1151–1164.

Eilbeck, J.C., Enolskii, V.Z., Kuznetsov, V.B., Leykin, D.V. (1993). Linear r-matrix algebra for systems separable in parabolic coordinates. *Phys. Lett.*, 180A, 208–214.

Eilbeck, J.C., Enolskii, V.Z., Kuznetsov, V.B., Tsiganov, A.V. (1994). Linear r-matrix algebra for classical separable systems. *J. Phys. A.: Math. Gen.*, 27, 567–578.

Euler, L. (1765). *Theoria motus corporum solidorum seu rigidorum*. A.F. Rose, Rostock, Greifswald.

Faddev, L.D. and Takhtajan, L.A. (2007). *Hamiltonian Methods in the Theory of Solitons*. Springer, Cham.

Fay, J. (1973). *Theta Functions on Riemann Surfaces*. Springer-Verlag, Cham.

Ferguson, W.E. (1980). The construction of Jacobi and periodic Jacobi matrices with prescribed spectra. *Math. Comp.*, 35, 1220–1230.

Flaschka, H. (1974). The Toda lattice I. Existence of integrals. *Phys. Rev. B*, 3(9), 1924–1925.

Fordy, A.P. (1991). The Hénon-Heiles system revisited. *Physica D*, 52, 204–210.

Garay, M. and van Straten, D. (2010). Classical and quantum integrability. *Mosc. Math. J.*, 10, 519–545.

Gardner, C.S., Greene, J.M., Kruskal, M.D., Miura, R.M. (1967). Method for solving the Korteweg-de Vries equation. *Phys. Rev. Lett.*, 19, 1095–1097.

Garnier, R. (1919). Sur une classe de systèmes différentiels abéliens déduits de la théorie des équations linéaires. *Ren. Circ. Math. Palermo*, 43(4), 155–191.

Gavrilov, L. and Angel Zhivkov, A. (1998). The complex geometry of the Lagrange top. *Enseign. Math.*, 44, 133–170.

Gavrilov, L., Ouazzani-Jamil, M., Caboz, R. (1992). Bifurcations des tores de Liouville du potentiel de Kolosoff $U = \rho + \frac{1}{\rho} - k\cos\varphi$. *CRAS*, 315(I), 289–294.

Gelfand, I.M. and Dickey, L. (1968). Family of Hamiltonian structures connected with integrable nonlinear differential equations. *Funct. Anal. Appl.*, 2, 92–93.

Gelfand, I.M. and Levitan, B.M. (1955). On the determination of a differential equation from its spectral function. *Amer. Math. Soc. Transl.*, 2(1), 253–304.

Gerd, R. and Matthias, S. (2013). *Differential Geometry and Mathematical Physics Part I: Manifolds, Lie Groups and Hamiltonian Systems*. Springer, Dordrecht.

Gervais, J.-L. (1985). Infinite family of polynomial functions of the Virasoro generators with vanishing Poisson bracket. *Phys. Lett. B*, 160(4–5), 277–278.

Goryachev, D. (1900). On the motion of a rigid material body about a fixed point in the case A=B=4C. *Mat. Sb.*, 21(3).

Grammaticos, B., Dorozzi, B., Ramani, A. (1983). Integrability of Hamiltonians with third and fourth-degree polynomial potentials. *J. Math. Phys.*, 24, 2289–2295.

Grammaticos, B., Dorozzi, B., Ramani, A. (1984). Hamiltonians with higher-order integrals and the "weak-Painlevé" concept. *J. Math. Phys.*, 25, 3470.

Griffiths, P.A. (1985). Linearizing flows and a cohomological interpretation of Lax equations. *Amer. J. Math.*, 107, 1445–1483.

Griffiths, P.A. and Harris, J. (1978). *Principles of Algebraic Geometry*. Wiley Interscience, Hoboken.

Guillemin, V. and Sternberg, S. (1984). *Symplectic Techniques in Physics*. Cambridge University Press, Cambridge.

Haine, L. (1983). Geodesic flow on $SO(4)$ and Abelian surfaces. *Math. Ann.*, 263, 435–472.

Haine, L. (1984). The algebraic complete integrability of geodesic flow on $SO(N)$ and Abelian surfaces. *Comm. Math. Phys.*, 94(2), 271–287.

Halphen, G.-H. (1888). *Traité des fonctions elliptiques et de leurs applications*. Gauthier-Villars, Paris.

Hénon, M. and Heiles, C. (1964). The applicability of the third integral of motion; some numerical experiments. *Astron. J.*, 69, 73–79.

Hess, W. (1890). Uber die Euler hen Bewegungsgleichungen und tlber eine neue par dare L(isung des Problems der Bewegung eines starren Korpers un einen festen punkt. *Math. Ann.*, 37(2), 153–181.

Hietarinta, J. (1984). Classical versus quantum integrability. *J. Math. Phys.*, 25, 1833–1840.

Hietarinta, J. (1985). How to construct integrable Fokker-Planck and electromagnetic Hamiltonians from ordinary integrable Hamiltonians. *J. Math. Phys.*, 26, 1970–1975.

Hietarinta, J. (1987). Direct methods for the search of the second invariant. *Phys. Rep.*, 147, 87–154.

Hitchin, N.J. (1983). On the construction of monopoles. *Comm. Math. Phys.*, 89(2), 145–190.

Hitchin, N.J. (1994). Stable bundles and integrable systems. *Duke Mathematical Journal*, 54(1), 91–114.

Hitchin, N.J. (2013). *Integrable Systems: Twistors, Loop Groups, and Riemann Surfaces*. Oxford University Press, Oxford.

Holmes, P.J. and Marsden, J.E. (1983). Horseshoes and Arnold diffusion for Hamiltonian systems on Lie groups. *Indiana Univ. Math. J.*, 32, 273–310.

Husson, E. (1906). Recherche des intégrales algébriques dans le mouvement d'un solide pesant autour d'un point fixe. *Ann. Fac. Sc. Univ. Toulouse*, Series 2, 8(1906), 73–152.

Jacobi, C. (1850). Sur la rotation d'un corps. *J. Reine Angew. Math.*, 39, 293–350.

Jacobi, C. (1969). *Vorlesungen über Dynamik, Königsberg Lectures of 1842–1843.* Chelsea Publishing Co., New York.

Kac, M. and van Moerbeke, P. (1975). On an explicitly soluble system of nonlinear differential equations related to certain Toda lattices. *Adv. in Math.*, 16, 160–169.

Kadomtsev, B.B. and Petviashvili, V.I. (1970). On the stability of solitary waves in weakly dispersing media. *Sov. Phys. Dokl.*, 15(6), 539–541.

Kalla, C. (2011). Fay's identity in the theory of integrable systems. Thesis, Institut de Mathématiques de Bourgogne, Dijon.

Kirchoff, G. (1876). *Vorlesungen über Mathematische Physik, Volume 1, Mechanik.* Teubner, Leipzig.

Knörrer, H. (1980). Geodesics on the ellipsoïd. *Invent. Math.*, 59, 119–143.

Knörrer, H. (1982). Geodesics on quadrics and a mechanical problem of C. Neumann. *J. Reine Angew. Math.*, 334, 69–78.

Koizumi, S. (1976). Theta relations and projective normality of abelian varieties. *Am. J. Math.*, 98, 865–889.

Kolossoff, G. (1903). Zur Rotation eines Körpers im Kowalewski'schen Falle. *Mathematische Annalen*, 56, 265–272.

Korteweg, D.J. and de Vries, G. (1895). On the change of form of long waves advancing in a rectangular canal and on a new type of long stationary waves. *Phil. Mag.*, 39, 422–443.

Kostant, B. (1979). The solution to a generalized Toda lattice and representation theory. *Adv. Math.*, 34(3), 195–338.

Kötter, F. (1892). Uber die Bewegung eines festen Körpers in einer Flüssigkeit I, II. *J. Reine Angew. Math.*, 109(51–81), 89–111.

Kötter, F. (1900). Die von Steklow und Lyapunov entdeckten intgralen Fälle der Bewegung eines Körpers in einen Flüssigkeit Sitzungsber. *König. Preuss. Akad. Wiss.*, 6, 79–87.

Krichever, I.M. (1976). Algebraic-geometric construction of Zakhorov-Shabat equations and their periodic solutions. *Sov. Math., Dokl.*, 17, 394–397.

Kowalewski, S. (1889). Sur le problème de la rotation d'un corps solide autour d'un point fixe. *Acta Math.*, 12, 177–232.

Lagrange, J.L. (1888). *Mécanique analytique*, Volume 11. Oeuvres de Lagrange, Gauthier-Villars, Paris.

Landau, L.D. and Lifshitz, E.M. (1935). On the theory of the dispersion of magnetic permeability in ferromagnetic bodies. *Phys. Zeitsch. der Sow.*, 8, 153–169.

Laurent-Gengoux, C., Miranda, E., Vanhaecke, P. (2011). Action-angle coordinates for integrable systems on Poisson manifolds. *Int. Math. Res. Not.*, 8, 1839–1869.

Lax, P. (1968). Integrals of nonlinear equations of evolution and solitary waves. *Comm. Pure Appl. Math.*, 21, 467–490.

Lesfari, A. (1986). Une approche systématique à la résolution du corps solide de Kowalewski. *C. R. Acad. Sc.*, Paris, 302(I), 347–350.

Lesfari, A. (1988). Abelian surfaces and Kowalewski's top. *Ann. Scient. École Norm. Sup.*, Paris, 21(4), 193–223.

Lesfari, A. (1999). Completely integrable systems: Jacobi's heritage. *J. Geom. Phys.*, 31, 265–286.

Lesfari, A. (2003). Le système différentiel de Hénon-Heiles et les variétés Prym. *Pacific J. Math.*, 212(1), 125–132.

Lesfari, A. (2007). Abelian varieties, surfaces of general type and integrable systems. *Beiträge zur Algebra und Geometrie*, 48(1), 95–114.

Lesfari, A. (2008a). Cyclic coverings of abelian varieties and the generalized Yang Mills system for a field with gauge groupe SU(2). *Int. J. Geom. Methods Mod. Phys.*, 5(6), 947–961.

Lesfari, A. (2008b). Fonctions et intégrales elliptiques. *Surv. Math. Appl.*, 3, 27–65.

Lesfari, A. (2009). Integrable systems and complex geometry. *Lobachevskii J. Math.*, 30(4), 292–326.

Lesfari, A. (2010). Théorie spectrale et problèmes non-linéaires. *Surv. Math. Appl.*, 5, 151–190.

Lesfari, A. (2011). Algebraic integrability: The Adler-van Moerbeke approach. *Regul. Chaotic Dyn.*, 16(3–4), 187–209.

Lesfari, A. (2013). Etude des équations stationnaire de Schrödinger, intégrale de Gelfand-Levitan et de Korteweg-de-Vries. Solitons et méthode de la diffusion inverse. *Aequat. Math.*, 85, 243–272.

Lesfari, A. (2014). Rotation d'un corps solide autour d'un point fixe. *Rend. Sem. Mat. Univ. Pol. Torino*, 72(1–2), 255–284.

Lesfari, A. (2015a). Géométrie et intégrabilité algébrique. *Rend. Mat. Appl.*, (7), 36(1–2), 27–76.

Lesfari, A. (2015b). *Introduction à la géométrie algébrique complexe*. Hermann, Paris.

Lesfari, A. (2019a). Réalisation du flot géodésique sur le groupe SO(n) comme flot sur des orbites de Kotant-Kirillov. *Surv. Math. Appl.*, 14, 231–259.

Lesfari, A. (2019b). Spectral theory, Jacobi matrices, continued fractions and difference operators. *Fundamental Journal of Mathematics and Applications*, 2(1), 63–90.

Lesfari, A. (2020). Generalized algebraic completely integrable systems. *Surv. Math. Appl.*, 15, 169–216.

Lesfari, A. (2021). *Géométrie symplectique, calcul des variations et dynamique hamiltonienne*. ISTE Editions Ltd, London.

Li, C.Z., He, J.S., Wu, K., Cheng, Y. (2010). Tau function and Hirota bilinear equations for the extended bigraded Toda Hierarchy. *J. Math. Phys.*, 51, 043514.

Liouville, R. (1896). Sur le mouvement d'un corps solide pesant suspendu par l'un de ses points. *Acta Math.*, XX, 239–284.

Lyapunov, A. (1894). On a property of the differential equations of the problem of motion of a rigid body having a fixed point. *Comm. Soc. Math. Kharkow*, 4(3), 123–140.

Manakov, S.V. (1976). Remarks on the integrals of the Euler equations of the n-dimensional heavy top. *Fund. Anal. Appl.*, 10(4), 93–94.

van Moerbeke, P. (1976). The spectrum of Jacobi matrices. *Invent. Math.*, 37, 45–81.

van Moerbeke, P. (1989). Introduction to algebraic integrable systems and their Painlevé analysis. *Theta Functions Bowdoin, Part 1, 107–131, Proc. Sympos. Pure Math., Vol. 49, Amer. Math. Soc.*, Providence.

van Moerbeke, P. (1994). Integrable foundations of string theory. In *Lectures on Integrable Systems*, Babelon, O., Cartier, P., Kosmann-Schwarzbach, Y., Verdier, J.-L. (eds). World Scientific, River Edge.

van Moerbeke, P. (2011). Random matrix theory and integrable systems. In *The Oxford Handbook of Random Matrix Theory*, Akemann, G., Baik, J., Di Francesco, P. (eds). Oxford University Press, Oxford.

van Moerbeke, P. and Mumford, D. (1979). The spectrum of difference operators and algebraic curves. *Acta Math.*, 143, 93–154.

McKean, H.P. and van Moerbeke, P. (1975). The spectrum of Hill's equation. *Invent. Math.*, 30, 217–274.

Mironov, A.E. (2010). Polynomial integrals of a mechanical system on a two-dimensional torus. *Izv. Math.*, 74, 805–817.

Mishchenko, A.S. and Fomenko, A.T. (1978). Euler equations on finite-dimensional Lie groups. *Math. USSR-Izv.*, 12(2), 371–389.

Moishezon, B.G. (1967). On n-dimensional compact varieties with n algebraically independent meromorphic functions. *Amer. Math. Soc. Transl.*, 63, 51–177.

Moser, J.K. (1965). On the volume elements on a manifold. *Trans. Amer. Math. Soc.*, 120, 286–294.

Moser, J.K. (1980). Geometry of quadrics and spectral theory. In *The Chern Symposium 1979*, Hsiang, W.-Y., KobayashiI, S., Singer, M., Wolf, J., Wu, H.-H., Weinstein, A. (eds). Springer, Cham.

Muir, T. (1960). *A Treatise on the Theory of Determinants*. Dover Publications, New York.

Mumford, D. (1967a). On the equations defining Abelian varieties I. *Invent. Math.*, 1, 287–354.

Mumford, D. (1967b). On the equations defining Abelian varieties II. *Invent. Math.*, 3, 75–135.

Mumford, D. (1967c). On the equations defining Abelian varieties III. *Invent. Math.*, 3, 215–244.

Mumford, D. (1983a). *Tata Lectures on Theta I*. Birkhaüser, Basel.

Mumford, D. (1983b). *Tata Lectures on Theta II*. Birkhaüser, Basel.

Nahm, W. (1981). All self-dual multi-monopoles for all gauge groups. CERNTH-3172. Presentation, Int. Summer Inst. on Theoretical Physics, Freiburg, West Germany, August 31–September 11.

Neumann, C. (1859). De problemate quodam mechanics, quod ad primam integralium ultraellipticorum classem revocatur. *J. Reine Angew. Math. (Crelles journal)*, 56, 46–63.

Novikov, S.P. (1974). The periodic problem for KdV equation. *F. Anal. Pril.*, 8, 53–66.

Painlevé, P. (1975). *Oeuvres, tomes 1, 2, 3*. Édition du CNRS, Paris.

Pelayo, A. (2017). Hamiltonian and symplectic symmetries: An introduction. *Bull. Amer. Math. Soc.*, 54, 383–436.

Perelomov, A.M. (1990). *Integrable Systems of Classical Mechanics and Lie Algebras*. Birkhäuser Verlag, Basel.

Piovan, L. (1992). Cyclic coverings of Abelian varieties and the Goryachev-Chaplygin top. *Math. Ann.*, 294, 755–764.

Poincaré, H. (1905). *Leçons de mécanique céleste*. Gauthier-Villars, Paris.

Poinsot, L. (1851). Théorie nouvelle de la rotation des corps. *Journal de mathématiques pures et appliquées, 1re série*, Volume 16, 9–129.

Ramani, A., Dorozzi, B., Grammaticos, B. (1982). Painlevé conjecture revisited. *Phys. Rev. Lett.*, 49, 1539–1541.

Ratiu, T. (1982). Euler-Poisson equations on Lie algebras and the N-dimensional heavy rigid body. *Amer. J. Math.*, 104, 409–448.

Ratiu, T. and van Moerbeke, P. (1982). The Lagrange rigid body motion. *Ann. Inst. Fourier*, 32(1), 211–234.

Ravoson, V., Gavrilov, L., Caboz, R. (1993). Separability and Lax pairs for Hénon-Heiles system. *J. Math. Phys.*, 34, 2385–2393.

Roekaerts, D. and Schwarz, F. (1987). Painléve analysis, Yoshida's theorems and direct methods in the search for integrable Hamiltonians. *J. Phys. A, Math. Gen.*, 20, L127–L133.

Rosemann, S. and Schöbel, K. (2015). Open problems in the theory of finite-dimensional integrable systems and related fields. *J. Geom. Phys.*, 87, 396–414.

Rudolph, G. and Schmidt, M. (2013). *Differential Geometry and Mathematical Physics, Part I. Manifolds, Lie Groups and Hamiltonian Systems.* Springer, Dordrecht.

Sato, M. (1989). The KP hierarchy and infinite-dimensional Grassmann manifolds. *Proc. of Sympos. Pure Math.*, 49, 51–66.

Sato, M. and Sato, Y. (1982). Soliton equations as dynamical systems on infinite dimensional Grassmann manifolds. *Lect. Notes in Num. Appl. Anal.*, 5, 259–271.

Scott-Russell, J. (1844). Report on waves. Report, 14th meeting of the British Association for Advancement of Science, 311-3-90.

Shiota, T. (1986). Characterization of Jacobian varieties in terms of soliton equations. *Invent. Math.*, 83, 333–382.

Siegel, C.L. (1955). Meromorphic funktionen anf kompakten mannigfaltigkeiten. *Nachrichten der Akademie der Wissenschaften in Göttingen, Math.-Phys. Klasse*, 4, 71–77.

Skrypnyk, T. (2019). Symmetric separation of variables for the Clebsch system. *J. Geom. Physics*, 135, 204–218.

Sokolov, V. (2017). Algebraic structures related to integrable differential equations. *Sociedade Brasileira de Mathematica, Ensaios Mathematicos*, 31, 1–108.

Steklov, V.A. (1893). Über die Bewegung eines festen Körper in einer Flüssigkeit. *Math. Ann.*, 42, 273–374.

Steklov, V.A. (1896). A certain case of motion of a heavy rigid body having a fixed point. *Trudy Otdel. Fiz. Nauk. Obsh. Lyubit. Estestvozn.*, 8(2), 19–21.

Sternberg, S. (1983). *Lectures on Differential Geometry*, 2nd edition. Chelsea, New York.

Symes, W. (1980). Systems of Toda type, inverse spectral problems and representation theory. *Invent. Math.*, 59, 13–53.

Teschil, G. (2000). *Jacobi Operators and Completely Integrable Nonlinear Lattices.* American Mathematical Society, Providence.

Toda, M. (1967). Wave propagation in anharmonic lattices. *J. Phys. Soc. of Japan*, 23, 501–506.

Tsiganov, A.V. (1999). The Lax pairs for the Holt system. *J. Phys. A, Math. Gen.*, 32(45), 7983–7987.

Valent, G. (2010). On a class of integrable systems with a cubic first integral. *Comm. Math. Phys.*, 299, 631–649.

Vanhaecke, P. (2001). *Integrable Systems in the Realm of Algebraic Geometry.* Springer-Verlag, Berlin.

Wadati, M. and Toda, M. (1972). The exact N-soliton solution of the Korteweg-de Vries equation. *J. Phys. Soc. Japan*, 32, 1403–1411.

Weil, A. (1983). *Euler and the Jacobians of Elliptic Curves.* Birkhauser Boston, Inc., Boston.

Weinstein, A. (1971). Symplectic manifolds and their Lagrangian submanifolds. *Adv. in Math.*, 6, 329–346.

Weinstein, A. (1977). Lectures on symplectic manifolds. *CBMS Regional Conference Series in Mathematics*, no. 29, AMS.

Whittaker, E.T. (1988). *A Treatise on the Analytical Dynamics of Particles and Rigid Bodies.* Cambridge University Press, Cambridge.

Wojciechowski, S. (1985). Integrability of one particle in a perturbed central quartic potential. *Physica Scripta*, 31, 433–438.

Zakharov, V.E., Manakov, S.V., Novikov, S.P., Pitaevskii, L.P. (1980). *Soliton Theory, Inverse Scattering Method.* Nauka, Moscow.

Ziglin, S.L. (1983a). Bifurcation of solutions and the nonexistence of first integrals in Hamiltonian mechanics I. *Funktsional. Anal. i Prilozhen.*, 16(3), 30–41, 96.

Ziglin, S.L. (1983b). Bifurcation of solutions and the nonexistence of first integrals in Hamiltonian mechanics II. *Funktsional. Anal. i Prilozhen.*, 17(1), 8–23.

Index

A

action, 49
  -variables, 74
adjoint
  orbit, 36
  wave function, 284
Adler–Kostant–Symes theorem, 104
algebra
  of differential operators, 275
  $\mathcal{W}$, 278
algebraic integrable system, 165
amplitude, 63
angle-variables, 74
angular momentum, 97
Arnold–Liouville theorem, 68, 73

B

Bäcklund transformation, 269
  auto-, 270
Baker–Akhiezer function, 273, 284
Bobylev–Steklov top, 90
Boussinesq equation, 243, 279
brachistochrone, 65
Burgers equation, 269, 270

C

Calogero
  –Moser system, 124, 273
  system, 123

Camassa–Holm equation, 243, 270
canonical
  basis, 2
  transformation, 55, 66
Carleman's condition, 138
Cartan homotopy formula, 18
Casimir functions, 74
Cauchy problem, 159
central
  element, 279
  force, 65
Clebsch's case, 91
coadjoint
  orbit, 36
  representation, 36
cocycle, 280
coisotropic space, 2
Cole–Hopf transformation, 269
commutator, 15
commute, 11, 27
complete, 73
completely integrable system, 67, 68, 74
constant of motion, 27
cyclic coordinate, 59

D

Darboux
  coordinates, 22
  theorem, 23

Davey–Stewartson equations, 270, 271
Dym
    equation, 271
    type equation, 272
dynamical system, 8

E

elliptic potential, 273
energy functional, 97
Euler
    –Arnold equation, 99, 107, 109
    –Lagrange equation, 51
    –Poinsot motion, 78
    differential equation, 85
    equation, 51
    top, 32, 78, 108, 173
extremal, 51

F

Fay
    differential identity, 286
    identity, 286
first integral, 27
fixed singularity, 160
Flaschka variables, 150
flow, 9
Fréchet derivative, 280
Frobenius theorem, 48
functions in involution, 27

G

Gardner
    equation, 269
    transformation, 269
Gelfand Dickey reduction, 277
generalized
    algebraic completely integrable systems, 221, 222
    periodic Toda systems, 213
generating function, 56
geodesic, 51
    flow, 96
    flow on an ellipsoid, 114
    flow on $SO(4)$, 33
    flow on $SO(n)$, 107, 210

Goryachev–Chaplygin top, 90, 121, 231
Griffiths theorem, 149
Gross–Neveu system, 214

H

Hénon–Heiles system, 31, 156, 191, 229
Hamilton
    –Jacobi equation, 57
    canonical equations, 31, 54
Hamiltonian (see also Jaynes–cummings–Gaudin Hamiltonian), 25, 31, 54, 67
    system, 25, 67
    vector field, 30
harmonic oscillator, 35, 60, 75, 100
heat equation, 269, 270
Heaviside function, 290
Heisenberg
    algebra, 31, 279, 280
    subalgebra, 280
Hesse–Appel'rot top, 90
Hirota
    bilinear equations, 289
    symbol, 289
Holt system, 125

I, J

integral
    action, 49
    curve, 9
interior product, 16
isospectral deformation, 104
isotropic space, 2
Jacobi
    elliptic function, 63
    geodesic flow on an ellipsoid, 114
    identity, 26, 31
    matrix, 129
    theorem, 58
Jaynes–cummings–Gaudin Hamiltonian, 126

K

Kac–van Moerbeke lattice, 212
Kadomtsev–Petviashvili equation, 243

Kaup–Kuperschdmit system, 123
KdV equation, 241
Kepler problem, 61, 100
kinetic energy, 97
Kirchhoff equations, 34, 91
Kolossof potential, 215
Korteweg–de Vries system, 123
 modified, 269
Kostant–Kirillov orbit, 36
Kowalewski
 matrix, 161
 top, 33, 83, 119, 175

L

Lagrange (*see also* system)
 equation, 51
 top, 82, 115, 153, 236
Lagrangian, 50, 53
 manifold, 5
 section, 6
 space, 2
Lax
 equation, 103
 pair, 103
left-invariant, 96
left angular velocity, 97
Legendre transformation, 52
Leibniz rule, 26
Levi–Civita symbol, 154
Lie
 bracket, 18
 derivative, 15, 16
light–cone coordinates, 126
Liouville (*see also* Sturm–Liouville
 equation)
 –Arnold–Adler–van Moerbeke
theorem, 171
 form, 4
 integrable system, 67, 74
 theorem, 64, 247
Lotka–Volterra system, 157
Lyapunov–Steklov's case, 93

M, N

Manakov geodesic flow, 109, 200
Mittag–Leffler problem, 148

Miura transformation, 269
mobile singularity, 160
moment of inertia operator, 98
Moser's lemma, 22
Nahm's equations, 154
Neumann problem, 114
Noether theorem, 28
normal order, 275
Novikov equation, 273

O, P

one-parameter group of diffeomorphism,
 9
orientable, 5
Parseval identity, 251
phase space, 31, 55
Poisson
 bracket, 15, 26, 28
 manifold, 27
 structure, 26, 27
 theorem, 27
pseudo-differential operator, 275
pull-back, 16

R, S

RDG integrable potential, 223, 225
rectification theorem, 48
regular value, 48
Riccati equation, 269
right angular velocity, 97
Sawada-Kotera system, 123
Schrödinger equations, 118, 125, 243, 269,
 271, 272
Schur polynomials, 288
Siegel half–space, 289
simple pendulum, 62
sine–Gordon equation, 125, 270, 271
special orthogonal group, 37
spectral
 curve, 104
 parameter, 103
spherical pendulum, 100
Sturm–Liouville equation, 257
Sutherland system, 124
symplectic
 basis, 2

diffeomorphism, 6
form, 2, 3
manifold, 3
matrix, 66
morphism, 6
space, 2
structure, 2, 3
transformation, 66
symplectomorphism, 6
system
    integrable by quadrature, 73
    of Lagrange equations, 28

T, V

tau function, 286
tensor, 98
Toda lattice, 150
trajectory, 9
trivial invariants, 74

variation, 50
variational problem, 49
vector field, 7
vertex operator, 287
Virasoro
    algebra, 280
    structure, 281
Volterra group, 276
volume form, 5

W, Y

wave
    function, 284
    operator, 283
weight-homogeneous, 161, 225
Yang–Mills
    equations, 119
    field, 93

Other titles from

in

Mathematics and Statistics

2022

DE SAPORTA Benoîte, ZILI Mounir
Martingales and Financial Mathematics in Discrete Time

2021

KOROLIOUK Dmitri, SAMOILENKO Igor
Random Evolutionary Systems: Asymptotic Properties and Large Deviations

MOKLYACHUK Mikhail
Convex Optimization: Introductory Course

POGORUI Anatoliy, SWISHCHUK Anatoliy, RODRÍGUEZ-DAGNINO Ramón M.
Random Motions in Markov and Semi-Markov Random Environments 1: Homogeneous Random Motions and their Applications
Random Motions in Markov and Semi-Markov Random Environments 2: High-dimensional Random Motions and Financial Applications

PROVENZI Edoardo
From Euclidean to Hilbert Spaces: Introduction to Functional Analysis and its Applications

2020

BARBU Vlad Stefan, VERGNE Nicolas
Statistical Topics and Stochastic Models for Dependent Data with Applications

CHABANYUK Yaroslav, NIKITIN Anatolii, KHIMKA Uliana
Asymptotic Analyses for Complex Evolutionary Systems with Markov and Semi-Markov Switching Using Approximation Schemes

KOROLIOUK Dmitri
Dynamics of Statistical Experiments

MANOU-ABI Solym Mawaki, DABO-NIANG Sophie, SALONE Jean-Jacques
Mathematical Modeling of Random and Deterministic Phenomena

2019

BANNA Oksana, MISHURA Yuliya, RALCHENKO Kostiantyn, SHKLYAR Sergiy
Fractional Brownian Motion: Approximations and Projections

GANA Kamel, BROC Guillaume
Structural Equation Modeling with lavaan

KUKUSH Alexander
Gaussian Measures in Hilbert Space: Construction and Properties

LUZ Maksym, MOKLYACHUK Mikhail
Estimation of Stochastic Processes with Stationary Increments and Cointegrated Sequences

MICHELITSCH Thomas, PÉREZ RIASCOS Alejandro, COLLET Bernard, NOWAKOWSKI Andrzej, NICOLLEAU Franck
Fractional Dynamics on Networks and Lattices

VOTSI Irene, LIMNIOS Nikolaos, PAPADIMITRIOU Eleftheria, TSAKLIDIS George
Earthquake Statistical Analysis through Multi-state Modeling (Statistical Methods for Earthquakes Set – Volume 2)

2018

AZAÏS Romain, BOUGUET Florian
Statistical Inference for Piecewise-deterministic Markov Processes

IBRAHIMI Mohammed
Mergers & Acquisitions: Theory, Strategy, Finance

PARROCHIA Daniel
Mathematics and Philosophy

2017

CARONI Chysseis
First Hitting Time Regression Models: Lifetime Data Analysis Based on Underlying Stochastic Processes
(Mathematical Models and Methods in Reliability Set – Volume 4)

CELANT Giorgio, BRONIATOWSKI Michel
Interpolation and Extrapolation Optimal Designs 2: Finite Dimensional General Models

CONSOLE Rodolfo, MURRU Maura, FALCONE Giuseppe
Earthquake Occurrence: Short- and Long-term Models and their Validation
(Statistical Methods for Earthquakes Set – Volume 1)

D'AMICO Guglielmo, DI BIASE Giuseppe, JANSSEN Jacques, MANCA Raimondo
Semi-Markov Migration Models for Credit Risk
(Stochastic Models for Insurance Set – Volume 1)

GONZÁLEZ VELASCO Miguel, del PUERTO GARCÍA Inés, YANEV George P.
Controlled Branching Processes
(Branching Processes, Branching Random Walks and Branching Particle Fields Set – Volume 2)

HARLAMOV Boris
Stochastic Analysis of Risk and Management
(Stochastic Models in Survival Analysis and Reliability Set – Volume 2)

KERSTING Götz, VATUTIN Vladimir
Discrete Time Branching Processes in Random Environment
(Branching Processes, Branching Random Walks and Branching Particle Fields Set – Volume 1)

MISHURA YULIYA, SHEVCHENKO Georgiy
Theory and Statistical Applications of Stochastic Processes

NIKULIN Mikhail, CHIMITOVA Ekaterina
Chi-squared Goodness-of-fit Tests for Censored Data
(Stochastic Models in Survival Analysis and Reliability Set – Volume 3)

SIMON Jacques
Banach, Fréchet, Hilbert and Neumann Spaces
(Analysis for PDEs Set – Volume 1)

2016

CELANT Giorgio, BRONIATOWSKI Michel
Interpolation and Extrapolation Optimal Designs 1: Polynomial Regression and Approximation Theory

CHIASSERINI Carla Fabiana, GRIBAUDO Marco, MANINI Daniele
Analytical Modeling of Wireless Communication Systems
(Stochastic Models in Computer Science and Telecommunication Networks Set – Volume 1)

GOUDON Thierry
Mathematics for Modeling and Scientific Computing

KAHLE Waltraud, MERCIER Sophie, PAROISSIN Christian
Degradation Processes in Reliability
(Mathematial Models and Methods in Reliability Set – Volume 3)

KERN Michel
Numerical Methods for Inverse Problems

RYKOV Vladimir
Reliability of Engineering Systems and Technological Risks
(Stochastic Models in Survival Analysis and Reliability Set – Volume 1)

2015

DE SAPORTA Benoîte, DUFOUR François, ZHANG Huilong
Numerical Methods for Simulation and Optimization of Piecewise Deterministic Markov Processes

DEVOLDER Pierre, JANSSEN Jacques, MANCA Raimondo
Basic Stochastic Processes

LE GAT Yves
*Recurrent Event Modeling Based on the Yule Process
(Mathematical Models and Methods in Reliability Set – Volume 2)*

2014

COOKE Roger M., NIEBOER Daan, MISIEWICZ Jolanta
*Fat-tailed Distributions: Data, Diagnostics and Dependence
(Mathematical Models and Methods in Reliability Set – Volume 1)*

MACKEVIČIUS Vigirdas
Integral and Measure: From Rather Simple to Rather Complex

PASCHOS Vangelis Th
*Combinatorial Optimization – 3-volume series – 2nd edition
Concepts of Combinatorial Optimization / Concepts and Fundamentals – volume 1
Paradigms of Combinatorial Optimization – volume 2
Applications of Combinatorial Optimization – volume 3*

2013

COUALLIER Vincent, GERVILLE-RÉACHE Léo, HUBER Catherine, LIMNIOS Nikolaos, MESBAH Mounir
Statistical Models and Methods for Reliability and Survival Analysis

JANSSEN Jacques, MANCA Oronzio, MANCA Raimondo
Applied Diffusion Processes from Engineering to Finance

SERICOLA Bruno
Markov Chains: Theory, Algorithms and Applications

2012

BOSQ Denis
Mathematical Statistics and Stochastic Processes

CHRISTENSEN Karl Bang, KREINER Svend, MESBAH Mounir
Rasch Models in Health

DEVOLDER Pierre, JANSSEN Jacques, MANCA Raimondo
Stochastic Methods for Pension Funds

2011

MACKEVIČIUS Vigirdas
Introduction to Stochastic Analysis: Integrals and Differential Equations

MAHJOUB Ridha
Recent Progress in Combinatorial Optimization – ISCO2010

RAYNAUD Hervé, ARROW Kenneth
Managerial Logic

2010

BAGDONAVIČIUS Vilijandas, KRUOPIS Julius, NIKULIN Mikhail
Nonparametric Tests for Censored Data

BAGDONAVIČIUS Vilijandas, KRUOPIS Julius, NIKULIN Mikhail
Nonparametric Tests for Complete Data

IOSIFESCU Marius *et al.*
Introduction to Stochastic Models

VASSILIOU PCG
Discrete-time Asset Pricing Models in Applied Stochastic Finance

2008

ANISIMOV Vladimir
Switching Processes in Queuing Models

FICHE Georges, HÉBUTERNE Gérard
Mathematics for Engineers

HUBER Catherine, LIMNIOS Nikolaos *et al.*
Mathematical Methods in Survival Analysis, Reliability and Quality of Life

JANSSEN Jacques, MANCA Raimondo, VOLPE Ernesto
Mathematical Finance

2007

HARLAMOV Boris
Continuous Semi-Markov Processes

2006

CLERC Maurice
Particle Swarm Optimization

Printed and bound by CPI Group (UK) Ltd, Croydon, CR0 4YY
17/01/2023
03181578-0001